Lecture Notes in Mathematics

Edited by A. Dold and B. Eckmann

993

Richard D. Bourgin

Geometric Aspects of Convex Sets with the Radon-Nikodým Property

Springer-Verlag
Berlin Heidelberg New York Tokyo 1983

Author

Richard D. Bourgin
Mathematics Department, Howard University
Washington, D.C. 20059, USA

AMS Subject Classifications (1980): (Primary) 46 B 22
(Secondary) 46 B 20, 46 G 10

ISBN 3-540-12296-6 Springer-Verlag Berlin Heidelberg New York Tokyo
ISBN 0-387-12296-6 Springer-Verlag New York Heidelberg Berlin Tokyo

Printed in Germany

Printing and binding: Beltz Offsetdruck, Hemsbach/Bergstr.
2146/3140-543210

To Leighton

PREFACE

Johann Radon [1913] established the theorem which bears his and Nikodým's name for measures of finite total variation on the Borel σ-field of Euclidean n-space; Otton Nikodým [1930] removed what are now known to be the unnecessarily restrictive assumptions Radon had imposed. Only three years elapsed from Nikodým's general theorem to the first published accounts dealing with the problem of differentiating Banach-valued absolutely continuous functions on [0,1]. There followed a spate of geometrical Banach space developments in the middle 1930s motivated by the search for conditions guaranteeing differentiability, the best known being Clarkson's introduction of uniformly convex spaces in 1936. After the results of Dunford, Pettis and Phillips in 1940 on the representability of linear operators on $L^1(\mu)$ as Bochner integrals, however, few direct substantive advances on Radon-Nikodým issues in Banach spaces were noted until the late 1960s. At that time M.A. Rieffel tied the Radon-Nikodým theorem in Banach spaces to the geometry via the notion of dentability. Also, S.D. Chatterji spelled out the intimate relationship between convergence of L^1-bounded Banach-valued martingales and the existence of Bochner integrable Radon-Nikodým derivatives. These results, in tandem with the concurrent geometrical advances of Lindenstrauss and others on the existence of extreme points for not necessarily compact closed bounded convex subsets of certain Banach spaces, spawned considerable research activity. A coherent picture of Banach spaces having the "Radon-Nikodým Property" developed from operator, martingale and geometric perspectives.

By the mid 1970s the broad strokes were in place and a period of reevaluation and refinement began. One of the upshots was the localization of many of the early theorems on Banach spaces to the corresponding ones for closed bounded convex sets. A second, expected improvement, was the tightening and reworking of several of the original arguments. What emerged was an elegant and comprehensive theory of sets with the Radon-Nikodým Property. In Chapters 2 through 6 of these notes this theory is described in detail, special attention being given to the geometrical aspects. By the late 1970s the Radon-Nikodým novelty had died down as the theory of such sets was by then fairly well understood. Emphasis shifted from the study of sets with the "RNP", which until then had flourished, to an analysis of their place in the larger functional analytic picture, especially their role in the structure theory of Banach spaces. The second half of these notes, Chapter 7, is a report on some aspects of this very active phase of the continuing Radon-Nikodým saga. Most of the topics discussed in Chapter 7, as listed in the Table of Contents, may be read independently of one another. (The parenthetical comments in the Table of Contents most often indicate subjects, in addition to those mentioned in section titles, which have been broached in the text. They are not to be construed as a full listing of the contents.)

I wish to thank each of the many people who has contributed in one way or another to this work. The notes would not have been completed without the timely receipt of manuscripts just written, the long discussions and the correspondence I have had on its contents. Several people provided detailed suggestions about portions of the first six chapters, many of which have been incorporated in the final version. They include G.A. Edgar, B. Epstein, S. Fitzpatrick and R.R. Phelps, to each of whom I extend my appreciation. Throughout this project I have had the benefit of the wisdom, advice, thoughtful comments and careful and elegant mathematical arguments of Isaac Namioka. His support has contributed significantly to making the work on these notes enjoyable. I am in his debt.

It is intended that anyone with a standard background in measure theory and basic functional analysis could, with a minimum of additional background as briefly outlined in the first part of Chapter 1, read large portions of these notes without undue difficulty. Arguments are presented in detail and, especially in the early chapters, some attempt has been made to predict later results on the basis of examples and preliminary theorems. A good example is afforded by Lévy's theorem in §1.3 which marks the dividing line between general martingale arguments substantively independent of the range space and those martingale arguments in which a knowledge of the range space is essential. The study of sets with the Radon-Nikodým Property begins in earnest in Chapter 2; the basic connections between martingale, measure theoretic and geometric notions is established by the end of §2.2. The sharpest geometric conclusions in these notes may be found in §§3.5, 3.6 and 3.7. Each of the next three chapters deals with a subtopic worthy of extended discussion. Although the Radon-Nikodým Property is centr to each of the sections of Chapter 7, the emphasis is often different than in the earlier chapters. In many of the sections, rather than discovering more about the structure of such sets, the RNP is used as a tool to study other objects. By virtue of this distinction, the range of background required for some of the sections of the last chapter is greater than that for the first six chapters, and in such cases I have provided references for additional background material. I have drawn on numerous sources, most of which have been credited in the text. The 1976-77 Rainwater seminar notes from the University of Washington were inadvertantly not credited in the text. They were used extensively while writing portions of Chapters 3 and 4.

Most notation is standard. However, it is worthwhile to mention here the use of "$\equiv$" in the text to denote equality in which one of the terms is defined by the equality. Thus, for example, $\mathbb{N}^+ \equiv \{1,2,3...\}$. With very few exceptions, discussions are limited to real Banach spaces and their subsets. Unless it is explicitly mentioned to the contrary, the word "subspace" means "closed linear subspace" in these notes.

TABLE OF CONTENTS

CHAPTER 1. BACKGROUND

The classical theorem of Radon and Nikodým contains conditions under which one measure varies smoothly with respect to another. In generalizing this smoothness criterion to the case of Banach-valued measures, one soon finds that the Banach space structure enters in a fundamental way. In fact, when appropriate conditions on the space are met, conclusions similar to those obtained in the classical case follow in somewhat the same manner, and it is therefore of interest to see exactly what these underlying spaces have in common which makes them so special. Historically this was done in a piecemeal fashion, with some of the "right" questions eluding everyone's attention, but recently a fairly unified picture has emerged.

In order to place the discussion on a more firm footing, we begin by stating one version of the Radon-Nikodým theorem which, in spite of its restrictive hypotheses, contains all the important features needed for subsequent generalizations. Recall first that a measurable space is a pair $(\Omega,\mathcal{F})$ with $\mathcal{F}$ a σ-algebra of subsets of the point set Ω. The triple $(\Omega,\mathcal{F},P)$ is a probability space if $(\Omega,\mathcal{F})$ is a measurable space and P is a nonnegative measure on $\mathcal{F}$ with $P(\Omega) = 1$. (All measures are countably additive unless the contrary is explicitly stated.)

Theorem 1.0.1 (Radon-Nikodým). Let $(\Omega,\mathcal{F},P)$ be a probability space and $m\colon \mathcal{F} \to \mathbb{R}$ a measure which is absolutely continuous with respect to P ($m \ll P$) and is of finite total variation ($|m|(\Omega) < \infty$). Then there is a P-integrable function $f\colon \Omega \to \mathbb{R}$, sometimes written $\frac{dm}{dP}$, such that

$$\int_A f\,dP = m(A) \quad \text{for each} \quad A \in \mathcal{F}\,.$$

If $\mathbb{R}$ in both instances above is replaced by a general Banach space X the resulting formal statement requires interpretation (what is a finite total variation, absolutely continuous X-valued measure, and what does $\int_A f\,dp$ mean?) even before its possible truth is considered. Section 1 deals with the problem of such interpretations while the machinery needed to analyze closed convex subsets of those spaces X for which the modified Radon-Nikodým statement above is true (with $\mathbb{R}$ replaced by X) is developed in the remainder of the chapter. All of this material will be used in the classification of the "Radon-Nikodým sets" begun in Chapter 2.

§1. Vector measures and vector integration (the Bochner integral).

There are very few real surprises in this section, and consequently no proofs are included. In fact, much of this material is easily predicted by direct analogy with the real situation. (One possible exception is the theorem of Hahn, Vitali, and Saks, which is stated as Theorem 1.1.5.) Sources which include details are Cohn [1980], Hille [1972], Hille and Phillips [1957], Dunford and Schwartz [1958] and Diestel and Uhl [1977].

NOTATION 1.1.1. $(X, \|\cdot\|)$ will always denote a real Banach space with dual X^*. (Ω, F, P) will always denote a probability space.

DEFINITIONS 1.1.2. Let F be a σ-algebra. A map $m: F \to X$ such that $m(\emptyset) = 0 \in X$ is called a vector measure (or, more precisely, an X-valued measure) if it is countably additive; that is, whenever $\{A_j : j = 1,2,\ldots\}$ is a sequence of pairwise disjoint sets in F then $m(\bigcup_{j=1}^{\infty} A_j) = \sum_{j=1}^{\infty} m(A_j)$. (Otherwise said, the partial sums $\sum_{j=1}^{n} m(A_j)$ are required to form a convergent sequence in the norm topology with limit $m(\bigcup_{j=1}^{\infty} A_j)$.) Such a measure is absolutely continuous with respect to P ($m << P$) if for $A \in F$ the condition $P(A) = 0$ implies $m(A) = 0$. It is of finite total variation ($|m|(\Omega) < \infty$) if there is a finite upper bound for the sums of the form $\sum_{j=1}^{k} \|m(A_j)\|$ as $\{A_j\}_{j=1}^{k}$ runs over all possible finite partitions of Ω with entries in F.

DEFINITION 1.1.3. Let (Ω, F, P) be a probability space. An X-valued measure m on (Ω, F) which is absolutely continuous with respect to P has average range defined by

$$AR(m) \equiv \{\frac{m(A)}{P(A)} : A \in F,\ P(A) > 0\}.$$

There is a complete metric space naturally associated with the probability space (Ω, F, P). Indeed, first define an equivalence relation $\sim$ between sets in F by

$$A \sim B \quad \text{if} \quad P(A \,\Delta\, B) = 0 \qquad A,B \in F .$$

$(A \,\Delta\, B \equiv (A \cap B^c) \cup (B \cap A^c)$ is the symmetric difference of A and B.) For $A \in F$

let $[A] \equiv \{A' \in \mathcal{F}: A \sim A'\}$ and define a metric d on $[\mathcal{F}] \equiv \{[A]: A \in \mathcal{F}\}$ as follows:

$$d([A],[B]) \equiv P(A \Delta B).$$

The following proposition is easily checked.

PROPOSITION 1.1.4. d is well defined. In fact, $([\mathcal{F}],d)$ is a complete metric space. (Embed $[\mathcal{F}]$ in $L^1(P)$ via the map $[A] \to \chi_A$, the characteristic function of A.) Moreover, if $\mathcal{B} \subset \mathcal{F}$ is an algebra of subsets of Ω and $\overline{[\mathcal{B}]}$ denotes the closure of $\{[A]: A \in \mathcal{B}\}$ in $([\mathcal{F}],d)$ then $\{A \in \mathcal{F}: [A] \in \overline{[\mathcal{B}]}\}$ is a σ-algebra containing $\mathcal{B}$ and contained in $\mathcal{F}$.

An X-valued measure on $\mathcal{F}$ may be naturally considered as a function on $[\mathcal{F}]$ precisely when it is absolutely continuous with respect to P. This fact may be applied in an elegant Baire category argument on $([\mathcal{F}],d)$ to prove the Hahn-Vitali-Saks theorem quoted here. See, for example, Dunford and Schwartz [1958], pages 158-9.

THEOREM 1.1.5 (Hahn, Vitali, Saks). Let $(\Omega,\mathcal{F},P)$ be a probability space and $\{m_n: n = 1,2,\ldots\}$ a sequence of X-valued measures on $\mathcal{F}$ each absolutely continuous with respect to P. If $\lim_n m_n(A)$ exists (and is denoted by $m(A)$) for each $A \in \mathcal{F}$ then m is an X-valued measure on $\mathcal{F}$. (That is, m is countably additive.)

We turn next to a discussion of the Bochner integral.

A Banach-valued integral defined by Dunford [1937] allows quite general functions to be integrated but yields a very weak theory of integration which is often of limited value. The Pettis integral is somewhat stronger, and the Bochner integral is at the other extreme. Here the integral is defined in such a way that the simple functions are L^1-dense in the space of Bochner integrable functions, and exactly because of this density, many of the theorems for the Bochner integral closely parallel their real counterparts.

Let $(\Omega,\mathcal{F},P)$ be a probability space and denote by $(\Omega,\mathcal{F}',P')$ its completion. A function $s: \Omega \to X$ is a simple function if s can be written in the form $\sum_{i=1}^{n} x_i\chi_{A_i}$ for distinct $x_i \in X$, $i = 1,\ldots, n$ and a finite partition $\{A_i\}_{i=1}^n$ of Ω chosen from $\mathcal{F}$. (When written in this way s is said to be in canonical form.) For any $A \in \mathcal{F}$ and simple function s as above define the Bochner integral of s over A by

$$\int_A s\,dP \;(= \int_A \sum_{i=1}^{n} x_i \chi_{A_i}\, dP) \equiv \sum_{i=1}^{n} x_i P(A \cap A_i)\;.$$

Generally, a function $f\colon \Omega \to X$ is said to be Bochner integrable (written $f \in L^1_X(\Omega, F, P)$ or sometimes $L^1_X(P)$ when no confusion arises) if there is a sequence $\{s_n\colon n = 1,2,\ldots\}$ of simple functions such that both

(1) $\lim_n s_n(\omega) = f(\omega)$ almost surely with respect to P (abbreviated [a.s.P]);

and

(2) $\lim_n \int_\Omega \|f(\omega) - s_n(\omega)\| dP'(\omega) = 0.$

(N.B. The terms "almost surely" and "almost everywhere" are used interchangeably throughout.) One can check that when f is Bochner integrable and $A \in F$ the $\lim_n \int_A s_n dP$ exists (in the norm sense) and is independent of the particular choice of sequence of simple functions satisfying (1) and (2). We thus define $\int_A f dP \equiv \lim_n \int_A s_n dP$ for any appropriate choice of s_n's. Before summarizing those properties of the Bochner integral needed in these notes some terminology is recalled.

DEFINITION 1.1.6. Let $f_i\colon \Omega \to X$ for $i = 1,\ldots,4$. If f_1 is the almost sure limit of a sequence of simple functions then it is said to be strongly F-measurable. If $F \circ f_2\colon \Omega \to \mathbb{R}$ is F-measurable for each $F \in X^*$ then f_2 is weakly F-measurable. Suppose that for some $A \in F$ both $P(A) = 0$ and $f_3(\Omega\backslash A)$ is a norm separable set. Then f_3 is almost separably valued. Finally, if $f_4^{-1}(O) \in F$ for each norm open set O in X then f_4 is F-Borel measurable.

It is not difficult to show that if $f\colon \Omega \to X$ is almost separably valued and F-Borel measurable then there is a sequence $\{s_n\colon\ n = 1,2,\ldots\}$ of simple functions which converges at each point of Ω such that both $\lim_n s_n(\omega) = f(\omega)$ [a.s.P] and $\|s_n(\omega)\| \leqq 2\|f(\omega)\|$ [a.s.P] for each n. (With a little care, the factor of 2 in this last inequality may be removed.) Thus, up to a subset of a set of P-measure zero, one version of f is $\lim_n s_n$, a separably valued, F-Borel measurable function. In the sequel two functions are identified when they disagree on a subset of a set of P-measure zero, thus allowing us to dispense with the completions F' and P' of F and P once and for all.

The next theorem provides useful criteria for determining whether or not a Banach-valued function on a probability space is Bochner integrable. It is due to Pettis and to Bochner.

THEOREM 1.1.7. Let (Ω,F,P) be a probability space. Then the following are equivalent.

(a) $f \in L^1_X(P)$.

(b) f is almost separably valued, weakly F'-measurable and $\int_\Omega \|f(\omega)\|dP'(\omega) < \infty$.

(c) f is strongly F-measurable and $\int_\Omega \|f(\omega)\|dP(\omega) < \infty$.

(d) f is almost separably valued, F'-Borel measurable and $\int_\Omega \|f(\omega)\|dP'(\omega) < \infty$.

The proof of the following theorem proceeds as in the real-valued case.

THEOREM 1.1.8. Let $f: \Omega \to X$ be Bochner integrable and $A \in F$. Then

(a) $\|\int_A fdP\| \leqq \int_A \|f(\omega)\|dP(\omega)$.

(b) $L^1_X(P)$ is a Banach space when equipped with the norm $\|f\|_1 \equiv \int_\Omega \|f(\omega)\|dP(\omega)$. The simple functions form a dense subspace of $L^1_X(P)$.

(c) (Dominated convergence): Suppose that $\{f_n: n = 1,2,...\}$ is a sequence in $L^1_X(P)$ and that $\|f_n(\omega)\| \leqq g(\omega)$ [a.s.P] for each n where g is P-integrable. If $\lim_n f_n(\omega) \equiv f(\omega)$ exists almost surely with respect to P in the norm sense then $f \in L^1_X(P)$ and both $\lim_n \|f - f_n\|_1 = 0$ and $\int fdP = \lim_n \int f_n dP$.

(d) $\int_A fdP$ is linear in f and countably additive in A.

(e) If $F \in X^*$ then $F(\int_A fdP) = \int_A F\circ fdP$. More generally, if Y is a Banach space and $T: X \to Y$ is a bounded linear operator then $T\circ f \in L^1_Y(P)$ and $T(\int_A fdP) = \int_A T\circ fdP$.

(f) If $\{f_n: n = 1,2,...\}$ is a sequence in $L^1_X(P)$ for which $\lim_n \|f_n - f\|_1 = 0$ then there is a subsequence $\{f_{n_i}: i = 1,2,...\}$ of $\{f_n: n = 1,2,...\}$ such that $\lim_i f_{n_i}(\omega) = f(\omega)$ [a.s.P].

§2. Expectations and conditional expectations.

Let (Ω,F,P) be a probability space and $A,B \in F$ with $P(A) > 0$. The probability that the "event" B happens given that the "event" A has already occured is written $P(B|A)$ and is given, as in intuition dictates it should be, by the formula

$P(B|A) \equiv \frac{P(A \cap B)}{P(A)}$. It is called the conditional probability of B given A. Suppose now that f and g are F-measurable simple functions written in canonical form, say $f \equiv \Sigma_{i=1}^{n} \alpha_i \chi_{A_i}$ and $g \equiv \Sigma_{j=1}^{k} \beta_j \chi_{B_j}$. Then $g^{-1}(\beta_j) = B_j$ and we temporarily adopt the sloppy but appealing notation $(g = \beta_j) \equiv \{\omega \in \Omega: g(\omega) = \beta_j\} = g^{-1}(\beta_j)$. Hence, for example, $P(g = \beta_j) = P(B_j)$. Let us further restrict g so that $P(g = \beta_j) > 0$ for each j. Common sense dictates that the average value, or expectation, of f should be

$$E[f] \equiv \sum_{i=1}^{n} \alpha_i P(f = \alpha_i) = \sum_{i=1}^{n} \alpha_i P(A_i) = \int_\Omega f dP$$

while the conditional expectation of f given that $g = \beta_j$ should be

$$E[f|g = \beta_j] \equiv \sum_{i=1}^{n} \alpha_i P(f = \alpha_i|g = \beta_j)$$

$$= \sum_{i=1}^{n} \alpha_i P(A_i|B_j) = \frac{1}{P(g = \beta_j)} \int_{(g=\beta_j)} f dP.$$

In fact, this same definition of conditional expectation will work for any f in $L^1(P)$ but there are problems when dealing with more general functions g, since $P(g = \beta_j)$ may be 0. They may be circumvented as follows.

The idea is to replace g by an appropriate σ-algebra G and then to interpret the symbol $E[f|G]$. (Cf. the latter part of 1.2.1.) Suppose then that g is a simple function as before and that $f \in L^1(P)$. Define $E[f|g]$ to be the real-valued function on Ω given by

$$E[f|g](\omega) \equiv \frac{1}{P(g = \beta_j)} \int_{(g=\beta_j)} f dP \quad \text{whenever} \quad g(\omega) = \beta_j .$$

(Thus, referring to the earlier notation, $E[f|g](\omega) = E[f|g = g(\omega)]$.) Let G denote the σ-algebra of all possible unions of the sets $(g = \beta_j)$, and let us write $E[f|G]$ henceforth in place of $E[f|g]$. Thus $E[f|G]$ is the function with constant value $\frac{1}{P(A)} \int_A f dP$ on each atom A of G. Observe that $E[f|G]$ is G-measurable and that for $A \in G$ we have $\int_A E[f|G](\omega) dP(\omega) = \int_A f dP$. This, then, is the motivation for the conditional expectation operator defined below.

DEFINITIONS 1.2.1. Let $f \in L^1(\Omega, F, P)$. Then the expectation of f is $E[f] \equiv \int_\Omega f dP$. If $G \subset F$ is a sub σ-algebra, the unique (up to almost sure equivalence) function g such that

1) g is G-measurable, and

2) $\int_A g dP = \int_A f dP$ for $A \in G$

is called the conditional expectation of f given G and is written $g \equiv E[f|G]$. For $f, h \in L^1(\Omega, F, P)$ define $E[f|h]$ to be $E[f|H]$ where H is the σ-algebra generated by $\{h^{-1}(O): O \text{ is open}\}$.

(To see that $E[f|G]$ exists let $m(A) \equiv \int_A f dP$ for $A \in G$. The measures m and P on the σ-algebra G satisfy the hypotheses of the Radon-Nikodým theorem (see 1.0.1) and the Radon-Nidodým derivative $g \equiv \frac{dm}{dP}$ so determined satisfies 1) and 2) of 1.2.1. Uniqueness may be verified directly.)

Note that $\frac{dm}{dP} = E[f|G]$ is nonnegative when $f \geqq 0$. This is part a) of the proposition below; the remaining parts of the proof are also straightforward.

PROPOSITION 1.2.2. Let $f, f_1, \ldots, f_n \in L^1(\Omega, F, P)$ and suppose that $G \subset F$ is a sub σ-algebra. Then

a) $E[f|G] \geqq 0$ [a.s.P] when $f \geqq 0$ [a.s.P];

b) $|E[f|G]| \leqq E[|f||G]$ [a.s.P]; and

c) $E[\sum_{i=1}^n c_i f_i | G] = \sum_{i=1}^n c_i E[f_i|G]$ [a.s.P].

Next, suppose that $f \equiv \sum_{i=1}^n x_i \chi_{A_i}$ is a strongly F-measurable X-valued simple function, written here in canonical form. If $G \subset F$ is a sub-σ-algebra, the conditional expectation of f given G, written $\mathbb{E}[f|G]$, is defined as might be expected:

$$\mathbb{E}[f|G] \equiv \sum_{i=1}^n x_i E[\chi_{A_i}|G].$$

A few observations are in order. (Each assumes that f is a strongly F-measurable X-valued simple function.)

a) $\mathbb{E}[f|G]$ is strongly G-measurable and Bochner integrable. Moreover, $\int_A f dP = \int_A \mathbb{E}[f|G] dP$ for each $A \in G$.

b) $\mathbb{E}[f|G]$ need not be a simple function even though f is.

c) $\mathbb{E}[f|G]$ is linear in f.

d) $\|\mathbb{E}[f|\mathcal{G}](\omega)\| \leqq E[\|f(\cdot)\| \; |\mathcal{G}](\omega)$ [a.s.P].

More generally, if $f \in L^1_X(\Omega,\mathcal{F},P)$ it may be approximated (in the norm of this space) by a sequence $\{s_n : n = 1,2,\ldots\}$ of simple functions. It is not difficult to show that $\{\mathbb{E}[s_n|\mathcal{G}]\}_{n=1}^{\infty}$ is Cauchy in $L^1_X(\Omega,\mathcal{F},P)$ and that its limit, which we shall denote by $\mathbb{E}[f|\mathcal{G}]$, is independent of the approximating sequence $\{s_n : n = 1,2,\ldots\}$. The following theorem summarizes some of the properties of $\mathbb{E}[f|\mathcal{G}]$.

<u>THEOREM</u> 1.2.3. Let $(\Omega,\mathcal{F},P)$ be a probability space and $\mathcal{G} \subset \mathcal{F}$ a sub σ-algebra. If $f \in L^1_X(\Omega,\mathcal{F},P)$ there exists a unique (up to almost sure equivalence) function, henceforth denoted by $\mathbb{E}[f|\mathcal{G}]$ and called the <u>conditional expectation of</u> f <u>given</u> $\mathcal{G}$, such that

a) $\mathbb{E}[f|\mathcal{G}] \in L^1_X(\Omega,\mathcal{G},P)$, and

b) $\int_A f dP = \int_A \mathbb{E}[f|\mathcal{G}]dP$ for each $A \in \mathcal{G}$.

The operator $\mathbb{E}$ is linear in f and satisfies the conditions:

1) $\|\mathbb{E}[f|\mathcal{G}](\omega)\| \leqq E[\|f(\cdot)\| \; |\mathcal{G}](\omega)$ [a.s.P], and

2) $\|\mathbb{E}[f|\mathcal{G}]\|_1 \leqq \|f\|_1$ for $f \in L^1_X(P)$.

If $\mathcal{H} \subset \mathcal{G}$ is another σ-algebra, then

3) $\mathbb{E}[\mathbb{E}[f|\mathcal{G}]|\mathcal{H}] = \mathbb{E}[\mathbb{E}[f|\mathcal{H}]|\mathcal{G}] = \mathbb{E}[f|\mathcal{H}]$ [a.s.P].

§3. <u>Lévy's theorem on the continuity of the conditional expectation operator.</u>

We will use Theorem 1.3.1 below to provide a first glimpse into the structure of spaces with the Radon-Nikodým Property (see §4). The theorem will then be applied in Chapter 2. (Lévy's original theorem [1937] p. 129 was stated for 0,1 - valued 'chance' variables. Doob [1940] Theorem 1.4 extended Lévy's result to general Random variables. Replacing $\mathbb{R}$ by X requires little change in the proof.)

<u>THEOREM</u> 1.3.1 (Lévy). Let $(\Omega,\mathcal{F},P)$ be a probability space, X a Banach space, and $f: \Omega \to X$ Bochner integrable. Suppose that $\mathcal{F}_1 \subset \mathcal{F}_2 \subset \ldots \subset \mathcal{F}$ is an increasing sequence of sub σ-algebras whose union generates $\mathcal{F}$. (That is, the smallest σ-algebra containing $\bigcup_{n=1}^{\infty} \mathcal{F}_n$ is $\mathcal{F}$.) Let $f_n \doteq \mathbb{E}[f|\mathcal{F}_n]$. Then $\lim_n f_n(\omega) = f(\omega)$ [a.s.P] and $\lim_n \|f_n - f\|_1 = 0$.

An important ingredient in the proof (which we deal with forthwith) is an inequality of the type commonly referred to as a maximal inequality.

<u>MAXIMAL INEQUALITY</u> 1.3.2. Let (Ω,F,P) and F_n, $n = 1,2,\ldots$ be as in 1.3.1. Suppose that $\{h_n: n = 1,2,\ldots\}$ is a sequence of nonnegative $L^1(\Omega,F,P)$ functions for which

a) h_j is F_j-measurable;

b) for $i \leqq n$ and $A \in F_i$, $\int_A h_i dP \leqq \int_A h_n dP$; and

c) $\sup_j \int_\Omega h_j dP < \infty$.

Then for each $c > 0$ we have

$$P\{\omega \in \Omega: \sup_j h_j(\omega) > c\} \leqq \frac{1}{c} \sup_j \int_\Omega h_j dP.$$

Note: conditions a) and b) may be rephrased as: $\{h_j: j = 1,2,\ldots\}$ is a submartingale. (See 1.4.2.)

Proof: Fix an integer $n > 0$. Let $A_1 \equiv \{\omega \in \Omega: h_1(\omega) > c\}$ and $A_i \equiv \{\omega: h_1(\omega) \leqq c,\ldots,h_{i-1}(\omega) \leqq c$, and $h_i(\omega) > c\}$ for $2 \leqq i \leqq n$. Observe that $A_i \in F_i$ for $i = 1,\ldots,n$ and that $A_i \cap A_j = \emptyset$ for $i \neq j$. Now

$$cP\{\omega: \sup_{j\leqq n} h_j(\omega) > c\} = cP(\bigcup_{i=1}^n A_i) = \sum_{i=1}^n cP(A_i) \leqq \sum_{i=1}^n \int_{A_i} h_i dP$$

$$\leqq \sum_{i=1}^n \int_{A_i} h_n dP \leqq \int_\Omega h_n dP \leqq \sup_j \int_\Omega h_j dP.$$

Consequently $P\{\omega: \sup_{j\leqq n} h_j(\omega) > c\} \leqq \frac{1}{c} \sup_j \int_\Omega h_j dP$ and the desired result follows by taking the limit as $n \to \infty$.

<u>Proof of 1.3.1.</u> For any $f \in L^1_X(\Omega,F,P)$ denote by f_n the function $\mathbb{E}[f|F_n]$ and let

$$F \equiv \{f \in L^1_X(\Omega,F,P). \text{ both a) } \|f_n(\omega) - f(\omega)\| \to 0 \text{ [a.s.P] as } n \to \infty,$$
$$\text{and b) } \|f_n - f\|_1 \to 0 \text{ as } n \to \infty\}.$$

We will establish that F is at once dense and closed in $L^1_X(P)$.

A) F is dense in $L^1_X(P)$: if $f \in L^1_X(\Omega,F_n,P)$ for some n then $\mathbb{E}[f|F_k] = f$ for each $k \geqq n$, whence $f \in F$. That is, $F \supset \bigcup_{n=1}^\infty L^1_X(\Omega,F_n,P)$. In particular $x\chi_A \in F$ for

$\Lambda \in \bigcup_{n=1}^{\infty} F_n$ and $x \in X$ and since $\bigcup_{n=1}^{\infty} F_n$ is dense in F (see Proposition 1.1.4) each $x\chi_B$ ($x \in X$, $B \in F$) is approximable arbitrarily well in $L_X^1(\Omega, F, P)$ by elements of F. Observe that F is a linear space which, by the argument just presented, contains a dense subset of the simple functions. Since the simple functions are dense in $L_X^1(P)$, it follows that F is also dense in $L_X^1(P)$.

B) F is closed in $L_X^1(\Omega, F, P)$: Suppose that f is in the closure of F. Given $\varepsilon > 0$ and $\delta > 0$ find $g \in F$ with $\|f - g\|_1 < \frac{1}{2}\varepsilon\delta$ and let g_n denote $\mathbb{E}[g|F_n]$. For almost all $\omega \in \Omega$ we have

$$\|f_n(\omega) - f_k(\omega)\| \leqq \|g_n(\omega) - g_k(\omega)\| + \|(f_n - g_n)(\omega) - (f_k - g_k)(\omega)\|$$

(1) $$\leqq \|g_n(\omega) - g_k(\omega)\| + \|\mathbb{E}[f - g|F_n](\omega)\| + \|\mathbb{E}[f - g|F_k](\omega)\|$$

$$\leqq \|g_n(\omega) - g_k(\omega)\| + 2 \sup_j E[\|f - g\| \,|F_j](\omega).$$

Let $h_j(\omega) \equiv E[\|f - g\| \,|F_j](\omega)$. It follows from (1) and the fact that $g \in F$ that

(2) $$\limsup_{n,k\to\infty} \|f_n(\omega) - f_k(\omega)\| \leqq 2 \sup_j h_j(\omega) \quad [\text{a.s.}P].$$

Note that $\{h_j : j = 1,2,\ldots\}$ satisfies the conditions of the maximal inequality 1.3.2. (See Theorem 1.2.3. Note that b) holds with "$\leqq$" replaced by "$=$".) Hence, from (2),

$$P\{\omega: \limsup_{n,k\to\infty} \|f_n(\omega) - f_k(\omega)\| > \varepsilon\} \leqq P\{\omega: 2 \sup_j h_j(\omega) > \varepsilon\}$$

(3) $$\leqq \frac{2}{\varepsilon} \sup_j \|h_j\|_1 = \frac{2}{\varepsilon} \sup_j \int_\Omega \|f(\omega) - g(\omega)\| dP(\omega) = \frac{2}{\varepsilon}\|f - g\|_1 < \delta.$$

Since both ε and δ are arbitrary, it follows that $\lim_n f_n(\omega) \equiv \hat{f}(\omega)$ exists [a.s.P].

We are now in a position to establish that $f \in F$. Using 1.1.4 as in part A), given $\varepsilon > 0$ it is possible to find a function $g \in L_X^1(\Omega, F_N, P)$ for some N such that $\|f - g\|_1 < \frac{\varepsilon}{2}$. When $n \geqq N$ it follows that

$$\|f - f_n\|_1 \leqq \|f - g\|_1 + \|g - \mathbb{E}[g|F_n]\|_1 + \|\mathbb{E}[g - f|F_n]\|_1$$

$$< \frac{\varepsilon}{2} + 0 + \|g - f\|_1 < \varepsilon.$$

That is $\lim_n \|f - f_n\|_1 = 0$. Thus some subsequence of the f_n's converges almost surely to f (cf. Theorem 1.1.8(f)), whence $f(\omega) = \lim_i f_{n_i}(\omega) = \hat{f}(\omega) = \lim_n f_n(\omega)$ [a.s.P].

§4. Banach valued martingales and their convergence: a preview.

We begin with a simple but important example.

EXAMPLE 1.4.1. Suppose that $f \in L^1_X(\Omega,F,P)$ and that $F_1 \subset F_2 \subset \ldots \subset F$ is an increasing sequence of sub σ-algebras of F. It is easy to check that the functions $f_n \equiv \mathbb{E}[f|F_n]$ satisfy:

a) f_n is strongly F_n-measurable for each n; and

b) $\int_A f_n dP = \int_A f_k dP$ whenever $n \leqq k$ and $A \in F_n$.

It turns out that the nature of the convergence of such a sequence of functions is typical of those of a large class of X-valued martingales (the L^1_X-bounded ones based on $\mathbb{N}$) if and only if X has the Radon-Nikodým Property. Consequently their study is quite appropriate here. Example 1.4.3 and the subsequent discussion elaborate on some of these statements. The next definition formalizes the important structure of the functions in Example 1.4.1 (listed in properties a) and b) there).

DEFINITION 1.4.2. Let $(B,<)$ be a directed set. (In these notes B will almost always be $\mathbb{N}$, the nonnegative integers.) Suppose that $(F_\beta)_{\beta\in B}$ is a net of sub σ-algebras of the σ-algebra F and that $F_\alpha \subset F_\beta$ whenever $\alpha < \beta$. Suppose that f_β is a function in $L^1_X(\Omega,F,P)$ for each β. If the net $(f_\beta)_{\beta\in B}$ satisfies the conditions

a) f_β is strongly F_β-measurable for each $\beta \in B$; and

b) $\int_A f_\alpha dP = \int_A f_\beta dP$ whenever $\alpha < \beta$ and $A \in F_\alpha$,

then the net $(f_\beta,F_\beta)_{\beta\in B}$ is said to be an X-valued martingale. (Observe that b) may be rewritten: $\mathbb{E}[f_\beta \mid F_\alpha] = f_\alpha$ [a.s.P] when $\alpha < \beta$.) When $X = \mathbb{R}$ and b) is replaced by the inequality $\int_A f_\alpha dP \leqq \int_A f_\beta dP$ for $\alpha < \beta$ and $A \in F_\alpha$ then $(f_\beta,F_\beta)_{\beta\in B}$ is called a submartingale. Suppose that $B = \mathbb{N}$, each F_n is generated by countably many atoms, and F is generated by $\bigcup_{n=1}^{\infty} F_n$. Then $(f_n,F_n)_{n=1}^{\infty}$ is called an elementary martingale. The X-valued martingale $(f_\beta,F_\beta)_{\beta\in B}$ is bounded if there exists a bounded set D in X such that $P\{\omega: f_\beta(\omega) \in D\} = 1$ for each β. It is L^1-bounded if $\{f_\beta\}_{\beta\in\mathcal{B}}$ is a bounded set in $L^1_X(\Omega,\mathcal{F},P)$.

Returning to Example 1.4.1 and its notation momentarily, note that $\lim_n f_n(\omega)$ exists [a.s.P] by Theorem 1.3.1. (In fact, 1.3.1 guarantees that if $\mathcal{F}_\infty$ is the σ-algebra generated by $\bigcup_{n=1}^{\infty} \mathcal{F}_n$ then $\lim_n f_n(\omega) = \mathbb{E}[f|\mathcal{F}_\infty](\omega)$ [a.s.P].) That such convergence behavior is not typical of X-valued martingales in general is the thrust of the following example.

EXAMPLE 1.4.3. Let $\Omega \equiv \prod_{i=1}^{\infty} \{-1,1\}$ denote the compact metric space of all sequences of -1's and +1's equipped with the product topology, and let us agree to denote the kth coordinate of a point $\omega \in \Omega$ by ω_k. Define P on the Borel σ-algebra of Ω (henceforth denoted by $\mathcal{F}$) to be the product measure each of whose factors assigns mass 1/2 to $\{-1\}$ and $\{+1\}$. Thus $(\Omega,\mathcal{F},P)$ is a probability space, and we take $\mathcal{F}_n \equiv \{A \times \prod_{i=n+1}^{\infty} \{-1,1\}: \quad A \subset \prod_{i=1}^{n} \{-1,1\}$ is arbitrary$\}$. The sets in $\mathcal{F}_n$ are then restricted in the first n coordinates only. Evidently $\bigcup_{n=1}^{\infty} \mathcal{F}_n$ generates $\mathcal{F}$.

Let $X \equiv c_0$, the Banach space of sequences of real numbers convergent to 0, with the usual coordinate operations and norm $\|(x_i)\| \equiv \max\{|x_i|: i = 1,2,\ldots\}$. Finally, we define functions $f_n: \Omega \to c_0$ as follows:

$$f_0(\omega) \equiv 0 \in c_0 \quad \text{for all } \omega;$$

while, for $n \geq 1$, f_n is the function which assumes the 2^n values $(\pm1,\pm1,\ldots,\pm1,0,0,\ldots)$ with equal probability, given by the formula

$$f_n(\omega) \equiv (\omega_1,\omega_2,\ldots\omega_n,0,0,\ldots) \in c_0 \quad \text{for all } \omega.$$

Observe that each f_n depends only on the first n coordinates of Ω and is hence strongly $\mathcal{F}_n$-measurable. To check that $(f_n,\mathcal{F}_n)_{n=1}^{\infty}$ is a martingale it suffices to show that $\int_A f_n dP = \int_A f_{n+1} dP$ where

$$A \equiv \{(\varepsilon_1,\varepsilon_2,\ldots, \varepsilon_n) \times \prod_{i=n+1}^{\infty} \{-1,1\}\} \qquad (\text{each } \varepsilon_i \text{ is } \pm1)$$

since each set in $\mathcal{F}_n$ is a finite union of such atoms, and the result for general

$k \geqq n$ follows from this one by induction. But for such an A we have $\int_A f_n dP = 2^{-n}(\varepsilon_1,\ldots,\varepsilon_n,0,0,\ldots)$ while

$$\int_A f_{n+1}dP = \int_{A_{-1}} f_{n+1}dP + \int_{A_{+1}} f_{n+1}dP$$

where $A_{\pm 1} \equiv \{(\varepsilon_1,\ldots,\varepsilon_n,\pm 1) \times \prod_{i=n+2}^{\infty} \{-1,1\}\}$. Hence

$$\int_A f_{n+1}dP = 2^{-(n+1)}(\varepsilon_1,\ldots,\varepsilon_n,-1,0,0,\ldots) + 2^{-(n+1)}(\varepsilon_1,\ldots,\varepsilon_n,+1,0,0,\ldots)$$

$$= 2^{-n}(\varepsilon_1,\ldots,\varepsilon_n,0,0,\ldots) = \int_A f_n dP.$$

This martingale is bounded, since $\|f_n(\omega)\| \leqq 1$ for each n and each ω. However, $\lim_n f_n(\omega)$ fails to exist for each ω. (Indeed, suppose that $\lim_n f_n(\omega) = (x_i)$ for some ω. Then each x_i would be ± 1 since for $k \geqq i$ the ith entry of $f_k(\omega)$ is ± 1. But then $\lim_i x_i \neq 0$ so $(x_i) \notin c_o$.) Thus, bounded elementary c_o-valued martingales need not converge at any point.

Let us analyze why 1.4.3 works. In order to go from the nth to the $(n+1)^{st}$ function f_{n+1}, we effectively constructed, for each point (x_i) in Range f_n, two points (y_i) and (z_i), far apart, whose midpoint is (x_i). So far, this is easy enough to do, no matter which Banach space X is used. However, the extra condition on this martingale is that it is bounded, which translates into the condition on (y_i) and (z_i) that their norms not be much larger (if at all) than that of (x_i). This last requirement places severe restrictions on X. Indeed, from this perspective one would <u>not</u> expect to be able to carry out a similar such construction in any Banach space in which the unit ball is, say, uniformly convex, since, in such a case, the norm of at least one of the points (y_i) and (z_i) would be much larger than that of (x_i). (A two dimensional picture is quite convincing.) On the other hand, in spaces like $L^1(\mu)$ (μ nonatomic) the unit ball again has the property that each point of the surface of the unit ball is the center of a large flat section of the unit ball. A second class of examples is furnished by C(Y) spaces for "most" compact Hausdorff spaces Y. (Although the unit ball of C(Y) may contain some extreme points, for many such Y it is not generated by its extreme points. Moreover, all points other than the extreme points lie in large flat sections of the unit ball.)

As expected, a similar construction is possible in these latter spaces.

The relevance of martingale convergence questions to those problems of concern to us stems from Corollary 2.3.7 which states in part that a Banach space X has the Radon-Nikodým Property if and only if bounded X-valued martingales based on $\mathbb{N}$ converge almost surely. Using this result prematurely, then, we might expect that uniformly convex spaces do have the Radon-Nikodým Property, while $L^1(\mu)$ (μ non-atomic), c_o, and most C(Y) spaces do not. These conjectures will be borne out in the course of these notes.

CHAPTER 2. BASIC RESULTS

In this chapter we begin the study of the geometry of those closed convex subsets of Banach spaces in which a vector-valued version of the Radon-Nikodým theorem holds. Section 1 covers the specific background needed to appreciate the basic characterizations of these sets, while the characterizations themselves appear in Section 2. The background material includes measure-theoretic, geometric, and martingale definitions and examples. The remainder of the chapter delves more carefully into some of the geometric aspects uncovered in the first two sections.

§1. The Radon-Nikodým Property, the Martingale Convergence Property, and subset s-dentability.

We begin with one of the common definitions of the Radon-Nikodým Property. (Others are mentioned after Example 2.1.2.)

DEFINITIONS 2.1.1. Let K be a closed bounded convex set in a Banach space X. Then K has the Radon-Nikodým Property for (Ω,F,P) if for each X-valued measure m on F which is absolutely continuous with respect to P and whose average range AR(m) is contained in K (cf. 1.1.3) there exists an $f \in L^1_X(\Omega,F,P)$ such that $m(A) = \int_A f dP$ for each $A \in F$. The set K is said to have the Radon-Nikodým Property (abbreviated RNP) if K has the Radon-Nikodým Property for each probability space (Ω,F,P). Finally, suppose that C is a closed convex (possibly unbounded) subset of X. (For example, C might be X itself.) Then C has the RNP if each of its closed bounded convex subsets has the RNP.

The above definition is applied directly in our first example.

EXAMPLE 2.1.2. Let $(\Omega,F,P) \equiv$ ([0,1], Lebesgue measurable sets, Lebesgue measure). Take $X \equiv L^1[0,1]$ and define $m: F \to X$ by $m(A) \equiv \chi_A$ for each $A \in F$. Note that both $m \ll P$ and $AR(m) \subset$ closed unit ball of X. We will prove that there is no function $f \in L^1_X(P)$ such that $m(A) = \int_A f dP$ for all $A \in F$ whence $L^1[0,1]$ does not have the RNP. Assume temporarily that such an f exists, pick $A \in F$, and choose $F \in L^\infty[0,1] = L^1[0,1]^*$. The notation (F,h) directly below denotes the value of $F \in X^*$ at $h \in X$. We have

$$\int_A (F,f(t))dt = (F,\int_A f(t)dt) = (F,\chi_A) = \int_A F(t)dt.$$

Hence $(F,f(t)) = F(t)$ for all $t \in [0,1]\backslash A(F)$ for some set $A(F) \in \mathcal{F}$ (possibly depending on F) with $P(A(F)) = 0$. Denote by $\{I_n : n = 1,2,\ldots\}$ a listing of all subintervals of $[0,1]$ with rational endpoints, and for each n let $F_n \equiv \chi_{I_n} \in L^\infty[0,1]$. Define $A \equiv \bigcup_{n=1}^{\infty} A(F_n)$ and pick any $x \in [0,1]\backslash A$. Then

$$\int_{I_n} f(x)(s)ds = \int F_n(s)f(x)(s)ds = F_n(x) = 0 \quad \text{whenever} \quad x \notin I_n .$$

This implies that $f(x)(s) = 0$ for almost all $s \in [0,1]$ as long as $x \in [0,1]\backslash A$, whence $f: [0,1] \to X$ vanishes almost surely. Since this is in contradiction to the hypothesis that $\int_A fdP = \chi_A \neq 0$ whenever $P(A) > 0$, the proof that $L^1[0,1]$ lacks the RNP is complete. (An easier proof of this fact, given the appropriate tools, appears in Corollary 3.3.7.)

We have listed below two alternative definitions of the RNP which sometimes appear in the literature. That each is equivalent to the one given in 2.1.1 follows by routine measure-theoretic arguments together with the fact that, as in the real case, when $m: \mathcal{F} \to X$ is a measure of finite total variation and $m \ll P$, then $\frac{dm}{d|m|} \frac{d|m|}{dP} = \frac{dm}{dP}$ [a.s.P] whenever $\frac{dm}{d|m|}$ exists.

a) X has the RNP if whenever m is an X-valued measure on a measurable space $(\Omega,\mathcal{F})$ which is of finite total variation, then there is an $f \in L^1_X(\Omega,\mathcal{F},|m|)$ such that $\int_A fd|m| = m(A)$ for $A \in \mathcal{F}$.

b) X has the RNP if each X-valued measure m on $\mathcal{F}$ which is of finite total variation and satisfies $m \ll P$ admits an $f \in L^1_X(\Omega,\mathcal{F},P)$ such that $\int_A fdP = m(A)$ for each $A \in \mathcal{F}$.

Early workers in this field (see, for example, Bochner [1933a] and [1933b], Clarkson [1936], Dunford and Morse [1936], Gelfand [1938], and Bochner and Taylor [1938]) were concerned with the Banach space property that each X-valued function of bounded variation on [0,1] be differentiable almost surely. It turns out that this property (known as the Gelfand-Fréchet property) is also equivalent to the RNP. The advances for many years had strong operator-theoretic overtones and were, in large part, stimulated by an important paper of Dunford and Pettis [1940]. Almost thirty years later the connections with other areas of mathematics began to appear, one of the first being the paper of S.D. Chatterji [1968] in which the relationship between martingale convergence and the Radon-Nikodým theorem is examined in detail. Before stating the martingale property which will be of interest to us, observe that when K is a closed bounded convex set in a Banach space X then

$$L^1_K(\Omega,\mathcal{F},P) \equiv \{f \in L^1_X(\Omega,\mathcal{F},P)\colon f(\omega) \in K \quad [\text{a.s.}P]\}$$

is also a closed bounded convex set.

DEFINITIONS 2.1.3. A closed bounded convex set $K \subset X$ has the Martingale Convergence Property for $(\Omega,\mathcal{F},P)$ if whenever $(f_n,\mathcal{F}_n)_{n=1}^{\infty}$ is a martingale for which $\bigcup_{n=1}^{\infty} \mathcal{F}_n$ generates $\mathcal{F}$ and $f_n \in L^1_K(\Omega,\mathcal{F}_n,P)$ for each n then there exists an $f \in L^1_K(\Omega,\mathcal{F},P)$ such that $\lim_n \|f_n(\omega) - f(\omega)\| = 0$ [a.s.P]. K has the Martingale Convergence Property (MCP) if it has the Martingale Convergence Property for each probability space $(\Omega,\mathcal{F},P)$. K has the elementary MCP if the above conditions are required to hold only for elementary martingales $(f_n,\mathcal{F}_n)_{n=1}^{\infty}$ which are K-valued and such that $\bigcup_{n=1}^{\infty} \mathcal{F}_n$ generates $\mathcal{F}$. Finally, a closed convex subset C of X has the MCP if each of its closed bounded convex subsets has the MCP.

In the new terminology, then, Example 1.4.3 demonstrates that the unit ball of c_o - and consequently c_o itself - lacks the MCP. Let us turn now to a purely geometric notion. It is convenient to introduce some notation first.

NOTATION 2.1.4. Let $D \subset X$. Then co(D) is the convex hull of D, $\overline{\text{co}}(D)$ denotes the norm-closure of the convex hull of D, and $\text{s-co}(D) \equiv \{\sum_{i=1}^{\infty} \alpha_i x_i\colon x_i \in D,\ \alpha_i \geq 0,\ \sum_{i=1}^{\infty} \alpha_i = 1\}$. (As usual, we require that $\sum_{i=1}^{\infty} \alpha_i x_i$ be the norm-limit of the Cauchy sequence of partial sums $\{\sum_{i=1}^{k} \alpha_i x_i\}_{k=1}^{\infty}$.) If $x \in X$ and $\varepsilon > 0$ then $U_\varepsilon(x) \equiv \{y \in X\colon \|y - x\| < \varepsilon\}$ and $U_\varepsilon[x] \equiv \{y \in X\colon \|y - x\| \leq \varepsilon\}$. If $K \subset X$ is convex then ex(K) denotes the set of extreme points of K.

If $D \subset \mathbb{R}^n$ is compact, a result of Carathéodory [1907] asserts that each point of $\overline{\text{co}}(D)$ is in fact a convex combination of at most n + 1 points of D. (In particular, then, $\text{co}(D) = \overline{\text{co}}(D)$.) Hence suppose that K is a compact convex set in

$\mathbb{R}^n$, $x \in ex(K)$, and $\varepsilon > 0$. Let $D \equiv K\backslash U_\varepsilon(x)$. Since x is extreme, $x \notin co(D)$ and thus $x \notin \overline{co}(D)$ by the Carathéodory result just quoted. The situation is more intricate for general Banach spaces X. If $D \subset X$ is closed bounded and convex then $co(D) \subset s\text{-}co(D) \subset \overline{co}(D)$ and each of these containments may be strict. Moreover, suppose that $K \subset X$ is a closed bounded convex set. If $x \in ex(K)$, $\varepsilon > 0$, and $D \equiv K\backslash U_\varepsilon(x)$ then it is immediate that $x \notin co(D)$ and, in fact, that $x \notin s\text{-}co(D)$. On the other hand, $x \in \overline{co}(D)$ is possible. (See, for example, 2.1.6.) In any case, it is intuitively appealing to think of a point x, extreme or not, of a convex set such that $x \notin co(K\backslash U_\varepsilon(x))$ (or $x \notin s\text{-}co(K\backslash U_\varepsilon(x))$, or $x \notin \overline{co}(K\backslash U_\varepsilon(x))$) as one around which K "bends". Upon reflection, one sees that this type of rotundity condition really has nothing to do with the convexity of K, and this is taken into account in Definition 2.1.5. The added generality is important, as will soon become evident.

DEFINITION 2.1.5. Let D be a bounded subset of X. Then D is s-dentable if for each $\varepsilon > 0$ there is a point $x_\varepsilon \in D$ such that $x_\varepsilon \notin s\text{-}co(D\backslash U_\varepsilon(x_\varepsilon))$. D is dentable (respectively, c-dentable) if for each $\varepsilon > 0$ there is a point $x_\varepsilon \in D$ such that $x_\varepsilon \notin \overline{co}(D\backslash U_\varepsilon(x_\varepsilon))$ (respectively, $x_\varepsilon \notin co(D\backslash U_\varepsilon(x_\varepsilon))$). Let C be a possibly unbounded subset of X. Then C is subset s-dentable if each bounded nonempty subset of C is s-dentable, and subset dentable (subset c-dentable) if each of its bounded nonempty subsets is dentable (respectively, c-dentable).

The example below illustrates some of these notions.

EXAMPLE 2.1.6. Let $X \equiv C[0,1]$, the Banach space of continuous functions on [0,1] with $\|f\| \equiv \max\{|f(t)|: t \in [0,1]\}$. Let K be its closed unit ball, $U_1[0]$. Observe that K has exactly two extreme points: namely, the functions identically +1 and identically -1. Thus K is s-dentable. (Take either extreme point to be f_ε in 2.1.5 for each $\varepsilon > 0$.) On the other hand, K is not dentable. Indeed, suppose that $f \in K$. For any integer $n > 0$ choose functions $f_1^n,\ldots,f_n^n$ in K so that $f_i^n(t) = f(t)$ for $t \notin [\frac{i-1}{n}, \frac{i}{n}]$ and $|f_i^n(t_i^n) - f(t_i^n)| > \frac{1}{2}$ for some point $t_i^n \in (\frac{i-1}{n}, \frac{i}{n})$. Then $\|f_i^n - f\| > \frac{1}{2}$ for $i = 1,\ldots,n$ and yet $\|\Sigma_{i=1}^n \frac{1}{n} f_i^n - f\| \leq \frac{2}{n}$, as is easily checked. It follows that $f \in \overline{co}(K\backslash U_{1/2}(f))$ since n was arbitrary. Hence K is not dentable (and thus neither is $C[0,1]$).

§2. The equivalences of RNP, MCP, and subset s-dentability for closed bounded convex sets.

Throughout this section K will denote a closed bounded convex subset of X.

THEOREM 2.2.1. If $K \subset X$ has the RNP for $(\Omega,\mathcal{F},P)$ then K has the MCP for $(\Omega,\mathcal{F},P)$.

Proof. Assume that K has the RNP for $(\Omega,\mathcal{F},P)$, that $\mathcal{F}_1 \subset \mathcal{F}_2 \subset \ldots$ is an increasing sequence of σ-algebras such that $\bigcup_{n=1}^{\infty} \mathcal{F}_n$ generates $\mathcal{F}$, and that $(f_n,\mathcal{F}_n)_{n=1}^{\infty}$ is a martingale taking its values in K. Let $m_n(A) \equiv \int_A f_n dP$ for each $A \in \mathcal{F}$. Then m_n is an X-valued measure on $\mathcal{F}$ and $m_n << P$. Moreover, $AR(m_n) \subset K$. (Indeed, for $A \in \mathcal{F}$, $P(A) > 0$, and for each $F \in X^*$ we have

$$F\left(\frac{m_n(A)}{P(A)}\right) = \frac{1}{P(A)}\int_A F\circ f_n dP \leqq \sup\{F(x)\colon x \in K\}.$$

Thus since K is a weakly closed convex set, $\frac{m_n(A)}{P(A)} \in K$.) Note that $\lim_n m_n(A)$ exists for each $A \in \bigcup_{n=1}^{\infty} \mathcal{F}_n$ since $(f_n,\mathcal{F}_n)_{n=1}^{\infty}$ is a martingale. In fact, this limit exists for all $A \in \mathcal{F}$ as we now demonstrate.

Let $M \equiv \sup\{\|x\|\colon x \in K\}$. Given $\varepsilon > 0$ and $A \in \mathcal{F}$ find $B \in \mathcal{F}_N$ for some N such that $P(A \Delta B) < \frac{\varepsilon}{2M}$. (Recall: $\bigcup_{n=1}^{\infty} \mathcal{F}_n$ is dense in $\mathcal{F}$ in the metric of 1.1.4.) If $n \geqq N$ then

$$\left\|\int_A f_n dP - \int_A f_N dP\right\| \leqq \left\|\int_B f_n dP - \int_B f_N dP\right\| + \int_{A\Delta B} \|f_n\| dP + \int_{A\Delta B} \|f_N\| dP$$

$$\leqq 2P(A \Delta B)M < \varepsilon .$$

Consequently $\{m_n(A)\colon n = 1,2,\ldots\}$ is a Cauchy sequence and $m(A) \equiv \lim_n m_n(A)$ exists for all $A \in \mathcal{F}$. The Hahn-Vitali-Saks Theorem (See 1.1.5) implies that m is a measure. Clearly $m << P$ and $AR(m) \subset \text{closure}\left(\bigcup_{n=1}^{\infty} AR(m_n)\right) \subset K$. Since K has the RNP for $(\Omega,\mathcal{F},P)$ there is an $f \in L^1_X(\Omega,\mathcal{F},P)$ such that $\int_A f dP = m(A)$ for each $A \in \mathcal{F}$. For $A \in \mathcal{F}_n$ we have

$$\int_A f dP = m(A) = m_n(A) = \int_A f_n dP.$$

Thus $\mathbb{E}[f|\mathcal{F}_n] = f_n$ [a.s.P], and Lévy's result (see 1.3.1) shows that

$\lim_n \|f_n(\omega) - f(\omega)\| = 0$ [a.s.P]. That is, K has the MCP for $(\Omega, \mathcal{F}, P)$.

H. Maynard [1973] introduced the notion of s-dentability when there were few firm links between geometry and integration theory, but several conjectures. In fact, his main result provided the first geometric characterization of spaces with the RNP. (Several important connections were established soon thereafter.) The proof of Theorem 2.2.2 is essentially Maynard's construction from his 1973 paper.

THEOREM 2.2.2 (Maynard). If K has the elementary MCP for ([0,1), Borel sets, Lebesgue measure) then K is subset s-dentable.

Proof. We will prove the contrapositive. Thus assume that $D \subset K$ is not s-dentable, and that $\varepsilon > 0$ has been found so that $x \in$ s-co$(D \setminus U_\varepsilon(x))$ for each $x \in D$. A nonconvergent martingale $(f_n, \mathcal{F}_n)_{n=1}^\infty$ will be constructed inductively.

At the 0th stage pick any point $x_{[0,1)} \in D$, let $f_0(t) \equiv x_{[0,1)}$ for each $t \in [0,1)$, and $\pi_0 \equiv \{[0,1)\}$. Define $\mathcal{F}_0$ to be the σ-algebra generated by π_0. Suppose now that π_n is a countable partition of [0,1) into half-open intervals, that $\mathcal{F}_n$ is the σ-algebra generated by π_n, and that $f_n \equiv \sum_{A \in \pi_n} x_A \chi_A$ for some choice of points $x_A \in D$. Temporarily fix $A \in \pi_n$. Since $x_A \in D$ we have $x_A \in$ s-co$(D \setminus U_\varepsilon(x_A))$. Consequently there are numbers t_j, $0 < t_j < \frac{1}{n+1}$ with $\Sigma\, t_j = 1$, and points $y_j \in D$ (not necessarily distinct) such that $\|x_A - y_j\| \geqq \varepsilon$ for each j and yet $\Sigma\, t_j y_j = x_A$. Partition A into half-open intervals, say $\{B_j\}$, one for each t_j, so that $P(B_j) = t_j P(A)$ for each j (P $\equiv$ Lebesgue measure). Now relabel y_j as x_{B_j} and let π_{n+1} be the collection of all such half-open intervals B_j as A ranges over π_n. Let $f_{n+1} \equiv \sum_{B \in \pi_{n+1}} x_B \chi_B$ and denote by $\mathcal{F}_{n+1}$ the σ-algebra generated by π_{n+1}. Evidently $\bigcup_{n=1}^\infty \mathcal{F}_n$ generates $\mathcal{F} \equiv$ Borel sets of [0,1), $(f_n, \mathcal{F}_n)_{n=1}^\infty$ is a K-valued elementary martingale, and $\lim_n f_n(t)$ fails to exist for each $t \in [0,1)$. Thus K does not have the elementary MCP for ([0,1), Borel sets, Lebesgue measure), which ends the proof.

The cycle begun in Theorems 2.2.1 and 2.2.2 is completed by Theorem 2.2.3 below. The following elementary discussion contains the heart of the argument found in the proof. Suppose, then, that $(\Omega, \mathcal{F}, P)$ and K are given and that $m: \mathcal{F} \to X$ is a vector measure of the form $m(A) \equiv \sum_{i=1}^n x_i P(A \cap B_i)$ for each $A \in \mathcal{F}$ (where each $x_i \in K$, and $\{B_i\}_{i=1}^n$ is a partition of Ω chosen from F with $P(B_i) > 0$ for each i). Note that when $A \in F$ and both $P(A) > 0$ and $A \subset B_i$ for some i then $\frac{m(A)}{P(A)} = x_i = \frac{m(B_i)}{P(B_i)}$. Let $f \equiv \sum_{i=1}^n x_i \chi_{B_i}$. Then for any $A \in F$ we have

$$\int_A f dP = \sum_{i=1}^n x_i P(A \cap B_i) = \sum_{i=1}^n m(A \cap B_i) = m(A).$$

Hence $f = \frac{dm}{dP}$. In general vector measures m such that $m << P$ and $AR(m) \subset K$ cannot be approximated by the simple ones with the above form, but when K is subset s-dentable such an approximation is possible. The detailed argument follows.

THEOREM 2.2.3. If K is subset s-dentable then K has the RNP.

The proof requires some preliminaries. Note that the next definition and lemma formalize the type of approximation just discussed. Through the end of the proof of 2.2.3, (Ω,F,P) will denote a fixed probability space and $m: F \to X$ a vector measure absolutely continuous with respect to P.

DEFINITION 2.2.4. For any $A \in F$ with $P(A) > 0$, $x \in X$, and $\varepsilon > 0$, the set A is said to be (x,ε)-pure if for each $B \in F$ with $B \subset A$ and $P(B) > 0$ we have $\left\|\frac{m(B)}{P(B)} - x\right\| < \varepsilon$. The set A is said to be ε-pure if it is (x,ε)-pure for some $x \in X$.

LEMMA 2.2.5. Assume that K is subset s-dentable. Let $m: F \to X$ be a measure absolutely continuous with respect to P such that $AR(m) \subset K$. The given $\varepsilon > 0$ there is a countable partition $\{B_i\}$ of Ω of sets chosen from F such that each B_i is ε-pure.

Proof. The first step is to produce an ε-pure subset of any prescribed set of positive measure. Thus suppose $A \in F$ and $P(A) > 0$. Let $D \equiv \{\frac{m(B)}{P(B)}: B \in F, B \subset A,$ and $P(B) > 0\}$. Since D is s-dentable, there is a point $x \in D$ such that $x \notin$ s-co$(D\setminus U_\varepsilon(x))$. Write $x \equiv \frac{m(A_o)}{P(A_o)}$ where $A_o \in F$, $P(A_o) > 0$, $A_o \subset A$. (This is the only place in the proof of 2.2.3 that the subset s-dentability hypothesis on K is used.) We claim that some subset B of A_o is (x,ε)-pure. If not, then each subset B of A_o such that $B \in F$, $P(B) > 0$, contains a subset C of B satisfying:

(1) $$P(C) > 0,\ C \in F \text{ and } \left\|x - \frac{m(C)}{P(C)}\right\| \geqq \varepsilon .$$

Let $\mathcal{C}$ be a maximal disjoint family of subsets C of A_o satisfying (1). Since $P(A_o) < \infty$, $\mathcal{C}$ is countable. Write $\mathcal{C} \equiv \{C_i: i = 1,2,...\}$ and let $C_o \equiv \bigcup_{i=1}^{\infty} C_i$. Then by the maximality of $\mathcal{C}$ and by the assumption, $P(A_o \backslash C_o) = 0$, and consequently $m(A_o \backslash C_o) = 0$. Hence

$$x = \frac{m(C_o)}{P(C_o)} = \sum_{i=1}^{\infty} \frac{P(C_i)}{P(C_o)} \frac{m(C_i)}{P(C_i)} .$$

Since $\frac{m(C_i)}{P(C_i)} \in D$ and $\left\| x - \frac{m(C_i)}{P(C_i)} \right\| \geqq \varepsilon$ for each i, this shows that $x \in \text{s-co}(D \backslash U_\varepsilon(x))$ contrary to our choice of x.

Now let $\mathcal{B}$ be a maximal disjoint family of ε-pure subsets of Ω. Then, as before, $\mathcal{B}$ is countable. Write $\mathcal{B} \equiv \{B_i: i = 1,2,...\}$ and let $B_o \equiv \bigcup_i B_i$. Then by the maximality of $\mathcal{B}$ and from what has been established in the first step of the proof, $P(\Omega \backslash B_0) = 0$. Since $m(\Omega \backslash B_0) = 0$ as well, the set $B_1' \equiv B_1 \cup (\Omega \backslash \bigcup_{i \geqq 2} B_i)$ is ε-pure. Thus the partition $\{B_1'\} \cup \{B_i\}_{i \geqq 2}$ satisfies the conditions of the lemma.

We are now in a position to prove Theorem 2.2.3.

<u>Proof of 2.2.3.</u> Suppose that $m: \mathcal{F} \to X$ is a measure with average range in K and that m is absolutely continuous with respect to P. Let $\{B_i(n): i = 1,2,...\}$ $n = 1,2,...$ be a sequence of countable partitions of Ω with entries in $\mathcal{F}$ chosen so that for each j and each $n \geqq 2$, $B_j(n) \subset B_i(n-1)$ for some i, and so that each $B_i(n)$ is $(x_i(n), \frac{1}{n})$ - pure for some $x_i(n) \in K$. (This can easily be accomplished by using 2.2.5 on each set of the $(n-1)^{\underline{st}}$ stage in order to construct the sets of the $n^{\underline{th}}$ stage.) Let $f_n \equiv \sum_i x_i(n) \chi_{B_i(n)}$. We will show $\{f_n: n = 1,2,...\}$ is Cauchy in $L^1_X(\Omega, \mathcal{F}, P)$. Assume that $k \geqq n$. Note that $B_j(k) \subset B_i(n)$ implies that

$$\|x_j(k) - x_i(n)\| \leqq \left\| x_j(k) - \frac{m(B_j(k))}{P(B_j(k))} \right\| + \left\| \frac{m(B_j(k))}{P(B_j(k))} - x_i(n) \right\| < \frac{1}{k} + \frac{1}{n} \leqq \frac{2}{n} .$$

Thus $\|f_n(\omega) - f_k(\omega)\| \leqq \frac{2}{n}$ for each $\omega \in \Omega$ and $k \geqq n$ and hence $\|f_n - f_k\|_1 \leqq \frac{2}{n}$ for $k \geqq n$.

Let f be the $L^1_X(P)$-limit of $\{f_n\}$. In order to prove that $\int_A f dP = m(A)$ for each $A \in \mathcal{F}$ first note that in the special case that $A \subset B_i(n)$ and $A \in \mathcal{F}$ then $\|m(A) - x_i(n)P(A)\| \leqq \frac{1}{n} P(A)$ whether $P(A) > 0$ or $P(A) = 0$. Hence, for general $A \in \mathcal{F}$ we have

$$\left\| \int_A f_n dP - m(A) \right\| = \left\| \int_A \sum_i x_i(n) \chi_{B_i(n)} dP - \sum_i m(A \cap B_i(n)) \right\|$$

$$= \left\| \sum_i [x_i(n) P(A \cap B_i(n)) - m(A \cap B_i(n)) \right\|$$

$$\leq \sum_i \frac{1}{n} P(A \cap B_i(n)) = \frac{1}{n} P(A).$$

Consequently

$$\left\| \int_A f dP - m(A) \right\| \leq \left\| \int_A f dP - \int_A f_n dP \right\| + \left\| \int_A f_n dP - m(A) \right\|$$

$$\leq \| f - f_n \|_1 + \frac{1}{n} P(A).$$

Since n is arbitrary we conclude that $\int_A f dP = m(A)$ for each $A \in F$, as was to be shown.

One way of viewing the proof of Theorem 2.2.3 is that $AR(m) \subset K$ is quite small (nearly compact, in fact) when K is subset s-dentable. The next result, which makes the above comment precise, is due to Rieffel [1967]. This paper is of the utmost historical importance: not only was the dentability notion introduced there, but questions posed in that paper served to direct the efforts of many people in fruitful ways. The proof of 2.2.6 which appears below uses a fact about weakly compact subsets of Banach spaces whose proof will be postponed until Chapter 3 (see 3.6.1): namely, each weakly compact convex subset of a Banach space has the RNP.

THEOREM 2.2.6 (Rieffel). Let (Ω, F, P) be a probability space, $K \subset X$ a closed bounded convex set, and $m: F \to X$ a vector measure such that both $m \ll P$ and $AR(m) \subset K$. Then the following are equivalent:

1) There is an $f \in L^1_X(\Omega, F, P)$ such that $\int_A f dP = m(A)$ for $A \in F$;
2) Given $\varepsilon > 0$ there is a set $\Omega_\varepsilon \in F$ such that:
   a) $P(\Omega \setminus \Omega_\varepsilon) < \varepsilon$ and
   b) $AR(m|_{\Omega_\varepsilon}) \equiv \{\frac{m(A)}{P(A)}: A \in F, P(A) > 0, A \subset \Omega_\varepsilon\}$ is relatively norm compact;
3) Given $\varepsilon > 0$ there is a set $\Omega_\varepsilon \in F$ such that:
   a) $P(\Omega \setminus \Omega_\varepsilon) < \varepsilon$ and
   b) $AR(m|_{\Omega_\varepsilon})$ is relatively weakly compact;
4) m locally has s-dentable average range. That is, given $A \in F$, $P(A) > 0$,

there is a set $B \in F$, $B \subset A$, and $P(B) > 0$, such that $AR(m|_B)$ is s-dentable.

Proof. 1 ⇒ 2 Let $\{s_n\}$ be a sequence of simple functions with $\lim_n s_n(\omega) = f(\omega)$ [a.s.P] and $\lim_n \|s_n - f\|_1 = 0$. Given $\varepsilon > 0$ it is possible by Egoroff's theorem to choose a set $\Omega_\varepsilon \in F$ such that both $P(\Omega\setminus\Omega_\varepsilon) < \varepsilon$ and $s_n \to f$ uniformly on Ω_ε. Let $m_n(A) \equiv \int_A s_n dP$ for $A \in F$, $A \subset \Omega_\varepsilon$. It is easy to see that

a) $AR(m_n|_{\Omega_\varepsilon}) \equiv \{\frac{m_n(A)}{P(A)} : A \in F,\ P(A) > 0,\ A \subset \Omega_\varepsilon\}$ is a finite-dimensional bounded set (which therefore contains a finite δ-net for each $\delta > 0$); and

b) If $A \in F$, $A \subset \Omega_\varepsilon$, and $P(A) > 0$, then

$$\left\|\frac{m(A)}{P(A)} - \frac{m_n(A)}{P(A)}\right\| \leq \sup_{\omega\in\Omega_\varepsilon} \|f(\omega) - s_n(\omega)\|.$$

Thus suppose $\delta > 0$. Pick n so large that $\sup_{\omega\in\Omega_\varepsilon} \|f(\omega) - s_n(\omega)\| < \frac{\delta}{2}$, and let $\{\frac{m_n(A_i)}{P(A_i)}: i = 1,\ldots,k\}$ be a finite $\frac{\delta}{2}$ - net in $AR(m_n|_{\Omega_\varepsilon})$. Then $\{\frac{m(A_i)}{P(A_i)}: i = 1,\ldots,k\}$ is a finite δ-net in $AR(m|_{\Omega_\varepsilon})$, whence this latter set is relatively norm-compact.

2 ⇒ 3 is obvious.

3 ⇒ 4 If for each $\varepsilon > 0$ the sets $AR(m|_{\Omega_\varepsilon})$ are relatively weakly compact, their closed convex hulls are weakly compact convex sets which, by Theorem 3.6.1, have the RNP. Hence each such set is subset s-dentable (put together 2.2.1 and 2.2.2 or, see 2.3.6) and condition 4 follows easily.

4 ⇒ 1 There is only one place in the proof of Theorem 2.2.3 in which the hypothesis "K is subset s-dentable" was used: namely, in showing that each subset of positive measure contains an ε-pure subset. (See paragraph one of the proof of 2.2.5.) In fact, the proof used only the formally weaker fact that each subset of positive measure contains a subset A of positive measure such that $D \equiv AR(m|_A)$ is s-dentable. But this is condition 4 of 2.2.6. Consequently statement 1 follows from 4 exactly as in the proof of 2.2.3.

The last result of this section strengthens Theorem 2.2.1. It is separated from the main cycle of results because it will not be needed subsequently. However, when dealing with the RNP from other perspectives, Theorem 2.2.7 below is quite important, and certainly it bears statement and proof here. The theorem is originally due to S.D. Chatterji [1968] whose proof was rather involved. The elegant one included here is due, I gather independently, to R.V. Chacon and L. Sucheston [1975], and to D.J.H. Garling [1976].

THEOREM 2.2.7 (Chatterji). If X has the RNP then each $L_X^1(\Omega,F,P)$-bounded martingale sequence on (Ω,F,P) converges almost surely.

Proof. Let $(F_n,F_n)_{n=1}^{\infty}$ be an X-valued martingale based on (Ω,F,P) with $\sup_n \|f_n\|_1 \equiv M < \infty$. By replacing F by the σ-algebra generated by $\bigcup_{n=1}^{\infty} F_n$ if necessary, assume that $\bigcup_{n=1}^{\infty} F_n$ generates F. We will use the following notation:

(a) $\|f_n\|: \Omega \to \mathbb{R}$ is the function defined by $\|f_n\|(\omega) \equiv \|f_n(\omega)\|$.

(b) If $\gamma: \Omega \to \mathbb{N}^+ \cup \{\infty\}$ is measurable then let f_γ be the function with domain $\{\omega: \gamma(\omega) < \infty\}$ given by $f_\gamma(\omega) \equiv f_{\gamma(\omega)}(\omega)$. Clearly f_γ is strongly measurable.

(c) For γ as in (b) and $k \in \mathbb{N}^+$ denote by $\gamma\wedge k$ the truncation of γ: $\gamma\wedge k(\omega) \equiv \min(k,\gamma(\omega))$.

(d) For γ as in (b), $\{\gamma = k\} \equiv \gamma^{-1}(k)$, $\{\gamma \geqq k\} \equiv \{\omega: \gamma(\omega) \geqq k\}$, etc. .

Fix $t > 0$ and define $\sigma: \Omega \to \mathbb{N}^+ \cup \{\infty\}$ as follows:

$$\sigma(\omega) \equiv \begin{cases} n & \text{if } \|f_i(\omega)\| < t \text{ for } i = 1,\dots,n-1 \text{ and } \|f_n(\omega)\| \geqq t; \\ \infty & \text{if } \|f_i(\omega)\| < t \text{ for each } i = 1,2,\dots . \end{cases}$$

Note that $\{\sigma = k\} \in F_k$ for each k.

We will prove that $(f_{\sigma\wedge k},F_k)_{k=1}^{\infty}$ is a martingale and that the corresponding sequence of measures $\{m_k\}$ given by $m_k(A) \equiv \int_A f_{\sigma\wedge k}\, dP$ is Cauchy. The Hahn-Vitali-Saks theorem guarantees that $m \equiv \lim_k m_k$ is also a measure and the RNP hypothesis provides an $L_X^1(P)$ function $g \equiv \frac{dm}{dP}$. The fact that $\mathbb{E}[g|F_k] = f_{\sigma\wedge k}$ is easily established. Observe that $f_n(\omega) = f_{\sigma\wedge n}(\omega)$ whenever $\sigma(\omega) = \infty$. In order to prove that $\{f_n\}$ converges almost surely, appeal is made to the maximal inequality 1.3.2. The details appear below.

Clearly $f_{\sigma\wedge k}$ is F_k-measurable. In order to prove that $(f_{\sigma\wedge k},F_k)$ is a martingale, recall that (f_n,F_n) is a martingale and observe that $A \cap \{\sigma > n\} \in F_n$ whenever $A \in F_n$. Thus for each such A,

$$\int_A f_{\sigma\wedge n}dP = \int_{A\cap\{\sigma\leqq n\}} f_{\sigma\wedge n}dP + \int_{A\cap\{\sigma>n\}} f_{\sigma\wedge n}dP$$

$$= \int_{A\cap\{\sigma\leqq n\}} f_{\sigma\wedge(n+1)}dP + \int_{A\cap\{\sigma>n\}} f_n dP$$

$$= \int_{A\cap\{\sigma\leqq n\}} f_{\sigma\wedge(n+1)}dP + \int_{A\cap\{\sigma>n\}} f_{n+1} dP$$

$$= \int_A f_{\sigma\wedge(n+1)}dP.$$

Thus $(f_{\sigma\wedge k}, F_k)$ is a martingale, and as indicated before, we let

$$m_k(A) \equiv \int_A f_{\sigma\wedge k}dP$$

define a sequence of X-valued measures on F. In order to prove that $\{m_k(A): k = 1,2,\ldots\}$ is Cauchy for each $A \in F$, we first show that the function $\omega \to \sup_k \|f_{\sigma\wedge k}(\omega)\|$ is integrable, and in this regard temporarily abbreviate $\sup_k \|f_{\sigma\wedge k}(\omega)\|$ by $h(\omega)$. Then h is a nonnegative measurable function and, from the definition of σ, $h(\omega) = \lim_k \|f_{\sigma\wedge k}(\omega)\|$ whenever $\sigma(\omega) < \infty$ and $h(\omega) \leqq t$ when $\sigma(\omega) = \infty$. An upper bound for $\int_\Omega hdP$ is established in two steps. Evidently

$$\text{(I)} \qquad \int_{\{\sigma=\infty\}} hdP \leqq t.$$

Next, from Fatou's lemma

$$\text{(II)} \qquad \int_{\{\sigma<\infty\}} hdP = \int_{\{\sigma<\infty\}} \lim \|f_{\sigma\wedge k}(\omega)\|dP(\omega)$$

$$\leqq \liminf_k \int_{\{\sigma<\infty\}} \|f_{\sigma\wedge k}\|dP \leqq \liminf_k \int_\Omega \|f_{\sigma\wedge k}\|dP$$

and it thus remains to estimate $\int_\Omega \|f_{\sigma\wedge k}\|dP$. Since

$$E[\|f_{n+1}\| \mid F_n] \geqq \|\mathbb{E}[f_{n+1} \mid F_n]\| = \|f_n\| \quad [a.s.P]$$

it follows that $\int_A \|f_n\|dP \leqq \int_A \|f_{n+1}\|dP$ for each $A \in F_n$ whence $(\|f_n\|, F_n)$ is a submartingale. Consequently

$$\int_\Omega \|f_{\sigma\wedge k}\|dP = \sum_{j=1}^{k-1} \int_{\{\sigma=j\}} \|f_j\|dP + \int_{\{\sigma\geqq k\}} \|f_k\|dP$$

$$\text{(III)} \qquad \leqq \sum_{j=1}^{k-1} \int_{\{\sigma=j\}} \|f_k\|dP + \int_{\{\sigma\geqq k\}} \|f_k\|dP$$

$$= \int_\Omega \|f_k\|dP \leqq M.$$

By putting (I), (II), and (III) together we obtain

$$\int_\Omega hdP \leqq M + t < \infty$$

and h is indeed integrable as claimed. Since h is integrable there is a $\delta > 0$ such that $\int_C hdP < \varepsilon/2$ whenever $P(C) < \delta$. Let $A \in F$ be arbitrary. Choose N and $B \in F_N$ such that $P(A \Delta B) < \delta$. Then for $k \geqq N$ we have

$$\|m_k(A) - m_N(A)\| \leqq \left\|\int_B f_{\sigma\wedge k}dP - \int_B f_{\sigma\wedge N}dP\right\| + \int_{A\Delta B} \|f_{\sigma\wedge k}\|dP + \int_{A\Delta B} \|f_{\sigma\wedge N}\|dP$$

$$\leqq 2\int_{A\Delta B} hdP < \varepsilon.$$

That is, $m(A) \equiv \lim_k m_k(A)$ exists for each $A \in F$, and by Theorem 1.1.5, m is an X-valued measure. It is clear that m is absolutely continuous with respect to P and that the total variation of m is bounded by $\|h\|_1$. Hence the conditions of the second alternative definition of the RNP are met (listed on the second page of this chapter) and so $g \equiv \frac{dm}{dP}$ exists. (Note that g depends on the choice of t, as do all the other functions defined thus far in the proof.) The function $f_{\sigma\wedge n}$ is F_n-measurable. Since $(f_{\sigma\wedge k}, F_k)$ is a martingale, for $A \in F_n$

$$\int_A f_{\sigma\wedge n}dP = m_n(A) = \lim_{k\geqq n} m_k(A) = m(A) = \int_A gdP.$$

Consequently $\mathbb{E}[g|F_n] = f_{\sigma\wedge n}$ [a.s.P] and Lévy's Theorem (1.3.1) asserts that $\{f_{\sigma\wedge k}: k = 1,2,\ldots\}$ converges almost surely to g.

Now let t vary. The maximal inequality 1.3.2 applied to the sequence $\{\|f_k\|\}$ shows that $P(\{\omega: \sigma(\omega) = \infty\})$ approaches 1 as $t \to \infty$. Since $f_{\sigma\wedge n}(\omega) = f_n(\omega)$ when $\sigma(\omega) = \infty$ it follows that $\{f_n\}$ converges almost surely as well, and the proof is complete.

§3. Dentable, s-dentable and c-dentable sets.

There is a useful and geometrically appealing characterization of dentability which sometimes appears as its definition. It is stated in Proposition 2.3.2 below.

DEFINITION AND NOTATION 2.3.1. Let D be a bounded subset of a Banach space X and suppose that $f \in X^*$, $f \neq 0$. Let

$$M(D,f) \equiv \sup\{f(x):\ x \in D\}.$$

If $\alpha > 0$ then the set

$$S(D,f,\alpha) \equiv \{x \in D:\ f(x) > M(D,f) - \alpha\}$$

is called the _slice of D determined by f and α_ (or more briefly, a slice of D).

PROPOSITION 2.3.2. A bounded subset D of X is dentable if and only if it has slices of arbitrarily small diameter.

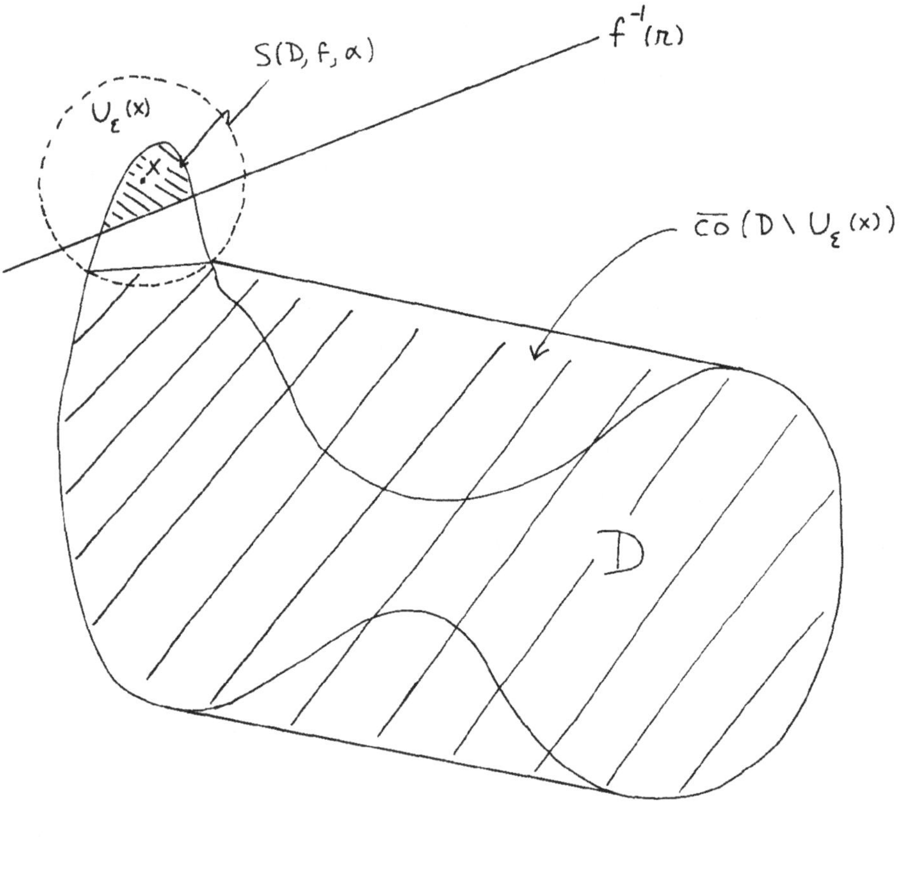

Proof of 2.3.2. (See diagram). If D is dentable and $\varepsilon > 0$ then choose $x \in D$ such that $x \notin \overline{co}(D\setminus U_\varepsilon(x))$. The Hahn-Banach theorem guarantees that there is an

$f \in X^*$, $\|f\| = 1$, such that $f(x) > r > M(\overline{co}(D\backslash U_\varepsilon(x)),f)$ for some $r \in \mathbb{R}$. Let $\alpha \equiv M(D,f) - r$. It is easy to check that $S(D,f,\alpha) \subset U_\varepsilon(x) \cap D$ and it consequently has diameter at most 2ε.

Conversely, suppose $\varepsilon > 0$ is prescribed and let $S(D,f,\alpha)$ be a slice of diameter less than ε. If $x \in S(D,f,\alpha)$ then

$$\overline{co}(D\backslash U_\varepsilon(x)) \subset \overline{co}(D\backslash S(D,f,\alpha)) \subset f^{-1}(-\infty,r]$$

where $r \equiv M(D,f) - \alpha$. Since $f(x) > r$ it follows that $x \notin \overline{co}(D\backslash U_\varepsilon(x))$ and the proof is complete.

COROLLARY 2.3.3 (Rieffel). Let D be a bounded subset of X. If $K \equiv \overline{co}(D)$ is dentable then so is D.

(Note: The converse of this result is also true, and is an easy consequence of 3.4.1.)

Proof. Observe that $M(D,f) = M(K,f)$ for $f \in X^*\backslash\{0\}$. Thus if $\alpha > 0$ and $0 \neq f \in X^*$ we have $S(D,f,\alpha) \subset S(K,f,\alpha)$. Hence if K has slices of arbitrarily small diameter, so does D.

The conclusion of 2.3.2 fails when "dentable" is replaced by "s-dentable" as is indicated next.

EXAMPLE 2.3.4. Let $D \equiv \{f \in C[0,1]: \|f\| < 1\}$ $(\equiv U_1(0))$. Then $K \equiv \overline{co}(D) = U_1[0]$ was shown to be s-dentable but not dentable in Example 2.1.6. Nevertheless, as the following lemma proves, D is not s-dentable. (In fact, D is not even c-dentable.)

LEMMA 2.3.5 (Davis-Phelps). Let $C \subset X$ be a closed bounded convex set whose interior, int C, is nonempty. If C is not dentable then int C is not c-dentable.

Proof. Since C is not dentable there is an $\varepsilon > 0$ such that $x \in \overline{co}(C\backslash U_\varepsilon(x))$ for each $x \in C$. If now $x \in$ int C, find $\delta > 0$ such that $U_\delta(x) \subset$ int C and pick $x_1,\ldots,x_n \in C$ and $t_1,\ldots t_n$ positive so that $\Sigma\, t_i = 1$, $\|x_i - x\| \geqq \varepsilon$ for each i, and $\|\sum_{i=1}^n t_i x_i - x\| < \min(\delta,\frac{\varepsilon}{2})$. Let $p \equiv x - \sum_{i=1}^n t_i x_i$ and note that

$x + p \in U_\delta(x) \subset \text{int } C$. Consequently $\frac{1}{2} x_i + \frac{1}{2}(x + p) \in \text{int } C$ for each i. Moreover

$$\|[\frac{1}{2} x_i + \frac{1}{2}(x + p)] - x\| \geqq \frac{1}{2}\|x_i - x\| - \frac{1}{2}\|p\| \geqq \frac{1}{2}\varepsilon - \frac{1}{4}\varepsilon = \frac{1}{4}\varepsilon.$$

Finally, $\sum_{i=1}^{n} t_i[\frac{1}{2} x_i + \frac{1}{2}(x + p)] = x$ so that $x \in \text{co}(\text{int } C\backslash U_{\varepsilon/4}(x))$ for each $x \in \text{int } C$. Thus int C is not c-dentable.

Lemma 2.3.5 first appeared in Davis and Phelps [1974] with a somewhat longer proof. Their paper contained the last link of a series of results beginning with Rieffel [1967] and continued in Maynard [1973] that X has the RNP if and only if it is subset dentable. In this paragraph the equivalence of the RNP and subset dentability conditions will be established for Banach spaces (since it is so easy now), even though Theorem 2.3.6 below establishes the considerably stronger result that the two notions are equivalent for closed convex sets. We follow the Davis-Phelps approach. Certainly if X is subset dentable then it is subset s-dentable. Conversely suppose that $D \subset X$ is a bounded nondentable subset. Then $D \cup -D$, and consequently by 2.3.3, $\overline{\text{co}}(D \cup -D)$, is nondentable. It is easy to check that $U_1[0] + \overline{\text{co}}(D \cup -D))$ fails to be dentable, and again by 2.3.3, $C \equiv \overline{\text{co}}(U_1[0] + \overline{\text{co}}(D \cup -D))$ is not dentable. Lemma 2.3.5 allows us to conclude that int C is not even c-dentable. That is, the notions of subset dentability, subset s-dentability, and subset c-dentability coincide for Banach spaces. (Moreover, from the above argument, it follows that X has the RNP if and only if the unit ball of each equivalent norm on X is dentable.)

The localized version of this problem (namely, whether the notions of subset dentability, s-dentability, and c-dentability are equivalent for a given closed bounded convex set K) is not amenable to solution as above, since Lemma 2.3.5 is not generally applicable. In a paper concurrent with Davis and Phelps [1974], Huff [1974] established the localized equivalence of subset s-dentability and subset dentability by modifying Maynard's construction (cf. 2.2.2) and using some integration-theoretic machinery. We shall present instead a purely geometric argument which establishes the equivalence of all three notions in a localized setting (and thus somewhat more than either Huff or Davis, Phelps). The possibility of a construction of the type contained below was first brought to my attention by I. Namioka. This also seems a good place to summarize the progress to this point.

THEOREM 2.3.6. Let K be a closed bounded convex subset of a Banach space. Then the following statements are equivalent:

1) K has the RNP.

2) Each closed bounded convex separable subset of K has the RNP.

3) K has the RNP for ([0,1], Borel sets, Lebesgue measure).

4) K has the MCP.

5) K has the MCP for ([0,1], Borel sets, Lebesgue measure).

6) K is subset s-dentable.

7) K is subset c-dentable.

8) Each countable subset of K is c-dentable. (In the language of §2.4, K contains no δ - bush for $\delta > 0$.)

9) K is subset dentable.

10) Each subset of K has slices of arbitrarily small diameter.

11) Each closed bounded convex subset of K is dentable.

Proof. $1 \Rightarrow 4$ is Theorem 2.2.1. $4 \Rightarrow 5$ is obvious. $5 \Rightarrow 6$ is Theorem 2.2.2, and $6 \Rightarrow 1$ is Theorem 2.2.3. Clearly $1 \Rightarrow 3$, while $3 \Rightarrow 5$ is again Theorem 2.2.1. $1 \Rightarrow 2$ is obvious. Suppose that 2 holds and that D is any countable subset of K. Then $K_1 \equiv \overline{co}(D)$ is a separable closed bounded convex set in K. Consequently K_1 has the RNP, whence D is s-dentable (and in particular, c-dentable) by virtue of the equivalence of conditions 1 and 6 (just established) for the set K. Hence $2 \Rightarrow 8$. Thus statements 1, 3, 4, 5, and 6 are equivalent, and $1 \Rightarrow 2 \Rightarrow 8$. Evidently $9 \Rightarrow 6$, $6 \Rightarrow 7$, and $7 \Rightarrow 8$. $11 \Rightarrow 9$ is contained in 2.2.3, while $9 \Rightarrow 11$ is obvious. $9 \Rightarrow 10$ and $10 \Rightarrow 9$ follow from 2.3.2. It therefore remains to establish $8 \Rightarrow 11$.

Thus suppose that C is a nondentable closed bounded convex subset of a Banach space. Then there is an $\varepsilon > 0$ such that $x \in \overline{co}(C \backslash U_{3\varepsilon}(x))$ for each x in C. Let δ be a positive number with $\delta \leqq \varepsilon$. If W is any nonvoid open (relative to C) subset of C of diameter at most ε and if $x \in W$ then there is a finite subset $\{y_1,\ldots,y_k\}$ of $C \backslash U_{3\varepsilon}(x)$ such that $W \cap co(\{y_1,\ldots,y_k\}) \neq \emptyset$. That is, there are nonnegative numbers λ_i, $i = 1,\ldots,k$ such that $\sum_{i=1}^{k} \lambda_i = 1$ and $\sum_{i=1}^{k} \lambda_i y_i \in W$. There are, consequently, sets V_i each of diameter at most δ and each open relative to C such that both $y_i \in V_i$ for $i = 1,\ldots,k$ and $\sum_{i=1}^{k} \lambda_i V_i \subset W$. Note that if $w \in W$ and $v \in V_i$ for any $i = 1,\ldots,k$ then

$$\varepsilon \leqq \|x - y_i\| - \|x - w\| - \|v - y_i\| \leqq \|w - v\|.$$

That is, distance $(W,V_i) \geqq \varepsilon$ for each i.

Pick any $x \in C$, let $W \equiv U_{\varepsilon/2}(x) \cap C$ and $\delta \equiv \varepsilon$. The above construction yields relatively open sets $V_1,\ldots,V_k$ each of diameter at most ε. Now let $\delta \equiv \frac{\varepsilon}{2}$ and repeat the above construction for each of the k sets $V_1,\ldots,V_k$. By continuing in this fashion it is evident that a sequence $\{A_n: n = 1,2,\ldots\}$ of finite families of nonvoid relatively open subsets of C will be produced satisfying

1) If $W \in A_n$ then diameter $W \leqq \frac{\varepsilon}{n}$;

2) If $W \in A_n$ then there is a nonnegative function λ_W on A_{n+1} such that $\Sigma\,\{\lambda_W(V)\colon V \in A_{n+1}\} = 1$ and $\Sigma\,\{\lambda_W(V)V\colon V \in A_{n+1}\} \subset W$;

3) If $W \in A_n$ and $V \in A_{n+1}$ and if $\lambda_W(V) \neq 0$ then distance $(W,V) \geqq \varepsilon$.

The final step of the argument is to replace each set W in A_k for $k = 1,2,\ldots$ by a carefully chosen point x_W while maintaining properties 2 and 3 above. In this regard, temporarily fix n and suppose that $k \leqq n$. For each $W \in A_k$ we define a subset $W(n)$ of W as follows: If $k = n$ then $W(n) \equiv W$ for all $W \in A_n$. Inductively assume that $k + 1 \leqq n$ and that $V(n)$ has already been defined for each $V \in A_{k+1}$. If $W \in A_k$ let $W(n) \equiv \Sigma\,\{\lambda_W(V)V(n)\colon V \in A_{k+1}\}$ where λ_W is any function on A_{k+1} as in 2 above. Now let n vary. It is easy to check that if $k \leqq n$ and $W \in A_k$ then both

$$\text{diameter } W(n) \leqq \frac{\varepsilon}{n} \quad \text{and} \quad W(n+1) \subset W(n).$$

Since C is complete it follows that $\bigcap_n \{\overline{W(n)}\colon k \leqq n\}$ is a singleton, say $\{x_W\}$ for each $W \in A_k$. Let $A \equiv \{x_W\colon W \in A_k,\ k = 1,2,\ldots\}$. The proof will be completed by showing that the countable set A is not c-dentable. First note that if $W \in A_k$ then $\Sigma\,\{\lambda_W(V)\overline{V(n)}\colon V \in A_{k+1}\} \subset \overline{W(n)}$ whenever $k + 1 \leqq n$. It follows that $\Sigma\,\{\lambda_W(V)x_V\colon V \in A_{k+1}\} = x_W$. Finally, 3) implies that $\|x_W - x_V\| \geqq \varepsilon$ if $\lambda_W(V) \neq 0$ and hence that $x_W \in \mathrm{co}(A \backslash U_\varepsilon(x_W))$ for each $x_W \in A$.

Since there is sometimes interest in conditions under which a Banach space X has the RNP (in contrast to one of the closed bounded convex subsets of X), we include the following result.

COROLLARY 2.3.7. Let X be a Banach space. Then the following statements are equivalent:

1) X has the RNP;

2) Whenever m is an X-valued measure on the measurable space (Ω,F) which is of finite total variation then there is an $f \in L^1_X(\Omega,F,|m|)$ such that $\int_A f d|m| = m(A)$ for each $A \in F$.

3) If $(f_n,F_n)_{n=1}^{\infty}$ is an X-valued martingale based on (Ω,F,P) (an arbitrary probability space) which is L^1_X-bounded then $\{f_n: n = 1,2,\ldots\}$ converges [a.s.P];

4) Any one of the 11 properties listed in Theorem 2.3.6 above holds for each closed bounded convex set $K \subset X$.

Proof. $1 \Rightarrow 2$ and $2 \Rightarrow 1$ are straightforward measure theory. (See the comments just after the end of Example 2.1.2. See also statement b there for yet another reformulation of the RNP for X.) $1 \Rightarrow 3$ comes from 2.2.7. Since 3) implies that K has the MCP for each closed bounded convex $K \subset X$, $3 \Rightarrow 1$ by Theorem 2.3.6. Finally, $1 \Rightarrow 4$ and $4 \Rightarrow 1$ follow directly from Theorem 2.3.6 and the definition of the RNP for Banach spaces.

§4. Bushes, trees, and nonhoriticultural applications.

Let $K \subset X$ be a closed bounded convex set which lacks the RNP. According either to the statement of $8 \Rightarrow 11$ of Theorem 2.3.6 or the construction employed in its proof, K must contain a countable set of points divided into "stages" which, together with some $\delta > 0$, satisfy the following conditions: Stage 0 contains a single point. If $x_1,\ldots,x_k$ are the points of the $n^{\underline{th}}$ stage, then the points of stage n+1 can be divided into finite sets (not necessarily disjoint), say $N_1,\ldots,N_k$, in such a way that $x_i \in co(N_i)$ and $\|x - x_i\| \geqq \delta$ for $x \in N_i$, $i = i,\ldots,k$. By joining x_i to the points of N_i for each i and each stage, a bush-like figure is obtained, and hence such a collection is called a _δ-bush_. Observe that δ-bushes are not c-dentable; thus the two statements which appear in part 8 of Theorem 2.3.6 are equivalent. A δ-bush which has exactly two branches at every point, each point of which is the midpoint of its branching points, is called a δ-tree. (Thus, formally, a δ-tree is a sequence $\{x_n : n = 1,2,\ldots\}$ such that $x_n = (1/2)(x_{2n} + x_{2n+1})$ and $\|x_{2n} - x_n\| = \|x_{2n+1} - x_n\| \geqq \delta$ for each $n = 1,2,\ldots$.) Consequently a closed bounded convex set which contains a δ-tree for some $\delta > 0$ lacks the RNP. The converse, though plausible, is false. Indeed, Bourgain and Rosenthal [1980a] have constructed a subspace X of $L^1[0,1]$ which lacks the RNP, which contains no bounded δ-tree for any $\delta > 0$, and which has the additional property that each L^1-bounded sequence in X has a subsequence converging in probability. (As will be seen in Chapter 4 - cf. Theorem 4.4.1 - X cannot be a dual space.) The details are presented in §7.3. See Theorem 7.3.2.

Several other tree-like conditions have been studied in other contexts. Perhaps the best known is the finite tree property of R.C. James [1972a]: A Banach space X has the _Finite Tree Property_ (FTP) if for each ε, $0 < \varepsilon < 1$ and each positive integer n there is a finite sequence $\{x_1,\ldots,x_{2^n-1}\}$ in the closed unit ball of X such that

$$x_j = (1/2)(x_{2j} + x_{2j+1}) \quad \text{and} \quad \|x_{2j} - x_j\| \ (= \|x_{2j+1} - x_j\|) \geqq \varepsilon$$

for $j = 1,\ldots,2^{n-1}-1$. In order to describe the connection between the RNP and the FTP we introduce some terminology originally due to James [1972a] and [1972b]. Say that a Banach space _Y is finitely representable in X_ if for each $\varepsilon > 0$ and finite dimensional subspace W of Y there is a 1-1 linear operator $T: W \to T(W) \subset X$ with $\|T\| \, \|T^{-1}\| \leqq 1 + \varepsilon$. Then X is said to be _super-reflexive_ (respectively, _super Radon-Nikodým_) if each Banach space Y finitely representable in X is reflexive (respectively, has the RNP). James [1972a] established that X is super-reflexive if and only if it lacks the FTP. Enflo [1972] characterized super-reflexive spaces as those which admit an equivalent uniformly convex norm, basing his work on James' FTP characterization of non super-reflexive spaces. Subsequently G. Pisier [1975], again starting from the work of James, significantly extended Enflo's theorem. An extraordinary conclusion from this paper is that if X is super-reflexive then there is an equivalent uniformly convex norm $\|\cdot\|$ on X whose modulus of uniform convexity $\delta(\varepsilon)$ (which then satisfies the conditions $\delta(\varepsilon) > 0$ and $(1/2)\|x + y\| \leqq 1 - \delta(\varepsilon)$ whenever $\|x\|, \|y\| \leqq 1$, $\|x - y\| \geqq \varepsilon$ and $0 < \varepsilon < 2$) has the property that for some $K > 0$ and q, $\delta(\varepsilon) \geqq K\varepsilon^q$ for each $\varepsilon \in (0,2)$. Moreover there is no least such q. One of the early results in Pisier [1975] is of direct interest here.

<u>PROPOSITION</u> 2.4.1. X is super-reflexive if and only if it is super-Radon-Nikodým.

Proof. Though it has not yet been done, it is an elementary matter to establish that reflexive spaces have the RNP (see Corollary 4.1.5). Thus if X is super reflexive and Y is finitely representable in X then Y is reflexive, hence has the RNP, whence X is super-Radon-Nikodým.

The more substantial half of the argument begins with the assumption that X is not super-reflexive. A Banach space Y will be produced which is finitely representable in X and which contains a bounded $\frac{1}{2}$-tree. Such a space of course lacks the RNP and thus X is not super-Radon-Nikodým. For each positive integer n choose $\{x_j(n): j = 1,\dots,2^n-1\}$ in the unit ball of X such that

$$x_j(n) = (1/2)(x_{2j}(n) + x_{2j+1}(n)) \quad \text{and} \quad \|x_{2j}(n) - x_j(n)\| \geqq 1/2$$

for $j = 1,\dots,2^{n-1}-1$. (This is possible by virtue of James' characterization.) If r is a positive integer and $t_1,\dots,t_r$ are rational numbers, the bounded sequence $\{\|\sum_{j=1}^{r} t_j x_j(k)\|: k = 1,2,\dots\}$ has a convergent subsequence. In fact, by diagonalization it is possible to find an increasing sequence $\{k_n: n = 1,2,\dots\}$ of positive integers such that whenever r is a positive integer and $t_1,\dots,t_r$ are rational numbers then $\lim_n \|\sum_{j=1}^{r} t_j x_j(k_n)\|$ exists. Observe that for any positive integer j

$$(1/2)x_{2j}(k_n) + (1/2)x_{2j+1}(k_n) - x_j(k_n) = 0$$

and

$$\|x_{2j}(k_n) - x_j(k_n)\| \; (= \|x_{2j+1}(k_n) - x_j(k_n)\|) \geqq 1/2$$

for all sufficiently large n.

Now let $\alpha_1, \alpha_2, \ldots$ be a sequence of symbols and let Z denote the vector space of finite linear combinations of the α_i's. For rational numbers $t_1, \ldots, t_r$ define $\|\sum_{j=1}^{r} t_j\alpha_j\|$ by

$$\|\sum_{j=1}^{r} t_j\alpha_j\| \equiv \lim_n \|\sum_{j=1}^{r} t_j x_j(k_n)\|.$$

By continuity it is easy to check that $\|\sum_{j=1}^{r} s_j\alpha_j\|$ is uniquely determined for $s_1, \ldots, s_r \in \mathbb{R}$ as the limit of "rational approximations". Then $(Z, \|\cdot\|)$ is a semi-normed space and the natural quotient space $\tilde{Z} \equiv Z/\{\alpha \in Z: \|\alpha\| = 0\}$ is a normed space for which the equations

$$(1/2)\tilde{\alpha}_{2j} + (1/2)\tilde{\alpha}_{2j+1} - \tilde{\alpha}_j = 0$$

and inequalities

$$\|\tilde{\alpha}_{2j} - \tilde{\alpha}_j\| \ (= \|\tilde{\alpha}_{2j+1} - \tilde{\alpha}_j\|) \geqq 1/2$$

hold for each $j = 1,2,\ldots$. Since $\|\tilde{\alpha}_j\| = \lim_n \|x_j(k_n)\| \leqq 1$ for each j, $\tilde{Z}$ contains a bounded 1/2-tree and thus so does Y ≡ completion of $\tilde{Z}$. Consequently Y lacks the RNP and it remains to show that $\tilde{Y}$ is finitely representable in X. It evidently suffices to establish that $\tilde{Z}$ is finitely representable in X. But each finite dimensional subspace is contained in the linear span of $\{\tilde{\alpha}_i: i \in F\}$ for some finite set F. If $\{\tilde{\alpha}_i: i \in G\}$ is a maximal linearly independent subset of $\{\tilde{\alpha}_i: i \in F\}$ for some subset G of F let T_n: linear span$(\{\tilde{\alpha}_i: i \in F\}) \to X$ be the linear map determined by the conditions $T_n(\tilde{\alpha}_i) \equiv x_i(k_n)$ for $i \in G$. Then T_n is well defined (at least for sufficiently large n) and since both T_n is 1-1 if n is large and $\lim_n \|T_n\| \|T_n^{-1}\| = 1$, the proof is complete.

Super-reflexivity has also been characterized in terms of the girth of the space by James and Schäffer [1972]. (For any two points x and y of norm one in a Banach space $(X, \|\cdot\|)$ let $\rho(x,y) \equiv \inf\{\text{length}(\underline{c}): \underline{c}$ is a rectifiable curve with endpoints x and y which lies entirely in the unit sphere of X$\}$. The <u>girth</u> $g(\|\cdot\|)$, is then $2\cdot \inf\{\rho(-x,x): \|x\| = 1\}$.) For X at least two dimensional, they establish that $g(\|\cdot\|) = 4$ for each equivalent norm $\|\cdot\|$ on X if and only if X is not super-reflexive.

Among the non super-reflexive spaces are those called flat by Harrell and Karlovitz [1970], [1972], [1974], [1975], and Karlovitz [1973]. ($(X, \|\cdot\|)$ is <u>flat</u> if for some rectifiable curve <u>c</u> in the unit sphere with antipodal endpoints, $g(\|\cdot\|) = 2$ length(<u>c</u>) $= 4$.) Although flatness is not an isomorphic invariant, it is true that if X is isomorphic to a flat space then neither X nor X* has the RNP. For recent accounts see Schäffer [1976], [1980] and [1981].

The weighted trees of A. Ho [1979], yet another variant on the tree theme, may be used to characterize the RNP. (Indeed, a Banach space has the RNP if and only if it does not have a weighted tree. See Section 7.6 for details.) In light of the Bourgain-Rosenthal example alluded to earlier (see also Chapter 7) perhaps Ho's result is a best possible positive result. Let us also mention briefly here that approximate trees and bushes of various sorts have been considered in several contexts. The construction of real trees and bushes from the approximate ones is a matter dealt with often in these notes. (See, for example, the proof of Theorem 2.3.6 and its more sophisticated sister, Theorem 3.7.2. This problem is also solved several times in Chapter 7.) We close this section with a plug for the approximate bushes which come from quasi martingales. This useful geometric tool was largely overlooked until the recent work of Kunen and Rosenthal [1982]. See §7.5, especially Theorem 7.5.3.

CHAPTER 3. EXTREME POINTS

Let K be a closed bounded convex subset of a Banach space. While it is possible that $ex(K) = \emptyset$, it is not difficult to establish that $ex(K) \neq \emptyset$ when K has the RNP (cf. 3.3.6). This result is the springboard for a detailed analysis of the extremal structure of such sets, the subject of the first part of this chapter. (Section 5 contains most of the end results of this discussion, though some of the background in sections 1-4 is of independent interest.) In direct contrast, section 7 deals with how "fuzzy" the "boundary" of certain subsets of K must be when K lacks the RNP.

Although for the purposes of this chapter it is immaterial whether or not K is a subset of a dual Banach space, such a distinction will be of crucial importance in the following two chapters. In order to prepare efficiently for the results of Chapters 4 and 5, then, various definitions, lemmas, etc. in this chapter will be reformulated in weak* versions. In all cases the proofs for the weak* versions require only minor, and most often obvious, modifications of the proofs provided here.

§1. Preliminary discussion.

Let K denote the closed unit ball of c_o. Then $ex(K) = \emptyset$, as the following argument shows. If $(x_i) \in K$ then there is an integer n with $|x_n| < \frac{1}{2}$. Let (y_i) and (z_i) be the points of K determined by the equations $x_i \equiv y_i \equiv z_i$ for $i \neq n$, $y_n \equiv x_n - \frac{1}{2}$, and $z_n \equiv x_n + \frac{1}{2}$. Then (y_i) and (z_i) belong to K, $\frac{1}{2}(y_i) + \frac{1}{2}(z_i) = (x_i)$, and consequently $(x_i) \notin ex(K)$.

As a further example of the same type recall (cf. 2.1.6) that the closed unit ball C of C[0,1] has exactly two extreme points. Evidently, then, $\overline{co}(ex(C)) \subsetneq C$. Let $C_1 \equiv \{f \in C\colon f(0) = 0\}$. Clearly C_1 is a closed bounded convex subset of C. Moreover, $ex(C_1) = \emptyset$. (Indeed, for $f \in C_1$ pick positive numbers a and ε such that $|f(t)| < 1 - \varepsilon$ for $t \in [0,a]$. Let $g(t) \equiv \begin{cases} \varepsilon \sin(\pi t/a) & 0 \leqq t \leqq a \\ 0 \text{ elsewhere} \end{cases}$. Then $f \pm g \in C_1$ whence $\frac{1}{2}(f + g) + \frac{1}{2}(f - g) = f \notin ex(C_1)$.) As Proposition 3.1.1 below demonstrates, the fact that C contains a closed bounded convex subset without extreme points follows from the fact that C is not generated by its extreme points. This result was first proved by Lindenstrauss [1966] by using the Bishop-Phelps theorem (see 3.3.1). The following proof was observed by Bourgin [1969] (or see Peck [1971], Lemma 1). Alternatively, see Asimow [1969].

<u>PROPOSITION</u> 3.1.1. Let C be a closed convex set in a Banach space X. If each closed bounded convex subset of C has at least one extreme point then each such set is the closed convex hull of its extreme points.

Proof. Let K be a closed bounded convex subset of C, let $K_1 \equiv \overline{co}(ex(K))$, and suppose that $K_1 \subsetneq K$. By the separation theorem there is an $f \in X^*$ such that $M(K_1,f) < r < M(K,f)$ for some $r \in \mathbb{R}$. Let $K_2 \equiv f^{-1}[r,\infty) \cap K$. Suppose that $y \in ex(K_2)$ (a nonempty set by hypothesis). Since $y \notin K_1$, $y \notin ex(K)$, and thus y is interior to a line segment $[x,z]$ in K with $f(x) < r < f(z)$. Let w be the endpoint of the ray from x through z intersected with K. Evidently $f(w) > r$, so that $w \notin ex(K)$. On the other hand, if w were interior to some nontrivial line segment in K (drawn with dotted lines in the diagram below) it would follow that $y \notin ex(K_2)$, a contradiction.

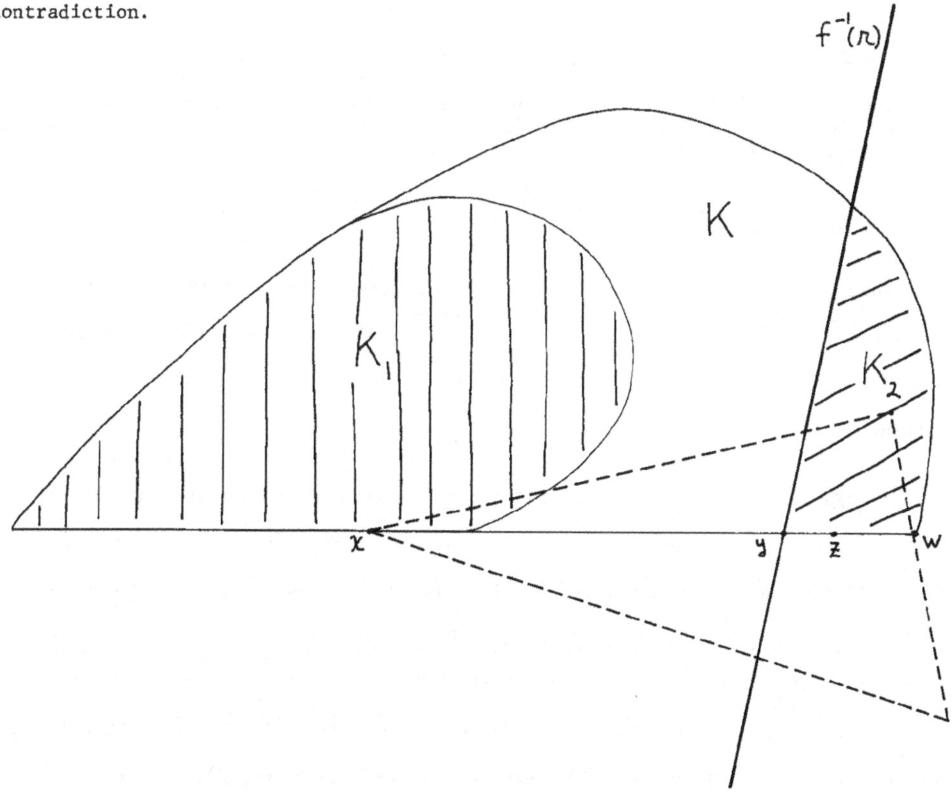

Observe that the same proof applies if the class of closed bounded convex subsets of C is replaced by $\mathcal{C}$, any class of linearly bounded closed convex subsets of a locally convex space such that the intersection of any member of $\mathcal{C}$ with a closed halfspace is again in $\mathcal{C}$. The Lindenstrauss proof no longer works in this general setting.

DEFINITION 3.1.2. Let C be a closed convex subset of a Banach space. Then C is said to have the Krein-Milman Property (abbreviated KMP) if each closed bounded convex subset K of C satisfies: $K = \overline{co}(ex(K))$. (Equivalently, C has the KMP if each closed bounded convex subset of C has an extreme point.)

Nontrivial examples of sets C with the KMP will be discussed in this chapter and the next as the connections between these sets and those with the RNP are analyzed. At this stage, though, let it suffice to note the following.

PROPOSITION 3.1.3. Reflexive Banach spaces have the KMP.

Proof. A closed bounded convex set K in a reflexive Banach space is weakly compact and convex. Thus $ex(K) \neq \emptyset$ by the classical Krein-Milman theorem, and the desired conclusion follows from 3.1.1.

§2. Exposed, strongly exposed, and support points.

Some of the most interesting geometric characterizations of the RNP for closed bounded convex sets are in terms of the existence of "sufficiently many" points or functionals of certain types. The first definitions of this section describe these distinguished points and linear functionals.

DEFINITIONS 3.2.1. Let K be a closed bounded convex subset of a Banach space X and suppose that $x \in K$. Then x is an exposed point of K if there is an $f \in X^*$ such that $f(x) > f(y)$ whenever $y \neq x$ and $y \in K$. The linear functional f is said to expose x. The point x is a strongly exposed point of K if there is an $f \in X^*$ which exposes x such that $\{S(K,f,\alpha): \alpha > 0\}$ is a neighborhood base for x in K in the norm topology (or equivalently, such that $\lim_{\alpha \to 0^+}(\text{norm-diameter } S(K,f,\alpha) = 0)$. Then f strongly exposes x. We will denote the set of exposed points of K by exp(K), the set of strongly exposed points of K by str exp(K), and the set of $f \in X^*$ which strongly expose some point of K by $\mathcal{SE}(K)$. If y is any point of a bounded subset D of X such that $f(y) = M(D,f)$ for some $f \in X^*\setminus\{0\}$ then y is a support point of D and f is a support functional of D. Let $\mathcal{S}(D) \equiv \{f \in X^*\setminus\{0\}$: f is a support functional of D$\}$.

DEFINITION 3.2.1 (w*). Let X* be a dual Banach space and $K \subset X^*$ a norm-closed bounded convex set. Then $x \in K$ is a weak*-strongly exposed point (written $x \in$ w*-str exp(K)) if there is an $f \in X$ which strongly exposes x. (Here and elsewhere we consider X as being naturally embedded in X**.) Denote by w*-$\mathcal{SE}(K)$ the set of weak*-strongly exposing functionals of K.

Two of the most useful results about compact convex sets from our perspective (besides the Krein-Milman theorem) are the Milman theorem [1947] and the Choquet theorem [1956]. Both are stated here for convenient reference. (See Choquet [1969], Phelps [1966], or Alfsen [1971] for details.)

THEOREM 3.2.2 (Milman). Let C be a compact convex set in a locally convex Hausdorff topological vector space, and suppose that D is a closed subset of C. If the closed convex hull of D equals C then $D \supset ex(C)$.

THEOREM 3.2.3 (Choquet). Let C be a compact convex metrizable subset of a locally convex Hausdorff topological vector space Y, and suppose that $x \in C$. Then there is a probability measure P on (C,Borel subsets of C) such that $P(ex(C)) = 1$ and $\int_C f dP = f(x)$ for each $f \in Y^*$.

After introducing some convenient notation, we turn to an example which illustrates some of the notions defined in 3.2.1 and 3.2.1 (w*).

NOTATION 3.2.4. a) If D is any subset of a dual Banach space, the notation w*cl(D) denotes the weak*-closure of D. Thus, for example, w*cl(co(D)) refers to the weak*-closed convex hull of the set D.

b) For any set D in a Banach space, let $\bar{D}$ denote the norm-closure of D.

EXAMPLE 3.2.5. Let X denote the Banach space c_0. Then $X^* = \ell_1$, the space of absolutely summable sequences of real numbers with $\|(x_i)\|_1 \equiv \Sigma_{i=1}^{\infty} |x_i|$. Moreover, $X^{**} = \ell_\infty$, the Banach space of bounded sequences with supremum norm. For each $n \geqq 1$ let δ_n denote the point (x_i) of ℓ_1 such that $x_n = 1$ and $x_j = 0$ for $j \neq n$. Let

$$D_1 \equiv \{\frac{1}{n}\delta_1 + \delta_n : n \geqq 2\} \cup \{\frac{1}{n}\delta_1 - \delta_n : n \geqq 2\} \quad \text{and} \quad K \equiv w^*cl(co(D_1)) \subset \ell_1 .$$

Since the weak*-limit of the sequence $\{\delta_n : n \geqq 2\}$ is $(0,0,\ldots) \equiv 0 \in \ell_1$ it follows that $D \equiv w^*cl(D_1) = D_1 \cup \{0\}$. The Milman theorem applies since K is weak*-compact and convex, and shows that $D \supset ex(K)$. That is, $ex(K) \subset D_1 \cup \{0\}$. The set K is weak* metrizable (since it is a bounded subset of the dual of the separable Banach space c_0), so the Choquet theorem is applicable as well. Since D is countable it follows that ex(K) is a countable set and thus each point of K may be written as a (possibly infinite) convex combination of the points of ex(K). (Note that the partial sums of such an infinite convex combination of points of D converge in norm to their limit since D is bounded.) But since 0 is the only point of D without a positive δ_1-coefficient (and therefore 0 cannot be written as a finite or infinite convex combination of other points of D) it must be an extreme point of K. Moreover, 0 is an exposed point of K since the linear functional $(-1,0,0,\ldots)$ attains its supremum exactly at 0. In order to see that 0 is not a strongly exposed point of K suppose to the contrary that $f \equiv (a_1,a_2,\ldots)$ strongly exposes 0. Then $f(\frac{1}{n}\delta_1 \pm \delta_n) < f(0) = 0$ and thus $\frac{1}{n}a_1 \pm a_n < 0$ for each $n \geqq 2$. It follows that $\lim_n a_n = 0$ (since $\{n|a_n|\}$ is bounded above by $|a_1|$). Finally, the equations $\lim_n f(\frac{1}{n}\delta_1 \pm \delta_n) = \lim_n (\frac{1}{n}a_1 \pm a_n) = 0$ imply that given any $\alpha > 0$ we have $\frac{1}{n}\delta_1 \pm \delta_n \in S(K,f,\alpha)$ for sufficiently large n. Since $\|(\frac{1}{n}\delta_1 + \delta_n) - (\frac{1}{n}\delta_1 - \delta_n)\|_1 = 2$ for each n, $\lim_{\alpha\to 0^+}$ (norm-diameter $S(K,f,\alpha)) \geqq 2$ and thus 0 is not a strongly exposed point.

Two final observations concerning 3.2.5 are in order. Note that each point of D_1 is a weak*-strongly exposed point of K. (The functionals

$$(0,\ldots,0,\pm 1,0,\ldots) \in c_0 \subset \ell_\infty$$

weak*-strongly expose the corresponding points $\frac{1}{n}\delta_1 \pm \delta_n$.) Thus K is the weak*-closed convex hull of its weak*-strongly exposed points. Moreover, even without mentioning the weak*-topology, conclusions stronger than those guaranteed by the Krein-Milman theorem hold. Indeed, as is easily checked, K is the norm-closed convex hull of its strongly exposed points. Neither of these conclusions about K is coincidental.

Before turning to a lemma of E. Bishop we mention several additional elementary examples.

(1) Let $K \equiv \overline{co}(\{\delta_n : n = 1,2,\ldots\}) \subset \ell_2$. Then 0 is an exposed point of this weakly compact convex set but a straightforward argument shows that 0 is not a strongly exposed point of K. (Indeed, each slice $S(K,f,\alpha)$ has diameter $\sqrt{2}$ if f exposes 0 in K.) There is a distinction to be made between this example and that appearing in 3.2.5 which will be examined in Chapter 7. 0 in this example is a strong extreme point of K (cf. 7.6.7) while 0 is not a strong extreme point of the set K in 3.2.5.

(2) Let $K \equiv \overline{co}(\{-\frac{1}{n}\delta_n : n = 1,2,\ldots\} \cup \{\frac{1}{4^n}(2\delta_n - \delta_{n+1}) : n = 1,2,\ldots\}) \subset \ell_1$. Then 0 is an extreme point of K. (Otherwise, by application of the Milman Theorem 3.2.2 and Choquet Theorem 3.2.3 as in Example 3.2.5 - note, K is compact - there would be nonnegative scalars s_n and t_n with $\sum_{n=1}^{\infty}(s_n + t_n) = 1$ such that

$$0 = \sum_{n=1}^{\infty}[s_n(-\frac{1}{n}\delta_n) + t_n(\frac{1}{4^n}(2\delta_n - \delta_{n+1}))].$$

For any $n \geq 2$ the coefficient of δ_n in the above sum must be 0. That is, $\frac{s_n}{n} + \frac{2t_n}{4^n} - \frac{t_{n-1}}{4^{n-1}} = 0$ and hence $t_n \geq 2t_{n-1}$ for each $n \geq 2$. It follows that $t_n = 0$ for each $n \geq 1$. But then $\sum_{n=1}^{\infty} s_n(-\frac{1}{n}\delta_n) = 0$ whence $s_n = 0$ for each n which contradicts the fact that $\sum_{n=1}^{\infty}(s_n + t_n) = 1$. Thus $0 \in ex(K)$.) Suppose now that $f \in \ell_1^* = \ell_\infty$ and that $f(0)$ $(=0) = \sup f(K)$. Write f as a bounded sequence (α_n) and apply f to each point used above to generate K. We obtain $-\alpha_n \leq 0$ and $2\alpha_n - \alpha_{n+1} \leq 0$ for each n. Hence $0 \leq 2\alpha_n \leq \alpha_{n+1}$ for each n and since (α_n) is bounded, $\alpha_n = 0$ for each n. Thus the only continuous linear functional which attains its supremum on K is the 0 functional, whence $0 \in K$ is an extreme point which fails to be a support point of K.

(3) In $\mathbb{R}^2$ let $K \equiv co(\{(x,y) \in \mathbb{R}^2 : x^2 + y^2 = 1\} \cup \{(1,-1)\})$. Then the point $(1,0)$ is an extreme point of K which is not an exposed point. Thus even finite dimensional compact convex sets may have non-exposed extreme points.

(4) Let C denote the Hilbert cube. (That is, $C \equiv \{(x_i) \in \ell_2 : |x_n| \leq \frac{1}{n}$ for each $n\}$.) Then C is a compact convex set with empty interior. (A direct argument, or use of the Baire category theorem shows that C even lacks core.) Thus each point of C is in its topological boundary. Nevertheless, it may be readily verified that $0 \in C$ is not a support point of C. Thus, although each support point of a closed bounded convex subset of a Banach space is in the boundary of that set, the converse need not hold in general. (Of course, if the closed bounded convex set in question has nonempty interior, then each boundary point is a support point by the Hahn-Banach theorem.)

The next lemma is geometrically clear, but warrants explicit statement since it is so often used in the sequel. Lemma 3.2.7 which follows it will be applied later on in this chapter.

LEMMA 3.2.6. Let D be a bounded subset of X and suppose that $\alpha > 0$ and $f \in X^*\backslash\{0\}$ are given. Then there is an $\varepsilon > 0$ such that

$$S(D,g,\tfrac{\alpha}{2}) \subset S(D,f,\alpha) \quad \text{whenever} \quad g \in X^* \quad \text{and} \quad \|f - g\| \leq \varepsilon.$$

Proof. Let $M \equiv \sup\{\|x\|: x \in D\}$ and suppose that we pick $0 < \varepsilon < \frac{\alpha}{4M}$. Take any functionals f and g in X* such that $\|f - g\| \leq \varepsilon$. If $y \in D$ and $g(y) > M(D,g) - \frac{\alpha}{2}$, then

$$\begin{aligned} f(y) \geq g(y) - |f(y) - g(y)| &> M(D,g) - \tfrac{\alpha}{2} - \varepsilon M \geq M(D,f) - \varepsilon M - \tfrac{\alpha}{2} - \varepsilon M \\ &> M(D,f) - \alpha. \end{aligned}$$

Consequently $S(K,f,\alpha) \supset S(K,g,\frac{\alpha}{2})$.

LEMMA 3.2.7 (Bishop). Let K be a closed bounded convex subset of a Banach space X. Suppose that for each $\delta > 0$ and $f \in X^*$ there is a $g \in X^*$ such that $\|f - g\| \leq \delta$ and g determines a slice of K of diameter at most δ. Then $K = \overline{co}(\text{str exp}(K))$. Moreover, $SE(K)$ is a dense G_δ-subset of X*.

Proof. For any given $\delta > 0$ let $\mathcal{O}_\delta \equiv \{f \in X^*: f$ determines a slice of K of diameter at most $\delta\}$. Then $\mathcal{O}_\delta$ is open by Lemma 3.2.6, and $\mathcal{O}_\delta$ is dense by hypothesis. Note that $\bigcap_{n=1}^{\infty} \mathcal{O}_{1/n}$ is a dense G_δ subset by the Baire category theorem, while, from the definition of the various sets involved, it is immediate that $SE(K) = \bigcap_{n=1}^{\infty} \mathcal{O}_{1/n}$. This establishes the second half of the lemma.

Suppose next that K properly contains $\overline{co}$(str exp(K)). Then use the separation theorem to produce a slice of K - say $S(K,f,\alpha)$ - which is disjoint from $\overline{co}$(str exp(K)). Since $SE(K)$ is dense in X* it is possible to find a functional $g \in SE(K)$ such that $S(K,g,\frac{\alpha}{2}) \subset S(K,f,\alpha)$ by 3.2.6. Evidently, if $x \in K$ is strongly exposed by g it must at once belong to $\overline{co}$(str exp(K)) and to $S(K,g,\frac{\alpha}{2})$, two disjoint sets. Thus the lemma is proved.

<u>LEMMA</u> 3.2.7 (w*). Let K be a weak*-compact convex subset of a dual Banach space X*. Suppose that for each $\delta > 0$ and $f \in X$ there is a $g \in X$ such that $\|f - g\| \leq \delta$ and g determines a slice of K of diameter at most δ. Then K = w*cl(co(w*-str exp(K))). Moreover, w*-$SE(K)$ is a dense G_δ subset of X.

Proof. Modify the proof of 3.2.7 as follows: 1) all functions belong to X; 2) replace "$SE(K)$" by "w*-$SE(K)$"; 3) replace "str exp(K)" by "w*-str exp(K)"; and 4) replace "$\overline{co}$" by "w*cl(co".

§3. <u>The density of support functionals. The RNP implies the KMP</u>.

Throughout this section and without further mention, K will denote a closed bounded convex subset of X.

<u>THEOREM</u> 3.3.1 (Bishop, Phelps [1961]). $S(K)$ is dense in X*.

The proof of this theorem is based on the following wonderfully simple geometric argument. Suppose that C is a convex cone in X with vertex 0 (that is, C is a convex set and nonnegative multiples of elements of C are again in C) and that the interior of C is nonempty. If $x_0 \in K$ is a point such that $(C + x_0) \cap K = \{x_0\}$ then the separation theorem guarantees that there is a $g \in X^*$, $\|g\| = 1$, such that $\sup g(K) = g(x_0) = \inf g(C + x_0)$. The "fatter" the cone C, the closer the hyperplane $g^{-1}(0)$ approximates the boundary of C (in a sense made precise below). Thus suppose that f is a given element of X* with $\|f\| = 1$. For any $\delta > 0$, $\delta < 1$, let $C \equiv C(f,\delta) \equiv \{x \in X: f(x) \geqq \delta\|x\|\}$. Then C is a closed convex cone with nonempty interior (and vertex 0), and for small $\delta > 0$ the boundary of C appears to closely approximate $f^{-1}(0)$. If x_o and g are chosen as above for this cone C, then $g^{-1}(0)$ and $f^{-1}(0)$ appear to be close together (and both are close to the boundary of C). (See the diagram.) The conclusion of Lemma 3.3.2 below is that, from such information, f and g (or possibly f and -g) are close together in X*. This outline of the Bishop-Phelps argument concludes with the observation that $g \in S(K)$.

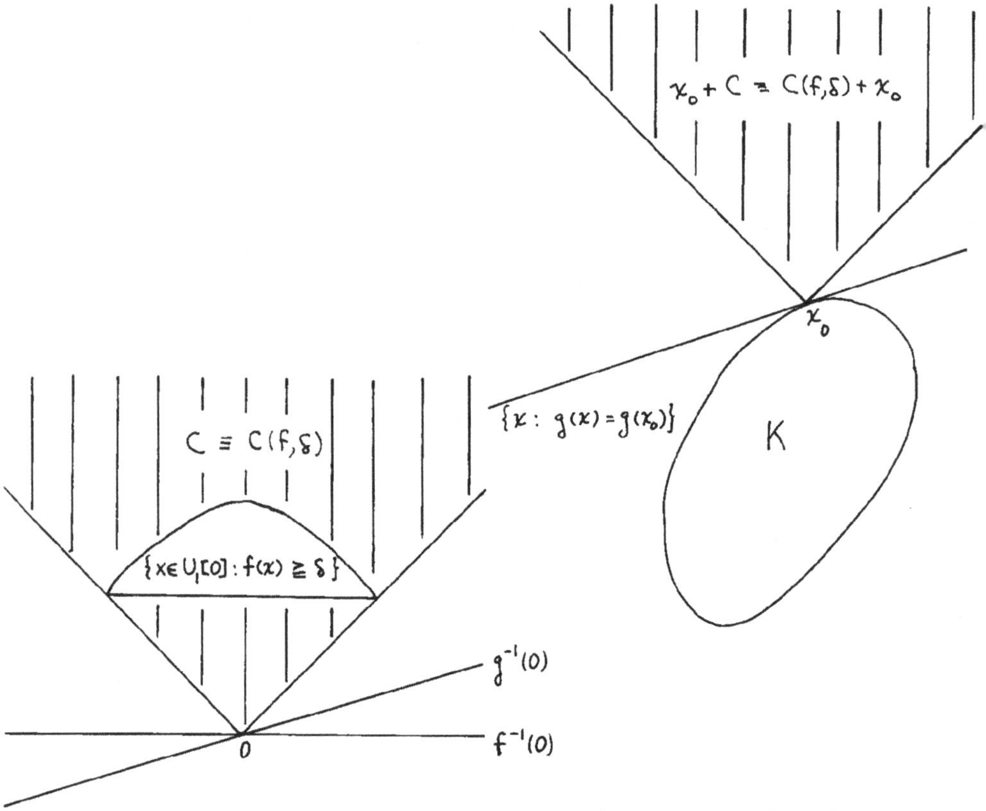

<u>LEMMA</u> 3.3.2. Let f and g be elements of X* of norm one, and fix $\varepsilon > 0$. If $f(x) = 0$ and $\|x\| \leqq 1$ together imply that $|g(x)| \leqq \frac{\varepsilon}{2}$ then either $\|f - g\| \leqq \varepsilon$ or $\|f + g\| \leqq \varepsilon$.

Proof. The restriction of g to the subspace $f^{-1}(0)$ is a linear functional of norm at most $\frac{\varepsilon}{2}$. Use the Hahn-Banach theorem to find an h in X* such that $\|h\| \leqq \frac{\varepsilon}{2}$

and $h = g$ on $f^{-1}(0)$. Then $g - h = 0$ on $f^{-1}(0)$ so that $g - h = tf$ for some $t \in \mathbb{R}$. Note that

$$\big|1 - |t|\big| = \big|\|g\| - \|g - h\|\big| \leqq \|h\| \leqq \frac{\varepsilon}{2}.$$

If $t \geqq 0$ then

$$\|f - g\| = \|(1 - t)f - h\| \leqq |1 - t| + \|h\| = \big|1 - |t|\big| + \|h\| \leqq \varepsilon$$

while, when $t < 0$ we have

$$\|f + g\| = \|(1 + t)f + h\| \leqq |1 + t| + \|h\| = \big|1 - |t|\big| + \|h\| \leqq \varepsilon.$$

The following technical consequence of 3.3.2 will be used in section 5.

<u>LEMMA</u> 3.3.3 Suppose that $f \in X^*$ and $\|f\| = 1$. For $t > 0$ denote by V_t the set $U_t[0] \cap f^{-1}(0)$. Assume that x_o and y are points of X such that $f(x_o) > f(y)$ and $\frac{2}{t}\|x_o - y\| \leqq 1$. If $g \in X^*$, $\|g\| = 1$, and $g(x_o) > M(y + V_t, g)$ then $\|f - g\| \leqq \frac{2}{t}\|x_o - y\|$.

Proof. In order to apply 3.3.2 assume that $f(x) = 0$ for a point x in X with $\|x\| \leqq 1$. Then $tx + y \in V_t + y$ and thus $tg(x) + g(y) < g(x_o)$. Let $\varepsilon \equiv \frac{2}{t}(g(x_o) - g(y))$. Then $g(x) \leqq \frac{\varepsilon}{2}$, and by symmetry it follows that $|g(x)| \leqq \frac{\varepsilon}{2}$. Consequently 3.3.2 guarantees that either $\|g - f\| \leqq \varepsilon = \frac{2}{t}(g(x_o) - g(y)) \leqq \frac{2}{t}\|x_o - y\|$ or that $\|g + f\| \leqq \frac{2}{t}(g(x_o) - g(y))$. But if $\|g + f\| \leqq \frac{2}{t}(g(x_o) - g(y))$ then

$$0 < g(x_o) - g(y) < g(x_o) - g(y) + f(x_o) - f(y) = (g + f)(x_o - y)$$

$$\leqq \|g + f\|\,\|x_o - y\| \leqq \frac{2}{t}(g(x_o) - g(y))\|x_o - y\|.$$

It would follow that $1 < \frac{2}{t}\|x_o - y\|$, a number less than one by assumption. Thus $\|g - f\| \leqq \frac{2}{t}\|x_o - y\|$ as was to be shown.

Let us continue with the discussion and notation introduced prior to the statement of Lemma 3.3.2. Observe that the equations

$$\sup g(K) = g(x_o) = \inf g(C(f,\delta) + x_o) \qquad (= \inf g(C(f,\delta)) + g(x_o))$$

imply that the restriction of g to the cone $C(f,\delta)$ is nonnegative. One might suspect from this that when $\delta > 0$ is small and $\|f\| = \|g\| = 1$ then $\|f - g\|$ is also small. The next lemma justifies this assertion.

LEMMA 3.3.4. Suppose that $0 < \varepsilon < 1$, that f and g are norm one elements of X^*, and that $0 < \delta < \frac{\varepsilon}{2 + \varepsilon}$. If g is nonnegative at each point of $C(f,\delta)$ then $\|g - f\| \leq \varepsilon$.

Proof. Suppose that $\|x\| \leq 1$ and $f(x) = 0$. We first show that $|g(x)| \leq \frac{\varepsilon}{2}$ so that 3.3.2 may be applied. To this end, choose $y \in X$, $\|y\| = 1$, so that $f(y) \geq \frac{\delta(2+\varepsilon)}{\varepsilon}$. Then

$$\delta\|y \pm \frac{2}{\varepsilon}x\| \leq \delta(1 + \frac{2}{\varepsilon}) = \frac{\delta(2+\varepsilon)}{\varepsilon} \leq f(y) = f(y \pm \frac{2}{\varepsilon}x)$$

and thus $y \pm \frac{2}{\varepsilon}x \in C(f,\delta)$. Since g is nonnegative on $C(f,\delta)$ it follows that $|\frac{2}{\varepsilon}g(x)| \leq g(y) \leq 1$, or, in the form employed in 3.3.2, $|g(x)| \leq \frac{\varepsilon}{2}$. Consequently, either $\|f - g\| \leq \varepsilon$ or $\|f + g\| \leq \varepsilon$ (by 3.3.2).

Pick any point $z \in X$, $\|z\| = 1$ with $f(z) > \max(\delta,\varepsilon)$. Then $f(z) > \delta\|z\|$ so that $z \in C(f,\delta)$. Hence $g(z) \geq 0$ and thus $\|f + g\| \geq (f + g)(z) \leq f(z) > \varepsilon$. Consequently we must have $\|f - g\| \leq \varepsilon$ as was to be shown.

Proof of 3.3.1. It clearly suffices to show that if $f \in X^*$, $\|f\| = 1$, and $0 < \varepsilon < 1$, then there is a $g \in S(K)$ with $\|f - g\| \leq \varepsilon$. Let $\delta \equiv \frac{1}{2} \cdot \frac{\varepsilon}{2+\varepsilon}$. Define a partial ordering $\leqslant$ on K as follows: for $x,y \in K$, say that $x \leqslant y$ if and only if $\delta\|y - x\| \leq f(y) - f(x)$. (Thus, $x \leqslant y$ if and only if $y - x \in C(f,\delta)$.)

Suppose that L is a linearly ordered subset of $(K,\leqslant)$. If x and y are in L and satisfy $x \leqslant y$, then $f(x) \leq f(y)$. Consequently the net $\{f(x): x \in L\}$ converges since it is monotonically nondecreasing in $\mathbb{R}$ and bounded above by $M(K,f)$. But since $\delta\|x - y\| \leq f(y) - f(x)$ for x and y as above, it follows that $\{x: x \in L\}$ is a norm-convergent net in K. The limit of this net is evidently an upper bound for L. Thus by Zorn's lemma, $(K,\leqslant)$ has a maximal element, say x_o.

If x is any point of $(C(f,\delta) + x_o) \cap K$ then $x - x_o \in C(f,\delta)$. Since both x and x_o are in K we have $x_o \leqslant x$, and hence $x = x_o$ by the maximality of x_o. That is,

$(C(f,\delta) + x_o) \cap K = \{x_o\}$. As indicated in the discussion, if $g \in X^*$, $\|g\| = 1$ satisfies

$$\sup g(K) = g(x_o) = \inf g(C(f,\delta) + x_o)$$

then $g \in S(K)$ and g is nonnegative on $C(f,\delta)$. Consequently $\|g - f\| \leq \varepsilon$ by 3.3.4.

The first results in the direction of Theorem 3.3.1 were obtained by R.R. Phelps in his doctoral thesis (Phelps [1957]). For further related results, see, for example, Phelps [1963], [1964], [1974b] and Bishop and Phelps [1963].

ℓ_1 was the first nontrivial example of a Banach space shown to have the KMP. This result, due to Lindenstrauss [1966] in 1966, was in part motivated by a question of Rieffel [1967] who was concerned with dentability questions related to Radon-Nikodým theorems. Several generalizations of Lindenstrauss' result in various directions followed soon thereafter. (See Asplund [1969], Bishop [1966], Bessaga and Pełczynski [1966], and Namioka [1967].) Based on a comparison of spaces with and without the RNP and with and without the KMP, J. Diestel raised the following question (when C itself is a Banach space) early in 1973:

PROBLEM 3.3.5. Let C be a closed convex subset of a Banach space. Does C have the RNP if and only if C has the KMP?

Later the same year, Lindenstrauss provided the first positive result connecting the RNP and the KMP (cf. Phelps [1974a]). (See §4.2 for other partial results.)

THEOREM 3.3.6 (Lindenstrauss). Let C be a closed convex subset of a Banach space X. If C has the RNP then C has the KMP.

Proof. Let K be a closed bounded convex subset of C and assume that C has the RNP. Then K and each of its subsets have slices of arbitrarily small norm-diameter (cf. 2.3.2). Construct a nested sequence of closed faces of K as follows: Let $K_o \equiv K$. If K_n has been constructed, let $S(K_n,f,\alpha)$ be a slice of K_n of diameter at most $2^{-(n+1)}$. Using Lemma 3.2.6 and the Bishop-Phelps theorem (3.3.1) find $g \in X^*$ such that $g \in \mathcal{S}(K_n)$ and $S(K_n,g,\frac{\alpha}{2}) \subset S(K_n,f,\alpha)$. Then let

$$K_{n+1} \equiv \{x \in K_n : g(x) = M(K_n, g)\}$$

and observe that K_{n+1} is a nonempty closed convex face of K_n. (That is, if $x, y \in K_n$ and $tx + (1-t)y \in K_{n+1}$ for some $t \in (0,1)$ then both x and y belong to K_{n+1}.) Moreover, the diameter of K_{n+1} is at most $2^{-(n+1)}$. Let $\{x_o\} \equiv \bigcap_{n=0}^{\infty} K_n$. Note that $\{x_o\}$ is a face of K since K_{n+1} is a face of K_n for each n. It follows that $x_o \in ex(K)$ and hence C has the KMP by Proposition 3.1.1.

<u>COROLLARY</u> 3.3.7. Each of the following Banach spaces lacks the KMP and hence lacks the RNP: c_o, $L^1(\mu)$ (μ finite, not purely atomic), ℓ_∞, $C(Y)$ (Y infinite compact Hausdorff.).

Proof. The closed unit ball of c_o lacks any extreme points. Given a not purely atomic finite measure μ, write it in the form $\mu_d + \mu_c$ where μ_d is a discrete measure and $0 \neq \mu_c$ assigns mass 0 to each atom. Then $\{f \in L^1(\mu): \int |f| d\mu_d = 0, \int |f| d\mu_c \leq 1\}$ has no extreme points. Next, the closed unit ball of c_o may be considered as a subset of ℓ_∞. Finally, if Y is infinite compact Hausdorff then there is a point $y_o \in Y$ which is not an isolated point. Consequently $\{f \in C(Y): \|f\| \leq 1$ and $f(y_o) = 0\}$ has no extreme points. (See the discussion near the beginning of section 1 for the case $Y = [0,1]$.)

§4. <u>Points of continuity of the identity map</u>.

In 1967 E. Asplund and I. Namioka provided a geometric proof of the Ryll-Nardzewski fixed point theorem. Their innovative idea was to show that many extreme points of, for example, a separable weakly compact convex subset C of a Banach space X are points at which the identity map id: (C, weak) → (C, norm) is continuous. Note that if $\varepsilon > 0$ and $x_o \in ex(C)$ is any such point of continuity then there is a slice of C containing x_o which has norm-diameter at most ε. (Indeed, by the continuity hypothesis there is a relative weak neighborhood V of x_o in C of norm-diameter at most ε. By the Milman theorem, cf. 3.2.2, $x_o \notin \overline{co}(C \setminus V)$ and the separation theorem guarantees the existence of a slice of the desired type.) Hence extreme points of such sets which are points of continuity are <u>denting points</u>. (A denting point x of D is a point such that $x \notin \overline{co}(C \setminus U_\varepsilon(x))$ for each $\varepsilon > 0$.) In a slightly later paper, Namioka [1967] examined the relevance of this same continuity idea to the kinds of problems dealt with in these notes. The following refinement of the Asplund-Namioka lemma is due to J. Bourgain [1978a].

<u>THEOREM</u> 3.4.1 (Asplund, Namioka; Bourgain). Suppose that $\varepsilon > 0$. Let J, K_o, and K_1 be closed bounded convex subsets of X such that

1) $J \subset \overline{co}(K_o \cup K_1)$
2) $K_o \subset J$ and diameter $(K_o) < \varepsilon$
3) $J \setminus K_1 \neq \emptyset$.

Then there is a slice of J which contains a point of K_o and is of diameter less than ε.

Proof. For each $r \in [0,1]$ let

$$C_r \equiv \{x \in X: \ x = (1 - \lambda)x_o + \lambda x_1,\ x_o \in K_o,\ x_1 \in K_1,\ r \leq \lambda \leq 1\}.$$

Note that $C_1 = K_1$ and that as r decreases from 1 to 0, the convex sets C_r increase from K_1 to the set $C_o \equiv co(K_o \cup K_1)$ whose closure contains J by hypothesis.

We first establish that if $0 < r \leq 1$ then $J \setminus \bar{C}_r \supset K_o \setminus \bar{C}_r \neq \emptyset$. Pick $f \in X^*$ such that $M(K_1,f) < M(J,f)$ (see hypothesis 3)). Then

$$M(J,f) \leq M(\bar{C}_o,f) = M(C_o,f) \leq \max\{M(K_o,f),\ M(K_1,f)\} \leq M(J,f).$$

On the other hand,

$$M(\bar{C}_r,f) = M(C_r,f) \leq \sup\{(1 - \lambda)M(K_o,f) + \lambda M(K_1,f): \lambda \in [r,1]\}$$

$$= (1 - r)M(J,f) + rM(K_1,f) < M(J,f) = M(K_o,f).$$

It is thus impossible that $\bar{C}_r \supset K_o$ since this would imply that $M(\bar{C}_r,f) \geq M(K_o,f)$.

The next step is to show that diameter $(J \setminus \bar{C}_r) < \varepsilon$ for r sufficiently close to 0. Since $J \subset \bar{C}_o$ we have $J \setminus \bar{C}_r \subset \bar{C}_o \setminus \bar{C}_r$. But since $\bar{C}_r$ is closed, it follows that $C_o \setminus \bar{C}_r$ is dense in $\bar{C}_o \setminus \bar{C}_r$. Thus if $w \in J \setminus \bar{C}_r$ and $\delta > 0$ there is a point $x \in C_o \setminus \bar{C}_r$ such that $\|w - x\| < \delta$. Pick $x_o \in K_o$, $x_1 \in K_1$ and $\lambda \in [0, r)$ such that $x = (1 - \lambda)x_o + \lambda x_1$. Then

$$\|w - x_0\| \leqq \|w - x\| + \|x - x_0\| < \delta + \lambda\|x_0 - x_1\| \leqq \delta + rM$$

where $M \equiv \sup\{\|y_0 - y_1\|: \ y_0 \in K_0, y_1 \in K_1\}$. It follows that

$$\text{diameter } (J\backslash\bar{C}_r) \leqq \delta + rM + \text{diameter } (K_0) + \delta + rM.$$

By picking $\delta > 0$ and $r > 0$ small enough, therefore (since diameter $(K_0) < \varepsilon$), we can force diameter $(J\backslash\bar{C}_r) < \varepsilon$. Fix one such r.

Pick any point $x_0 \in K_0$ such that $x_0 \in J\backslash\bar{C}_r$. (See paragraph 2 of this proof.) Then by the separation theorem, x_0 is contained in a slice of J disjoint from $\bar{C}_r$. This slice necessarily has diameter less than ε, and the proof is complete.

<u>THEOREM</u> 3.4.1 (w*). Suppose that $\varepsilon > 0$. Let J, K_0, and K_1 be weak*-compact convex subsets of a dual Banach space X* such that

1) $J \subset co(K_0 \cup K_1)$ $(= w^*cl(co(K_0 \cup K_1)))$;
2) $K_0 \subset J$ and diameter $(K_0) < \varepsilon$; and
3) $J\backslash K_1 \neq \emptyset$.

Then there is a weak*-slice of J which contains a point of K_0 and is of diameter less than ε.

(<u>Note</u>: a weak*-slice of J is a slice of J determined by an $f \in X$.)

Proof. In this setting all the sets C_r of the proof of 3.4.1 are already weak*-compact. Consequently, the proof of 3.4.1 may be simplified in several places. Since all sets involved are weak*-compact, the slice chosen at the end of the proof of 3.4.1 may be chosen to be a weak*-slice.

Evidently there are close ties between dentability, points of continuity of the identity, and the existence of extreme points. That a certain degree of caution need be exercised, however, is made clear from the following result of Bourgain and Rosenthal [1980b]:

There is a Banach space X which fails the RNP such that whenever D is a closed bounded subset of X then id: (D,weak) $\to$ (D,norm) has a point of continuity. Information about spaces having the "point of continuity property" described by Bourgain and Rosenthal above may be found in the preprint <u>Topological Properties of Banach Spaces</u> by G.A. Edgar and R.F. Wheeler.

§5. Category and strongly exposed point characterizations of the RNP.

The work of this section is, in its present form, largely due to J. Bourgain and C. Stegall. However, many of the central ideas as well as earlier versions of some of the results come from others, notably R. R. Phelps, E. Bishop, E. Asplund, and I. Namioka. In any case, the results contained below provide powerful and elegant geometric characterizations of the RNP which point out unequivocally the special nature of the sets which possess this property. A few preliminary results are needed.

LEMMA 3.5.1. Let D and C be closed bounded convex subsets of X. Let $J \equiv \overline{co}(D \cup C)$ and suppose that $f(x) = M(J,f) > M(C,f)$ for some $f \in X^*$ and $x \in J$. Then $x \subset D$.

Proof. There is a sequence $\{x_n\} \subset co(D \cup C)$ such that $\lim_n \|x_n - x\| = 0$. Suppose that $x_n = \lambda_n d_n + (1 - \lambda_n)c_n$ where $0 \leqq \lambda_n \leqq 1$, $d_n \in D$, and $c_n \in C$. Without loss of generality assume that $\lambda \equiv \lim_n \lambda_n$ exists. If $\lambda < 1$ then

$$M(J,f) = f(x) = \lim_n [\lambda_n f(d_n) + (1 - \lambda_n)f(c_n)] \leqq \lambda M(D,f) + (1 - \lambda)M(C,f)$$

$$< M(J,f).$$

Thus $\lambda = 1$. Since C is bounded it follows that $\lim_n \|x - d_n\| = 0$ and since D is closed, $x \in D$.

The connection between the RNP and the closed convex hull of two convex sets, one of which has slices of arbitrarily small diameter, was hinted at in Theorem 3.4.1. It is made explicit in the following.

PROPOSITION 3.5.2. Let C and K be closed bounded convex subsets of X and let $J \equiv \overline{co}(C \cup K)$. Suppose that K has the RNP, that $K \backslash C \neq \emptyset$ and that $\varepsilon > 0$. Then there is a slice of J of diameter less than ε which contains a point of K.

Proof. Let $D \equiv \{x \in J: \text{ there is an } f \in X^* \text{ such that } f(x) = M(J,f) > M(C,f)\}$. Then $D \subset K$ by 3.5.1. Clearly $\overline{co}(D \cup C) \subset J$. Suppose momentarily that $\overline{co}(D \cup C)$ is a proper subset of J. Then there is a slice S of J such that $S \cap \overline{co}(D \cup C) = \emptyset$. Use Lemma 3.2.6 and the Bishop-Phelps Theorem (3.3.1) to find an $f \in S(J)$ and $\alpha > 0$ with $S(J,f,\alpha) \subset S$. Then f supports J at some point x. Clearly $x \notin \overline{co}(D \cup C) \supset D$ and yet $f(x) = M(J,f) > M(C,f)$ so $x \in D$. That is, we must have $J = \overline{co}(D \cup C)$. In particular $D \neq \emptyset$ since $K \setminus C \neq \emptyset$ by hypothesis.

Since K has the RNP the set $D \subset K$ is dentable and thus there is a point $x_o \in D$ such that $x_o \notin \overline{co}(D \setminus U_{\varepsilon/3}(x_o))$. Let $K_o \equiv \overline{co}(U_{\varepsilon/3}(x_o) \cap D)$ and $K_1 \equiv \overline{co}[(D \setminus U_{\varepsilon/3}(x_o)) \cup C]$. Then the hypotheses of 3.4.1 are met by J, K_o, and K_1 as we show next.

1) It is clear that $J = \overline{co}(K_o \cup K_1)$ since $J = \overline{co}(D \cup C)$.

2) Evidently diameter $(K_o) \leqq 2\varepsilon/3 < \varepsilon$ and $K_o \subset \overline{co}(D) \subset J$.

3) We will prove that $x_o \in J \setminus K_1$. Indeed, if $x_o \in K_1$, let $D_1 \equiv \overline{co}(D \setminus U_{\varepsilon/3}(x_o))$ and note that $K_1 = \overline{co}(D_1 \cup C)$. Since $x_o \in D$, there is an $f \in X^*$ such that $f(x_0) = M(J,f) > M(C,f)$ and since $x_o \in K_1 \subset J$ we have $f(x_0) = M(K_1,f) > M(C,f)$. Thus Lemma 3.5.1 applies (with D_1 in place of D and K_1 instead of J) and x_o thus belongs to D_1. Since this contradicts the choice of x_o, it follows that $x_o \in J \setminus K_1 \neq \emptyset$.

Theorem 3.4.1 guarantees that there is a slice of J of diameter less than ε which contains a point of K_o. Since $K_o \subset K$, the proof is complete.

The following terminology is needed in order to state the weak*-version of 3.5.2.

DEFINITION 3.5.3. Let A be an arbitrary bounded subset of a dual Banach space X*. Then A is weak*-dentable if it contains weak*-slices of arbitrarily small norm-diameter. A subset D of a dual Banach space is subset weak*-dentable if each of its bounded subsets is weak*-dentable.

Just as in 2.3.2, a bounded set $A \subset X^*$ is weak*-dentable if and only if for each $\varepsilon > 0$ there is a point $x_\varepsilon \in A$ such that $x_\varepsilon \notin w^*cl(co(A \setminus U_\varepsilon(x_\varepsilon)))$.

PROPOSITION 3.5.2 (w*). Let C and K be weak*-compact convex subsets of a dual Banach space X* and let $J \equiv co(C \cup K)$. Suppose that $\varepsilon > 0$, that K is subset weak*-dentable, and that $K \setminus C \neq \emptyset$. Then there is a weak*-slice of J of diameter less than ε which contains a point of K.

Proof. Follow the proof of 3.5.2 with the following alterations: J as defined above is already weak*-compact and convex. All linear functionals are chosen from X (rather than from X**). Replace "co" by "w*cl(co" throughout. Since the sets in question in the first paragraph of the proof of 3.5.2 are weak*-compact there is no need to apply 3.2.6 or 3.3.1. (The functional which defines the slice will suffice.) The point x_o chosen in the second paragraph (which satisfies the condition $x_o \notin w^*cl(co(D\setminus U_{\varepsilon/3}(x_o)))$ exists since K is subset weak*-dentable. In paragraph 3, apply 3.4.1 (w*) rather than 3.4.1 in the appropriate place.

We are finally in a position to characterize the RNP in terms of strongly exposed points. (See §7.5 for an alternative proof of this theorem.)

THEOREM 3.5.4 (Phelps; Bourgain). Let K be a closed bounded convex subset of X and assume that K has the RNP. Then $K = \overline{co}(\text{str exp}(K))$. Moreover, $SE(K)$ is a dense G_δ subset of X*.

Proof. The conclusions of this theorem coincide with those of Bishop's lemma (3.2.7) and it therefore suffices to check that the above hypotheses imply those of

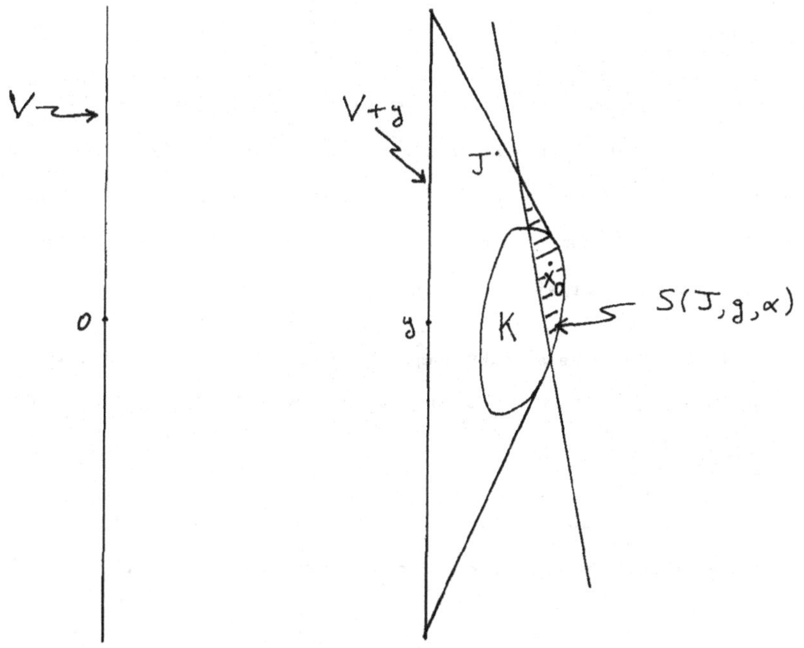

3.2.7. Thus suppose that $0 < \delta < 1$ and that $f \in X^*$, $\|f\| = 1$. Since K is bounded there is a point $y \in X$ such that $f(y) < f(x) - 1$ for each $x \in K$. We will use the following notation: $M \equiv \sup\{\|x - y\|: x \in K\}$; $t \equiv 2M/\delta$; $V \equiv f^{-1}(0) \cap U_t[0]$; and $C \equiv V + y$. Note that whenever $z \in C$ and $x \in K$ then $f(z) = f(y) < f(x) - 1$ so that $K \cap C = \emptyset$. In particular, $K \setminus C \neq \emptyset$. Consequently Proposition 3.5.2 asserts that, for $J \equiv \overline{co}(C \cup K)$, there is a slice $S(J,g,\alpha)$ for some $g \in X^*$, $\|g\| = 1$, which is of diameter less than δ and which contains a point x_o of K.

Observe first that $S(J,g,\alpha) \cap C = \emptyset$. (Indeed, if $w \in S(J,g,\alpha) \cap C$ then

$$1 > \delta > \text{diameter } S(J,g,\alpha) \geqq \|x_o - w\| \geqq f(x_o) - f(w) = f(x_o) - f(y) > 1,$$

a contradiction.) Secondly, $M(K,g) > M(C,g)$ since otherwise we would have $M(J,g) = M(C,g)$ and hence each slice of J determined by g would contain points of C. Since this is not the case for the slice $S(J,g,\alpha)$ it follows that $M(K,g) > M(C,g)$. Hence $M(K,g) = \max\{M(K,g), M(C,g)\} = M(J,g)$.

Since $K \subset J$ it follows that $S(K,g,\alpha) \subset S(J,g,\alpha)$ and hence diameter $S(K,g,\alpha) < \delta$. Finally $\frac{2}{t}\|x_o - y\| \leqq 2M/t = \delta < 1$ so that Lemma 3.3.3 shows that $\|g - f\| \leqq \frac{2}{t}\|x_o - y\| \leqq \delta$. That is, the hypotheses of Bishop's lemma are satisfied, and the theorem is proved.

The idea of separating (with a slice of small diameter) a large flat section of a hyperplane from a point (V from x_o in the terminology of 3.5.4) in the search for strongly exposed points was first employed by Phelps [1974a]. He proved that the conclusion of 3.5.4 holds for each closed bounded convex subset of a Banach space with the RNP. Also see Bourgain [1977] and Phelps [1983].

THEOREM 3.5.4 (w*). Let K be a weak* compact convex subset of X* and assume that K is subset weak*-dentable. Then K = w*cl(co(w*-str exp(K))). Moreover, w*-SE(K) is a dense G_δ subset of X.

Proof. Mimic the proof of 3.5.4, using the weak* versions of 3.2.7 and 3.5.2. Of course all functionals must be chosen from X instead of from X*.

The next characterization of sets with the RNP was communicated to us by Namioka who first heard this result presented orally by Bourgain with Stegall attending. It represents the first successful attempt at determining an important class of closed bounded convex subsets of a Banach space X in terms of the distribution of support functionals in X* since the theorem of James [1964] ($K \subset X$ is weakly compact if and only if $S(K) = X^*$). (Of course, $S(K)$ is automatically dense in X* by virtue of the Bishop-Phelps theorem 3.3.1.)

THEOREM 3.5.5 (Bourgain, Stegall). Let $C \subset X$ be a closed convex set and suppose that $S(K) \subset X^*$ is of the second Baire category for each closed bounded convex set $K \subset C$. Then C has the RNP.

Proof. If C lacks the RNP then there is a separable closed bounded convex subset K of C which is not dentable by Theorem 2.3.6 part 2.

Choose $\delta > 0$ such that each slice of K has diameter at least 3δ. Pick a countable dense set $\{x_i\}$ in K, and let $U_i \equiv K \cap U_\delta[x_i]$ and $\mathcal{O}_i \equiv \{g \in X^*:$ g determines a slice of K disjoint from $U_i\}$. Observe that if $g \in \bigcap_i \mathcal{O}_i$ and $x \in K$ then $x \in U_j$ for some j whence $g(x) < M(K,g)$. That is, $X^* \setminus \bigcap_i \mathcal{O}_i \supset S(K)$.

It thus remains to prove that each $\mathcal{O}_i$ is both open and dense in X*. That $\mathcal{O}_i$ is open is immediate from Lemma 3.2.6. To check on the density of $\mathcal{O}_i$, suppose that $f \in X^*$ and $0 < \varepsilon < 1$. Since $\mathcal{O}_i$ is closed under multiples by positive scalars, we henceforth assume that $\|f\| = 1$. Pick $y \in X$ such that $f(x) - f(y) > 0$ for $x \in K$ and let $M \equiv \sup\{\|x - y\|: x \in K\}$. Choose $t \geqq 2M/\varepsilon$, let $V = f^{-1}(0) \cap U_t[0]$, and denote by C the set V + y. Clearly $U_i \setminus C \neq \emptyset$. Next observe that by Theorem 3.4.1, $J \equiv \overline{co}(U_i \cup C)$ contains a slice S such that diameter (S) $< 3\delta$ and $S \cap U_i \neq \emptyset$ since diameter (U_i) is at most 2δ. But if $K \subset J$ then, since $U_i \subset K$ by definition, $S \cap K$ would be a slice of K of diameter less than 3δ which contradicts the choice of δ. Consequently there is a point $x_o \in K\setminus J$. Choose $g \in X^*$, $\|g\| = 1$, such that $g(x_o) > M(J,f)$ and note that from

$$\frac{2}{t}\|x_o - y\| \leqq \frac{2}{t}M \leqq \varepsilon < 1$$

and Lemma 3.3.3, we have $\|g - f\| \leqq \varepsilon$. Moreover g evidently determines a slice of K disjoint from U_i, whence $g \in \mathcal{O}_i$. That is, $\mathcal{O}_i$ is dense and the proof is complete.

The following question arises naturally in connection with 3.5.5.

PROBLEM 3.5.6. Does there exist a closed bounded convex set K in a Banach space X which is not dentable such that $S(K)$ is of the second Baire category?

The answer is "no" when K is separable, as is shown in the proof of 3.5.5. In addition Talagrand observed that the answer is also "no" when K is the unit ball of a C(Y) space, Y infinite compact Hausdorff.

The main results of this section may be combined in the following.

COROLLARY 3.5.7. Let C be a closed convex subset of a Banach space. Then the following are equivalent:

1) C has the RNP;

2) Each closed bounded convex subset K of C satisfies:

$$K = \overline{co}(\text{str exp}(K));$$

3) $S(K)$ is of the second Baire category for each closed bounded convex subset K of C.

Proof. If K is a closed bounded convex set then the following facts are immediate: a) $SE(K) \subset S(K)$; and b) if $x_o \in \text{str exp}(K)$ then $x_o \notin \overline{co}(K \setminus U_\varepsilon(x_o))$ for each $\varepsilon > 0$. (In particular, K is dentable.) Consequently, 2 $\Rightarrow$ 1 follows from b and Theorem 2.3.6, while 1 $\Rightarrow$ 3 is a consequence of a and Theorem 3.5.4. 3 $\Rightarrow$ 1 is the content of Theorem 3.5.5 while 1 $\Rightarrow$ 2 is contained in Theorem 3.5.4.

Finally, it might seem likely (in view of 3.5.7, Part 3) that if K were a closed bounded convex dentable set, then $S(K)$ must necessarily be of the second Baire category. The following example shows this to be false.

EXAMPLE 3.5.8 (Bourgin). Let U denote the closed unit ball of c_o and for each $n = 1,2,\ldots$ let $\delta_n \equiv (0,\ldots,0,1,0,\ldots) \in c_o$ be the point with exactly one 1 in the nth place. Given numbers t_i such that $1 > t_1 > t_2 > \ldots > 0$ and $\lim_i t_i = 0$ define a set $K \subset c_o$ as follows:

$$K \equiv \overline{co}(U \cup \bigcup_{i=1}^{\infty} [t_iU + 2\delta_i]).$$

A direct argument or appropriate application of Theorem 3.4.1 shows that K is dentable. Suppose next that $f \in S(K)$. Now

$$M(K,f) = \sup\{M(U,f),\ t_iM(U,f) + f(2\delta_i):\ i = 1,2,\dots\}$$

and since $\lim_{i\to\infty} [t_iM(U,f) + f(2\delta_i)] = 0$ there are numbers $r > 0$, $\varepsilon > 0$, and $N \in \mathbb{N}^+$ such that

$$M(K,f) = r = \max\{M(U,f),\ t_iM(U,f) + f(2\delta_i):\ i = 1,\dots,N\}$$

$$> r - \varepsilon \geqq \sup\{t_iM(U,f) + f(2\delta_i):\ i \geqq N + 1\}.$$

To see that $K_1 \equiv co(U \cup \bigcup_{i=1}^{N} [t_iU + 2\delta_i])$ is closed, pick any point $x \in co(w^*cl(U) \cup \bigcup_{i=1}^{N} [w^*cl(t_iU + 2\delta_i)]) \cap c_o$. (This latter set is the intersection of a weak* compact convex subset of ℓ_∞ with c_o, and it certainly contains the embedded set K_1.) Pick points w, u_i, $v_i \in w^*cl(U)$, and nonnegative numbers α, and λ_i, $i = 1,\dots,N$ such that $\alpha + \sum_{i=1}^{N} \lambda_i = 1$ and $x = \alpha w + \sum_{i=1}^{N} \lambda_i(t_iu_i + 2\delta_i)$. Both x and δ_i are points of c_o, $i = 1,\dots,N$, and hence $y \equiv \alpha w + \sum_{i=1}^{N} \lambda_it_iu_i \in c_o$. Moreover, $\|y\| \leqq 1$ since $0 < t_i \leqq 1$ for each i. That is, $y \in U$. Consequently

$x = (1 - \sum_{i=1}^{N} t_i\lambda_i)y + \sum_{i=1}^{N} \lambda_i(t_iy + 2\delta_i) \in K_1$. Hence K_1 is closed. Let

$K_2 \equiv \overline{co}(\bigcup_{i=N+1}^{\infty} [t_iU + 2\delta_i])$. Then $K = \overline{co}(K_1 \cup K_2)$ and $\{x \in K:\ f(x) = M(K,f)\} = \{x \in K_1:\ f(x) = M(K,f)\}$ by Lemma 3.5.1. Thus, if $f \in S(K)$ then $f \in S(K_1)$ and consequently $f \in S(t_iU + 2\delta_i)$ (or $f \in S(U)$) for some $i = 1,\dots,N$. Thus $f \in S(U)$ and hence $S(K) \subset S(U)$, a set of the first Baire category (either by direct argument, or by appeal to the proof of Theorem 3.5.5, since U is separable.)

§6. Weakly compact sets.

Lindenstrauss [1963] showed that each weakly compact convex separable subset of a Banach space is the closed convex hull of its strongly exposed points. Troyanski [1971] removed the separability restriction. Both results were, in some sense, incidental corollaries of somewhat deeper structure theoretic results of these authors. In the time period between these two results, Rieffel [1967] noted that Lindenstrauss had in fact shown that separable weakly compact convex sets in Banach spaces are dentable (and thus have the RNP). (See Theorem 2.3.6, part 11.) Later, Maynard [1973] observed that the RNP was a separably determined property (see Theorem 2.3.6, part 2) and that consequently weakly compact convex subsets of Banach spaces, separable or not, have the RNP. A direct geometric proof of the Lindenstrauss-Troyanski result was given by Bourgain [1976] somewhat before his more general Theorem 3.5.4. We follow Bourgain's ideas below.

<u>THEOREM</u> 3.6.1 (Lindenstrauss; Troyanski). Let K be a weakly compact convex subset of a Banach space X. Then K has the RNP. Consequently $K = \overline{co}(\text{str exp}(K))$.

Proof. Suppose first that K is a separable weakly compact convex subset of X and that K is not dentable. Let $\{x_i\}$ be a norm dense subset of K and fix $\varepsilon > 0$. Let $D \equiv$ weak-closure $(ex(K)) \subset K$, and note that $D \subset \bigcup_{i=1}^{\infty} U_{\varepsilon/3}[x_i]$. Since D is a compact Hausdorff space in the weak topology and since $D \cap U_{\varepsilon/3}[x_i]$ is a weakly closed subset of D for each i, the Baire category theorem guarantees that there is a weakly open subset V of X such that

$$\emptyset \neq V \cap D \subset U_{\varepsilon/3}[x_i] \cap D \quad \text{for some} \quad i.$$

Let $K_o \equiv \overline{co}(V \cap D)$ and $K_1 \equiv \overline{co}(K\backslash V)$. Observe that

1) $K \subset \overline{co}(K_o \cup K_1)$;

and

2) diameter $(K_o) < \varepsilon$.

Now choose $x \in ex(K) \cap V$. (Since $V \cap D \neq \emptyset$, such a point necessarily exists.) Then $x \notin$ weak-closure $(K\backslash V)$, and the Milman theorem (cf. 3.2.2) guarantees that $x \notin K_1$. That is,

3) $x \in K\backslash K_1 \neq \emptyset$.

Theorem 3.4.1 thus applies, and there is a slice of K of diameter less than ε. Since $\varepsilon > 0$ is arbitrary, K is dentable.

Note that conditions 2 and 11 of Theorem 2.3.6 may be combined to give the criterion: K has the RNP if and only if each closed bounded convex separable subset of K is dentable. The conclusion of the first half of 3.6.1 thus follows from the above paragraphs. Finally, the fact that $K = \overline{co}(\text{str exp}(K))$ is a consequence of 3.5.4.

§7. Sets which lack the RNP.

Thus far we have concentrated on positive results about closed convex sets with the RNP. In taking the opposite tack, Huff and Morris asked how badly behaved some subset of a closed convex set C must be if C lacks the RNP. (Clearly not all subsets of such a set can be expected to exhibit "unpleasant" characteristics.) Some of their answers are startling. The main ideas here stem from work of R.E. Huff and P.D. Morris [1976], although in many cases, the results presented owe their present form to J. Bourgain [1978a]. In addition, I. Namioka has streamlined some of Bourgain's constructions.

It is convenient to prove the main result of this section in terms introduced in the following definitions.

DEFINITIONS 3.7.1. a) Let A be a finite set. A real-valued function λ on A is called a positive charge on A if $\lambda(x) > 0$ for each $x \in A$ and $\Sigma \{\lambda(x): x \in A\} = 1$.

b) Let D be a bounded subset of a Banach space X and $f \in X^*$, $f \neq 0$. For $\alpha > 0$, the slice $S(D,f,\alpha)$ is called a pointed slice if there is an $x_0 \in D$ such that $f(x_0) = M(D,f)$. If $S(D,f,\alpha)$ is a pointed slice, then any point $y \in D$ for which $f(y) = M(D,f)$ is called a slicing point of the pointed slice. (There may, of course, be more than one slicing point of a pointed slice.)

The main result of this section appears below. Its proof will require some preliminary work.

THEOREM 3.7.2 (Huff, Morris; Bourgain). Let $C \subset X$ be a closed convex set which lacks the RNP. Then there is a weakly closed and weakly discrete infinite bounded subset W of C such that

(a) W has no extreme points (i.e. for each $x \in W$, $x \in co(W\setminus\{x\})$); and

(b) whenever $f \in X^*$ attains its supremum on W then f is constant on W. (That is, if $f \in X^*$ and there is a point $x \in W$ such that $f(x) = M(W,f)$ then $f(x) = f(y)$ for each $y \in W$.)

The following lemmas will be used.

LEMMA 3.7.3. Let K_1 and K be closed bounded convex subsets of X and suppose K_1 is a proper subset of K. Then there exist $y \in K$, $g \in X^*$, and $\beta > 0$, such that

$$g(y) = M(K,g) > M(K_1,g) + \beta.$$

Proof. Strictly separate some point of $K \setminus K_1$ from K_1, use Lemma 3.2.6, and apply the Bishop-Phelps theorem (3.3.1).

LEMMA 3.7.4. Let K be a closed bounded convex nondentable subset of X. Then there is an $\varepsilon > 0$ such that whenever D_0 and D_1 are subsets of K for which

(1) diameter $(D_0) \leqq \varepsilon$; and

(2) $K = \overline{co}(D_0 \cup D_1)$,

then $K = \overline{co}(D_1)$.

Proof. Pick $\varepsilon > 0$ so that each slice of K has diameter at least 2ε. Let $K_i \equiv \overline{co}(D_i)$ $i = 0,1$. Of course, diameter $(K_0) \leqq \varepsilon$. If also $K \setminus K_1 \neq \emptyset$ then the hypotheses of Theorem 3.4.1 would be met and there would exist a slice of K of diameter less than 2ε. From this impossibility we conclude that $K = \overline{co}(D_1)$.

LEMMA 3.7.5. Let $K \subset X$ be a closed bounded convex nondentable set. Then there is an $\varepsilon > 0$ with the following property: given any

$B \subset K$ of norm diameter at most ε;

$S(K,f,\alpha)$ a pointed slice of K with slicing point x_0;

$\mathcal{O}$ a norm neighborhood of x_0;

there exist a finite number of pointed slices $S_1,\ldots,S_n$ with corresponding slicing points $x_1,\ldots,x_n$, such that

(a) $\bar{S}_i \subset \bar{S}(K,f,\alpha) \setminus B$ $\quad i = 1,\ldots,n$; and

(b) $\mathcal{O} \cap co(\{x_1,\ldots,x_n\}) \neq \emptyset$.

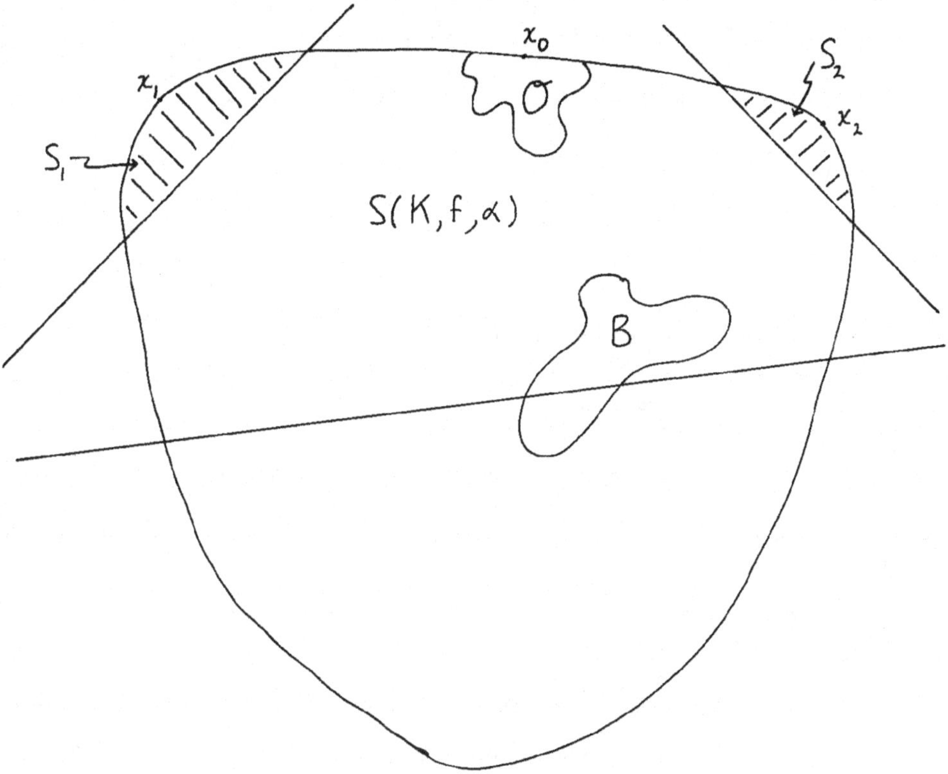

Proof. For our given set K choose $\varepsilon > 0$ as in Lemma 3.7.4. Let

$$D \equiv \{y \in K\colon \text{ there is a } g \in X^* \text{ with } g(y) = M(K,g) > \sup\{g[(K\backslash\bar{S}(K,f,\alpha)) \cup B]\}.$$

Lemma 3.7.3 and the definition of D yield

$$K = \overline{co}[(K\backslash\bar{S}(K,f,\alpha)) \cup B \cup D].$$

But B has diameter at most ε and consequently Lemma 3.7.4 applies. Thus

$$K = \overline{co}[(K\backslash\bar{S}(K,f,\alpha)) \cup D].$$

In particular, then, since $K\backslash\bar{S}(K,f,\alpha)$ is already convex, there are points $z_n \in K\backslash\bar{S}(K,f,\alpha)$ and $w_n \in co(D)$ as well as numbers $\beta_n \in [0,1]$ such that

$$x_o = \lim_n (\beta_n z_n + (1 - \beta_n)w_n).$$

Since

$$M(K,f) = f(x_o) = \lim_n [\beta_n f(z_n) + (1 - \beta_n) f(w_n)]$$

$$\leqq \lim [\beta_n (M(K,f) - \alpha) + (1 - \beta_n) M(K,f)] = M(K,f) - \alpha \lim_n \beta_n$$

it follows that $\lim_n \beta_n = 0$. Hence there is an N such that $w_N \in \mathcal{O}$. If $w_N = \sum_{i=1}^{n} \lambda_1 x_i$ ($x_i \in D$, $\lambda_i > 0$, $i = 1,\ldots,n$, and $\sum_{i=1}^{n} \lambda_i = 1$) choose pointed slices $S_1,\ldots,S_n$ corresponding to $x_1,\ldots,x_n$ such that condition (a) of this lemma is satisfied. (Observe that, from the definition of D, $\bar{B} \cap D = \emptyset$.)

We are now ready to prove Theorem 3.7.2.

Proof of 3.7.2. We will obtain W as the union of a sequence $\{A_i: i = 1,2,\ldots\}$ of finite subsets of C such that if $x \in A_i$ then there is a positive charge λ on A_{i+1} such that $x = \Sigma \{\lambda(y)y: y \in A_{i+1}\}$.

By Theorem 2.3.6, parts 2 and 11, if C is not dentable it contains a closed bounded convex separable nondentable subset K. Choose $\varepsilon > 0$ for this set K as guaranteed by Lemma 3.7.4 and observe that since K is separable, there is a sequence of subsets B_i of K, each of diameter at most ε, such that $K = \bigcup_{i=1}^{\infty} B_i$. By induction we will construct a sequence $\{A_n: n = 1,2,\ldots\}$ and corresponding collections $\{S_{\mathcal{O}}: \mathcal{O} \in A_n\}$ subject to the following conditions:

(1) A_n is a finite set for each n. The elements of A_n are relatively norm-open subsets of K;

(2) For each $\mathcal{O} \in A_n$, $S_{\mathcal{O}}$ is a pointed slice of K;

(3) If $\mathcal{O} \in A_n$ then diameter $(\mathcal{O}) \leqq 1/n$;

(4) For each $\mathcal{O} \in A_n$ there is a positive charge $\lambda_{\mathcal{O}}$ on the finite set A_{n+1} such that $\Sigma \{\lambda_{\mathcal{O}}(V)V: V \in A_{n+1}\} \subset \mathcal{O}$;

(5) $\mathcal{O} \subset S_{\mathcal{O}}$ and $\mathcal{O}$ contains a slicing point of $S_{\mathcal{O}}$;

(6) $\bar{S}_{\mathcal{O}} \cap B_n = \emptyset$ for each $\mathcal{O} \in A_n$;

(7) For each $V \in A_{n+1}$, $\bar{S}_V \subset \bar{S}_{\mathcal{O}}$ for some $\mathcal{O} \in A_n$.

To begin, pick any pointed slice S of K such that $\bar{S} \cap B_1 = \emptyset$. (This is easy to do using Lemma 3.7.3 since $\overline{co}(B_1) \neq K$ by 3.7.4.) Let x be a slicing point of S. A_1 then contains the single element $U_{1/2}(x) \cap S$, and we let the collection

$\{S_O: O \in A_1\}$ be the set whose sole element is S. Suppose next that $A_1,\ldots,A_n$ and corresponding collections of pointed slices have been constructed satisfying (1) - (7) for all appropriate indices. Let $A_n = \{O_1,\ldots,O_k\}$ and for convenience we will denote S_{O_i} by S_i, $i = 1,\ldots,k$. Condition (5) guarantees that for each $i = 1,\ldots,k$ there is a slicing point x_i of S_i such that $x_i \in O_i$. Now apply 3.7.5 to each triple B_{n+1}, S_i, O_i, $i = 1,\ldots,k$ (in place of B, S(K,f,α), O in the notation of 3.7.5). Denote the collection of pointed slices and slicing points so produced by $\{T_j: j = 1,\ldots,r\}$ and $\{y_j: j = 1,\ldots r\}$. It follows directly from 3.7.5 that

(A) for each $j = 1,\ldots,r$, $\bar{T}_j \subset \bar{S}_i$ for some $i = 1,\ldots,k$;

(B) $\bar{T}_j \cap B_{n+1} = \emptyset$ for each $j = 1,\ldots,r$;

(C) $O_i \cap \mathrm{co}(\{y_j: j = 1,\ldots,r\}) \neq \emptyset$ for each $i = 1,\ldots,k$.

Condition (C) implies that (for each $i = 1,\ldots,k$) there is a positive charge λ_i on the finite set $\{y_j: j = 1,\ldots,r\}$ such that $\Sigma\, \{\lambda_i(y_j)y_j: j = 1,\ldots,r\} \in O_i$.

By choosing sufficiently small relatively norm-open sets V_j containing y_j we can guarantee that

(D) $\Sigma\, \{\lambda_i(y_j)V_j: j = 1,\ldots,r\} \subset O_i$ for each $i = 1,\ldots,k$. By further restricting the size of the V_j's if necessary, we may assume further that, for each $j = 1,\ldots,r$ we have

(E) diameter $(V_j) \leq \frac{1}{n+1}$; and

(F) $V_j \subset \overline{T_j}$. (This condition is easy to satisfy since V_j is a neighborhood of $y_j \in T_j$, a relatively open set.)

Let $A_{n+1} \equiv \{V_j: j = 1,\ldots,r\}$, $\lambda_{O_i}(V_j) \equiv \lambda_i(y_j)$ for $i = 1,\ldots,k$, $j = 1,\ldots,r$, and $S_{V_j} \equiv T_j$. Then properties (E), (D), (F), (B), and (A), correspond to inductive conditions (3), (4), (5), (6), and (7) respectively. Since (1) and (2) are immediate for the (n + 1)st stage, the induction is complete.

For each $n \geq 1$ define sets $O(n)$ for $O \in \bigcup_{i=1}^{n} A_i$ in the following inductive manner: If $O \in A_n$ let $O(n) \equiv O$, and if $O(n)$ has been defined for each element of $\bigcup_{i=j+1}^{n} A_i$ let

(*) $$O(n) \equiv \sum \{\lambda_O(V)V(n): V \in A_{j+1}\} \quad \text{for} \quad O \in A_j.$$

An easy argument establishes that for $n \leqq j$ and $\mathcal{O} \in \mathcal{A}_n$ we have

$$\text{diameter } \mathcal{O}(j) \leqq \frac{1}{j} \quad \text{and} \quad \mathcal{O}(j) \subset \mathcal{O}(n) = \mathcal{O}.$$

$$\begin{array}{llll}
\mathcal{A}_1 & \mathcal{O}(1) & \mathcal{O}(2) & \mathcal{O}(3) \ \ldots \\
 & & \uparrow & \uparrow \\
\mathcal{A}_2 & & \mathcal{O}(2) & \mathcal{O}(3) \ \ldots \\
 & & & \uparrow \\
\mathcal{A}_3 & & & \mathcal{O}(3) \ \ldots \\
\vdots & & &
\end{array}$$

For each $\mathcal{O} \in \mathcal{A}_n$ let x be the unique point in the intersection $\bigcap_{j=n}^{\infty} \overline{\mathcal{O}(j)}$. Since $\bar{S}_{\mathcal{O}}$ is closed and $\overline{\mathcal{O}(j)} \subset \bar{\mathcal{O}} \subset \bar{S}_{\mathcal{O}}$ by inductive condition (5), it follows that $x_{\mathcal{O}} \in \bar{S}_{\mathcal{O}}$ for each $\mathcal{O}$. For each n let $A_n \equiv \{x_{\mathcal{O}}: \mathcal{O} \in \mathcal{A}_n\}$, and let $W \equiv \bigcup_{n=1}^{\infty} A_n$. It is easy to check that (6) and (7) imply that some $A_p \cap B_n = \emptyset$ for $p \geqq n$ and thus W is infinite, for otherwise $W \subset \bigcup_{n=1}^{k} B_n$ for some k. Moreover, it follows from (*) that

$$\sum \{\lambda_{\mathcal{O}}(V)\overline{V(j)}: V \in \mathcal{A}_{n+1}\} \subset \overline{\mathcal{O}(j)} \quad \text{for} \quad n + 1 \leqq j,\ \mathcal{O} \in \mathcal{A}_n .$$

Consequently, for $\mathcal{O} \in \mathcal{A}_n$,

$$(**) \qquad x_{\mathcal{O}} = \sum \{\lambda_{\mathcal{O}}(V)x_V: V \in \mathcal{A}_{n+1}\},$$

which establishes that the sequence $\{A_n: n = 1,2,\ldots\}$ has the properties claimed for it in the first sentence of this proof.

It thus remains to prove that W is weakly closed and weakly discrete (since W is automatically bounded, being a subset of K, and since conclusions (a) and (b) follow easily from (**)). First note that K is weakly closed. Hence it suffices to show that W is a relatively weakly closed, weakly discrete subset of K. If $z \in K$, then $z \in B_m$ for some $m \geqq 1$. By inductive properties (6) and (7), if $r \geqq m$ and if $V \in \mathcal{A}_r$ then $x_V \in \bar{S}_V \subset \cup\{\bar{S}_{\mathcal{O}}: \mathcal{O} \in \mathcal{A}_m\}$ so that $\bigcup_{r=m}^{\infty} A_r \subset \cup\{\bar{S}_{\mathcal{O}}: \mathcal{O} \in \mathcal{A}_m\}$. It is also true that $B_m \cap \cup\{\bar{S}_{\mathcal{O}}: \mathcal{O} \in \mathcal{A}_m\} = \emptyset$. Let $Z \equiv \cap\{K\backslash\bar{S}_{\mathcal{O}}: \mathcal{O} \in \mathcal{A}_m\}$. Then Z contains z, Z is relatively weakly open, and $Z \cap W \subset \bigcup_{j=1}^{m-1} A_j$. In particular $Z \cap W$ is finite. Hence W is both weakly closed and weakly discrete. Thus the proof of Theorem 3.7.2 is complete.

We shall call a point x_o of a convex set $C \subset X$ a <u>strongly extreme point</u> if there is a neighborhood base for x_o in C in the weak topology which consists of slices. (Recall the elementary fact that each extreme point of a compact convex set in a locally convex space is strongly extreme. Easy examples show that this is no longer the case for closed bounded convex sets in Banach spaces even when they have the RNP. Indeed, the point 0 in Example 3.2.5 is not strongly extreme.) The following corollary (at least that portion detailing the relationship between the RNP and strongly extreme points) was first noted by C. Stegall.

<u>COROLLARY</u> 3.7.6. If $C \subset X$ is a closed convex set then the following are equivalent:

1) C has the RNP;

2) Each closed bounded convex subset of C is the closed convex hull of its strongly extreme points;

3) Each closed bounded convex subset of $C \subset X \subset X^{**}$ contains an extreme point of its weak*-closure in X^{**};

4) Each weakly closed bounded subset of C contains an extreme point of its weak*-closed convex hull in X^{**};

5) Each weakly closed bounded subset of C contains an extreme point of its convex hull;

6) Each closed bounded subset of C contains an extreme point of its closed convex hull.

Proof. $1 \Rightarrow 2$ follows from Theorem 3.5.4 since each strongly exposed point is strongly extreme. $2 \Rightarrow 3$ is direct since any strongly extreme point of a closed bounded convex set $K \subset X$ is an extreme point of $w^*cl(K) \subset X^{**}$. $3 \Rightarrow 4$: If D is weakly closed and bounded in X and if $K \equiv \overline{co}(D)$ then since $w^*cl(K) \subset X^{**}$ is weak*-compact and convex, each of its extreme points is a strongly extreme point in the weak*-sense. By hypothesis, there is a point $x_o \in K$ such that $x_o \in ex(w^*cl(K))$. It follows that x_o is a strongly extreme point of K, and since D is weakly closed, x_o must belong to D. $4 \Rightarrow 5$ is obvious, while $5 \Rightarrow 1$ is part of Theorem 3.7.2. Thus 1 - 5 are equivalent. Evidently $6 \Rightarrow 5$. It remains, therefore, to establish that $1 \Rightarrow 6$. If D is a closed bounded subset of C then $K \equiv \overline{co}(D)$ has a strongly exposed point (from Theorem 3.5.4). Such a point necessarily belongs to D.

The final results of this section are similar to those contained in 3.7.2 and 3.7.6, but for the case $C = X$. While it might seem that this special case deserves no special attention, the global case yields distinct conclusions (which do not seem to follow immediately from the results obtained thus far). An elementary example due to R.R. Phelps [1974b] serves to introduce the global nature of the next results.

EXAMPLE 3.7.7 (Phelps). For each positive integer n and choice of n-tuple $(\varepsilon_1, \varepsilon_2, \ldots, \varepsilon_n)$ (where each ε_i is ± 1) let $x(\varepsilon_1, \ldots, \varepsilon_n) \equiv \frac{n}{n+1}(\varepsilon_1, \ldots, \varepsilon_n, 0, 0, \ldots) \in c_o$. The norm of the difference of any two such points of c_o is at least $\frac{2}{3}$ so that the collection of all such points, D, is a norm-closed set. Clearly $D \subset U_1(0)$ (the open unit ball of c_o). Suppose next that $f \in \ell_1 = c_o^*$. It is easy to check that $\sup f(D) = \|f\|$; consequently the separation theorem guarantees that $\overline{co}(D) = U_1[0]$. Finally, if $f \in \ell_1$ and there is a point $y \in D$ such that $f(y) = M(D,f)$ it follows that $f = 0$.

THEOREM 3.7.8 (Huff, Morris). Let X be a Banach space which lacks the RNP. Then there is an equivalent norm ρ for X and a closed subset D of the open ρ-unit ball such that $\overline{co}(D)$ is the closed ρ-unit ball.

Proof. Let $\|\cdot\|$ be an equivalent norm on X whose closed unit ball is nondentable. (The argument presented after the proof of Lemma 2.3.5 shows that such a norm exists.) We denote the $\|\cdot\|$-open unit ball by U, its closure by $\bar{U}$, and $\|\cdot\|$-ε balls by $U_\varepsilon[x]$. By Lemma 3.7.4 there is an $\varepsilon > 0$ such that whenever $D_o, D_1 \subset \bar{U}$, diameter $(D_o) \leq 2\varepsilon$, and $\bar{U} = \overline{co}(D_o \cup D_1)$, then $\bar{U} = \overline{co}(D_1)$. Pick any finite set $\{x_1, \ldots, x_n\} \subset U$. Then by sequential application of the above version of 3.7.4, we see that

$$\bar{U} = \overline{co}(U \backslash U_\varepsilon[x_1]) = \overline{co}(U \backslash \bigcup_{i=1}^{2} U_\varepsilon[x_i]) = \ldots = \overline{co}(U \backslash \bigcup_{i=1}^{n} U_\varepsilon[x_i]).$$

Let $V \equiv U \backslash \bigcup_{i=1}^{n} U_\varepsilon[x_i]$. Then V is an open subset of U and $\overline{co}(V) = \bar{U}$. We shall now prove that $U = co(V)$. Indeed, suppose that $x_o \in V$ and that x is any point of U.

Then for some $\delta > 0$, $U_\delta(x_o) \subset V$. Pick r, $0 < r < 1$, and $y \in \bar{U}$, such that $x = ry + (1 - r)x_o$. Let $y' \in co(V)$ be any point such that $\|y - y'\| < \frac{\delta(1 - r)}{r}$.

Then

$$p \equiv x_o + \frac{r}{1 - r}(y - y') \in U_\delta(x_o) \subset V$$

whence

$$x = (1 - r)x_o + ry = ry' + (1 - r)p \in (r)co(V) + (1 - r)V \subset co(V).$$

Consequently $co(V) = U$ as claimed.

If X is separable, pick a countable dense subset $\{z_i\}_{i=1}^{\infty}$ of U, and define finite subsets $\tilde{F}_1, \tilde{F}_2, \ldots,$ of U by induction: Let $\tilde{F}_1 \equiv \{z_1\}$. Suppose next that $\tilde{F}_1, \ldots, \tilde{F}_n$ have been found. By virtue of the construction in the first paragraph, there is a finite set $\tilde{F}_{n+1}$ such that

a) $\tilde{F}_n \cup \{z_{n+1}\} \subset co(\tilde{F}_{n+1})$; and

b) $\tilde{F}_{n+1} \subset U \setminus \bigcup_{i=1}^{n} U_\varepsilon[\tilde{F}_i]$

(where by $U_\varepsilon[\tilde{F}_i]$ we mean $\{y: \|x - y\| \leqq \varepsilon\}$ for some $x \in \tilde{F}_i$). Note that if $x \in \tilde{F}_n$, $y \in \tilde{F}_k$, $n \neq k$, then $\|x - y\| \geqq \varepsilon$ whence $\tilde{D} \equiv \bigcup_{n=1}^{\infty} \tilde{F}_n$ is a norm-closed subset of U. Since $co(\tilde{D}) \supset \{z_i\}_{i=1}^{\infty}$ it follows that $\overline{co}(\tilde{D}) = U$. Thus the proof is complete when X is separable.

If $(X, |||\cdot|||)$ is a nonseparable Banach space which lacks the RNP it contains (as a consequence of Theorem 2.3.6, part 8) a separable subspace Y which lacks the RNP. Let $\|\cdot\|$ denote an equivalent norm on Y whose closed unit ball $\bar{U}$ is not dentable, construct the finite sets $\tilde{F}_n \subset U$ such that $\|x - y\| \geqq \varepsilon$ whenever $x \in \tilde{F}_n$, $y \in \tilde{F}_k$, $n \neq k$, whose union $\tilde{D}$ is closed and satisfies: $\overline{co}(\tilde{D}) = \bar{U}$. Now let K denote the $|||\cdot|||$-closed unit ball of X and observe that since $|||\cdot|||$ and $\|\cdot\|$ are equivalent norms on Y there is a $\delta > 0$ such that $|||x - y||| \geqq \delta$ whenever $x \in \tilde{F}_n$, $y \in \tilde{F}_k$, $n \neq k$. For each n let $F_n \equiv \tilde{F}_n + \frac{\delta}{3}(1 - \frac{1}{n + 1})K$. Then F_n is a norm closed set and it is easy to check that if $w \in F_n$, $v \in F_k$, and $n \neq k$, then $|||w - v||| \geqq \frac{\delta}{3}$. Consequently $D \equiv \bigcup_{n=1}^{\infty} F_n$ is closed. Also, $co(D)$ is dense in $\bar{U} + \frac{\delta}{3}K$ (since $co(\tilde{D})$ is dense in $\bar{U}$). The set $\bar{U} + \frac{\delta}{3}K$ is a bounded convex subset of X with nonempty interior which is symmetric about 0. Consequently, its Minkowski functional ρ defines an equivalent norm on X. The proof of Theorem 3.7.8 is thus complete since it is evident that

$D \subset \{x: \rho(x) < 1\}$ while $\overline{co}(D) = \{x: \rho(x) \leqq 1\}$.

<u>COROLLARY</u> 3.7.9 (Huff, Morris). The following are equivalent for a Banach space X:

1) X has the RNP;

2) $S(B) \neq \emptyset$ whenever B is a closed bounded subset of X;

3) $S(B)$ contains a dense G_δ-subset of X* whenever B is a closed bounded subset of X.

Proof. Let X have the RNP and suppose that B is a closed bounded subset of X. Denote its closed convex hull by C. Then $SE(C)$ is a dense G_δ subset of X* by Theorem 3.5.4. Moreover, it is easy to check that $SE(C) \subset S(B)$. Consequently $1 \Rightarrow 3 \Rightarrow 2$. Since not $1 \Rightarrow$ not 2 follows directly from Theorem 3.7.8, the proof is complete.

CHAPTER 4. DUAL SPACES

The material on dual spaces with the RNP is the stuff of fairy tales: everything within reason works as it should. There is one main construction leading to one main theorem (both due to Stegall [1975]), and numerous consequences. (Having made such a sweeping statement let it immediately be added that some of these consequences are far from obvious.) In fact, a number of the most important results about dual spaces follow from a predecessor of Stegall's - the well known and historically important Dunford-Pettis theorem. This and related results (which will be derived as consequences of some of the material of Chapter 2) will be discussed before unfolding the localized version of Stegall's construction.

§1. The Dunford-Pettis theorem.

The Dunford-Pettis theorem below captures the essence of the spirit of the Radon-Nikodým property. Although it is an immediate corollary of Stegall's theorem, it is presented here along with a sample of its consequences to emphasize how much of the theory follows from this beautiful (and elementary) result. It will follow as a consequence of the first proposition which originally appeared in Bourgin [1978].

PROPOSITION 4.1.1. Let $D \subset X^*$ be a weak* compact set and suppose that D is separable. Then D is subset s-dentable.

The proof of 4.1.1 is presented below after introducing some notation which will see much use in the following chapters.

NOTATION 4.1.2. For any Banach space $(X, \|\cdot\|)$ let

(a) $U(X)$ denote the closed unit ball of X and recall (cf. 2.1.4) that for $x \in X$ and $\varepsilon > 0$, $U_\varepsilon(x)$ denotes the open ε-ball centered at x, while $U_\varepsilon[x]$ denotes its closure;

(b) $S(X)$ denote the unit sphere of X, i.e. $S(X) \equiv \{x \in X: \|x\| = 1\}$.

(c) For $x \in X$ let $\hat{x} \in X^{**}$ be the image of x under the natural embedding (thus $\hat{x}(f) \equiv f(x)$ for $f \in X^*$) and let $\hat{X} \equiv \{\hat{x}: x \in X\}$.

Finally, for any subset $A \subset X$

(d) span(A) denotes the smallest linear subspace (not necessarily closed) containing A, while $\overline{\text{span}}(A)$ denotes its norm closure.

Proof of 4.1.1. If $\{f_i\}$ is a countable dense subset of D and for each i $\{x_{i,j}: j = 1,2,\ldots\}$ is a sequence in U(X) such that $\lim_j f_i(x_{i,j}) = \|f_i\|$ then the separable subspace

$$Y \equiv \overline{\text{span}}(\{x_{i,j}: i,j = 1,2,\ldots\})$$

has the following obvious property: $id^*: X^* \to Y^*$ is an affine isometry and weak*-weak* homeomorphism when restricted to $D \subset X^*$ (where $id: Y \to X$ is the identity embedding). Thus from now on assume that D is a weak* compact norm separable subset of Y^* with Y separable, and let $\{y_k\}_{k=1}^{\infty}$ be a countable dense subset of the unit sphere of Y, S(Y).

Suppose that some subset A of D were not s-dentable. Precisely as in the proof of Theorem 2.2.2 a Y-valued martingale (f_n, F_n) could be constructed based on $([0,1), F, \lambda)$ (F = Lebesgue measurable sets; λ = Lebesgue measure) such that

(a) $\bigcup_{n=1}^{\infty} \text{Range}(f_n)$ is a countable subset of A, and

(b) there is an $\varepsilon > 0$ such that $\|f_{n+1}(t) - f_n(t)\| \geqq \varepsilon$.

for each $t \in [0,1)$ and $n = 1,2,\ldots$.

Since $\mathbb{R}$ has the RNP, whenever $F \in Y^{**}$ the martingale $(F \circ f_n, F_n)$ converges pointwise at each point of a set of Lebesgue measure one. Let Z_F denote the exceptional set where this martingale fails to converge so that $\lambda(Z_F) = 0$. Observe that if for some t, $\{f_n(t): n = 1,2,\ldots\}$ is not weak* convergent then there is a k such that $\{f_n(t)(y_k): n = 1,2,\ldots\}$ is not convergent and hence

$$\text{weak*} \lim_n f_n(t) \equiv f(t)$$

exists except on a subset Z of $\bigcup_{k=1}^{\infty} Z_{\hat{y}_k}$. That is, f has been defined on $[0,1)\backslash Z$, a set of measure one, and its definition may be extended to all of $[0,1)$ by letting f have constant value $a \in A$ on Z for some fixed a.

Observe that f is separably valued and that $\text{Range}(f) \subset A$, a bounded set. Thus in order to show that f is Bochner integrable it suffices to prove that $F \circ f$ is measurable for each $F \in Y^{**}$ by Theorem 1.1.7. First of all, if $y \in Y$ the map $f(\cdot)(y)$ is F-measurable since it is the almost sure pointwise limit of the sequence $\{f_n(\cdot)(y)\}$. Thus $f^{-1}(H) \in F$ for each weak* open half space $H \subset Y^*$. Since D is separable, each norm open subset of D is a countable union of weak* compact sets (closed balls intersected with D in fact) so the weak* and norm Borel subsets of D coincide. Moreover, since Y is separable, D is weak* metrizable and hence each weak* open subset of D is the intersection with D of a countable union of finite intersections of weak* open half spaces. From the last two sentences it follows that $f^{-1}(V) \in F$ for each relatively norm open subset V of D. Thus if $F \in Y^{**}$ and W is an open subset of $\mathbb{R}$ then

$$(F \circ f)^{-1}(W) = f^{-1}(F^{-1}(W) \cap D) \in F.$$

Hence f is weakly measurable and consequently Bochner integrable from above.

Suppose now that $E \in F_n$ for some n. Then for $y \in Y$ it follows from the fact that $(f_n(\cdot)(y), F_n)$ is a martingale, together with Theorem 1.1.8(e) that

$$\left(\int_E f d\lambda\right)(y) = \int_E f(t)(y) d\lambda(t) = \int_E f_n(t)(y) d\lambda(t) = \left(\int_E f_n d\lambda\right)(y).$$

Consequently $\int_E f d\lambda = \int_E f_n d\lambda$. Thus $\mathbb{E}[f \mid F_n] = f_n$ [a.s.λ] and Lévy's Theorem 1.3.1 shows that $\{f_n\}$ is pointwise convergent almost surely to $\mathbb{E}[f \mid B]$ (B = Borel sigma algebra of $[0,1)$). Since this contradicts assumption (b) of the martingale $\{f_n\}$ it follows that A is s-dentable. The proof is complete.

The classical Dunford-Pettis theorem [1940] follows immediately.

<u>THEOREM</u> 4.1.3. (Dunford, Pettis). Separable dual spaces have the RNP.

Proof. If X^* is a separable dual space then $U(X^*)$ is weak* compact and norm separable. Hence $U(X^*)$ is subset s-dentable and the desired conclusion follows from Theorem 2.3.6.

A second consequence, below, of Proposition 4.1.1 also follows easily. While it at first appears to be significantly stronger than the proposition itself, a result of Haydon [1976a] shows that it is not. The proof below of Haydon's result is due to Bourgain [1978b].

THEOREM 4.1.4. Let K be a weak* compact convex set in X* for which ex(K) is a norm separable set. Then K has the RNP, K itself is norm separable, and $K = \overline{\text{co}}(\text{ex}(K))$.

Proof. Once Haydon's theorem is established (that under the above hypotheses, K itself is separable) the other conclusions follow from Proposition 4.1.1 followed by, say, Theorems 2.3.6 and 3.3.6. Let $\{f_i\}$ be a countable norm dense subset of ex(K), pick $\varepsilon > 0$ and suppose that $f \in K$. Note that $\bigcup_{i=1}^{\infty} (U_\varepsilon[f_i] \cap K)$ is a weak* F_σ - subset of K which contains ex(K). Consequently the Bishop-deLeeuw generalization of the Choquet theorem (see, for example, Phelps [1966] page 24) yields a probability measure μ on K with barycenter f such that $\mu(\bigcup_{i=1}^{\infty} (U_\varepsilon[f_i] \cap K)) = 1$. For each n let

$$B_n \equiv U_\varepsilon[f_n] \setminus \bigcup_{i=1}^{\infty} U_\varepsilon[f_i].$$

If $\mu(B_n \cap K) = 0$ let μ_n be the point mass at f_n. Otherwise let

$$\mu_n \equiv \frac{1}{\mu(B_n \cap K)} \mu|_{B_n \cap K}$$

(i.e. μ_n is the normalized restriction of μ to $B_n \cap K$.) The barycenter g_n of μ_n lies in $U_\varepsilon[f_n]$ in either case and

$$f = \sum_{n=1}^{\infty} \mu(B_n \cap K) g_n \quad \text{(the series is norm convergent).}$$

Evidently $\sum_{n=1}^{\infty} \mu(B_n \cap K) f_n$ lies within ε of f since for each n, $\|f_n - g_n\| \leq \varepsilon$ and hence for some k

$$\left[\sum_{n=1}^{k} \mu(B_n \cap K)\right]^{-1} \sum_{n=1}^{k} \mu(B_n \cap K) f_n$$

is within 2ε of f. That is, convex combinations of $\{f_i\}_{i=1}^{\infty}$ are norm dense in K. Hence K is separable and the proof is complete.

COROLLARY 4.1.5. Reflexive spaces have the RNP.

Proof. If Y is a separable subspace of the reflexive space X then Y = Y** so Y is a separable dual. Thus Y has the RNP and the desired conclusion follows since the RNP is a separably determined property (see, for example, Theorem 2.3.6 part (2) or part (8)).

Of course, Corollary 4.1.5 is also an immediate consequence of Theorem 3.6.1. The proof we presented of this corollary, though, points up what remains as the main technique for showing that spaces have the RNP. While the next two corollaries are direct, their use is pervasive.

COROLLARY 4.1.6. A Banach space X has the RNP if each separable subspace of X is isomorphic to a subspace of a separable dual space.

Proof. The proof is a consequence of Theorem 4.1.3 followed by Theorem 2.3.6 part (2).

Uhl [1972] observed a variation of the above corollary in the case of dual spaces.

COROLLARY 4.1.7. Let X* be a dual space. Then X* has the RNP if each separable subspace of X has separable dual.

Proof. Let Y be a separable subspace of X*. It suffices to show that Y is isomorphic to a subspace of a separable dual by the previous corollary. If $\{f_i\}$ is a countable dense subset of S(Y) and if for each i, $\{x_{i,j}\}_{j=1}^{\infty}$ is a sequence in U(X) such that $\lim_j f_i(x_{i,j}) = \|f_i\| = 1$ then for each $f \in Y$,

$$\|f\| = \sup\{f(x): x \in U(W)\} \quad \text{where } W \equiv \overline{\text{span}}(\{x_{i,j}: i,j = 1,2,\ldots\}).$$

The map $f \to f|_W$ from Y to W* is an isometric isomorphism, and since W* is separable by hypothesis (since W is separable) the proof is complete.

Originally Stegall [1975] established Corollary 4.1.8 below using the powerful tools of section 4.2. In that paper he also showed by example that it cannot be strengthened to characterize weak* compact convex sets with the RNP. The elementary proof given here is from Bourgin [1978].

COROLLARY 4.1.8. Let K be a weak* compact convex subset of X*. Then K has the RNP if the weak* closure of each countable subset of K is norm separable.

Proof. Suppose that K lacks the RNP. Then there is a countable subset D of K which fails to be dentable by Theorem 2.3.6 parts (8) and (9). Let $C \equiv w^*cl(co(D))$. By the Milman Theorem 3.2.2 $ex(C) \subset w^*cl(D)$. If $w^*cl(D)$ were separable then ex(C) would be separable as well and C would have the RNP by Theorem 4.1.4. Thus D would be dentable by Theorem 2.3.6 part (9). This impossibility completes the proof.

We turn next to the identification of some important classes of spaces with the RNP.

COROLLARY 4.1.9. For any set $\Gamma \neq \emptyset$, $\ell_1(\Gamma)$ has the RNP.

Proof. Let $Y \subset \ell_1(\Gamma)$ be a separable subspace. Then there is a countable subset $\Gamma_0 \subset \Gamma$ such that $y(\gamma) = 0$ if $\gamma \in \Gamma \backslash \Gamma_0$. But $\ell_1(\Gamma_0)$ is isomorphic to a separable dual space (c_0* if Γ_0 is infinite, $\mathbb{R}^n$ if Γ_0 is finite) and hence Y has the RNP. Thus $\ell_1(\Gamma)$ has the RNP as well.

Recall that X is a weakly compactly generated Banach space (abbreviated WCG) if X is the closed linear span of some weakly compact set. It is easily seen that reflexive spaces and separable spaces are WCG, and it may be shown that C(K) is WCG for K a weakly compact subset of a Banach space. For background, see, for example, Amir and Lindenstrauss [1968] and Diestel [1975]. The proof of the following result is due to P.D. Morris, and the result itself is due to Kuo [1975].

COROLLARY 4.1.10. Suppose that the dual space X* is isomorphic to a subspace of some WCG space Z. Then X* has the RNP.

Proof. Let Y be a separable subspace of X. By Corollary 4.1.7 it suffices to show that Y* is separable. By hypothesis, there is a linear map $I: Y^* \to Z$ which maps Y* isomorphically into Z. Choose a countable dense subset $\{y_i\}_{i=1}^{\infty}$ of S(Y), let e_i denote the $i^{\underline{th}}$ unit vector of ℓ_1, and let $T(e_i) \equiv y_i$ for each i. It is an elementary argument to show that T may be (uniquely) extended to a continuous linear operator of ℓ_1 onto Y. By the open mapping theorem T is open; hence the adjoint $T^*: Y^* \to \ell_\infty$ is an isomorphic embedding. The continuous linear map

$R \equiv T^* \circ I^{-1}$: $T(Y^*) \to \ell_\infty$ can be extended to a continuous linear map $\bar{R}: Z \to \ell_\infty$. (The argument goes as follows: for each n, $\hat{e}_n \circ R$ is a continuous linear functional on $I(Y^*)$. Let $f_n \in Z^*$ be a norm preserving extension of $\hat{e}_n \circ R$, and define $\bar{R}$ by $\bar{R}(z)(n) \equiv f_n(z)$ for each n.) The continuous linear image $\bar{R}(Z)$ of Z is itself WCG. Of course the weak and weak* topologies must coincide on any weakly compact subset of ℓ_∞ (since one is formally weaker than the other) and since weak* compact sets in ℓ_∞ are metrizable and separable (since $\ell_1^* = \ell_\infty$) such a set is weakly (hence norm) separable. It follows that $\bar{R}(Z)$ is separable. This shows that $T^*(Y^*) \subset \bar{R}(Z)$ is separable and since T^* is an isomorphism, Y^* is separable.

Before beginning Stegall's construction (see §4.2) let us briefly examine an approach recently taken by Bourgain and Rosenthal which yields the Dunford-Pettis Theorem 4.1.3 along the way. The next definition is due to Lotz, Peck and Porta [1979].

DEFINITION 4.1.11. Let $T: X \to Y$ be a one-one continuous linear operator from X into the Banach space Y. If $T(U(X))$ is a closed subset of Y then *T is said to be a semi-embedding* and *X semi-embeds in Y*.

The results below appear early in the Bourgain and Rosenthal paper [1983] and form a small selection of the interesting uses to which semi-embeddings may be put.

LEMMA 4.1.12. Let X be a separable Banach space. Then X* semi-embeds in ℓ_2 ($\equiv$ separable Hilbert space).

Proof. Let $\{x_n: n = 1,2,\ldots\}$ be a countable dense subset of $U(X)$. Define $T_1: \ell_2 \to X$ by $T_1((a_i)) \equiv \sum_{i=1}^{\infty} \frac{a_i}{i} x_i$ for $(a_i) \in \ell_2$. Clearly T_1 has dense range, and consequently $T_1^*: X^* \to \ell_2^*$ is one-one. Of course T_1^* is weak*-weak* continuous, so $T_1^*(U(X^*))$ is weak* (hence norm) closed in $\ell_2^* \approx \ell_2$. Thus $T \equiv T_1^*$ is a semi-embedding of X* into ℓ_2.

THEOREM 4.1.13. Let K be a separable closed bounded convex subset of a Banach space X and suppose that $T: X \to Y$ is a continuous linear operator into the Banach space Y. If the restriction of T to K is one-one and if $T(K)$ is closed in Y then K has the RNP if $T(K) \subset Y$ has the RNP.

Proof. Let λ denote Lebesgue measure on $[0,1]$ and $\mathcal{F}$ the Borel σ-algebra on $[0,1]$. Let $\{\mathcal{F}_n: n = 1,2,\ldots\}$ be an increasing sequence of sub σ-algebras of $\mathcal{F}$ whose union generates $\mathcal{F}$ and suppose that $(f_n,\mathcal{F}_n)$ is a K-valued martingale on $([0,1],\mathcal{F},\lambda)$. According to Theorem 2.3.6 part (5), K will have the RNP if such a martingale is almost surely convergent. Observe that $(T\circ f_n,\mathcal{F}_n)$ is a $T(K)$-valued martingale (by Theorem 1.1.8(e)) and hence, since $T(K)$ has the RNP by hypothesis, there is a $g \in L^1_{T(K)}(\lambda)$ such that $\lim_n T\circ f_n = g$ [a.s.λ]. Since $\{t \in [0,1]: g(t) \notin T(K)\}$ is a Borel set of λ-measure 0, g may be modified on this set if necessary so that the resulting function takes each of its values in $T(K)$. Let $f \equiv (T|_K)^{-1}\circ g$. Then $f: [0,1] \to K$ is separably valued and bounded. Let B be any Borel subset of K. Because K is complete separable metric and T is continuous and one-one with metric range, a theorem of Lusin (see, for example, Cohn [1980] p. 276, Theorem 8.3.7) asserts that $T(B)$ is also a Borel set. But then $g^{-1}(T(B)) \in \mathcal{F}$. That is, $f^{-1}(B) \in \mathcal{F}$ for each Borel set B in K and hence f is Bochner integrable by Theorem 1.1.7(d). Now for any k, $\mathbb{E}[g|\mathcal{F}_k] = T(f_k)$ [a.s.λ] since the martingale $(T(f_n),\mathcal{F}_n)$ converges to g inside the bounded set $T(K)$. Consequently

$$T(\mathbb{E}[f|\mathcal{F}_k]) = \mathbb{E}[T(f)|\mathcal{F}_k] = \mathbb{E}[g|\mathcal{F}_k] = T(f_k) \quad [\text{a.s.}\lambda]$$

and since T is one-one on K, $\mathbb{E}[f|\mathcal{F}_k] = f_k$ [a.s.λ] for each k. The proof of the theorem is completed by an appeal to Lévy's Theorem 1.3.1 which asserts that $\lim_k \mathbb{E}[f|\mathcal{F}_k] = f$ [a.s.λ] (and hence $\lim_k f_k = f$ [a.s.λ] as was to be shown).

Suppose now that X^* is a separable dual space. Then X is separable so X^* semi-embeds in ℓ_2 by Lemma 4.1.12. Since ℓ_2 has the RNP (a direct proof of this fact is not too difficult from scratch) it follows from Theorem 4.1.13 that $U(X^*)$ has the RNP. That is, we have just provided another proof of the Dunford-Pettis Theorem 4.1.3. (A more direct measure theoretic proof of 4.1.13 appears in Bourgain and Rosenthal [1983], Theorem 1.1', also. This proof does not rely on the use of martingales.)

Let $\mathcal{R}$ denote the smallest class of separable Banach spaces with the properties that $\ell_2 \in \mathcal{R}$ and $X \in \mathcal{R}$ whenever X is a separable Banach space and for some $Y \in \mathcal{R}$, X semi-embeds in Y. Then $\mathcal{R}$ contains the class of (subspaces of) separable dual spaces by the above argument. Moreover, the space of McCartney and O'Brien [1980] (cf. §7.1) provides an example of an element of $\mathcal{R}$ which does not embed in a separable dual (cf. Proposition 1.4 of Bourgain and Rosenthal [1983]) and hence $\mathcal{R}$ strictly includes the class of subspaces of separable duals. Evidently each element of $\mathcal{R}$ has the RNP and it might seem hopeful that $\mathcal{R}$ is exactly the class of separable Banach spaces with the RNP. Unfortunately, Delbaen [1983] has shown that no separable infinite dimensional $\mathcal{L}_\infty$ space is in $\mathcal{R}$, and since Bourgain and Delbaen [1980] (cf. §7.2) have produced an example of such a space with the RNP, $\mathcal{R}$ is a proper subclass of the separable RNP spaces. (Other examples of spaces in $\mathcal{R}$ which do not embed in separable duals have been given by Johnson and Lindenstrauss [1980].)

§2. The main results in a localized setting.

Stegall's construction (see, for example, its culmination in Lemma 4.2.12(a)) is quite flexible and various modifications have found use in a number of connections. The construction presented here is a streamlined version of the one given by Stegall [1975] which allows localized conclusions while Stegall originally restricted his attention to the case of dual Banach spaces with and without the RNP. The spirit of his construction has not been seriously altered, though, nor has the spirit of the conclusions which can be drawn. In their present form the next pages for the most part follow some unpublished notes of Fitzpatrick [1977] and of Namioka [1981]. Some notation and conventions are gathered below which will save us energy later on.

NOTATION 4.2.1. For any set A, $|A|$ denotes the cardinality of A. If $(T,\mathcal{T})$ is a topological space, dens(T) denotes the smallest cardinal number for which there is a $\mathcal{T}$-dense set of that cardinality. It is called the density character of $(T,\mathcal{T})$. Each cardinal number is identified with the first ordinal number of that cardinality.

The first lemma sets the stage for dealing with a Banach space and its dual of differing density characters.

LEMMA 4.2.2. Let X be an infinite dimensional Banach space, ξ an infinite cardinal number and $D \subset S(X)$ a subset such that $\text{dens}(D) \geqq \xi$. Then there are $M > 0$, nets $\{x_\alpha : \alpha < \xi\}$ in D and $\{f_\alpha : \alpha < \xi\}$ in X* such that $\|f_\alpha\| < M$ for each $\alpha < \xi$ and

$$f_\alpha(x_\beta) = \begin{cases} 0 & \text{if} \quad \beta < \alpha \\ 1 & \text{if} \quad \alpha = \beta \qquad \alpha,\beta < \xi \\ ? & \text{if} \quad \beta > \alpha \end{cases}$$

Proof. Choose δ, $0 < \delta < 1$ so that for some subset D_0 of D, $|D_0| \geqq \xi$ and $\|x - x'\| > \delta$ whenever x and x' are distinct elements of D_0. Pick any $M > 2/\delta$. The proof is by induction. Pick x_0 in D_0 and f_0 in X* such that $f_0(x_0) = 1$, $\|f_0\| < M$. If for some $\gamma < \xi$, $\{x_\alpha : \alpha < \gamma\}$ and $\{f_\alpha : \alpha < \gamma\}$ have already been chosen which satisfy the conditions of the lemma for $\alpha,\beta < \gamma$, let $Y \equiv \overline{\text{span}}(\{x_\alpha : \alpha < \gamma\})$ and observe that $\text{dens}(Y) \leqq |\gamma| < \xi$. Thus there are $|\gamma|$ balls, each of radius $\delta/4$ which cover Y and yet by expanding each of these balls to have radius $\delta/2$ without changing the centers, the new collection of balls fails to cover D_0 because each expanded ball can contain at most one point of D_0. Thus there is an $x_\gamma \in D_0$ such that distance $(x_\gamma, Y) \geqq \delta/2$. Hence there is an $f_\gamma \in X^*$ such that $\|f_\gamma\| \leqq 2/\delta < M$ while both $f_\gamma|_Y = 0$ and $f_\gamma(x_\gamma) = 1$. This completes the inductive step and hence the proof.

The next definition and lemma generalize well known information about condensation points (the case $\eta = \aleph_0$ below) in topological spaces.

DEFINITION 4.2.3. Let η be a cardinal number and S a subset of a topological space. Then $x \in S$ is an η-condensation point of S if for each neighborhood V of x, $|S \cap V| > \eta$.

LEMMA 4.2.4. Let X be an infinite dimensional Banach space and $A \subset U(X^*)$ a subset such that $|A| > \text{dens}(X)$. Then there is a subset A_0 of A such that both

(a) $|A \backslash A_0| \leqq \text{dens}(X)$;

(b) For each $f \in A_0$ and each weak* neighborhood V of f, $|A_0 \cap V| > \text{dens}(X)$.

(That is, each point of A_o is a dens(X)-weak* condensation point of A_o.)

Proof. Observe first that there is a base for $U(X^*)$ in the weak* topology of cardinality at most dens(X). (Indeed, if T is a total subset of X with $|T| =$ dens(X), $(U(X^*), \text{weak*})$ may be considered a subset of $\Pi \equiv \Pi\{I_t: t \in T\}$ in the product topology, where each $I_t \equiv [-1,1]$. But Π has a base of neighborhoods of the form $\bigcap_{i=1}^{n} P_{t_i}^{-1}(B_i)$ where $P_{t_i}: \Pi \to I_{t_i}$ is the natural projection and $\{B_i: i = 1,2,\ldots\}$ is a countable base for $[-1,1]$. The cardinality of this base for Π is at most $\aleph_0 \cdot |T| = |T| =$ dens(X).)

In the remainder of the proof denote dens(X) by η. Let

$$A_o \equiv \{f \in A: f \text{ is an } \eta\text{-weak* condensation point of } A\}$$

and let V be a base for the weak* topology on $U(X^*)$ with $|V| \leqq \eta$. If $f \in A\backslash A_o$ there is a $V_f \in V$ with $f \in V_f$ and $|A \cap V_f| \leqq \eta$. Clearly $\{V_f: f \in A\backslash A_o\}$ covers $A\backslash A_o$ so

$$|A\backslash A_o| \leqq |\cup \{g: g \in V_f \cap A \text{ for some } f \in A\backslash A_o\}| \leqq \eta \cdot |V| \leqq \eta \cdot \eta = \eta.$$

This establishes (a). Evidently for each $f \in A_o$ and weak* neighborhood V of f, $|A\backslash V| > \eta$. But

$$\eta < |A \cap V| = \max\{|(A\backslash A_o) \cap V|, |A_o \cap V|\} \leqq \max\{\eta, |A_o \cap V|\}.$$

Thus $\eta < |A_o \cap V|$ whenever V is a weak* neighborhood of $f \in A_o$. This completes the proof.

It will be convenient to fix the hypothesis and notation before delving into the heart of Stegall's construction.

HYPOTHESIS AND NOTATION 4.2.5. Let X be a nonzero dimensional Banach space and denote dens(X) by η. Let η^+ denote the next larger cardinal number (i.e. $\eta^+ \equiv \aleph_{\gamma+1}$ if $\eta = \aleph_\gamma$). Assume that $D \subset S(X^*)$ is a subset such that dens(D) $\geqq \eta^+$. Then by Lemma 4.2.2 for some $M > 0$ there are nets $\{g_\alpha: \alpha < \eta^+\}$ in D and $\{G_\alpha: \alpha < \eta^+\}$ in X^{**} with $\|G_\alpha\| < M$ for each $\alpha < \eta^+$ and

$$G_\alpha(g_\beta) = \begin{cases} 0 & \text{if } \beta < \alpha \\ 1 & \text{if } \beta = \alpha \\ ? & \text{if } \beta > \alpha \end{cases} \qquad \alpha,\beta < \eta^+.$$

Let $A \equiv \{g_\alpha\}_{\alpha<\eta^+}$.

<u>LEMMA</u> 4.2.6. In addition to the hypothesis and notation of 4.2.5, assume that $\delta > 0$ and that $(A_1,\ldots,A_n)$ is a finite sequence of subsets of A <u>not necessarily distinct</u>, such that $|A_i| > \eta$ for $i = 1,\ldots,n$. Then there are subsets $B_i \subset A_i$ $i \leq i \leq n$ and elements $x_i \in X$ $1 \leq i \leq n$ such that

(a) $|B_i| > \eta$ for $1 \leq i \leq n$;

(b) $w^*cl(B_i) \cap w^*cl(B_j) = \emptyset$ for $i \neq j$;

(c) $|f(x_i) - \chi_{w^*cl(B_i)}(f)| < \delta$ for $f \in \bigcup_{j=1}^{n} w^*cl(B_j)$ $(1 \leq i \leq n)$;

(d) $\|x_i\| < M$ for $1 \leq i \leq n$;

(e) $w^*cl(B_i) \subset U(X^*)\backslash(1-\delta)U(X^*)$ for $1 \leq i \leq n$.

Proof. We first prove a prelemma: Under the hypotheses of 4.2.6 there are subsets $B_i \subset A_i$ $1 \leq i \leq n$ and an element x_1 of X such that (a), (b), (e) above and (c'), (d') directly below all hold.

(c') $|f(x_1) - \chi_{w^*cl(B_1)}(f)| < \delta$ for $f \in \bigcup_{i=1}^{n} w^*cl(B_i)$;

(d') $\|x_1\| < M$.

(N.B. (c') and (d') are (c) and (d) when $i = 1$.) Suppose momentarily that the prelemma has been established and that $B_i \subset A_i$ $1 \leq i \leq n$ and $x_1 \in X$ are produced satisfying (a), (b), (c'), (d') and (e). Then by applying the prelemma to $(B_2,B_1,B_3,\ldots,B_n)$ (in place of $(A_1,A_2,\ldots,A_n)$) we see that by shrinking the B_i's if necessary we can choose $x_2 \in X$ so that (a), (b), (e) still hold and now (c) and (d) hold for $i = 1,2$. (In fact when the prelemma is applied to $(B_2,B_1,B_3,\ldots,B_n)$, subsets $C_i \subset B_i$ $1 \leq i \leq n$ are produced which satisfy (a), (b), (e) and (c), (d) for $i = 1,2$. Rather than call them C_i's we refer to them as the "shrunken" B_i's.) Now apply the prelemma once again, this time to $(B_3,B_1,B_2,B_4,\ldots,B_n)$. By shrinking the B_i's further one can choose x_3 in X so that (a), (b), (e) continue to hold and (c) and (d) hold for $i = 1,2,3$. By repeating this process a total of $n - 1$ times we arrive at subsets $B_i \subset A_i$ and elements $x_i \in X$ $1 \leqq i \leqq n$ which satisfy (a), (b), (c), (d) and (e) of the lemma. Hence it remains to prove the prelemma.

Suppose that A' is a subset of A with $|A'| > \eta$ and $\alpha < \eta^+$ is any ordinal. Then there is an ordinal β, $\alpha < \beta < \eta^+$, such that g_β is an η-weak* condensation point of A'. (Indeed, let $A'' \equiv A' \cap \{g_\beta : \alpha < \beta < \eta^+\}$. Then $|A''| > \eta$ and any η-weak* condensation point of A'' - of which there are many by Lemma 4.2.4 - will be a g_β as sought for.) Thus there are ordinals $\alpha_n < \alpha_{n-1} <\ldots< \alpha_1 < \eta^+$ such that g_{α_j} is an η-weak* condensation point of A_j. Hence $G_{\alpha_1}(g_{\alpha_1}) = 1$ and $G_{\alpha_1}(g_{\alpha_j}) = 0$ for $2 \leqq j \leqq n$. The next lemma (sometimes called Helly's theorem) establishes the elementary fact that under these conditions there is an $x_1 \in X$, $\|x_1\| < M$ such that $g_{\alpha_1}(x_1) = 1$ and $g_{\alpha_j}(x_1) = 0$ for $2 \leqq j \leqq n$. Thus there are weak* closed neighborhoods $V_1,\ldots,V_n$ of $g_{\alpha_1},\ldots,g_{\alpha_n}$ respectively such that $V_i \cap V_j = \emptyset$ for $i \neq j$ and

$$|g(x_1) - 1| < \delta \text{ for } g \in V_1 \quad\text{while}\quad |g(x_1)| < \delta \quad\text{for}\quad g \in \bigcup_{j=2}^{n} V_j .$$

Because (X*, weak*) is completely regular, these weak* neighborhoods may also be chosen to be disjoint from the weak* closed set $(1 - \delta)U(X^*)$. Let $B_i \equiv V_i \cap A_i$. Then $|B_i| \geqq \eta^+$ (since g_{α_i} is an η-weak* condensation point of A_i). Moreover, $B_i \subset A_i \subset A \subset U(X^*)$. Hence $w^*cl(B_i) \subset U(X^*)$. Since $w^*cl(B_i) \subset V_i$, a weak* closed set disjoint from $(1 - \delta)U(X^*)$, it follows that

$$w^*cl(B_i) \subset U(X^*)\backslash(1 - \delta)U(X^*) \quad \text{for} \quad 1 \leqq i \leqq n.$$

Properties (b) and (c') are also immediate, while (d') follows from the construction. Thus, once Helly's theorem is proved, the proof of this lemma will be complete. (See Shirey [1980] for a generalization of the next result.)

<u>LEMMA</u> 4.2.7. (Helly's theorem). Let $0 \neq F \in X^{**}$ and suppose that $f_1,\ldots,f_n$ are in X^*. Then given $\varepsilon > 0$ there is an $x \in X$, $\|x\| < \|F\| + \varepsilon$ with $F(f_i) = f_i(x)$ for $i = 1,\ldots,n$.

Proof. Let $T: X^{**} \to \mathbb{R}^n$ be the map $T(G) \equiv (G(f_1),\dots,G(f_n))$ for $G \in X^{**}$. Then T is weak* continuous. Let $E \equiv T(X^{**})$. Since $\hat{X}$ is weak* dense in X^{**}, $T(\hat{X}) = E$ as well (since E is finite dimensional). Now abbreviate $U(X)$ by U and $U(X^{**})$ by U^{**}. Then $T(\hat{U})$ is a neighborhood of 0 in E, and since $\|F\|\hat{U}$ is weak* dense in $\|F\|U^{**}$ we have

$$T(\|F\|U^{**}) \subset \overline{T(\|F\|\hat{U})} \subset T(\|F\|\hat{U}) + \varepsilon T(\hat{U}) = T((\|F\| + \varepsilon)\hat{U}).$$

But $F \in \|F\|U^{**}$ and thus there is an $x \in (\|F\| + \varepsilon)U$ such that $T(F) = T(\hat{x})$.

The standard base for the topology of the Cantor set (together with the Cantor set itself) forms the canonical Haar system as defined below. In the next few lemmas such a system will be placed in the unit sphere of any dual space without the RNP.

DEFINITION 4.2.8. Let Δ be a compact Hausdorff space. A Haar system of compact subsets of Δ is a sequence $\{\Delta_n: n = 1,2,\dots\}$ of compact nonempty subsets of X such that

$$\Delta_1 = \Delta,\ \Delta_n = \Delta_{2n} \cup \Delta_{2n+1}, \text{ and } \Delta_{2n} \cap \Delta_{2n+1} = \emptyset \quad \text{for } n = 1,2,\dots .$$

Observe that for each $k = 1,2,\dots$ $\{\Delta_n: 2^k \leqq n < 2^{k+1}\}$ is a partition of Δ and that each Δ_n is both open and closed in Δ. The heart of Stegall's construction appears in the next lemma and in Lemma 4.2.12(a).

LEMMA 4.2.9. Along with the hypothesis and notation of 4.2.5, let $\{\delta_k: k = 1,2,\dots\}$ be a sequence of positive numbers tending to 0. Then there is a Haar system $\{\Delta_n: n = 1,2,\dots\}$ of weak* compact subsets of $S(X^*) \cap w^*cl(D)$ and a sequence $\{x_n: n = 1,2,\dots\}$ in X such that for each n

$\|x_n\| < M$ and

$|f(x_n) - \chi_{\Delta_n}(f)| < \delta_{k+1}$ for $2^k \leqq n < 2^{k+1}$ and $f \in \Delta \equiv \Delta_1$.

Proof. The bulk of the proof comes in constructing a sequence $\{A_n: n = 1,2,\dots\}$ of subsets of A (see 4.2.5 for the definition of A) and a sequence $\{x_n: n = 1,2,\dots\}$ in X such that

$\|x_n\| < M$ and $|A_n| > \eta$ for each n;

$A_{2n} \cup A_{2n+1} \subset A_n$ for each n;

$w^*cl(A_{2n}) \cap w^*cl(A_{2n+1}) = \emptyset$ for each n;

$w^*cl(A_n) \subset U(X^*)\backslash(1 - \delta_{k+1})U(X^*)$ if $2^k \leqq n < 2^{k+1}$;

$|f(x_n) - \chi_{w^*cl(A_n)}(f)| < \delta_{k+1}$ for $2^k \leqq n < 2^{k+1}$ and $f \in w^*cl(A_n)$ for some n, $2^k \leqq n < 2^{k+1}$.

This construction is by induction and it requires Lemma 4.2.6 at each inductive step.

Step 0: Apply Lemma 4.2.6 to the finite sequence (A) and to $\delta \equiv \delta_1$. Obtain $A_1 \subset A$, $x_1 \in X$ such that

$\|x_1\| < M$;

$|A_1| > \eta$;

$|f(x_1) - \chi_{w^*cl(A_1)}(f)| < \delta_1$ for $f \in w^*cl(A_1)$;

$w^*cl(A_1) \subset U(X^*)\backslash(1 - \delta_1)U(X^*)$.

Step k: Suppose that sets $A_1, A_2, \ldots, A_{2^k-1}$ and points $x_1, x_2, \ldots, x_{2^k-1}$ have already been constructed. Apply Lemma 4.2.6 to the 2^k-tuple

$$(A_{2^{k-1}}, A_{2^{k-1}}, A_{2^{k-1}+1}, A_{2^{k-1}+1}, \ldots, A_{2^k-1}, A_{2^k-1})$$

with δ replaced by δ_{k+1} to obtain A_i and x_i, $2^k \leqq i < 2^{k+1}$ such that

$A_{2n} \subset A_n$ and $A_{2n+1} \subset A_n$ for $2^{k-1} \leqq n < 2^k$;

$w^*cl(A_{2n}) \cap w^*cl(A_{2n+1}) = \emptyset$ for $2^{k-1} \leqq n < 2^k$;

$\|x_i\| < M$ for $2^k \leqq i < 2^{k+1}$;

$|A_i| > \eta$ for $2^k \leqq i < 2^{k+1}$;

$w^*cl(A_i) \subset U(X^*)\backslash(1 - \delta_{k+1})U(X^*)$ for $2^k \leqq i < 2^{k+1}$; and

$|f(x_i) - \chi_{w^*cl(A_i)}(f)| < \delta_{k+1}$ for $2^k \leqq i < 2^{k+1}$ and $f \in w^*cl(A_i)$ for some i, $2^k \leqq i < 2^{k+1}$.

These are precisely the properties required, and therefore the inductive construction is complete.

The proof of Lemma 4.2.9 is now an easy matter. Indeed, let

$$\Delta \equiv \bigcap_{k=1}^{\infty} \bigcup_{n=2^k}^{2^{k+1}-1} w^*cl(A_n) \quad \text{and} \quad \Delta_n \equiv \Delta \cap w^*cl(A_n).$$

Note that $\Delta \subset S(X^*)$. In fact,

$$\Delta \subset w^*cl(A_1) \cap S(X^*) \subset w^*cl(A) \cap S(X^*) \subset w^*cl(D) \cap S(X^*).$$

Furthermore, $\Delta_n \neq \emptyset$ for any n by compactness. The other conditions placed on Δ are similarly straightforward to verify, so the proof of Lemma 4.2.9 is complete.

Let $\{\Delta_n : n = 1,2,\ldots\}$ be a Haar system of compact subsets of a compact Hausdorff space, and suppose that for each sequence $\{n_k : k = 1,2,\ldots\}$ of positive integers tending to ∞ then $\left|\bigcap_{k=1}^{\infty} \Delta_{n_k}\right| \leqq 1$. It is not difficult to check that Δ is homeomorphic to the Cantor set and that the homeomorphism can be constructed so that the usual Haar system for the Cantor set is the image of the Haar system $\{\Delta_n : n = 1,2,\ldots\}$. We thus obtain the following result which will find use in the next section.

<u>COROLLARY</u> 4.2.10. Suppose that, in addition to the hypothesis of Lemma 4.2.9, we also assume that X is separable. Then Δ may be taken to be weak* homeomorphic to the Cantor set in the conclusions of 4.2.9.

Proof. From the preceding remarks, it suffices to guarantee that the Haar system produced in Lemma 4.2.9 can be chosen so that $\left|\bigcap_{k=1}^{\infty} \Delta_{n_k}\right| \leqq 1$ for each increasing sequence $\{n_k : k = 1,2,\ldots\}$ of positive integers. But if X is separable, $U(X^*)$ is weak* metrizable, so the weak* closed neighborhoods V_i of the proof of 4.2.6 may be chosen to be of arbitrarily small weak* diameter, and hence the conclusion of 4.2.6 may be strengthened to include the extra condition:

$$w^*\text{-diameter}(B_i) < \delta \quad \text{for} \quad 1 \leqq i \leqq n$$

as well when X is separable. Finally, in the inductive proof of 4.2.9, use the new full strength of 4.2.6 at each stage. At the kth step, then, the 2^k sets $A_{2^k},\ldots,A_{2^{k+1}-1}$ constructed will also have the property that w*-diameter$(A_i) < \delta_{k+1}$ for $i = 2^k,\ldots,2^{k+1} - 1$. The fact that the intersection of a sequence of distinct sets from $\{\Delta_n: n = 1,2,\ldots\}$ is either empty or contains exactly one point is an immediate consequence.

Since the Cantor set may be represented as those $x \in [0,1]$ which have a ternary expansion with only 0's and 2's, it is evidently homeomorphic to $\Pi\{0,2\}$, a countable product of two point sets with the product topology, and the image of the natural Haar system for the Cantor set becomes the Haar system which, at stage k, has the 2^k disjoint compact sets constructed by fixing a 0 or a 2 in each of the first k coordinates and leaving the remaining coordinates unrestricted. Suppose that $\{\Delta_n: n = 1,2,\ldots\}$ is a Haar system of compact subsets of some compact Hausdorff space which is homeomorphic to the Cantor set (and the Δ_n's are the images of the canonical Haar system of the Cantor set). By "pulling back" the product measure on $\Pi\{0,2\}$ whose factors assign mass 1/2 to each of $\{0\}$ and $\{2\}$, via the above mentioned homeomorphisms (cf. 4.3.7) a probability measure μ on $\Delta \equiv \Delta_1$ is constructed with the property that $\mu(\Delta_n) = 2^{-k}$ for $k = 1,2,\ldots$ and $2^k \leqq n < 2^{k+1}$. This process is generalized below.

LEMMA 4.2.11. Let Δ be a compact Hausdorff space and $\{\Delta_n: n = 1,2,\ldots\}$ a Haar system of compact subsets of Δ. Then there is a probability measure μ on the Borel subsets of Δ such that $\mu(\Delta_n) = 2^{-k}$ for $2^k \leqq n < 2^{k+1}$ and $k = 1,2,\ldots$.

Proof. Let $Y \equiv \mathrm{span}(\{\chi_{\Delta_n} : n = 1,2,\ldots\})$ (not necessarily closed) and $F: Y \to \mathbb{R}$ the unique linear map determined by the conditions $F(\chi_{\Delta_n}) \equiv 2^{-k}$ when $2^k \leqq n < 2^{k+1}$. Observe that F is well defined since $\chi_{\Delta_{2n}} + \chi_{\Delta_{2n+1}} = \chi_{\Delta_n}$ for each n. If $f \in Y$ then for k sufficiently large, f has the representation

$$f = \sum_{n=2^k}^{2^{k+1}-1} \alpha_n \chi_{\Delta_n}$$

for some choice of scalars α_n. If $f \geqq 0$ then since the Δ_n's $(2^k \leqq n < 2^{k+1})$ are pairwise disjoint, each $\alpha_n \geqq 0$. Hence $F(f) \geqq 0$. Since $F(1) = 1$ and considering $Y \subset C(\Delta)$, $F \in Y^*$ and $\|F\| = 1$. Let $F_1 \in C(\Delta)^*$ be a norm one extension of F. Then

$\|F_1\| = F_1(1) = 1$ so $F_1 \geqq 0$ and the Riesz representation theorem guarantees that there is a probability measure μ on the Borel subsets of Δ such that $\int f d\mu = F_1(f)$ for each $f \in C(\Delta)$. In particular, then,

$$\mu(\Delta_n) = F_1(\chi_{\Delta_n}) = 2^{-k} \quad \text{when} \quad 2^k \leqq n < 2^{k+1} .$$

As the final lemma of this section makes abundantly clear, any dual space satisfying the conclusions of Lemma 4.2.9 is very far from having the RNP. Part (b) is due to Huff and Morris [1975].

LEMMA 4.2.12. Let $M > 0$, $\{\delta_k: k = 1,2,\ldots\}$ a sequence of positive numbers with $\lim_k \delta_k = 0$ and $\sup_k \delta_k \equiv \alpha < 1/2$, and $\{\Delta_n: n = 1,2,\ldots\}$ a Haar system of weak* compact subsets of $S(X^*)$ such that for each n there is an $x_n \in X$, $\|x_n\| < M$ with $|f(x_n) - \chi_{\Delta_n}(f)| < \delta_{k+1}$ whenever $f \in \Delta \equiv \Delta_1$ and $2^k \leqq n < 2^{k+1}$. Then there is a sequence $\{f_n\}$ in $L \equiv w^*cl(co(\Delta))$ such that

(a) $\{f_n: n = 1,2,\ldots\}$ is a bounded $(1 - 2\alpha)(2M)^{-1}$-tree. (That is to say, $\frac{1}{2} f_{2n} + \frac{1}{2} f_{2n+1} = f_n$ and $\|f_{2n} - f_n\| = \|f_{2n+1} - f_n\| \geqq (1 - 2\alpha)(2M)^{-1}$ for each $n = 1,2,\ldots$. See §2.4.)

(b) If $K \equiv \{f \in L: \lim_n f(x_n) = 0\}$ then K is a nonempty norm closed bounded convex set with no extreme points.

Proof. Let $e: X \to C(\Delta)$ denote the evaluation map: namely,

$$e(y)(f) \equiv f(y) \quad \text{for} \quad y \in X,\ f \in \Delta .$$

Let μ be a probability measure on the Borel subsets of Δ such that $\mu(\Delta_n) = 2^{-k}$ whenever $2^k \leqq n < 2^{k+1}$ (Lemma 4.2.11) and for each such n and k let

$$\mu_n \equiv 2^k \mu|_{\Delta_n} \in C(\Delta)^* .$$

(By $\mu|_{\Delta_n}$ we mean the measure whose value at the Borel set B is $\mu(B \cap \Delta_n)$.) Let $f_n \equiv e^*(\mu_n)$ for each n. Observe first that $f_n \in L$ $(\equiv w^*cl(co(\Delta)))$ for each n since otherwise, for some $y \in X$

$$\inf\{f(y): f \in \Delta\} > f_n(y) = e^*(\mu_n)(y) = \int_\Delta f(y)\, d\mu_n(f) \geqq \inf\{f(y): f \in \Delta\}$$

since μ_n is a probability measure.

In order to establish part (a) of the lemma note that $\{f_n : n = 1,2,\ldots\}$ is a tree since $e^*(\mu_n) = f_n$ and $\mu_n = (1/2)\mu_{2n} + (1/2)\mu_{2n+1}$ for each n. The proof is completed by showing that (for $\alpha \equiv \sup \delta_k < 1/2$)

$$\|f_{2n} - f_n\| = \|f_{2n+1} - f_n\| \geqq (1 - 2\alpha)(2M)^{-1} \quad \text{for each } n.$$

For each n

$$\|f_{2n+1} - f_n\|M \geqq (f_n - f_{2n+1})(x_{2n}) = e^*(\mu_n)(x_{2n}) - e^*(\mu_{2n+1})(x_{2n})$$

$$= \int_\Delta f(x_{2n})d\mu_n(f) - \int_\Delta f(x_{2n})d\mu_{2n+1}(f)$$

$$\text{(I)} \qquad = \frac{1}{2}\int_\Delta f(x_{2n})d(\mu_{2n} - \mu_{2n+1})(f)$$

$$= \frac{1}{2}\int_{\underline{\Delta}} [f(x_{2n}) - \chi_{\Delta_{2n}}(f)]\, d(\mu_{2n} - \mu_{2n+1})(f)$$

$$+ \frac{1}{2}\int_\Delta \chi_{\Delta_{2n}}(f)d(\mu_{2n} - \mu_{2n+1})(f).$$

Now $|f(x_{2n}) - \chi_{\Delta_{2n}}(f)| < \alpha$ by definition of α whence

$$\text{(II)} \qquad \frac{1}{2}\int_{\Delta^-} [f(x_{2n}) - \chi_{\Delta_{2n}}(f)]\, d(\mu_{2n} - \mu_{2n+1})(f) \leqq \frac{1}{2}\alpha\cdot 2 = \alpha$$

while

$$\text{(III)} \qquad \frac{1}{2}\int_\Delta \chi_{\Delta_{2n}}(f)d(\mu_{2n} - \mu_{2n+1})(f) = \frac{1}{2}\,\mu_{2n}(\Delta_{2n}) = \frac{1}{2}\,.$$

Thus from (I), (II) and (III), $\|f_{2n+1} - f_n\|M \geqq \frac{1}{2} - \alpha$. That is, $\|f_{2n+1} - f_n\| \geqq (1 - 2\alpha)(2M)^{-1}$, which proves (a) of the lemma.

In order to prove Lemma 4.2.11(b) first observe that K is both convex and bounded. Suppose next that $f \in \bar{K}$. Then $f \in L$ and in order to show that K is closed it suffices to show that $\lim_n f(x_n) = 0$. Now for any $\varepsilon > 0$ pick $g \in K$ such that $\|f - g\| < \varepsilon(2M)^{-1}$. Choose N so that $|g(x_n)| < \varepsilon/2$ for $n \geqq N$ and note that

$$|f(x_n)| \leqq \|f - g\|\cdot\|x_n\| + |g(x_n)| < \varepsilon(2M)^{-1}M + \varepsilon/2 = \varepsilon$$

so that $\lim_n f(x_n) = 0$ and $f \in K$. Hence K is a closed bounded convex subset of L.

It remains to show that $K \neq \emptyset$ and that $\mathrm{ex}(K) = \emptyset$. Let $e: X \to C(\Delta)$ be as in the proof of part (a). Then for each n, $\|e(x_n) - \chi_{\Delta_n}\|_\infty < \delta_k$ provided $2^{k-1} \leqq n < 2^k$ (where $\|\cdot\|_\infty$ denotes the sup norm on $C(\Delta)$). The adjoint map $e^*: C(\Delta)^* \to X^*$ is weak*-weak* continuous and it sends the point mass δ_f at $f \in \Delta$ to f itself. Let $P(\Delta)$ denote the set of probability measures on the Borel subsets of Δ, and note that

$$e^*(P(\Delta)) = e^*(w^*cl(co(\{\delta_f: f \in \Delta\}))) = w^*cl(co(\Delta)) = L.$$

Let $\nu \in P(\Delta)$. Then $e^*(\nu) \in K$ precisely when

$$0 = \lim_n e^*(\nu)(x_n) = \lim_n \int_\Delta e(x_n)d\nu = \lim_n \int_\Delta \chi_{\Delta_n} d\nu = \lim_n \nu(\Delta_n).$$

Let $M \equiv \{\nu \in P(\Delta): \lim_n \nu(\Delta_n) = 0\}$. Then $P(\Delta) \cap e^{*-1}(K) = M$ and hence $e^*(M) = K$. Since $M \neq \emptyset$ (for instance, μ in the proof of part (a) of this lemma is in M) it follows that $K \neq \emptyset$. In order to show that $\mathrm{ex}(K) = \emptyset$ first observe that if $g_1, g_2 \in L$ and $(1/2)(g_1 + g_2) \in K$ then $g_1, g_2 \in K$. (In the vernacular, then, K is an extremal subset of L. To see this, let $\nu_1, \nu_2 \in P(\Delta)$ satisfy $e^*(\nu_1) = g_1$ and $e^*(\nu_2) = g_2$. Then since $e^*((1/2)(\nu_1 + \nu_2)) \in K$, $(1/2)(\nu_1 + \nu_2) \in M$, i.e. $(1/2)\nu_1(\Delta_n) + (1/2)\nu_2(\Delta_n) \to 0$ as $n \to \infty$. Since $\nu_1(\Delta_n) \geqq 0$ and $\nu_2(\Delta_n) \geqq 0$ for each n it follows that $\nu_1(\Delta_n) \to 0$ and $\nu_2(\Delta_n) \to 0$ as $n \to \infty$. That is, $\nu_1, \nu_2 \in M$. Hence $g_1, g_2 \in K$.) Now suppose that $f_o \in \mathrm{ex}(K)$. Then from above, $f_o \in \mathrm{ex}(L)$. Then there is a $\nu_o \in \mathrm{ex}(P(\Delta))$ such that $e^*(\nu_o) = f_o$. (This is a standard argument: $e^{*-1}(f_o) \cap P(\Delta)$ is extremal in $P(\Delta)$. Hence if ν_o is an extreme point of $e^{*-1}(f_o) \cap P(\Delta)$ - and there are such extreme points since this set is weak* compact and convex - then ν_o satisfies both conditions.) By Milman's Theorem 3.2.2, $\nu_o = \delta_f$ for some $f \in \Delta$. But then $\nu_o \notin M$ since $f \in \Delta_n$ for infinitely many n so that $\limsup_n \nu_o(\Delta_n) = 1$. Thus $e^*(\nu_o) = f_o \notin K$. This contradiction proves part (b).

Here, finally, is the main theorem. Its proof is fairly straightforward after all this work. Part (c) was first stated by Fitzpatrick [1977] while (b) ⇒ (a) (at least in the global case) is due to Huff and Morris [1975]. The global version of (a) ⇒ (c) (i.e. the case $C \equiv U(X^*)$) is Stegall's fundamental contribution [1975], and our proof of not(g) ⇒ not(c) uses Stegall's Cantor like proof [1978a] which, in a slightly different context provided the last link in the original proof of Theorem 5.2.12. The argument in not(g) $\Rightarrow$ not(h) below goes back to Namioka [1967], and the properties of the identity map on C in parts (i) and (j) have been studied by Michael and Namioka [1976]. (They call such a map, appropriately enough, "barely continuous".)

THEOREM 4.2.13. Let $C \subset X^*$ be a weak* compact convex set. Then the following are equivalent.

(a) C has the RNP.

(b) C has the KMP.

(c) If Y is a separable subspace of X then $C|_Y$ is separable. (Recall: $C|_Y \equiv \{f|_Y: f \in C\} \subset Y^*$ is the set of restrictions to Y of the elements of C.)

(d) For each subspace Y of X, $\text{dens}(C|_Y) \leqq \text{dens}(Y)$.

(e) C contains no δ-tree for any $\delta > 0$.

(f) For each weak* compact convex subset K of C, $K = \text{w*cl}(\text{co}(\text{w*-str exp}(K)))$.

(g) C is subset weak* dentable.

(h) Each weak* compact subset D of C has relatively weak* open subsets of arbitrarily small diameter.

(i) For each weak* compact subset D of C there is a point of continuity of the identity map id: $(D, \text{weak*}) \to (D, \|\cdot\|)$.

(j) For each norm closed subset D of C there is a point of continuity of the identity map id: $(D, \text{weak*}) \to (D, \|\cdot\|)$.

Proof.

(a) $\Rightarrow$ (b) is Theorem 3.3.6.

not(d) $\Rightarrow$ (not(e) and not(b)) Assume first that C is a weak* compact convex subset of $S(X^*)$ and that Y is a subspace of X such that both $\text{dens}(C|_Y) > \text{dens}(Y)$ and $C|_Y \subset S(Y^*)$. Let $D \equiv C|_Y$. Thus D is a weak* compact convex subset of $S(Y^*)$ and $\text{dens}(D) > \eta \equiv \text{dens}(Y)$ - i.e. the hypothesis of 4.2.5 is satisfied with Y in place of X. By Lemma 4.2.9 there is a Haar system $\{\Delta_n': n = 1,2,...\}$ of weak* compact subsets of $S(Y^*) \cap \text{w*cl}(D) = D$ and there are points $x_n \in Y$, $\|x_n\| < M$ for some $M > 0$ and all $n = 1,2,...$ such that

$$|f(x_n) - \chi_{\Delta'_n}(f)| < 2^{-(k+3)} \quad \text{whenever} \quad 2^k \leqq n < 2^{k+1} \quad \text{for} \quad f \in \Delta' .$$

For each n let $\Delta_n \equiv \{f \in C: f|_Y \in \Delta'_n\}$. It is easy to check that $\{\Delta_n : n = 1,2,...\}$ is a Haar system of weak* compact subsets of C and that for each n, $|f(x_n) - \chi_{\Delta_n}(f)| < 2^{-(k+3)}$ whenever $2^k \leqq n < 2^{k+1}$, $f \in \Delta$. Thus Lemma 4.2.12 applies. Part (a) of that lemma establishes not(e) of this theorem for C as above while (b) of 4.2.12 establishes not(b) of this theorem for such C.

For the case of general weak* compact convex $C \subset X^*$ and subspace $Y \subset X$ with $\text{dens}(C|_Y) > \text{dens}(Y)$ proceed as follows. (We assume without loss of generality that $C \subset U(X^*)$.) Let $X_1 \equiv X \times \mathbb{R}$ with $\|(x,r)\| \equiv \|x\| + |r|$. Then X_1^* is isomorphic to $X^* \times \mathbb{R}$ and the norm on X_1^* is given by $\|(f,r)\| \equiv \max\{\|f\|, |r|\}$. Let $T: X^* \to X_1^*$ be the map $T(f) \equiv (f,1) \in X_1^*$ for $f \in X^*$. Then T is an affine norm-norm and weak*-weak* homeomorphism onto its range. Thus $T(C) \doteq C_1$ is a weak* compact convex subset of $S(X_1^*)$. Moreover, $Y_1 \equiv Y \times \mathbb{R} \subset X_1$ is a subspace with $\text{dens}(Y_1) = \text{dens}(Y)$ and $\text{dens}(C_1|_{Y_1}) = \text{dens}(C|_Y) > \text{dens}(Y_1)$. Finally, if $(f,1) \in C_1$ then since $(0,1) \in Y_1$ has norm one,

$$1 \geqq \|(f,1)|_{Y_1}\| \geqq (f,1)(0,1) = 1.$$

Hence $C_1|_{Y_1} \subset S(Y_1^*)$ as well. From what has already been established in the first paragraph of this part of the theorem, C_1 contains a bounded δ-tree $\{(f_n,1): n = 1,2,...\}$ for some $\delta > 0$ and a closed bounded convex subset K_1 with $\text{ex}(K_1) = \emptyset$. Then $\{f_n : n = 1,2,...\}$ is a bounded δ-tree in C and $K \equiv T^{-1}(K_1)$ is a closed bounded convex subset of C with no extreme points.

<u>(c) ⇒ (a)</u> Let $B \equiv \{h_i : i = 1,2,...\}$ be a countable subset of C and let $A \subset U(X)$ be a countable set with

$$\sup_{x \in A} (h_i(x) - h_j(x)) = \|h_i - h_j\| \quad \text{for all} \quad i,j = 1,2,\ldots .$$

Let $Y \equiv \overline{\text{span}}(A)$ and note that Y is separable. By hypothesis $C|_Y$ is separable and since $C|_Y$ is also weak* compact and convex, $C|_Y$ is subset s-dentable by Proposition 4.1.1. But for id: $Y \to X$ the inclusion map, id*: $X^* \to Y^*$ is an affine isometry on B and $\text{id}^*(B) = B|_Y \subset C|_Y$ is s-dentable. Hence B itself is s-dentable. It follows that C has the RNP by Theorem 2.3.6 parts (1) and (8).

<u>(d) ⇒ (c)</u> This is trivial.

<u>(a) ⇒ (e)</u> Any set containing a bounded δ-tree for some $\delta > 0$ fails to be subset dentable and hence lacks the RNP by Theorem 2.3.6 parts (1) and (9).

Thus (a), (b), (c), (d) and (e) are equivalent.

<u>(g) ⇒ (f)</u> This follows from Theorem 3.5.4(w*).

<u>(g) ⇒ (a)</u> This is immediate since weak* dentable sets are dentable. Hence C is subset dentable and the conclusion follows from parts (1) and (9) of Theorem 2.3.6.

<u>(j) ⇒ (i)</u> is clear.

<u>(i) ⇒ (h)</u> If $f_o \in D$ is a point of continuity of the identity map then f_o has relatively weak* open neighborhoods of arbitrarily small norm diameter.

<u>not(g) ⇒ not(h)</u> Choose $A \subset C$ such that A is not weak* dentable and let $J \equiv w^*cl(co(A))$. Then J fails to be weak* dentable as well (since weak* slices of J have nonempty intersection with A) and hence there is an $\varepsilon > 0$ such that each weak* slice of J has norm diameter exceeding 2ε. Let $D \equiv w^*cl(ex(J))$. We show now that each weak* open subset W of X* such that $W \cap D \neq \emptyset$ satisfies diameter $(W \cap D) > \varepsilon$ (thus establishing not(h)). If this were not the case there is a weak* open set W in X* with $W \cap D \neq \emptyset$ and diameter $(W \cap D) \leqq \varepsilon$. Let $K_o \equiv w^*cl(co(W \cap ex(J)))$. Note that $K_o \neq \emptyset$ by choice of W and that diameter $(K_o) \leqq \varepsilon < 2\varepsilon$. (If $f \in W \cap D$ then $U_\varepsilon[f] \supset W \cap D$. Hence $U_\varepsilon[f] \supset K_o$. It follows that for each $g \in K_o$, $U_\varepsilon[g] \supset W \cap D$. Hence $U_\varepsilon[g] \supset K_o$.) Moreover, if $K_1 \equiv w^*cl(co(C\backslash W))$ (a possibly empty set) then K_i $(i = 0,1)$ is a weak* compact convex subset of J and

$$J = w^*cl(co(K_o \cup K_1)) = co(K_o \cup K_1).$$

Because there is an $f \in W \cap ex(J)$, $f \notin D\backslash W$ and hence by the Milman Theorem 3.2.2, $f \notin K_1$. Thus $J\backslash K_1 \neq \emptyset$. It follows from Theorem 3.4.1(w*) that there is a weak* slice of J of diameter less than 2ε and since this is impossible, diameter $(W \cap D) > \varepsilon$ as was to be shown.

<u>(f) ⇒ (j)</u> Let D be a norm closed subset of C and let $K \equiv w^*cl(co(D))$. Pick any $f_o \in w^*\text{-str}\ exp(K)$ and suppose that $x_o \in X$ weak* strongly exposes K at f_o. Now

$S(K,x_o,\alpha)$ is a relatively weak* open subset of K containing f_o when $\alpha > 0$, and diameter $(S(K,x_o,\alpha)) \to 0$ as $\alpha \to 0$. Thus f_o is a point of continuity of the identity map id: $(K,\text{weak*}) \to (K,\|\cdot\|)$. It is easy to check that $f_o \in D$. (Otherwise for some $\varepsilon > 0$, $U_\varepsilon(f_o) \cap D = \emptyset$ and hence for some $\alpha > 0$, $S(K,x_o,\alpha) \cap D = \emptyset$. But then

$$f_o \notin \{f \in X^*: f(x_o) \leqq M(x_o,K) - \alpha\} \supset K.)$$

Thus f_o is a point of continuity of id: $(D,\text{weak*}) \to (D,\|\cdot\|)$ as well.

Thus (f), (g), (h), (i) and (j) are equivalent and (g) $\Rightarrow$ (a). The following implication, then, is enough to establish the theorem.

<u>not(g) $\Rightarrow$ not(c)</u> Suppose that D (a weak* compact subset of C) and $\varepsilon > 0$ have been chosen such that diameter $(V) > \varepsilon$ whenever V is a nonempty relatively weak* open subset of D. The ensuing argument is treelike, originally due to Stegall [1978a] and depends on the following observation:

(IV) If V is a nonempty relatively weak* open subset of D then there are nonempty relatively weak* open subsets V_1 and V_2 of V along with some $x \in S(X)$ such that whenever $f_i \in w^*cl(V_i)$ $i = 1,2$, then $|(f_1 - f_2)(x)| \geqq \varepsilon/3$.

(Indeed, given a nonempty relatively weak* open subset V of D find $g_1, g_2 \in V$ with $\|g_1 - g_2\| > \varepsilon$. This is possible by hypothesis. Next pick $x \in S(X)$ such that $g_1(x) - g_2(x) > \varepsilon$ and let $V_1 \equiv \{h \in X^*: h(x) > g_1(x) - \varepsilon/3\} \cap V$ and $V_2 \equiv \{h \in X^*: h(x) < g_2(x) + \varepsilon/3\} \cap V$. Clearly V_1 and V_2 have the desired properties.)

Inductively apply (IV) with $W_1 \equiv D$ to create a sequence $\{W_n: n = 1,2,\ldots\}$ of nonempty relatively weak* open subsets of D together with a sequence $\{x_n: n = 1,2,\ldots\}$ in S(X) such that for each n

(a) $W_{2n} \cap W_{2n+1} \subset W_n$; and

(b) If $f \in w^*cl(W_{2n})$ and $g \in w^*cl(W_{2n+1})$ then $|(f - g)(x_n)| \geqq \varepsilon/3$.

Let $Y \equiv \overline{\text{span}}(\{x_n: n = 1,2,\ldots\})$ and

$$\Delta \equiv \bigcap_{k=1}^{\infty} \bigcup_{n=2^k}^{2^{k+1}-1} w^*cl(W_n).$$

Then $|\Delta|$ is at least that of the continuum and $\Delta \subset C$. Let $\{n_k: k = 1,2,\ldots\}$ and $\{m_j: j = 1,2,\ldots\}$ be two distinct increasing sequences of integers such that $n_1 = m_1 = 1$ and

$$n_{k+1} = \begin{cases} 2n_k & \text{or} \\ 2n_k+1 \end{cases} \quad \text{for each } k \text{ and} \quad m_{j+1} = \begin{cases} 2m_j & \text{or} \\ 2m_j+1 \end{cases} \quad \text{for each } j.$$

(Clearly there are uncountably many such sequences.) Then

$$\bigcap_{k=1}^{\infty} w^*cl(W_{n_k}) \neq \emptyset \quad \text{and} \quad \bigcap_{j=1}^{\infty} w^*cl(W_{m_j}) \neq \emptyset.$$

But if f belongs to the first intersection and g to the second then there is an i such that $f \in w^*cl(W_{2i})$ and $g \in w^*cl(W_{2i+1})$ (or visa versa) and hence $|(f - g)(x_i)| \geqq \varepsilon/3$. It follows that $\|f|_Y - g|_Y\| \geqq \varepsilon/3$ whence $\Delta|_Y$ contains an uncountable subset, each two points of which are norm separated by at least $\varepsilon/3$. Thus $C|_Y \supset \Delta|_Y$ is not separable. The proof of Theorem 4.2.13 is complete.

It should be remarked that the assumption that C be weak* compact (rather than just closed and bounded) is necessary in 4.2.13 (a) => (c). Indeed, let $X \equiv C[0,1]$ and for each $t \in [0,1]$ let $\delta_t \in U(X^*)$ be the point mass at t. If $C \equiv \overline{co}(\{\pm\delta_t: t \in [0,1]\})$ then C is affinely isometric with $U(\ell_1([0,1]))$ as may easily be checked, and thus C has the RNP by Corollary 4.1.9. Nevertheless, since X itself is separable, $C|_Y = C$, a nonseparable set. We close this section with some elementary consequences of Theorem 4.2.13, the first of which was noted by Stegall [1981] and by Fitzpatrick [1977], the second two by Stegall [1981] and [1975] and the last by Edgar [1977] (in global form).

<u>COROLLARY</u> 4.2.14. Let $K \subset X^*$ be weak* compact and convex. Then K has the RNP if and only if $T^*(K) \subset Y^*$ has the RNP for each $T: Y \to X$ a continuous linear map on the Banach space Y.

Proof. One direction is trivially verified by taking Y = X and T = identity on X. Conversely, if $K \subset X^*$ has the RNP and $T: Y \to X$ is a continuous linear operator then $T^*(K)$ has the RNP if and only if $T^*(K)|_Z$ is separable whenever $Z \subset Y$ is a separable subspace. Denote the closure of the linear space $T(Z) \subset X$ by Z_1. Then Z_1

is separable and hence $K|_{Z_1}$ is separable. Observe that if $f \in K$ and $f|_{Z_1} = 0$ then $f(T(z)) = 0$ for each $z \in Z$ whence $T^*(f)|_Z = 0$. Thus the map $f|_{Z_1} \to T^*(f)|_Z$ from $K|_{Z_1}$ to $T^*(K)|_Z$ is well defined. It is easy to check that it is both norm-norm continuous and surjective as well. Thus $T^*(K)|_Z$ is separable since $K|_{Z_1}$ is separable, which completes the proof.

On the basis of 4.2.14 it seems reasonable to conjecture that for a weak* compact convex set $K \subset X^*$ and $R: K \to Y^*$ a weak*-weak* affine homeomorphism, K has the RNP if and only if $R(K)$ does. This is, however, false, as the following elementary example points out. Let $T: c_o \to \ell_1$ be the map $T((x_i)) \equiv (2^{-i}x_i)$ for $(x_i) \in c_o$. Then T is a continuous 1-1 linear operator with dense range so $T^*|_{U(\ell_\infty)}: U(\ell_\infty) \to \ell_1$ is an affine weak*-weak* homeomorphsm onto its range. Observe that $T^*(U(\ell_\infty))$ is a weak* compact convex subset of ℓ_1 and thus has the RNP while $U(\ell_\infty)$ does not have the RNP (since for example, it contains $U(c_o)$ a closed bounded convex set with no extreme points).

<u>COROLLARY</u> 4.2.15. If $K \subset X^*$ is weak* compact and convex then K has the RNP if and only if whenever Y is a separable Banach space and $T: Y \to X$ is a continuous linear operator then $T^*(K)$ is a norm separable subset of Y^*.

Proof. If K has the RNP and $T: Y \to X$ then $T^*(K) \subset Y^*$ has the RNP by Corollary 4.2.14. If Y itself is separable then $T^*(K)|_Y$ is separable by Theorem 4.2.13 and the desired conclusion follows since $T^*(K)|_Y = T^*(K)$. Conversely, if K lacks the RNP then there is a separable subspace Y of X such that $K|_Y$ is nonseparable. Let $T: Y \to X$ be the identity embedding. Then $T^*(K) = K|_Y$ is a nonseparable subset of Y^*.

For any Banach space Z, a bounded <u>biorthogonal system</u> in $Z \times Z^*$ is a bounded subset B of $Z \times Z^*$ such that if (x,f) and (y,g) are distinct elements of B then $g(x) = f(y) = 0$ and $f(x) = g(y) = 1$. The next result (at least the first part of it) appears in Stegall [1975].

<u>COROLLARY</u> 4.2.16. Let Y be a separable Banach space and $C \subset Y^*$ a nonseparable weak* compact convex set. Then C contains a weakly discrete closed subset of the

power of the continuum. If, in addition, $C \subset S(Y^*)$ then there is a bounded biorthogonal system in $C \times Y^{**}$ of the cardinality of the continuum.

Proof. Assume first that $C \subset S(Y^*)$. According to Corollary 4.2.10 there is a Haar system $\{\Delta_n : n = 1,2,\ldots\}$ in C of weak* compact sets, an $M > 0$ and a sequence $\{y_n : n = 1,2,\ldots\}$ in $MU(Y)$ such that Δ is weak* homeomorphic to the Cantor set and

$$|f(y_n) - \chi_{\Delta_n}(f)| < 2^{-(k+3)} \quad \text{whenever} \quad f \in \Delta \quad \text{and} \quad 2^k \leqq n < 2^{k+1}.$$

Pick any $f \in \Delta$. Then there is a unique sequence $\{n_k : k = 1,2,\ldots\}$ of positive integers such that $2^k \leqq n_k < 2^{k+1}$ for each k and $f \in \Delta_{n_k}$ for each k. Let $F_f \in Y^{**}$ be any weak* cluster point of the sequence $\{\hat{y}_{n_k} : k = 1,2,\ldots\}$ in Y^{**}. Then $\|F_f\| \leqq M$ and since

$$|\hat{y}_{n_k}(f) - 1| = |f(y_{n_k}) - \chi_{\Delta_{n_k}}(f)| < 2^{-(k+3)} \quad \text{for each } k$$

it follows that $F_f(f) = 1$. On the other hand, for any $g \in \Delta$, if $g \neq f$ then $g \notin \Delta_{n_k}$ for all sufficiently large k. Hence for each such k

$$|\hat{y}_{n_k}(g)| = |\hat{y}_{n_k}(g) - \chi_{\Delta_{n_k}}(g)| < 2^{-(k+3)}$$

whence $F_f(g) = 0$. It follows that $\{(f,F_f) : f \in \Delta\}$ is a bounded biorthogonal system in $C \times Y^{**}$ of cardinality $|\Delta| = 2^{\aleph_0}$ when $C \subset S(Y^*)$.

If $C \subset Y^*$ is not necessarily a subset of $S(Y^*)$, an appropriate multiple of the norm on Y^* places C in its unit ball. Then embed C in $S(Y^* \times \mathbb{R})$ via the map $f \to (f,1)$ (see the second paragraph of the proof of Theorem 4.2.13). Now produce a bounded biorthogonal system $\{(f_r,1), (F_r,t_r) : r \in \mathbb{R}\}$ in $(C \times \{1\}) \times (Y^* \times \mathbb{R})^*$ as in the previous paragraph. Then $F_r(f_s) = -t_r$ for r,s distinct real numbers and $F_r(f_r) + t_r = 1$. Hence $F_r(f_r - f_s) = 1$ for each pair of distinct numbers r and s. It follows that $\{f_r : r \in \mathbb{R}\} \subset C$ is a weakly discrete and closed subset of C of the power of the continuum.

Recall that a topological space (D,T) is <u>Lindelöf</u> if each open cover has a countable subcover. Talagrand [1975] established that each WCG Banach space is Lindelöf in its weak topology. Thus the next result, a localized version of a result of Edgar, yields a generalization of Corollary 4.1.10 (since if (D,T) is Lindelöf, so is each of its closed subsets in the relative topology). Note that the next corollary generalizes a result of Saab [1980a] on weakly K-analytic sets since K-analytic sets are always Lindelöf. (See 7.8.7(b) and 7.8.8 for details.)

COROLLARY 4.2.17. Let $C \subset X^*$ be a weak* compact convex set and suppose that C is Lindelöf in its weak topology. Then C has the RNP.

Proof. If C lacks the RNP then there is a separable subspace Y of X such that $C|_Y$ is not separable by Theorem 4.2.13, and hence by Corollary 4.2.16, there is a weakly discrete and closed subset in $C|_Y$ of the power of the continuum. That subset is thus not Lindelöf, and since it is weakly closed in $C|_Y$, $C|_Y$ fails to be weakly Lindelöf. Hence C is not weakly Lindelöf (since $id^*: C \to C|_Y$ is surjective and weak-weak continuous where $id: Y \to X$ denotes the identity embedding, and the continuous image of a Lindelöf space is Lindelöf). This contradiction establishes the corollary

§3. Measurable sets.

The main results of this section are essentially due to Edgar [1977] and are of quite a different sort than those which have appeared earlier in these notes. The problems of identifying σ-algebras generated (in some sense) by various of the natural topologies on a given subset of a Banach space and of obtaining a useful handle on the support of (a large class of) measures on such a σ-algebra are important ones when dealing with detailed information about measures on Banach spaces. (See, for example, Chapter 6 where a general integral representation theory for closed bounded convex sets with the RNP is developed.) These problems are studied in this and the next section.

Many of the technical lemmas needed here (or their proofs) are well known, and for this reason their proofs have been postponed until the end of the section. (Nevertheless the statements of the lemmas appear early on for convenient referencing in the proof of the main result, Theorem 4.3.11.) All topological spaces are assumed to be Hausdorff. Observe that the weak* topology on a dual space is completely regular (since it is a subset of a product of metric spaces in the product topology) and hence each subset of a dual space is completely regular in the weak* topology. Similarly, each subset of (X,weak) is completely regular.

NOTATION 4.3.1. Let $(D,\mathcal{T})$ be a topological space.

(a) $\mathcal{B}(D,\mathcal{T})$ (abbreviated $\mathcal{B}(D)$ when no confusion arises) denotes the Borel σ-algebra on D. That is, $\mathcal{B}(D,\mathcal{T})$ is the smallest σ-algebra containing the $\mathcal{T}$-open sets.

(b) Baire $(D,\mathcal{T})$ (or Baire(D) when possible) denotes the smallest σ-algebra containing all sets of the form $f^{-1}(\mathcal{O})$ for $\mathcal{O}$ open in $\mathbb{R}$ and $f: D \to \mathbb{R}$ $\mathcal{T}$-continuous.

N.B. In general Baire(D) $\subsetneq \mathcal{B}(D)$ although the two coincide when $(D,\mathcal{T})$ is metrizable as is easily checked.

The next definition contains some slightly nonstandard terminology. (Indeed, some authors use "regular" in place of "tight" below, and restrict the use of "tight" to measures defined on the Baire sets which are approximable by compact subsets as are the measures of 4.3.2. The terminology "Radon Measure" is often used in the literature in place of our "tight Measure". The excellent source Schwartz [1973] contains an accounting of most of the "well known" information about such measures.) Tight measures will appear in Chapter 6 as well, there defined on subsets of a Banach space in the norm topology, and in that context (since Baire(D) = $\mathcal{B}(D)$ when D is metrizable) the terminology is standard.

DEFINITION 4.3.2. Let $(D,\mathcal{T})$ be a topological space.

(a) Let μ be a nonnegative measure on $\mathcal{B}(D,\mathcal{T})$ with $\mu(D) = 1$. (Such a measure is called a probability measure.) Say that μ is tight if for each $A \in \mathcal{B}(D)$ and $\varepsilon > 0$ there is a $\mathcal{T}$-compact set $K \subset A$ such that $\mu(K) > \mu(A) - \varepsilon$.

(b) Denote the collection of all tight probability measures on $\mathcal{B}(D,\mathcal{T})$ by $\mathcal{P}_t(D,\mathcal{T})$ (or more simply $\mathcal{P}_t(D)$ if possible without confusion).

In order to be prepared to deal with Baire functions as well as Baire sets it will be convenient to introduce the following.

NOTATION 4.3.3. Let $(D,\mathcal{T})$ be a topological space.

(a) By $C_b(D)$ we shall mean the Banach space of continuous bounded real valued functions on D with the supremum norm.

(b) Let Brefct$(D,\mathcal{T})$ (Brefct(D) if possible) denote the smallest collection of real valued functions on D which includes $C_b(D)$ and which is closed under sequential pointwise limits. (That is, if $\{f_n : n = 1,2,\ldots\}$ is a sequence in Brefct(D), and if $\lim_n f_n(x) \equiv f(x)$ exists for each $x \in D$, then $f \in$ Brefct(D).) Thus, the functions in Brefct(D) are precisely the Baire functions on D.

The connection between Baire(D) and Brefct(D) is partially spelled out in the first lemma. Recall: the proofs of the lemmas will appear after the proof of 4.3.11

LEMMA 4.3.4. Let $(D,\mathcal{T})$ be a normal topological space. If $B \in$ Baire(D) then $\chi_B \in$ Brefct(D).

Since Baire sets are nice and so are compact sets even when they are not Baire sets, so too is the next lemma.

LEMMA 4.3.5. Let $(D,\mathcal{T})$ be a completely regular topological space. If $C \subset D$ is σ-compact (that is, if C is a countable union of compact sets) then for any $\mu \in \mathcal{P}_t(D)$ there is a set $B \in$ Baire(D) such that $B \supset C$ and $\mu(B) = \mu(C)$.

The full strength of the next lemma will be used in the proof of Theorem 4.3.11.

LEMMA 4.3.6. Let $D \subset X^*$ be a norm closed bounded and separable subset. Then D is a weak* $K_{\sigma\delta}$ set. (That is, D is a countable intersection of countable unions of weak* compact sets.) In particular, $D \in \mathcal{B}(X^*,\text{weak*})$. Moreover $\mathcal{B}(D,\text{norm}) = \mathcal{B}(D,\text{weak*})$ and $\mathcal{B}(D,\text{norm}) = D \cap \mathcal{B}(X^*,\text{weak*})$.

There are numerous situations when measures on one space need to be "pushed over" onto another space. (For example, change of variable theorems may be viewed in this way.) The most natural way to do this in most cases is formalized next.

DEFINITION 4.3.7. Let $(E_1,\mathcal{F}_1)$ and $(E_2,\mathcal{F}_2)$ be two measurable spaces and suppose that $\psi: E_1 \to E_2$ is an $\mathcal{F}_1 - \mathcal{F}_2$ measurable function. (That is, assume that $\psi^{-1}(A) \in \mathcal{F}_1$ for each $A \in \mathcal{F}_2$.) If μ is a finite measure on $\mathcal{F}_1$ then define $\psi(\mu)$ to be the measure on $\mathcal{F}_2$ given by $\psi(\mu)(A) \equiv \mu(\psi^{-1}(A))$ for each $A \in \mathcal{F}_2$. $\psi(\mu)$ is called the distribution of μ by ψ.

Although the Borel σ-algebra appears naturally when dealing with measures on topological spaces it is in a very real sense artificially small from a measure theory perspective. An enlarged σ-algebra is defined below which still preserves its ties with the Borel σ-algebra (and hence to the topology) but is suitably enlarged to include those sets which cause no measure theoretic problems. CAUTION: Definition 4.3.8 will be quite misleading if the word "tight" is ignored. This could cause a misinterpretation of Theorem 4.3.11.

DEFINITION 4.3.8. Let $(D,\mathcal{T})$ be a topological space and let μ be a probability measure on $\mathcal{B}(D)$. Denote the μ-completion of $\mathcal{B}(D)$ by $\mathcal{M}_\mu$. Thus

$$\mathcal{M}_\mu \equiv \{A \subset D\colon \text{There are } A_1, A_2 \in \mathcal{B}(D) \text{ with } A_1 \subset A \subset A_2 \text{ and } \mu(A_2 \setminus A_1) = 0\}$$

The σ-algebra of universally measurable sets with respect to the tight measures on $\mathcal{B}(D,\mathcal{T})$, denoted by $\mathcal{U}_t(\mathcal{B}(D,\mathcal{T}))$ (or $\mathcal{U}_t(\mathcal{B}(D))$) is defined by

$$\mathcal{U}_t(\mathcal{B}(D,\mathcal{T})) \equiv \bigcap_{\mu \in \mathcal{P}_t(D,\mathcal{T})} \mathcal{M}_\mu$$

The problem with definition 4.3.8 is that in general there is no reason to believe that $\mathcal{U}_t(\mathcal{B}(D,\mathcal{T}))$ coincides with

(V) $\bigcap \mathcal{M}_\mu$ the intersection being indexed by all probability measures μ on $\mathcal{B}(D,\mathcal{T})$

and yet the σ-algebra of (V) (abbreviated $\mathcal{U}(\mathcal{B}(D,\mathcal{T}))$ or $\mathcal{U}(\mathcal{B}(D))$ - cf. Definition 6.2.3) is commonly known as the universally measurable σ-algebra with respect to $\mathcal{B}(D,\mathcal{T})$. On the other hand, it is obvious that $\mathcal{U}_t(\mathcal{B}(D)) \supset \mathcal{U}(\mathcal{B}(D))$ in general so that important classes of non Borel sets known to be in $\mathcal{U}(\mathcal{B}(D))$ also belong to $\mathcal{U}_t(\mathcal{B}(D))$. There may be some consolation when $(D,\mathcal{T})$ is completely metrizable. Namely, it is consistent (for such $(D,\mathcal{T})$) with the Zermelo-Fraenkel axioms + the axiom of choice that each probability measure on $(D,\mathcal{T})$ is tight. Hence it is consistent with (ZF + AC) that $\mathcal{U}_t(\mathcal{B}(D)) = \mathcal{U}(\mathcal{B}(D))$ when $(D,\mathcal{T})$ is completely metrizable. (Combine Marczewski and Sikorski [1948] and Solovay [1967].) In any case, it is $\mathcal{U}_t(\mathcal{B}(D))$ of interest to us in this and the following section, and the penultimate lemma of this sequence provides some useful technical information about it.

LEMMA 4.3.9. Let D and E be topological spaces. Suppose that $\psi: D \to E$ is continuous and surjective.

(a) If $A \in \mathcal{U}_t(\mathcal{B}(E))$ then $\psi^{-1}(A) \in \mathcal{U}_t(\mathcal{B}(D))$.

Moreover assume either that

(1) D is compact and that, as above, ψ is continuous and surjective; or

(2) D is complete metric, E is completely regular, and in addition to being continuous and surjective, ψ is an open map as well.

Then both the following statements are true.

(b) If $\lambda \in \mathcal{P}_t(E)$ there is a $\nu \in \mathcal{P}_t(D)$ such that $\psi(\nu) = \lambda$;

(c) If for some $A \subset E$ we have $\psi^{-1}(A) \in \mathcal{U}_t(\mathcal{B}(D))$ then $A \in \mathcal{U}_t(\mathcal{B}(E))$.

There is one important case in which it is true that $\mathcal{U}_t(\mathcal{B}(D))$ and $\mathcal{U}(\mathcal{B}(D))$ coincide. Indeed, the following elementary result was recorded by Oxtoby and Ulam [1939].

LEMMA 4.3.10. Let M be a complete separable metric space. If μ is a probability measure on M then $\mu \in \mathcal{P}_t(M)$. (That is, μ is tight.)

Here is the main theorem. Edgar [1977] stated and proved it for dual spaces with the RNP although his proof carries over to the local setting below with little trouble.

THEOREM 4.3.11 (Edgar). Let K be a weak* compact convex subset of a dual Banach space X*. Then the following are equivalent.

(a) K has the RNP;

(b) If $\mu \in \mathcal{P}_t(K,\text{weak*})$ then there is a norm separable closed bounded convex subset K_1 of K with $K_1 \in \mathcal{B}(K,\text{weak*})$ and $\mu(K_1) = 1$. That is, μ has norm separable support.

(c) $\mathcal{U}_t(\mathcal{B}(K,\text{weak*})) = \mathcal{U}_t(\mathcal{B}(K,\text{norm}))$.

Proof.

(a) ⇒ (b) Suppose that $\mu \in \mathcal{P}_t(K,\text{weak*})$ and that $A \in \mathcal{B}(K,\text{weak*})$. Let $m(A)$ be that real valued function on X whose value at $x \in X$ is

$$m(A)(x) \equiv \int_A \hat{x}(f)\,d\mu(f).$$

(Note that $f \to \hat{x}(f)$ is weak* continuous so $m(A)(x)$ really makes sense.) Evidently $m(A) \in X^*$ and $m(A) \in \mu(A)K$. (Otherwise, since $\mu(A)K$ is weak* compact and convex there is an $x \in X$ such that $\hat{x}$ strictly separates $m(A)$ from $\mu(A)K$. This leads to the contradiction

$$m(A)(x) > \sup\{\mu(A)\hat{x}(f): f \in K\} \geqq \mu(A)\int_A \hat{x}(f)d\frac{\mu}{\mu(A)}(f) = m(A)(x)$$

when $\mu(A) \neq 0$. The case $\mu(A) = 0$ is trivial.) Also observe that m is finitely additive. Thus $m: \mathcal{B}(K,\text{weak*}) \to X^*$ is finitely additive, $m \ll \mu$ and $AR(m) \subset K$. But for any pairwise disjoint sequence $\{A_n: n = 1,2,\ldots\}$ in $\mathcal{B}(K,\text{weak*})$ it is clear that

$$\|m(\bigcup_{i=1}^{\infty} A_i) - \sum_{i=1}^{\infty} m(A_i)\| = \lim_n \|m(\bigcup_{i=1}^{\infty} A_i) - \sum_{i=1}^{n-1} m(A_i)\|$$

$$= \lim_n \|m(\bigcup_{i=n}^{\infty} A_i)\| \leqq \lim_n \mu(\bigcup_{i=n}^{\infty} A_i)[\sup\{\|f\|: f \in K\}] = 0$$

and hence m is countably additive. Since K has the RNP there is a Bochner integrable function $\psi \in L^1_K(K,\mathcal{B}(K,\text{weak*}),\mu)$ such that

$$m(A) = \int_A \psi(f)\,d\mu(f) \qquad \text{for} \quad A \in \mathcal{B}(K,\text{weak*})$$

We may as well modify ψ on a μ-null set if necessary and assume that ψ is $\mathcal{B}(K,\text{weak*}) - \mathcal{B}(K,\text{norm})$ measurable and is separably valued (cf. Theorem 1.1.7). Thus there is a norm closed bounded convex separable subset K_1 of K such that $\mu(\{f: \psi(f) \in K_1\}) = 1$.

Let $x \in X$. Then for any $A \in \mathcal{B}(K,\text{weak*})$, by using Theorem 1.1.8(e) we obtain

$$\int_A \hat{x}(f)d\mu(f) = m(A)(x) = \hat{x}(m(A)) = \hat{x}(\int_A \psi(f)d\mu(f)) = \int_A \hat{x} \circ \psi(f)d\mu(f)$$

whence $\hat{x} = \hat{x} \circ \psi$ [a.s.μ] for each $x \in X$. Next observe that

$$\mathcal{A} \equiv \{f: K \to \mathbb{R}: f = f \circ \psi \text{ [a.s.}\mu]\}$$

is an algebra of functions which contains a separating family of weak* continuous functions on K (namely $\{\hat{x}: x \in X\}$) and the constant functions (obviously). Hence $\mathcal{A}$

contains a dense subset of $C(K)$ by the Stone-Weierstrass theorem (K in the weak* topology of course). But if $\{f_n : n = 1,2,\dots\}$ is a sequence in A such that for each $x \in K$, $\lim_n f_n(x) \equiv f(x)$ exists then it is clear that $f \in A$ as well. In particular, then, $A \supset C(K)$ and A is closed under sequential pointwise limits. By Lemma 4.3.4, then, for any $B \in \text{Baire}(K,\text{weak*})$, $\chi_B \in A$. (Note that $(K,\text{weak*})$ is compact Hausdorff, hence normal, so the hypotheses of 4.3.4 are met.) Consequently, if $B \in \text{Baire}(K,\text{weak*})$ and $B \supset K_1$ then

$$
\begin{aligned}
\mu(B) &= \int_K \chi_B(f)\,d\mu(f) = \int_K \chi_B \circ \psi(f)\,d\mu(f) \\
&= \int_K \chi_{\psi^{-1}(B)}(g)\,d\mu(g) \qquad \text{(VI)} \\
&= \mu(\{g: \psi(g) \in B\}) \geqq \mu(\{f: \psi(f) \in K_1\}) \\
&= 1.
\end{aligned}
$$

Thus $\mu(B) = 1$ whenever $B \in \text{Baire}(K,\text{weak*})$, $B \supset K_1$.

Now suppose that C is a weak* σ-compact subset of K and that $C \supset K_1$. Then $\mu(C) = \mu(B)$ for some Baire set $B \supset C$ by Lemma 4.3.5 whence $\mu(C) = \mu(B) = 1$ from (VI). Since K_1 is a weak* $K_{\sigma\delta}$ set (by Lemma 4.3.6) it follows that there is a decreasing sequence $\{C_n\}$ of weak* σ-compact subsets of K with $\bigcap_{n=1}^{\infty} C_n = K_1$ and thus $\mu(K_1) = \lim_n \mu(C_n) = 1$. This establishes (b) of 4.3.11.

<u>(b) $\Rightarrow$ (c)</u> Evidently $U_t(\mathcal{B}(K,\text{weak*})) \subset U_t(\mathcal{B}(K,\text{norm}))$. Conversely, let $\mu \in P_t(K,\text{weak*})$. By hypothesis there is a norm closed bounded convex separable set $K_1 \subset K$ which is in $\mathcal{B}(K,\text{weak*})$ and satisfies $\mu(K_1) = 1$. By Lemma 4.3.6, $\mathcal{B}(K_1,\text{norm}) = \mathcal{B}(K_1,\text{weak*})$. The rest of this proof is bookeeping. Define $\tilde{\mu}$ on $\mathcal{B}(K,\text{norm})$ by

$$\tilde{\mu}(A) \equiv \mu(K_1 \cap A) \quad \text{for} \quad A \in \mathcal{B}(K,\text{norm})$$

and observe that $\tilde{\mu} \in P_t(K,\text{norm})$ by Lemma 4.3.10. Thus suppose that $A \in U_t(\mathcal{B}(K,\text{norm}))$. Then for any $\mu \in P_t(K,\text{weak*})$ and $\tilde{\mu}$ as constructed above in $P_t(K,\text{norm})$ find $A_1, A_2 \in \mathcal{B}(K,\text{norm})$ with

$$A_1 \subset A \subset A_2 \quad \text{and} \quad \tilde{\mu}(A_2 \backslash A_1) = 0.$$

Let $A_1' \equiv A_1 \cap K_1$ and $A_2' \equiv (A_2 \cap K_1) \cup (K \setminus K_1)$. Then

$$A_1' \subset A_1 \subset A \subset A_2 \subset A_2' \quad \text{and} \quad A_1', A_2' \in \mathcal{B}(K, \text{weak*})$$

(since $\mathcal{B}(K_1, \text{norm}) = \mathcal{B}(K_1, \text{weak*})$ and $K_1 \in \mathcal{B}(K, \text{weak*})$) and $\tilde{\mu}(A_2' \setminus A_1') = 0$. Thus $\mu(A_2 \setminus A_1) = 0$ so that $A \in \mathcal{U}_t(\mathcal{B}(K, \text{weak*}))$ as desired.

(c) $\Rightarrow$ (a). If K lacks the RNP there is, by Theorem 4.2.13, a separable subspace Y of X such that $K|_Y$ is not separable. Exactly as in the second paragraph of the proof of Theorem 4.2.13 assume that $K \subset U(X^*)$ and let $T: X^* \to X^* \times \mathbb{R}$ be the isometry $T(f) \equiv (f,1)$ for $f \in X^*$. If $X^* \times \mathbb{R}$ is endowed with the max norm, it follows that $T(K) \subset S((X \times \mathbb{R})^*)$ and $T(K)|_{Y_1}$ is not separable for $Y_1 \equiv Y \times \mathbb{R}$. Corollary 4.2.10 applies to the weak* compact convex set $T(K)|_{Y_1} \subset S(Y_1^*)$. Hence there is an $M > 0$ and a Haar system $\{\Delta_n': n = 1,2,\ldots\}$ of weak* compact subsets of $T(K)|_{Y_1}$ together with a sequence $\{y_n': n = 1,2,\ldots\}$ in $MU(Y_1)$ such that both $|f(y_n') - \chi_{\Delta_n'}(f)| < 2^{-(k+3)}$ whenever $f \in \Delta' \equiv \Delta_1'$ and $2^k \leqq n < 2^{k+1}$, and $|\bigcap_{k=1}^{\infty} \Delta_{n_k}'| \leqq 1$ for any increasing sequence $\{n_k: k = 1,2,\ldots\}$ of integers (cf. the discussion preceding 4.2.10). In particular, if $f, g \in \Delta'$ and $f \neq g$ then there is a k and there are n and j, $n \neq j$ such that $2^k \leqq n, j < 2^{k+1}$ and both $f \in \Delta_n$ and $g \in \Delta_j$. Thus

$$\begin{aligned} \|f - g\| &\geqq M^{-1}|f(y_n') - g(y_n')| \\ &\geqq M^{-1}(|\chi_{\Delta_n'}(f) - \chi_{\Delta_n'}(g)| - |f(y_n') - \chi_{\Delta_n'}(f)| - |g(y_n') - \chi_{\Delta_n'}(g)|) \\ &\geqq M^{-1}(1 - 2^{-(k+3)} - 2^{-(k+3)}) \geqq 3(4M)^{-1} . \end{aligned}$$

Thus Δ' is a norm closed set each two points of which have distance at least $3(4M)^{-1}$ from one another. It is clear that $K|_Y$ and $T(K)|_{Y_1}$ are norm-norm isometric and weak*-weak* homeomorphic. Let $\Delta_n \subset K|_Y$ correspond to $\Delta_n' \subset T(K)|_{Y_1}$ under this isometry. Then $\{\Delta_n: n = 1,2,\ldots\}$ is a Haar system of weak* compact sets and each pair of distinct points in $\Delta \equiv \Delta_1$ is norm separated by at least $3(4M)^{-1}$. It follows that each subset of Δ is in $\mathcal{U}_t(\mathcal{B}(K|_Y, \text{norm}))$. Moreover, one of the conclusions of Corollary 4.2.10 is that $\{\Delta_n': n = 1,2,\ldots\}$ is a Haar system homeomorphic term by term to the usual Haar system of the Cantor set, and hence the same is true of $\{\Delta_n: n = 1,2,\ldots\}$.

Let us assume for now that a subset L of Δ has been found such that $L \in \mathcal{U}_t(\mathcal{B}(\Delta,\text{norm}))\setminus\mathcal{U}_t(\mathcal{B}(\Delta,\text{weak*}))$. (See the next paragraph for details.) If id: $Y \to X$ denotes the identity embedding then $id^*|_K: K \to K|_Y$ is weak*-weak* continuous and surjective. Thus $id^*|_K^{-1}(L) \notin \mathcal{U}_t(\mathcal{B}(K,\text{weak*}))$ by Lemma 4.3.9(c) under hypothesis (1) (since it is clear that $\mathcal{U}_t(\mathcal{B}(\Delta,\text{weak*})) = \mathcal{U}_t(\mathcal{B}(K|_Y,\text{weak*})) \cap \Delta$). Now L is a closed subset of Y* and hence $L \in \mathcal{U}_t(\mathcal{B}(Y^*,\text{norm}))$. Lemma 4.3.9(c) under hypothesis (2) applies here and shows that $id^{*-1}(L) \in \mathcal{U}_t(\mathcal{B}(X^*,\text{norm}))$ since $id^*: X^* \to Y^*$ is an open map by the Hahn-Banach theorem. Since $K \in \mathcal{B}(X^*,\text{norm})$ we have $\mathcal{U}_t(\mathcal{B}(K,\text{norm})) = \mathcal{U}_t(\mathcal{B}(X^*,\text{norm}) \cap K$ and hence $id^{*-1}(L) \cap K \in \mathcal{U}_t(\mathcal{B}(K,\text{norm}))$. Finally, since $id^{*-1}(L) \cap K = id^*|_K^{-1}(L)$ it follows that

$$id^*|_K^{-1}(L) \in \mathcal{U}_t(\mathcal{B}(K,\text{norm}))\setminus\mathcal{U}_t(\mathcal{B}(K,\text{weak*}))$$

as was to be shown.

It thus suffices to produce $L \in \mathcal{U}_t(\mathcal{B}(\Delta,\text{norm}))\setminus\mathcal{U}_t(\mathcal{B}(\Delta,\text{weak*}))$ and since Δ is closed in Y* and discrete in the norm topology, each subset of Δ belongs to $\mathcal{U}_t(\mathcal{B}(\Delta,\text{norm}))$. Thus it suffices to find $L \notin \mathcal{U}_t(\mathcal{B}(\Delta,\text{weak*}))$. Let $h: \Delta \to C$ be a homeomorphism ($C \equiv$ Cantor set), let $\psi: C \to [0,1]$ be continuous and surjective, and let λ denote Lebesgue measure on $[0,1]$. (An example of a ψ is the map which assigns the point with $n^{\underline{th}}$ ternary digit a_n for a point $._3a_1a_2\ldots$ with each a_i either 0 or 2, the point whose $n^{\underline{th}}$ digit in its binary expansion is $a_n/2$.) By Lemma 4.3.9(b) there is a $\mu \in \mathcal{P}_t(\Delta)$ such that $\psi \circ h(\mu) = \lambda$. Let D be a subset of $[0,1]$ with inner Lebesgue measure zero and outer Lebesgue measure one. (See, for example, Cohn [1980], Proposition 1.4.9, p. 33.) If E is a compact subset of $(\psi \circ h)^{-1}(D)$ then

$$E \subset (\psi\circ h)^{-1}(\psi\circ h(E)) \subset (\psi\circ h)^{-1}(D)$$

so that

$$\mu(E) \leqq \mu((\psi\circ h)^{-1}(\psi\circ h)(E)) = \lambda(\psi\circ h(E)) = 0.$$

Similarly, when $E \subset \Delta\setminus(\psi\circ h)^{-1}(D)$ is compact then $\mu(E) = 0$. Consequently $(\psi\circ h)^{-1}(D)$ fails to be μ-measurable in particular, $(\psi\circ h)^{-1}(D) \notin \mathcal{U}_t(\mathcal{B}(\Delta,\text{weak*}))$. This completes the proof of Theorem 4.3.11.

The proof of the lemmas follows.

Proof of 4.3.4. Let $G \equiv \{B \subset D: \chi_B \in \text{Brefct}(D)\}$. For $a \in \mathbb{R}$ and $f \in C_b(D)$ there is a continuous function $g_n: D \to \mathbb{R}$ such that $0 \leqq g_n \leqq 1$, $g_n = 0$ on $f^{-1}(-\infty, a - \frac{1}{n}]$ and $g_n = 1$ on $f^{-1}[a,\infty)$ by Urysohn's lemma. Hence $\chi_{f^{-1}[a,\infty)}(x) = \lim_n g_n(x)$ for each $x \in D$ whence

$$G \supset \{f^{-1}[a,\infty): \quad a \in \mathbb{R}, \; f \in C_b(D)\}.$$

If $\{A_n: n = 1,2,\ldots\}$ is a monotonically increasing sequence of subsets of D with $A_n \in G$ for each n then $\chi_{\cup A_n}$ is the pointwise limit of $\{\chi_{A_n}: n = 1,2,\ldots\}$ so $\bigcap_{n=1}^{\infty} A_n \in G$ as well. By the monotone class theorem, then (see, for example, Halmos [1950]) it suffices to show that G is a ring.

Observe that if $f,g \in \text{Brefct}(D)$ then $f + g \in \text{Brefct}(D)$ as well. (A proof of this fact goes as follows: if $f \in C_b(D)$ then $\{g: f + g \in \text{Brefct}(D)\}$ is closed under sequential pointwise limits and includes $C_b(D)$. Now $\{f: f + g \in \text{Brefct}(D)$ for each $g \in \text{Brefct}(D)\}$ is itself a collection closed under pointwise sequential limits and it includes $C_b(D)$ from above. It thus includes Brefct(D).) In a similar manner, $rf \in \text{Brefct}(D)$ $(r \in \mathbb{R},\ f \in \text{Brefct}(D))$ and $fg \in \text{Brefct}(D)$ if $f, g \in \text{Brefct}(D)$. (That is, Brefct(D) is an algebra.) Thus if $B \in G$ then $\chi_B \in \text{Brefct}(D)$ and since $1 \in \text{Brefct}(D)$, so is $1 - \chi_B = \chi_{B^c}$. That is, $B^c \in G$ if $B \in G$. Finally, if $B_1, B_2 \in G$ then $\chi_{B_1 \cap B_2} = \chi_{B_1}\chi_{B_2} \in \text{Brefct}(D)$ whence $B_1 \cap B_2 \in G$. From the above remarks this completes the proof of 4.3.4.

Proof of 4.3.5. If C is compact and $\mu \in P_t(D)$ let

$$a \equiv \inf\{\mu(B): B \supset C,\ B \in \text{Baire}(D)\}.$$

By taking $B_n \in \text{Baire}(D)$, $B_n \supset C$ and $\mu(B_n) < a + n^{-1}$ the set $\bigcap_{n=1}^{\infty} B_n \equiv B \in \text{Baire}(D)$ satisfies $B \supset C$ and $\mu(B) = a$. If $a = \mu(C)$ we are done so assume that $a > \mu(C)$. Because μ is tight there is a compact set $C_1 \subset B\backslash C$ with $\mu(C_1) > 0$. Since (D,T) is completely regular there is a continuous function $f: D \to \mathbb{R}$ such that $f|_C = 1$ and $f|_{C_1} < 1/2$ while $0 \leqq f \leqq 1$ on D. Then $C \subset f^{-1}(1/2,\infty) \cap B \subset B \cap C_1^{\,c}$ so

$$\mu(C) \leq \mu(f^{-1}(1/2,\infty) \cap B) < a.$$

Since $f^{-1}(1/2,\infty) \cap B$ is a Baire set this contradicts the choice of a. Thus $a = \mu(C)$ when C is compact.

If $C = \bigcup_{n=1}^{\infty} C_n$ where each C_i is compact then for each n there is a Baire set $B_n \supset C_n$ with $\mu(B_n) = \mu(C_n)$. Then $\bigcup_{i=1}^{\infty} B_i$ belongs to Baire(D), $\cup B_i \supset C$ and it is elementary to check that $\mu(\cup B_i) = \mu(C)$.

<u>Proof of 4.3.6.</u> Assume that $D \subset U(X^*)$ and that $\{f_i: i = 1,2,...\}$ is a norm dense subset of D. Let $D_n \equiv \{f_1,...,f_n\}$ for each n. Now

$$D = \bigcap_{j=1}^{\infty} (D + \frac{1}{j} U(X^*)) = \bigcap_{j=1}^{\infty} (\bigcup_{n=1}^{\infty} D_n + \frac{1}{j} U(X^*)) = \bigcap_{j=1}^{\infty} \bigcup_{n=1}^{\infty} (D_n + \frac{1}{j} U(X^*)) \supset D.$$

Since $D_n + \frac{1}{j} U(X^*)$ is weak* compact for each n, D is a weak* $K_{\sigma\delta}$ set.

Each norm open ball in X* is a countable union of weak* compact sets (closed balls of smaller radius) so any norm open subset of X* which can be written as a countable union of open balls is weak* Borel. Since D is separable, each norm relatively open set in D is the intersection with D of a countable union of open balls. Since $D \in \mathcal{B}(X^*,\text{weak*})$ from above, it follows that each relatively norm open subset of D belongs to $\mathcal{B}(D,\text{weak*})$ and hence $\mathcal{B}(D,\text{norm}) \subset \mathcal{B}(D,\text{weak*})$. Since the reverse inclusion is obvious, it follows that $\mathcal{B}(D,\text{norm}) = \mathcal{B}(D,\text{weak*})$. The last statement follows from this and the fact that $D \in \mathcal{B}(X^*,\text{weak*})$.

<u>Proof of 4.3.9.</u>

(a). Let $\mu \in P_t(D)$. Then $\psi(\mu) \in P_t(E)$ as may be easily checked since ψ is continuous. Hence if $A \in \mathcal{U}_t(\mathcal{B}(E))$ there are sets $A_1, A_2 \in \mathcal{B}(E)$ such that $A_1 \subset A \subset A_2$ and $\psi(\mu)(A_2 \backslash A_1) = 0$. Now $\psi^{-1}(A_1), \psi^{-1}(A_2) \in \mathcal{B}(D)$ and $\psi^{-1}(A_1) \subset \psi^{-1}(A) \subset \psi^{-1}(A_2)$. Moreover

$$\mu(\psi^{-1}(A_2)\backslash\psi^{-1}(A_1)) = \psi(\mu)(A_2\backslash A_1) = 0.$$

Thus $\psi^{-1}(A) \in \mathcal{U}_t(\mathcal{B}(D))$.

(b). (Under hypothesis (1)). Suppose that $\lambda \in P_t(E)$. Let $T_\psi: C(E) \to C(D)$ be the map $T_\psi(f) \equiv f \circ \psi$ for $f \in C(E)$. Thus T_ψ is a linear isometry of C(E) into C(D) and for $g \in T_\psi(C(E))$ let $F(g) \equiv \int_E T_\psi^{-1}(g)d\lambda$. Thus F is a linear functional on the closed subspace $T_\psi(C(E))$ of $C(D)$ which is nonnegative and satisfies $\|F\| = F(1) = \lambda(E) = 1$. By the Hahn-Banach theorem F has a norm preserving extension to an element of $C(D)^*$, say F'. Since $\|F'\| = 1 = F(1) = F'(1)$ it follows that $F' \geqq 0$ and by the Riesz representation theorem there is a $\nu \in P_t(D)$ with $\int_D g d\nu = F(g)$ for each $g \in C(D)$. Consequently, for $f \in C(E)$

$$\int_E f d\lambda = F(T_\psi(f)) = \int_D T_\psi(f)d\nu = \int_D f \circ \psi d\nu$$

and it follows that $\lambda = \psi(\nu)$.

(b) (Under Hypothesis (2)). If $\lambda \in P_t(E)$ and $K \subset E$ is a compact set, define a sequence $\{C_n: n = 1,2,\ldots\}$ of subsets of D inductively as follows. Let $C_0 \equiv D$ and if C_{n-1} has already been defined, for each $y \in K$ choose $x_y \in \text{interior}(C_{n-1})$ with $\psi(x_y) = y$. Let V_y be a closed neighborhood of x_y such that $V_y \subset \text{interior}(C_{n-1})$ and $\text{diameter}(V_y) \leqq 1/n$. Because ψ is open and K is compact there are finitely many points $y_1,\ldots y_k$ in K with $\bigcup_{i=1}^{k} \psi(\text{interior}(V_{y_i})) \supset K$. Let $C_n \equiv \bigcup_{i=1}^{k} V_{y_i}$. Clearly $C_n \subset C_{n-1}$, $\psi(\text{interior}(C_n)) \supset K$ and C_n has a finite $1/n$ net. Thus $C \equiv \cap C_n$ is closed and totally bounded, hence compact (since D is complete). To see that $\psi(C) \supset K$ as well, fix $y \in K$ and for each n choose $x_n \in C_n$ such that $\psi(x_n) = y$. This is possible because $\psi(C_n) \supset K$ for all n. We claim that the sequence $\{x_n: n = 1,2,\ldots\}$ has a Cauchy subsequence. Since C_1 is a finite union of sets of diameter at most one, there is a subsequence $\{x_{n,1}: n = 1,2,\ldots\}$ of $\{x_n: n = 1,2,\ldots\}$ such that the diameter of the set $\{x_{n,1}\}_{n=1}^{\infty}$ is at most one. The sequence $\{x_{n,1}: n = 1,2,\ldots\}$ is eventually in C_2 and C_2 is a finite union of sets of diameter at most $1/2$. Hence there is a subsequence $\{x_{n,2}: n = 1,2,\ldots\}$ of $\{x_{n,1}: n = 1,2,\ldots\}$ such that the diameter of $\{x_{n,2}\}_{n=1}^{\infty}$ is at most $1/2$. Continuing this process we obtain a subsequence $\{x_{n,k+1}: n = 1,2,\ldots\}$ of $\{x_{n,k}: n = 1,2,\ldots\}$ such that the diameter of $\{x_{n,k+1}\}_{n=1}^{\infty} \leqq (k+1)^{-1}$. We claim that $\{x_{n,n}: n = 1,2,\ldots\}$ is Cauchy. Indeed, if $p, q \geq N$ then both $x_{p,p}$ and $x_{q,q}$ are terms in the sequence $\{x_{n,N}: n = 1,2,\ldots\}$. Hence the distance between $x_{p,p}$ and $x_{q,q}$ is at most $1/N$. Let $x \equiv \lim_n x_{n,n}$. Then $x \in C$ and since ψ is continuous at x, $\psi(x) = y$. Hence $\psi(C) \supset K$ as claimed above, and it follows that $\psi(C \cap \psi^{-1}(K)) = K$. That is, each compact subset of E is the image under ψ of a compact subset of D.

Define a sequence $\{k_n : n = 1,2,\ldots\}$ of positive integers and a sequence $\{K_n : n = 1,2,\ldots\}$ of compact sets in E as follows. Let k_1 be the smallest positive integer such that for some compact set K_1 in E, $\lambda(K_1) > 1/k_1$. If $\lambda(K_1) = 1$, stop. Otherwise, assume that $k_1,\ldots,k_{n-1}$ and $K_1,\ldots,K_{n-1}$ have been chosen and that $\lambda(\bigcup_{i=1}^{n-1} K_i) < 1$. Find the smallest integer k_n such that for some compact set $K_n \subset E\setminus \bigcup_{i=1}^{n-1} K_i$, $\lambda(K_n) > 1/k_n$. Note that if K is compact and $K \subset E\setminus \bigcup_{i=1}^{n-1} K_i$ then $\lambda(K) \leqq (k_n - 1)^{-1}$. Since

$$1 \geqq \lambda(\cup K_i) = \sum \lambda(K_i) \geqq \sum 1/k_i$$

it follows that $\lim_i k_i = \infty$. Hence if K is a compact subset of $E\setminus \bigcup_{i=1}^{\infty} K_i$ (and hence a compact subset of $E\setminus \bigcup_{i=1}^{n-1} K_i$ for each n)

$$\lambda(K) \leqq \limsup_n (k_n - 1)^{-1} = 0.$$

Now let C_n be a compact subset of D such that $\psi(C_n) = K_n$ for each n and let $\lambda_n \equiv \lambda|_{K_n}$. Then $(\lambda(K_n))^{-1}\lambda_n \in P_t(K_n)$ and since $\psi(C_n) = K_n$ it follows from part (b) of this lemma that there is a ν_n such that $(\lambda(K_n))^{-1}\nu_n \in P_t(C_n)$ and $\psi(\nu_n) = \lambda_n$. Let $\nu \equiv \Sigma \nu_n$. Evidently $\nu \in P_t(D)$ and $\psi(\nu) = \lambda$.

(c) (under either set of hypotheses). Suppose that $A \subset E$ and that $\psi^{-1}(A) \in U_t(B(D))$. Pick any $\lambda \in P_t(E)$. Then by (b) (or (d)) there is a $\nu \in P_t(D)$ such that $\psi(\nu) = \lambda$. Thus given $\varepsilon > 0$ there are compact sets $K_1, K_2 \subset D$ with $K_1 \subset \psi^{-1}(A) \subset K_2{}^c$ and $\nu(K_2{}^c\setminus K_1) \leqq \varepsilon$. Of course $\psi(K_1)$, $\psi(K_2)$ are compact in E and $\psi(K_1) \subset A \subset \psi(K_2)^c$. Moreover

$$\lambda(\psi(K_2)^c\setminus\psi(K_1)) = \nu(\psi^{-1}(\psi(K_2)^c)\setminus\psi^{-1}(\psi(K_1))) \leqq \nu(K_2{}^c\setminus K_1) \leqq \varepsilon$$

since $\psi^{-1}(\psi(K_2)^c)\setminus\psi^{-1}(\psi(K_1)) \subset K_2{}^c\setminus K_1$. It follows that $A \in U_t(B(E))$. This completes the proof of Lemma 4.3.9.

Proof of 4.3.10. Given $\varepsilon > 0$ there are finitely many closed balls, say $B_1(1),\ldots,B_1(k_1)$ in M each of radius one such that $\mu(M_1) > 1 - \varepsilon/2$ where $M_1 \equiv \bigcup_{i=1}^{k} B_1(i)$. Similarly there are finitely many closed balls $B_2(1),\ldots,B_2(k_2)$

in the metric space M_1, each of radius 1/2 such that $\mu(M_2) > 1 - \varepsilon/2$ where $M_2 \equiv \bigcup_{i=1}^{k_2} B_2(i)$. Continuing in this fashion, there is a decreasing sequence of Borel subsets $\{M_n\}$ of M such that $\mu(M_n) > 1 - \varepsilon/2$ for each n and M_n has a finite 1/n net. Denote the closure of $\cap M_n$ by K. Evidently K is complete and totally bounded, hence compact, and

$$\mu(K) \geqq \liminf_n \mu(M_n) \geqq 1 - \varepsilon/2.$$

Now let

$$\mathcal{A} \equiv \{A \in \mathcal{B}(M)\text{: For each } \delta > 0 \text{ there are a closed set } C \text{ and an open set } \mathcal{O} \text{ in } M \text{ with } C \subset A \subset \mathcal{O} \text{ and } \mu(\mathcal{O}\backslash C) < \delta\}.$$

Because each open set in M is an F_σ set, it is straightforward to check both that $\mathcal{A}$ contains the open sets and is a σ-algebra. Hence $\mathcal{A} = \mathcal{B}(M)$. To complete the proof of this lemma, pick any $A \in \mathcal{B}(M)$. For the $\varepsilon > 0$ of the first paragraph, according to the alternative description of $\mathcal{B}(M)$ as $\mathcal{A}$, there is a closed set $C \subset A$ with $\mu(A) < \mu(C) + \varepsilon/2$. If K is the compact set constructed in the first paragraph, then $\mu(C) < \mu(C \cap K) + \varepsilon/2$ and hence $\mu(C \cap K) > \mu(A) - \varepsilon$. But $C \cap K$ is compact. Since $\varepsilon > 0$ was arbitrary, the proof is complete.

§4. Global results.

Several results are presented here, only some of which are global translates of local theorems of earlier sections. Theorem 4.4.1 below includes the main theorem of Stegall's basic dual space paper [1975] on which so much of this chapter has been based (namely, (a) ⇒ (c) of the next theorem).

THEOREM 4.4.1. For a dual space X* the following are equivalent.

(a) X* has the RNP;

(b) X* has the KMP;

(c) If Y is a separable subspace of X then Y* is separable;

(d) X* contains no bounded δ-tree for any $\delta > 0$;

(e) For each weak* compact convex subset K of X*

$$K = w^*cl(co(w^*\text{-str } exp(K)));$$

(f) Each bounded subset of X* is weak* dentable;

(g) If Y is a subspace of X then dens(Y) = dens(Y*);

(h) $\mathcal{U}_t(\mathcal{B}(X^*,\text{norm})) = \mathcal{U}_t(\mathcal{B}(X^*,\text{weak}^*))$.

Proof. (a) ⟺ (b) ⟺ (d) ⟺ (e) ⟺ (f) are immediate from Theorem 4.2.13 applied to the set $C \equiv U(X^*)$. Observe that if Y is a separable subspace of X and $C \equiv U(X^*)$ then Y* is separable if and only if $C|_Y$ (= U(Y*)) is separable. Thus (a) ⟺ (c) follows from Theorem 4.2.13 as well. In a similar manner, (a) ⟺ (g) is a consequence of Theorem 4.2.13.

(a) ⟺ (h) In the following two sentences consistently provide X* with the norm topology, or consistently provide it with the weak* topology. $\mathcal{U}_t(\mathcal{B}(U(X^*)) = \mathcal{U}_t(\mathcal{B}(X^*)) \cap U(X^*)$ is easy to establish. It follows that $\mathcal{U}_t(\mathcal{B}(X^*))$ consist of all countable unions of sets in $\bigcup_{n=1}^{\infty} \mathcal{U}_t(\mathcal{B}(nU(X^*)))$. Thus it is not difficult to see that X* has the RNP if and only if

$$\begin{aligned}\mathcal{U}_t(\mathcal{B}(X^*,\text{norm})) &= \{\text{all countable unions from } \bigcup_{n=1}^{\infty} \mathcal{U}_t(\mathcal{B}(nU(X^*),\text{norm}))\}\\ &= \{\text{all countable unions from } \bigcup_{n=1}^{\infty} \mathcal{U}_t(\mathcal{B}(nU(X^*),\text{weak}^*))\} \text{ by 4.3.11}\\ &= \mathcal{U}_t(\mathcal{B}(X^*,\text{weak}^*)).\end{aligned}$$

This completes the proof of the theorem.

Say that a dual Banach space X* has <u>Property (**)</u> provided the weak* and the norm topologies agree on S(X*). The following results are due to Edgar [1977].

<u>PROPOSITION</u> 4.4.2. (a) Let X be a Banach space such that the norm and weak topologies agree on S(X). Then $\mathcal{B}(X,\text{weak}) = \mathcal{B}(X,\text{norm})$.

(b) Let X* be a dual Banach space with Property (**). Then $\mathcal{B}(X^*,\text{weak}^*) = \mathcal{B}(X^*,\text{norm})$.

(N.B. An alternative proof of 4.4.2, even more direct than the one here, appears in the proof of 7.7.11. That Corollary also contains further results along these lines.)

Proof. The proof for (b) below may be mimicked to provide a proof of (a). Thus suppose that X* has Property (**). For any set $A \subset X^*$ let w*-int(A) denote the interior of A in the weak* topology. Since $\mathcal{B}(X^*,\text{weak*}) \subset \mathcal{B}(X^*,\text{norm})$ it suffices to show that $V \in \mathcal{B}(X^*,\text{weak*})$ for each norm open set V in X*. In this regard, let

$$V' \equiv \bigcup_{\substack{r>0 \\ r \text{ rational}}} U_r[0] \cap \text{w*-int}[V \cup (X^* \setminus U_r[0])].$$

Evidently $V' \in \mathcal{B}(X^*,\text{weak*})$. The proof is completed by demonstrating that $V = V'$. Clearly $V' \subset V$. Conversely, suppose that $f \in V$ and pick $\varepsilon > 0$ such that $U_\varepsilon[f] \subset V$. If there were a net $\{f_\alpha\colon \alpha \in A\}$ in $U_{\|f\|}[0]\setminus U_{\varepsilon/2}[f]$ which converged weak* to f then $\lim_\alpha \|f_\alpha\| = \|f\|$ and hence the net $\{\|f_\alpha\|^{-1}\|f\| f_\alpha\colon \alpha \in A\}$ would converge weak* to f as well. But this latter net is eventually outside $U_{\varepsilon/4}[f]$ (since $\| \|f_\alpha\|^{-1}\|f\|f_\alpha - f_\alpha\| \to 0$) while always lying inside $\|f\|\, S(X^*)$ (a place where the weak* and norm topologies agree by hypothesis). This is clearly impossible, whence

$$f \in \text{w*-int}(U_{\varepsilon/2}[f] \cup (X^*\setminus U_{\|f\|}[0])).$$

Let W denote the set $\text{w*-int}(U_{\varepsilon/2}[f] \cup (X^*\setminus U_{\|f\|}[0]))$. Then W is weak* open and $f \in W$. Hence there is a rational number $r > \|f\|$ satisfying

$$f \in r\|f\|^{-1}W \quad\text{and}\quad r\|f\|^{-1}\, U_{\varepsilon/2}[f] \subset U_\varepsilon[f].$$

Consequently

$$f \in r\|f\|^{-1}\, W \subset \text{w*-int}(U_\varepsilon[f] \cup (X^*\setminus U_r[0]))$$

$$\subset \text{w*-int}(V \cup (X^*\setminus U_r[0])).$$

Thus

$$f \in U_r[0] \cap \text{w*-int}[V \cup (X^*\setminus U_r[0])] \subset V'$$

and the proof is complete.

Suppose that X* is a dual space. Then $\mathcal{P}_t(X^*,\text{norm}) \subset \mathcal{P}_t(X^*,\text{weak*})$ since each weak* Borel set is a norm Borel set and each norm compact set is weak* compact. If X* has Property (**) then $\mathcal{B}(X^*,\text{norm}) = \mathcal{B}(X^*,\text{weak*})$ (4.4.2(b)) and from the discussion preceding Lemma 4.3.9 about tight measures, it is consistent with the Zermelo-Fraenkel axioms plus the axiom of choice to assume that each probability measure on $\mathcal{B}(X^*,\text{norm})$ is tight. In particular, it is consistent to assume that $P_t(X^*,\text{weak}^*) \subset P_t(X^*,\text{norm})$. Hence it is consistent with (ZF + AC) that $\mathcal{U}_t(\mathcal{B}(X^*,\text{norm})) = \mathcal{U}_t(\mathcal{B}(X^*,\text{weak}^*))$ when X* has Property (**). However it does not follow from this line of reasoning that these two σ-algebras are the same - only, as just mentioned, that it is consistent that they are the same. Similarly, when Theorem 4.3.11(c) is invoked, we cannot conclude on the basis of this reasoning that X* has the RNP if it has Property (**) - only that it is consistent that X* has the RNP if it has Property (**). This points up the danger of misreading 4.3.11 as discussed near definition 4.3.8. Nevertheless it is a result of Namioka and Phelps [1975] that X* (definitely) has the RNP under these hypotheses.

THEOREM 4.4.3. If X* has Property (**) then X* has the RNP. (In particular, if X* is locally uniformly convex then X* has the RNP.)

Proof. Let Y be a separable subspace of X. Then U(Y*) is weak* compact and metrizable, hence weak* separable. Let D be a countable weak* dense subset of U(Y*). If $f \in S(Y^*)$ there is a sequence in D, say $\{f_n : n = 1,2,\ldots\}$ which converges weak* to f. Since $\|f\| \leqq \liminf_n \|f_n\|$ it follows that $\lim_n \|f_n\| = 1$. Now let $g_n \in X^*$ be a norm preserving extension of $\|f_n\|^{-1} f_n$. Then $\{g_n : n = 1,2,\ldots\}$ is a sequence in $S(X^*)$ which must have a weak* cluster point $g \in U(X^*)$. If $\{g_{n_\alpha} : \alpha \in A\}$ is a subnet converging weak* to g then for each $x \in Y$,

$$g(x) = \lim_\alpha g_{n_\alpha}(x) = \lim_\alpha f_{n_\alpha}(x) = f(x)$$

since $\{f_n\}$ converges weak* to f. That is, g is a norm preserving extension of f. By Property (**) g belongs to the norm closure of $\{g_n\}_{n=1}^\infty$ and thus $g|_Y$ belongs to the norm closure of $\{g_n|_Y\}_{n=1}^\infty$. That is, f belongs to $\bar{D}$ for each $f \in S(Y^*)$. Consequently Y* itself is separable (since $\bar{D}$ is) and X* has the RNP by Theorem 4.4.1.

Both parts of 4.4.2 need some special conditions on the Banach space in question in order to establish the respective conclusions. (Indeed, Talagrand [1978] has shown that ℓ_∞ is a Banach space in which the weak and norm Borel σ-algebras differ. Moreover, if $X^* \equiv L^\infty[0,1] = L^1[0,1]^*$, the set

$$A \equiv \{\chi_{[0,t]}: 0 \leq t \leq 1\} \subset X^*$$

is discrete and closed in X^* in the norm topology while being homeomorphic to [0,1] in the weak* topology. Thus there is a subset of A in $\mathcal{B}(X^*,\text{norm})\setminus\mathcal{B}(X^*,\text{weak}^*)$ since each subset of A is in $\mathcal{B}(X^*,\text{norm})$.) What is perhaps surprising, in view of the aforementioned ℓ_∞ example, is the following result of L. Schwartz. (See Schwartz [1973], p. 162.)

Theorem 4.4.4. For any Banach space X, if $\mu \in \mathcal{P}_t(X,\text{weak})$ then there is a norm separable closed convex subset C of X with $\mu(C) = 1$. Consequently $\mathcal{U}_t(\mathcal{B}(X,\text{norm})) = \mathcal{U}_t(\mathcal{B}(X,\text{weak}))$.

Proof. First recall that if K is a weakly compact convex subset of a Banach space then K has the RNP (Theorem 3.6.1). Thus $(\hat{K},\text{weak}^*) \subset (X^{**},\text{weak}^*)$ is a weak* compact convex set with the RNP. If $\mu \in \mathcal{P}_t(K,\text{weak})$ and if $\hat{\mu}$ is its image under the embedding $x \to \hat{x}$ then $\hat{\mu} \in \mathcal{P}_t(\hat{K},\text{weak}^*)$. Thus there is a separable norm closed bounded convex set in $(\hat{K},\text{weak}^*)$, say $\hat{K}_1$, such that $\hat{\mu}(\hat{K}_1) = 1$ (Theorem 4.3.11(b)). Consequently there is a separable closed bounded convex subset K_1 of K such that $\mu(K_1) = 1$. It follows that whenever $\nu \in \mathcal{P}_t(X,\text{weak})$ there is a separable subspace L of X (hence $L \in \mathcal{B}(X,\text{weak})$) such that $\nu(L) = 1$. The remainder of the proof exactly mimics the proof of (b) $\Rightarrow$ (c) in Theorem 4.3.11, using the elementary fact that $\mathcal{B}(L,\text{weak}) = \mathcal{B}(L,\text{norm})$ since L is separable.

CHAPTER 5. ASPLUND SPACES AND THE GSP

Past chapters dealt with properties of sets with the RNP, many of which reflect the plethora of various types of "sharp" extreme points. In general one expects to find a duality between convexity notions on the one hand and smoothness conditions on the other and there is no exception in this case. The extraordinary results of Chapters 3 and 4 are here dualized to produce powerful smoothness criteria. The duality is especially pleasing when dealing with Banach spaces with the RNP and in fact, the early work done by E. Asplund [1968] (on spaces he called strong differentiability spaces and now known as Asplund spaces) preceded the relatively recent resurgence of interest in the Banach space version of the Radon-Nikodým theorem and its many consequences. Even more recently Stegall [1981] has managed to localize the duality (which heretofore was restricted to RNP Banach spaces and their duality counterparts, Asplund spaces). His work, based on techniques and results of Grothendieck [1955], appears in various forms throughout this chapter, though the approach we have taken leans heavily on Fitzpatrick's thesis [1980]. The localized theory remains somewhat unsatisfactory in the sense that it is not possible at this stage to tell whether or not a set has the "dual RN property" (called the GSP) by just knowing about the set itself; information about the Banach space in which it sits seems to be required. (This contrasts disappointingly with properties like the RNP and the KMP which at least in some equivalent reformulation only require information about the set itself. In any case, the added generality of GSP sets (over Asplund spaces) yields important information about the RNP and introduces an interesting class of sets in its own right. See also the brief remarks at the end of this chapter (§5.9) for some recent information concerning an approach due to O. I. Reynov.

§1. Subdifferentials and the Asplund property (AP).

The GSP may be described in terms of a differentiability criterion which generalizes those of Fréchet and Gateaux. That criterion has been studied in detail by Asplund and Rockafellar [1969].

DEFINITION 5.1.1. Let $D \subset X$ be a bounded set and $F: X \to \mathbb{R}$ any function. Then F is D-differentiable at $x \in X$ if there is an $f \in X^*$ such that

$$\lim_{t \to 0^+} \sup_{d \in D} \left| \frac{F(x+td) - F(x)}{t} - f(d) \right| = 0 .$$

If such an f exists it is called a D-gradient of F at x.

Observe that when the set of directions D is the unit ball of X, the condition that F be D-differentiable at x may be more classically described by saying that F is Fréchet differentiable at x while F is Gateaux differentiable at x in the direction d if and only if F is $\{d,-d\}$-differentiable at x. The next definitions formalize the objects of central interest in this chapter.

DEFINITION 5.1.2. (a) Let $D \subset X$ be a bounded set. Then D has the Asplund Property (AP) if and only if each convex continuous function $F: X \to \mathbb{R}$ is D-differentiable on a residual subset of X. (Recall: $W \subset X$ is residual if it contains a dense G_δ subset of X. F is convex if for each $t \in [0,1]$ and $x,y \in X$ we have $F(tx + (1-t)y) \leq tF(x) + (1-t)F(y)$.) (b) X is an Asplund space if each convex continuous $F: X \to \mathbb{R}$ is Fréchet differentiable on a dense G_δ set in X. (Equivalently: X is an Asplund space if and only if $U(X)$ has the GSP. See Corollary 5.6.10.)

Fitzpatrick [1980] used a definition quite different than the one in 5.1.2(a) in his treatment and Stegall [1981] introduced GSP sets with another definition still. (These alternatives will be considered in due course.) Fitzpatrick did show them all equivalent. It is reasonably well known that each continuous convex function on $\mathbb{R}$ is differentiable on a dense G_δ subset of $\mathbb{R}$. This fact may be parlayed into showing that finite sets have the AP. Much stronger results will soon follow.

Let $F: X \to \mathbb{R}$ be continuous and convex and denote by epi(F) the epigraph of F:

$$\operatorname{epi}(F) \equiv \{(x,r) \in X \times \mathbb{R}: F(x) \leq r\} .$$

Thus epi(F) is a closed convex set in $X \times \mathbb{R}$ with nonempty interior. Associate

$x \in X$ with the collection of elements of X^* whose graphs in $X \times \mathbb{R}$ are translate of supporting hyperplanes to epi(F) at $(x,F(x))$. (Such an association is always nonempty by the Hahn-Banach theorem.) This possibly multivalued mapping from X to X^* provides the main link in the duality theory below, and may be analytically described as follows: We describe first a natural 1-1 correspondence between the elements $\bar{f}$ of $(X \times \mathbb{R})^*$ such that $\bar{f}(0,1) = 1$ and those in X^* . Indeed, given an $\bar{f} \in (X \times \mathbb{R})^*$ with $\bar{f}(0,1) = 1$, for each $x \in X$ find the unique number $f(x)$ such that $\bar{f}(x,f(x)) = 0$. Then $f \in X^*$ as may easily be checked and the correspondence we have in mind associates f and $\bar{f}$. (The inverse of this correspondence maps $f \in X^*$ to $\bar{f} \in (X \times \mathbb{R})^*$ given by $\bar{f}(x,s) \equiv s - f(x)$.)

Suppose now that $x \in X$ and that $\bar{f}$ supports epi(F) at $(x,F(x))$. That is, suppose that $\bar{f}(x,F(x)) < \bar{f}(y,s)$ for each (y,s) in the interior of epi(F) and in particular then, $\bar{f}(x,F(x)) \leqq \bar{f}(y,F(y))$ for each $y \in X$. Note that $(x,F(x) + 1)$ is in the interior of epi(F) so that $\bar{f}(0,1) > 0$. Normalize $\bar{f}$ so that, in addition to its other properties, $\bar{f}(0,1) = 1$. Then for $f \in X^*$ corresponding to $\bar{f}$ as above and for each $y \in X$ we have

$$\begin{aligned} 0 &\leqq \bar{f}(y,F(y)) - \bar{f}(x,F(x)) - \bar{f}(y,f(y)) + \bar{f}(x,f(x)) \\ &= \bar{f}(0,F(y)-F(x)-f(y)+f(x)) \ . \end{aligned}$$

Since $\bar{f}(0,1) > 0$ it follows that

$$(1) \qquad f(y) - f(x) \leqq F(y) - F(x) \quad \text{for each } y \in X \ .$$

Conversely, if $f \in X^*$ satisfies (1) and $\bar{f} \in (X \times \mathbb{R})^*$ corresponds to f as above then $\bar{f}$ supports epi(F) at $(x,F(x))$ as may be verified by reversing the last few lines. Let us now formalize the key notions implicitly introduced thus far.

DEFINITION 5.1.3. Let $F: X \to \mathbb{R}$ be convex and continuous. Then the subdifferential of F , ∂F , is the (possibly multivalued) map on X to X^* defined at $x \in X$ by:

$$\partial F(x) \equiv \{f \in X^*: F(y) - F(x) \geqq f(y) - f(x) \text{ for each } y \in X\} \ .$$

Observe that in the notation we established in the previous paragraph, when $F: X \to \mathbb{R}$ is convex and continuous and $x \in X$, then $f \in \partial F(x)$ if and only if $\bar{f}$ supports epi(F) at $(x,F(x))$. In particular, $\partial F(x) \neq \emptyset$ for each $x \in X$.

When $G: \mathbb{R} \to \mathbb{R}$ is convex and continuous its left and right sided derivatives exist at each $t \in \mathbb{R}$ ($G^{(\ell)}(t)$ and $G^{(r)}(t)$, respectively), each is monotone non-decreasing, and $G^{(r)}(t) \leq G^{(\ell)}(s)$ whenever $t < s$. Moreover, if $F: X \to \mathbb{R}$ is convex and continuous and both $f \in \partial F(x)$ and $g \in \partial F(y)$ then

$$f(y - x) \leq F(y) - F(x) \leq g(y - x)$$

whence $(g - f)(y - x) \geq 0$ whenever $x,y \in X$, $f \in \partial F(x)$, $g \in \partial F(y)$. Thus subdifferentials of continuous convex functions are special cases of monotone operators as defined below.

<u>DEFINITION</u> 5.1.4. Let $T: X \to X^*$ be a multivalued mapping.

(a) T is a <u>monotone operator</u> on X if $(g - f)(y - x) \geq 0$ whenever $f \in T(x)$, $g \in T(y)$, $x,y \in X$.

(b) T is <u>maximal monotone</u> if it is monotone and if its graph $Gr(T) \equiv \{(x,f) \in X \times X^*: x \in X, f \in T(x)\}$ is not properly contained in the graph of any other monotone operator on X .

(c) T is <u>locally bounded</u> at $x \in X$ if there is a neighborhood V of x such that $T(V) \equiv \cup T(y)$ (where the union is over y in V) is bounded in X^* .

(d) The <u>domain</u> of a monotone operator T , $D(T)$, is the set $\{x \in X: T(x) \neq \emptyset\}$.

Parts of the next proposition are due to Minty [1964].

<u>PROPOSITION</u> 5.1.5. Let $F: X \to \mathbb{R}$ be continuous and convex. Then ∂F is maximal monotone with domain X .

Proof. The previous discussion established that ∂F is monotone with domain X and it remains to show that ∂F is maximal monotone. A simple translation shows that, if it were not, there would be a continuous convex function $G: X \to \mathbb{R}$ with $G(0) = 0$, a point $y \neq 0$ in X and an element $f \in X^*$ such that $G(y) < f(y)$ (hence $f \notin \partial G(0)$) and yet the operator whose graph is $Gr(\partial G) \cup \{(0,f)\}$ is still monotone. Let $\tilde{G}(t) \equiv G(ty)$ and $\tilde{f}(t) \equiv f(ty)$ for $t \in \mathbb{R}$ so that $\tilde{G}: \mathbb{R} \to \mathbb{R}$ is convex and continuous, $\tilde{G}(0) = 0$ and $\tilde{G}(1) < \tilde{f}(1)$. Since the right handed derivative

$\tilde{G}^{(r)}(t)$ determines a support line for $\mathrm{epi}(\tilde{G})$ at $(t,\tilde{G}(t))$, not only is $\tilde{G}^{(r)}(0) <$ $\tilde{f}(1)$ (since the contrary forces $\tilde{G}(1) \geqq \tilde{f}(1)$) but, for $t_o > 0$ sufficiently close to 0, $\tilde{G}^{(r)}(t_o) < \tilde{f}(1)$ as well. Fix any such $t_o > 0$ and let $\tilde{g}(t) \equiv \tilde{G}^{(r)}(t_o)t$ for $t \in \mathbb{R}$. Then $\tilde{g} \in \partial\tilde{G}(t_o)$. In $X \times \mathbb{R}$ the line

$$L \equiv \{(ty,s) \in X \times \mathbb{R}:\ s,t \in \mathbb{R},\ s - \tilde{g}(t) = G(t_o y) - \tilde{g}(t_o)\}$$

lies everywhere below $\mathrm{epi}(G)$ and contains the point $(t_o y, G(t_o y))$ of $\mathrm{epi}(G)$. Thus there is an extension of the linear functional $\hat{g}$ defined on $\{ty: t \in \mathbb{R}\} \times \mathbb{R}$ given by $\hat{g}(ty,s) \equiv s - \tilde{g}(t)$ to an element $\bar{g}$ of $(X \times \mathbb{R})^*$ whose hyperplane $\{(x,s): \bar{g}(x,s) = G(t_o y) - \tilde{g}(t_o)\}$ still supports $\mathrm{epi}(G)$ at $(t_o y, G(t_o y))$. Let $g \in X^*$ correspond to $\bar{g}$ as usual: $\bar{g}(x,g(x)) = 0$ for all $x \in X$. Then as per the discussion prior to Definition 5.1.3, $g \in \partial G(t_o y)$. Note that $g(t_o y) = \tilde{g}(t_o)$ (since $\hat{g}(t_o y, g(t_o y)) = 0$) and hence

$$\begin{aligned}(g - f)(t_o y - 0) &= \tilde{g}(t_o) - t_o f(y)\\ &= (\tilde{G}^{(r)}(t_o) - \tilde{f}(1))t_o < 0\,.\end{aligned}$$

This inequality contradicts the monotonicity assumption of the supposedly monotone extension of F, and the proof is complete.

§2. A characterization of the Asplund property: the Stegall property (SP).

For any bounded set $D \subset X$ let $\mathrm{aco}(D)$ denote the set $\mathrm{co}(D \cup -D)$, $\overline{\mathrm{aco}}(D)$ its norm closure, and ρ_D the seminorm on X^* given by

$$\begin{aligned}\rho_D(f) &\equiv \sup\{|f(x)|:\ x \in D\} \qquad (f \in X^*)\\ &= \sup\{f(x):\ x \in \mathrm{aco}(D)\}\\ &= \sup\{f(x):\ x \in \overline{\mathrm{aco}}(D)\}\,.\end{aligned}$$

One of the principal results of this section is the following.

THEOREM 5.2.1. (Fitzpatrick). A bounded set $D \subset X$ has the Asplund Property if and only if

(*) The seminormed space (X^*, ρ_A) is separable whenever $A \subset D$ is countable.

A bounded set $D \subset X$ will henceforth be said to have the Stegall Property (SP) if it satisfies (*) in the theorem above. (See Theorems 4.4.1(c) and 4.2.13(c) for the genesis of this terminology.) The importance of Theorem 5.2.1 lies in the ease with which the SP may be applied in a variety of situations as shall presently be seen; in fact, Fitzpatrick [1980] actually used a minor variant of the SP (see 5.2.2.(b)) as his definition of the CSP. The proof of 5.2.1 will take some time and along the way several important results in their own right will emerge. It should be emphasized that once the proof of the equivalence of the AP, the SP and the GSP is complete, intermediate results which include hypotheses about sets having the SP could be rephrased entirely in terms of their being sets with the GSP. Some elementary observations are gathered in the next proposition for convenient reference.

PROPOSITION 5.2.2. Let $D \subset X$ be a bounded set.

(a) If $D_1 \subset X$ then $\rho_D = \rho_{D_1}$ if and only if $\overline{\text{aco}}(D) = \overline{\text{aco}}(D_1)$.

(b) D has the SP if and only if (X^*,ρ_A) is separable for each norm separable set $A \subset D$.

(c) The SP is an hereditary condition. That is, if D has the SP so does each subset of D .

(d) D has the SP if and only if $\overline{\text{aco}}(D)$ has the SP.

(e) If X and Y are Banach spaces and $T: X \to Y$ is a continuous linear operator then the adjoint $T^*: (Y^*,\rho_{T(D)}) \to (X^*,\rho_D)$ is an isometry onto its range.

Proof.

(a) This follows immediately from the definition in conjunction with the separation theorem.

(b) If $A \subset D$ is separable and $B \subset A$ is a countable dense subset of A then $\overline{\text{aco}}(B) = \overline{\text{aco}}(A)$. Hence from part (a), (X,ρ_B) is separable if and only if (X,ρ_A) is separable.

(c) is obvious.

(d) follows directly from parts (a) and (c).

(e) If $f \in Y^*$ then

$$\rho_D(T^*(f)) = \sup\{|T^*(f(x))|: x \in D\} = \sup\{|f(T(x))|: x \in D\} = \rho_{T(D)}(f) .$$

The SP is directly linked by duality to the RNP as the next result indicates.

THEOREM 5.2.3. Let $T: X \to Y$ be a continuous linear operator. Then $T(U(X))$ has the SP if and only if $T^*(U(Y^*))$ has the RNP.

Proof. Note that if $B \subset T(U(X))$ is countable then there is a countable set $A \subset U(X)$ with $T(A) = B$. Each statement below is easily seen to be equivalent to the next.

$T(U(X))$ has the SP.

(Y^*, ρ_B) is separable for each countable $B \subset T(U(X))$.

$(Y^*, \rho_{T(A)})$ is separable for each countable $A \subset U(X)$.

$(T^*(Y^*), \rho_A)$ is separable for each countable $A \subset U(X)$. (See 5.2.2(e).)

$(T^*(U(Y^*)), \rho_A)$ is separable for each countable $A \subset U(X)$.

$(T^*(U(Y^*)), \rho_{U(Z)})$ is separable for each separable subspace Z of X.

$T^*(U(Y^*))|_Z$ is separable for each separable subspace Z of X.

$T^*(U(Y^*))$ has the RNP. (See Theorem 4.2.13.)

Theorem 5.2.3 has consequently been proved.

The particular case of Theorem 5.2.3 in which $X = Y$ and T is the identity on X shows that $U(X)$ has the SP if and only if X^* has the RNP. Combining this with the anticipated Theorem 5.2.1 (and Corollary 5.6.10) yields (slightly prematurely) the important result that X is an Asplund space if and only if X^* has the RNP. Thus we are naturally led to seek other local duality connections between the AP (or, again modulo 5.2.1, the SP) and RNP results for subsets of dual spaces. An excellent example of such a result appears next. (See Theorem 4.4.1(f).)

THEOREM 5.2.4. (Fitzpatrick). A bounded set $D \subset X$ has the SP if and only if each bounded subset of X^* has weak* slices of arbitrarily small ρ_D-diameter.

Proof. As in the previous chapter, let $\ell_1(D)$ denote the Banach space of (countably nonzero) absolutely summable real valued functions (w_d) on D with norm $\|(w_d)\| \equiv \sum_{d \in D} |w_d|$, and let $T: \ell_1(D) \to X$ be the continuous linear operator defined at (w_d) by

$$T((w_d)) \equiv \sum_{d\in D} w_d d .$$

If $d \in D$, the element of $\ell_1(D)$ which is 1 at d and 0 elsewhere is mapped by T to d, and since $U(\ell_1(D))$ is the closed convex hull of such points (and their negatives) it follows that

$$aco(D) \subset T(U(\ell_1(D))) \subset \overline{aco}(D) .$$

Suppose that D has the SP and that $\varepsilon > 0$. Then $\overline{aco}(D)$ also has the SP (Proposition 5.2.2(d)) and hence so does its subset $T(U(\ell_1(D)))$. Let C be a bounded subset of X^* and suppose that $K \equiv w^*cl(co(C))$. Theorem 5.2.3 shows that the weak* compact convex set $T^*(K)$ has the RNP and Theorem 4.2.13(g) applied to $T^*(C) \subset T^*(K)$ yields a weak* slice of $T^*(C)$, say $S(T^*(C),w,\alpha)$ for some $w \in \ell_1(D)$ and $\alpha > 0$, of norm diameter less than ε. Now T^* is an isometry between $(X^*,\rho_{T(U(\ell_1(D)))})$ and $(T^*(X^*),\rho_{U(\ell_1(D))})$ by Proposition 5.2.2(e). Since the closure of $T(U(\ell_1(D)))$ coincides with $\overline{aco}(D)$ it follows that $\rho_{T(U(\ell_1(D)))} = \rho_{\overline{aco}(D)} = \rho_D$ and hence that T^* is an isometry between (X^*,ρ_D) and $(T^*(X^*),\text{norm})$. Because T^* maps $S(C,T(w),\alpha)$ into $S(T^*(C),w,\alpha)$, a set of norm diameter less than ε, it follows that $S(C,T(w),\alpha)$ has ρ_D-diameter less than ε as was to be shown.

Conversely, if D does not have the SP then (for T as above) $T^*(U(X^*))$ lacks the RNP by Theorem 5.2.3 since $T(U(\ell_1(D))) \supset D$. Theorem 4.2.13(g) applied to $T^*(U(X^*))$ states that there is an $\varepsilon > 0$ and a nonempty subset C of $T^*(U(X^*))$ all of whose weak* slices have diameter greater than ε. Let $K \equiv w^*cl(co(C))$. Evidently each weak* slice of K has diameter exceeding ε as well.

Since T^* is weak*-weak* continuous, $T^*(U(X^*))$ is a weak* compact convex set containing K. Thus the set

$$W \equiv T^{*-1}(K) \cap U(X^*) \subset X^*$$

is a weak* compact convex set and $T^*(W) = K$. A straightforward Zorn's lemma argument establishes that there is a minimal weak* compact convex set $V \subset W$ with $T^*(V) = K$. The proof will be completed by showing that each weak* slice of V has ρ_D-diameter exceeding ε. Let S be a weak* slice of V. Then $V \setminus S$ is a proper weak* compact convex subset of V and the minimality of V guarantees that $T^*(V \setminus S)$ is a proper weak* compact convex subset of K. There is thus a weak* slice S' of K which misses $T^*(V \setminus S)$ and hence $T^*(S) \supset S'$. The slice S' has diameter greater than ε by choice of K and hence the diameter of $T^*(S)$ is greater than ε as well. In the first half of the proof it was established that T^* is an isometry between (X^*, ρ_D) and $(T^*(X^*), \text{norm})$. Hence the ρ_D-diameter of S exceeds ε and the proof of Theorem 5.2.4 is complete.

The geometric characterization of the SP in Theorem 5.2.4 sets the stage for making the key connections between the SP and the AP. In fact, the proof of Theorem 5.2.1 will follow as an elementary consequence of the next theorem.

THEOREM 5.2.5. (Fitzpatrick). Let K be a bounded, absolutely convex (i.e., $K = \text{aco}(K)$) subset of X. (K is not assumed to be closed.) Then the following are equivalent.

(a) K has the SP.

(b) K has the AP.

(c) Each continuous convex function on X is K-differentiable at some point of X.

The proof of Theorem 5.2.5 will be broken into a series of four lemmas, the first of which provides information in addition to that already obtained in Proposition 5.1.5 about subdifferentials of continuous convex functions.

LEMMA 5.2.6. Let $F: X \to \mathbb{R}$ be convex and continuous. Then

(a) ∂F is locally bounded at each point of X (see 5.1.4(c));

(b) $\partial F(x)$ is a weak* compact convex set for each $x \in X$;

(c) the map $x \to \partial F(x)$ is norm-weak* upper semicontinuous.

(That is, for each $x \in X$ and weak* open $V \supset \partial F(x)$ there is a norm open set U with $x \in U$ such that $\partial F(y) \subset V$ for $y \in U$.)

(d) F is locally Lipschitz. That is, for each $x \in X$ there are a $\delta > 0$ and an $M > 0$ such that

$$|F(x+y) - F(x)| < M\,\|y\| \qquad \text{if } \|y\| < \delta .$$

Proof.

(a) If $x \in X$ pick $\varepsilon > 0$ such that $|F(x) - F(y)| < 1$ when $y \in U_\varepsilon(x)$. Suppose that $z \in U_{\varepsilon/2}(x)$, $f \in \partial F(z)$ and that $a \in X$, $\|a\| < \varepsilon/2$. Then

$$f(a) = f(z+a) - f(z) \leq F(z+a) - F(z) \leq |F(z+a) - F(x)| + |F(x) - F(z)| < 2$$

since $\|z + a - x\| < \varepsilon$. It follows that $\|f\| < 2(2/\varepsilon) = 4/\varepsilon$ whenever $f \in \partial F(z)$ for $\|z - x\| < \varepsilon/2$.

(b) It follows from part (a) that $\partial F(x)$ is a bounded set for each x. If (f_α) is a weak* convergent net in $\partial F(x)$ with limit $f \in X^*$ then for each $y \in X$

$$f(y) - f(x) = \lim_\alpha (f_\alpha(y) - f_\alpha(x)) \leq F(y) - F(x)$$

whence $f \in \partial F(x)$. Thus $\partial F(x)$ is weak* compact. The convexity of this set is also straightforward.

(c) Suppose temporarily that ∂F were not norm-weak* upper semicontinuous at x. Then for some weak* open set $V \supset \partial F(x)$ there is a sequence $\{y_n : n = 1,2,\ldots\}$ in X with $\lim_n \|y_n - x\| = 0$ and an $f_n \in \partial F(y_n)$ such that $f_n \notin V$ for each n. By the local boundedness of ∂F, $\{f_n : n = 1,2,\ldots\}$ is bounded and hence it has a weak* convergent subnet $(f_{n_\alpha})_\alpha$ with limit $f \in X^*$. Then $f \notin V$, and in particular, $f \notin \partial F(x)$. But for any $z \in X$ we have

$$\begin{aligned}
f(z) - f(x) &= \lim_\alpha (f_{n_\alpha}(z) - f_{n_\alpha}(x)) \\
&\leq \limsup_\alpha (f_{n_\alpha}(z) - f_{n_\alpha}(y_{n_\alpha})) + \limsup_\alpha (f_{n_\alpha}(y_{n_\alpha}) - f_{n_\alpha}(x)) \\
&\leq \limsup_\alpha (F(z) - F(y_{n_\alpha})) + \limsup_\alpha (\sup_n \|f_n\|)\, \|y_{n_\alpha} - x\| \\
&= F(z) - F(x) .
\end{aligned}$$

Thus $f \in \partial F(x)$, establishing a contradiction and thus the desired conclusion.

(d) Since ∂F is locally bounded at x by part (a) there are $\delta > 0$ and $M > 0$ such that if $\|y\| < \delta$ and $g \in \partial F(x+y)$ then $\|g\| < M$. Then for any such y and for any $f \in \partial F(x)$,

$$f(y) \leq F(x+y) - F(x) \leq g(y) .$$

Consequently $-M\|y\| \leq F(x+y) - F(x) \leq M\|y\|$ as was to be shown.

The definition below helps make the statement of the lemma following it clearer.

DEFINITION 5.2.7. Let $T: X \to X^*$ be a monotone operator and suppose that $D \subset X$ is a bounded set. If x is in the domain of T then T is D-continuous at x if whenever $\{x_n : n = 1,2,\dots\}$ is a sequence in X with $\lim_n \|x_n - x\| = 0$, $f_n \in T(x_n)$ for each n and $f \in T(x)$, then $\lim_n \rho_D(f_n - f) = 0$.

LEMMA 5.2.8. Suppose that $D \subset X$ is bounded and that D has the SP. If $F: X \to \mathbb{R}$ is convex and continuous then ∂F is D-continuous on a dense G_δ subset of X.

Proof. For each n let O_n be the union of all open sets V in X such that the ρ_D-diameter of $\partial F(V) \equiv \bigcup_{x \in V} F(x)$ is less than $1/n$. Then each O_n is open and it is straightforward to check that the set of points of D-continuity of F is exactly $\bigcap_n O_n$. Thus the lemma will be proved once it is shown that each O_n is dense in X.

Pick any positive integer n, any $x \in X$, and any $\varepsilon > 0$. Because ∂F is locally bounded there is a $\delta > 0$, $\delta \leq \varepsilon$, such that $C \equiv \partial F(U_\delta(x))$ is a bounded set. Theorem 5.2.4 applied to C guarantees the existence of a weak* slice $S \equiv S(C,z,\alpha)$ of C whose ρ_D-diameter is less than $1/n$. Let $f \in S$. Then by construction there is a $y \in U_\delta(x)$ such that $f \in \partial F(y)$. Let $\beta > 0$ be so small that the point $w \equiv y + \beta z$ is in $U_\delta(x)$. We will show that $w \in O_n$. If $g_w \in \partial F$ then by the monotonicity of ∂F we have $0 \leq (g_w - f)(w - y) = \beta(g_w - f)(z)$ and thu

(2) $$g_w(z) \geq f(z) > M(C,\hat{z}) - \alpha .$$

It follows from the norm-weak* upper semicontinuity of the subdifferential map (Lemma 5.2.6(c)) that there is a norm open set W with $w \in W \subset U_\delta(x)$ and

(3) $$h(z) > M(C,\hat{z}) - \alpha \quad \text{whenever} \quad h \in \partial F(W) .$$

Since $W \subset U_\delta(x)$ it follows that $\partial F(W) \subset C$, and even more, from (3), that $\partial F(W) \subset$ But the ρ_D-diameter of S is less than $1/n$; hence the ρ_D-diameter of $\partial F(W)$ is le than $1/n$. That is, $W \subset O_n$. Since $w \in U_\delta(x)$ and $\delta \leq \varepsilon$, $\|w - x\| < \varepsilon$. Thus O_n is dense as was to be shown.

The next result (and its notation) will be used in several places in this chapter.

LEMMA 5.2.9. Let C be a bounded set in X^* and let $\lambda_C(x) \equiv M(C,\hat{x})$ for each $x \in X$. Then $\lambda_C = \lambda_{w^*cl(co(C))}$ is convex and continuous. Moreover for $g \in C$ and $y \in X$ the following are equivalent:

(a) $g(y) = \lambda_C(y)$;

(b) $g \in \partial\lambda_C(y)$.

Proof. It is easy to check that λ_C is both convex and continuous on X. If for some $y \in X$ and $g \in C$, $g \in \partial\lambda_C(y)$ then $\lambda_C(z) - \lambda_C(y) \geqq g(z) - g(y)$ for each $z \in X$. Since $\lambda_C(0) = 0$ it follows that $g(y) \geqq \lambda_C(y)$. But $g \in C$ so that automatically $g(y) \leqq M(C,\hat{y}) \equiv \lambda_C(y)$ and thus $g(y) = \lambda_C(y)$. Conversely, if $g \in C$ and $g(y) = \lambda_C(y)$ for some $y \in X$ then for each $z \in X$

$$\lambda_C(z) - \lambda_C(y) = M(C,\hat{z}) - g(y) \geqq g(z) - g(y)$$

and thus $g \in \partial\lambda_C(y)$.

The final lemma of this sequence ties together the D-continuity of ∂F with the D-differentiability of F .

LEMMA 5.2.10. Let K be a bounded convex subset of X and suppose that $F: X \to \mathbb{R}$ is convex and continuous. Then

(a) If ∂F is K-continuous at x then F is K-differentiable at x .

(b) Let C be a weak* compact convex set in X^* and assume that K is absolutely convex and bounded. Then λ_C (see 5.2.9 for this notation) is K-differentiable at x if and only if $\lim_n \rho_K(f_n - f) = 0$ whenever $f \in \partial\lambda_C(x)$, $f_n \in C$ and $\lim_n f_n(x) = \lambda_C(x)$.

Proof.

(a) Pick any $f \in \partial F(x)$. If f were not a K-gradient of F at x then there would be a decreasing sequence $\{t_n: n = 1,2,\dots\}$ of positive numbers with $\lim_n t_n = 0$ and a sequence $\{k_n: n = 1,2,\dots\}$ in K such that for some $\varepsilon > 0$

$$\left|\frac{F(x + t_n k_n) - F(x)}{t_n} - f(k_n)\right| \geqq \varepsilon \quad \text{for each } n . \tag{4}$$

Let $x_n \equiv x + t_n k_n$ so that $\lim_n \|x_n - x\| = 0$ and suppose that $f_n \in \partial F(x_n)$. The

$$\rho_K(f_n - f) \geq f_n(k_n) - f(k_n) = \frac{f_n(x + t_n k_n) - f_n(x)}{t_n} - f(k_n)$$

(5) $$\geq \frac{F(x + t_n k_n) - F(x)}{t_n} - f(k_n)$$

$$= \frac{F(x + t_n k_n) - F(x) - (f(x + t_n k_n) - f(x))}{t_n} \geq 0 \ .$$

It follows from (4) and (5) that $\rho_K(f_n - f) \geq \varepsilon$ for each n which contradicts the K-continuity of ∂F at x.

(b) Suppose that $\lim_n \rho_K(f_n - f) = 0$ whenever $f \in \partial\lambda_C(x)$, $f_n \in C$ and $\lim_n f_n(x)$ $\lambda_C(x)$. As in part (a) (see (5) and its aftermath) if $f \in \partial\lambda_C(x)$ and f is not a K-gradient of λ_C at x then there is a sequence $\{t_n : n = 1,2,\ldots\}$ converging do to 0 and a sequence $\{k_n : n = 1,2,\ldots\}$ in K such that for some $\varepsilon > 0$

$$\frac{\lambda_C(x + t_n k_n) - \lambda_C(x)}{t_n} - f(k_n) \geq \varepsilon \qquad \text{for each } n \ .$$

(Note that no absolute values are needed since $f \in \partial\lambda_C(x)$.) Pick any $f_n \in C$ suc that $f_n(x + t_n k_n) = \lambda_C(x + t_n k_n)$. Then $f_n \in \partial\lambda_C(x + t_n k_n)$ by Lemma 5.2.9 and

(6) $$\begin{aligned}\lambda_C(x) = M(C,\hat{x}) &\geq f_n(x) = f_n(x + t_n k_n) - f_n(t_n k_n) \\ &\geq \lambda_C(x + t_n k_n) - t_n (\sup_{g \in C} \|g\|)(\sup_{k \in K} \|k\|) \ .\end{aligned}$$

But $\lim_n \lambda_C(x + t_n k_n) = \lambda_C(x) = M(C,\hat{x})$ since λ_C is continuous. Hence from (6), $\lambda_C(x) = \lim_n f_n(x)$. By hypothesis, then, $\lim_n \rho_K(f_n - f) = 0$. On the other hand, again as in part (a)

$$\rho_K(f_n - f) \geq f_n(k_n) - f(k_n) \geq \frac{\lambda_C(x + t_n k_n) - \lambda_C(x)}{t_n} - f(k_n) \geq \varepsilon$$

for each n which is impossible. Thus f must be a K-gradient of λ_C at x.

Conversely, suppose λ_C is K-differentiable at x with gradient g and that the sequence $\{f_n : n = 1,2,\ldots\}$ in C satisfies $\lim_n f_n(x) = \lambda_C(x)$. Let $f \in \partial\lambda_C$

Observe first that $f(k) = g(k)$ for each $k \in K$. (Indeed,

$$f(x) = f(2x) - f(x) \leqq \lambda_C(2x) - \lambda_C(x) = \lambda_C(x) = \lambda_C(x) - \lambda_C(0)$$
$$\leqq f(x) - f(0) = f(x) .$$

Thus $f(x) = \lambda_C(x)$. Hence

$$g(k) = \lim_{t\to 0^+} \frac{\lambda_C(x+tk) - \lambda_C(x)}{t} \geqq \limsup_{t\to 0^+} \frac{f(x+tk) - f(x)}{t} = f(k) .$$

Since $K = \mathrm{aco}(K)$, $g(-k) \geqq f(-k)$ as well and thus $f = g$ on K. Note also that $f \in C$ by the separation theorem since $f(y) \leqq \lambda_C(y) - \lambda_C(x) + f(x) = \lambda_C(y)$ for each $y \in X$. Consequently $f \in \partial\lambda_C(x)$ is a K-gradient of λ_C at x and f belongs to C. Thus given $\varepsilon > 0$ it is possible to find a $t_o > 0$ such that

$$0 \leqq \frac{\lambda_C(x + t_o k) - \lambda_C(x)}{t_o} - f(k) < \varepsilon/2 \quad \text{for each } k \in K .$$

Pick N so that $|f_n(x) - f(x)|$ $(= |f_n(x) - \lambda_C(x)|) < t_o(\varepsilon/2)$ whenever $n \geqq N$. Then for each $k \in K$ and each $n \geqq N$

$$f_n(k) - f(k) \leqq f_n(k) - \frac{\lambda_C(x + t_o k) - \lambda_C(x)}{t_o} + \frac{\varepsilon}{2}$$
$$= \frac{f_n(x + t_o k) - \lambda_C(x + t_o k)}{t_o} + \frac{\lambda_C(x) - f_n(x)}{t_o} + \frac{\varepsilon}{2}$$
$$\leqq 0 + t_o\varepsilon/(2t_o) + \varepsilon/2 = \varepsilon .$$

That is, $f_n(k) - f(k) \leqq \varepsilon$ for each $k \in K$. But $K = \mathrm{aco}(K)$. It follows that $|f_n(k) - f(k)| \leqq \varepsilon$ for each $k \in K$ and hence that $\rho_K(f_n - f) \leqq \varepsilon$ whenever $n \geqq N$ Thus $\lim_n \rho_K(f_n - f) = 0$. The proof of Lemma 5.2.10 is complete.

The proof of Theorem 5.2.5 is now an easy matter.

Proof of 5.2.5.

(a) $\Longrightarrow$ (b) If $F\colon X \to \mathbb{R}$ is convex and continuous then ∂F is K-continuous on a dense G_δ set in X by Lemma 5.2.8 which implies, by Lemma 5.2.10(a), that F is

K-differentiable on a dense G_δ set in X .

(b) $\Longrightarrow$ (c) is obvious.

(c) $\Longrightarrow$ (a) If C is weak* compact and convex in X^* let $\lambda_C: X \to \mathbb{R}$ be defined as usual: $\lambda_C(x) \equiv M(C,\hat{x})$ for $x \in X$. If λ_C is K-differentiable at x then Lemma 5.2.10(b) may be rephrased to state that x determines weak* slices of C of arbitrarily small ρ_K-diameter. Now suppose that $A \subset X^*$ is bounded and set $C \equiv$ w*cl(co(A)) . Then any weak* slice of C must contain points of A and hence each bounded subset of X^* has weak* slices of arbitrarily small ρ_K-diameter. Theorem 5.2.4 then guarantees that K has the SP.

The fact that the SP and the AP are equivalent (Theorem 5.2.1) may now be established.

Proof of 5.2.1. Suppose first that D has the SP. Then $\overline{\text{aco}}(D)$ has the SP by Proposition 5.2.2.(d). By Theorem 5.2.5, $\overline{\text{aco}}(D)$ has the AP and consequently so does D .

Conversely, if D has the AP let $A \subset D$ be a countable subset, say $A \equiv \{a_n\}_{n=1}^\infty$ and let $K \equiv \overline{\text{aco}}(\{n^{-1}a_n\}_{n=1}^\infty)$. We will prove that A has the SP. (In order to check that the compact set K has the AP just observe that the restriction map from (X^*,ρ_K) into $(C(K)$, sup norm) is an isometry. Since K is compact and metrizable $C(K)$ is separable. Hence so is (X^*,ρ_K).)

Let $F: X \to \mathbb{R}$ be convex and continuous and define $\tilde{F}: X \to \mathbb{R}$ by $\tilde{F}(x) \equiv F(-x)$ for $x \in X$. Then there are residual sets G_1 and G_2 in X such that F and $\tilde{F}$ are A-differentiable on G_1 and G_2 respectively, and there is a residual set G_3 of X such that F is K-differentiable on G_3 . Let $G \equiv G_1 \cap (-G_2) \cap G_3$. Then G is residual. If $x \in G$ there are $f,g,h \in X^*$ such that

$$(7)\qquad \limsup_{t\to 0^+ a\in A} \left|\frac{F(x+ta) - F(x)}{t} - f(a)\right| = 0 ;$$

$$(8)\qquad \limsup_{t\to 0^+ a\in A} \left|\frac{F(x-ta) - F(x)}{t} - g(a)\right| = 0 \qquad (\text{since } \tilde{F}(-x+ta) = F(x-ta)) ; \text{ and}$$

$$(9)\qquad \limsup_{t\to 0^+ k\in K} \left|\frac{F(x+tk) - F(x)}{t} - h(k)\right| = 0 .$$

But (9) implies that for each $a \in A \cup -A$ we have

$$h(a) = \lim_{t\to 0^+} \frac{F(x + ta) - F(x)}{t} .$$

It follows that f in (7) may be replaced by h , and g in (8) may be replaced by $-h$. That is,

$$(10) \qquad \lim_{t\to 0^+} \sup_{a\in A\cup -A} \left| \frac{F(x + ta) - F(x)}{t} - h(a) \right| = 0 .$$

It is routine that (10) still holds when $A \cup -A$ is replaced by its convex hull, $\mathrm{aco}(A)$, and thus F is $\mathrm{aco}(A)$-differentiable on G . That is, we have just shown that $\mathrm{aco}(A)$ has the AP, and Theorem 5.2.5 states that $\mathrm{aco}(A)$ must therefore satisfy the SP. Hence A has the SP as well. The proof that D has the SP now reduces to the observation that a set has the SP if and only if each of its countable subsets has the SP.

As a matter of record we list below some of the results obtained thus far in terms of the recently justified terminology.

THEOREM 5.2.11. Let D be a bounded set in X . Then D has the AP if and only if any of the following hold:

(a) Each countable subset of D has the AP.

(b) (X^*,ρ_A) is separable for each separable subset A of D .

(c) Each subset of D has the AP.

(d) $\overline{\mathrm{aco}}(D)$ has the AP.

(e) Each bounded subset of X^* has weak* slices of arbitrarily small ρ_D-diameter.

(f) ∂F is D-continuous on a dense G_δ subset of X for each convex continuous function F on X .

Moreover, for any Banach space Y ,

(g) If $T: X \to Y$ is a continuous linear operator then $T(U(X))$ has the AP if and only if $T^*(U(Y^*))$ has the RNP.

Proof. (a) is obvious and (b)-(g) are translations of results appearing with mildly different terminology earlier in this section.

<u>THEOREM</u> 5.2.12. Let X be a Banach space. Then X is an Asplund space if and only if X^* has the RNP.

Proof. Apply Theorem 5.2.11(g) with $Y \equiv X$ and $T \equiv$ identity on X. The only point requiring clarification is that if $F: X \to \mathbb{R}$ is convex and continuous and F is known to be $U(X)$-differentiable on a dense set then that dense set is automatical a G_δ set. (This is Corollary 5.6.10, whose proof may be read with no further background.)

A few remarks are in order concerning the level of generality of the results of this section. For example, parts of 5.2.6(a) and (c) may be stated for (maximal) mo tone operators (with some technical assumptions about the domains). More importantl Lemma 5.2.8 is a special case of the fact that monotone operators are K-continuous o residual subsets of the interior of their domains (K bounded). Lemma 5.2.10, too, has been stripped down since the converse of 5.2.10(a) is also true. The following theorem is an easy consequence of these results in their added generality: Let K be a bounded subset of X with $\overline{\text{span}}(K) = X$. If K has the AP and $T: X \to X^*$ is monotone then T is singlevalued on a residual subset of the interior of its domain (Indeed, by the generalized version of 5.2.8 there is a residual subset of the interi of T's domain on which T is K-continuous. But the ρ_K-topology on X^* is Hausdo by choice of X, and thus T must be singlevalued at least on that residual set where it is K-continuous.)

§3. <u>Operator results and factorization.</u>

Davis, Figiel, Johnson, and Pełczynski [1974] proved that whenever $T: X \to Y$ a weakly compact linear operator than there are a reflexive Banach space R and continuous linear operators $T_1: X \to R$ and $T_2: R \to Y$ such that $T = T_2 \circ T_1$. (That is, weakly compact operators <u>factor</u> through reflexive spaces.) Moreover, if $W \subset X$ is a weakly compact set then there are a reflexive space R_1 and a continuous 1-1 linear operator $T': R_1 \to X$ such that $T'(U(R_1)) \supset W$. In this section analogues o these results are established for sets with the AP.

In the sequel Theorem 5.2.1 will be used so frequently that its references will be understood without specific invocation. Some preliminaries are necessary.

LEMMA 5.3.1. Let $T: X \to Y$ be a continuous linear operator and suppose that $D \subset X$ is a bounded set.

(a) If D has the AP then so does $T(D) \subset Y$.

(b) If T is an isomorphism of X into Y then $D \subset X$ has the AP if and only if $T(D) \subset Y$ has the AP.

Proof. (a) Given a countable set $B \subset T(D)$ there is a countable set $A \subset D$ such that $T(A) = B$. Since (X^*, ρ_A) is separable by hypothesis and $T^*: (Y^*, \rho_B) \to (X^*, \rho_A)$ is an isometry onto its range (Proposition 5.2.2(e)) it follows that (Y^*, ρ_B) is separable and hence that $T(D) \subset Y$ has the AP.
(b) One direction was just established in part (a). If now $T(D) \subset Y$ has the AP and $A \subset D$ is countable then $(Y^*, \rho_{T(A)})$ is separable and hence so is $(T^*(Y^*), \rho_A)$ by Proposition 5.2.2(e). Since T is an isomorphism, $T^*(Y^*) = X^*$ and thus (X^*, ρ_A) is separable. It follows that D has the AP.

Let $\{X_n : n = 1,2,\ldots\}$ be a sequence of Banach spaces. Then $(\oplus \sum X_n)_{c_o}$ denotes the Banach space of sequences (x_n) with $x_n \in X_n$ for each n and $\lim_n \|x_n\| = 0$. The norm is $\|(x_n)\| \equiv \max_n \|x_n\|$. Its dual space is isometrically isomorphic to $(\oplus \sum X_n^*)_{\ell_1}$ the space of sequences (f_n) with $f_n \in X_n^*$ for each n and $\|(f_n)\| \equiv \sum \|f_n\| < \infty$.

LEMMA 5.3.2. Let $\{X_n : n = 1,2,\ldots\}$ be a sequence of Banach spaces and suppose that D is a bounded subset of $(\oplus \sum X_n)_{c_o}$. Then D has the AP if and only if $P_n(D) \subset X_n$ has the AP for each n (where $P_n : (\oplus \sum X_n)_{c_o} \to X_n$ denotes the $n^{\underline{th}}$ coordinate projection).

Proof. If $D \subset (\oplus \sum X_n)_{c_o} \equiv X$ has the AP so does $P_n(D) \subset X_n$ for each n by Lemma 5.3.1(a). Conversely, suppose that $A \subset D$ is a countable set and that $P_n(D) \subset X_n$ has the AP for each n. Then $A_n \equiv P_n(A)$ has the AP for each n and thus there is a countable set $W_n \subset X_n^*$ which is ρ_{A_n}-dense in X_n^*. It suffices to show that the countable set

$$W \equiv \{ \sum_{i=1}^{N} P_i^*(g_i) : g_i \in W_i \text{ for } 1 \leq i \leq N;\ N = 1,2,\ldots \} \subset X^*$$

is ρ_A-dense in X^*. If $h \in X^*$ then $h \equiv (h_i)$ where $h_i \in X_i^*$ for each i and $\sum \|h_i\| < \infty$. Given $\varepsilon > 0$ choose N so that $\sum_{i=N+1}^{\infty} \|h_i\| < \varepsilon$, and let $g_i \in W_i$ be chosen so that $\rho_{A_i}(h_i - g_i) < \varepsilon/N$ for $1 \leq i \leq N$. Observe that whenever $g \in X$ then $\rho_A(P_k^*(g)) = \rho_{A_k}(g)$ so that

$$\rho_A\left(h - \sum_{i=1}^{N} P_i^*(g_i)\right) \leq \rho_A\left(\sum_{i=N+1}^{\infty} P_i^*(h_i)\right) + \sum_{i=1}^{N} \rho_A(P_i^*(h_i - g_i))$$

$$\leq \sum_{i=N+1}^{\infty} (\sup\{ \|d\| : d \in D\}) \cdot \|h_i\| + \sum_{i=1}^{N} \rho_{A_i}(h_i - g_i)$$

$$\leq \varepsilon(1 + \sup\{ \|d\| : d \in D\}) .$$

This completes the proof of Lemma 5.3.2.

LEMMA 5.3.3. The sum and the union of a finite number of bounded sets each satisfying the AP again have the AP.

Proof. For each n let $X_n \equiv X$ and define $T_n : (\oplus \sum X_i)_{c_o} \to X$ by $T_n((x_i)) \equiv \sum_{i=1}^{n} x_i$. Suppose now that $D_1,\ldots,D_n$ are bounded sets in X each having the AP. By Lemma 5.3.2 and Theorem 5.2.10(d) the set D has the AP where

$$D \equiv \{(d_1,\ldots,d_n,0,0,\ldots) : d_i \in \overline{aco}(D_i),\ i = 1,\ldots,n\} \subset (\oplus \sum X_i)_{c_o} .$$

Consequently

$$T_n(D) = \overline{aco}(d_1) + \overline{aco}(D_2) + \ldots + \overline{aco}(D_n) \supset (D_1 + \ldots + D_n) \cup \left(\bigcup_{i=1}^{n} D_i\right)$$

has the AP by Lemma 5.3.1. It follows that $D_1 + \ldots + D_n$ and $D_1 \cup \ldots \cup D_n$ as subsets of a set with the AP, have the AP.

The final preliminary lemma helps set the stage for a slight variant of the Davis-Figiel-Johnson-Pełczynski construction alluded to earlier. (See the proof of Theorem 5.3.5.)

LEMMA 5.3.4. Let $\{D_n\}$ be a sequence of bounded subsets of X each with the AP and let $\{t_n\}$ be a sequence of positive numbers such that $\lim_n t_n = 0$. Then $D \equiv \bigcap_{n=1}^{\infty} (D_n + t_n U(X))$ has the AP.

Proof. Let $A \subset D$ be countable. For each n pick a countable set $A_n \subset D_n$ such that $A \subset A_n + t_n U(X)$. Then A_n has the AP and hence there is a countable set $W_n \equiv \{f_{n,j}: j = 1,2,\dots\}$ which is ρ_{A_n}-dense in $U(X^*)$ for each n. For every $h \in X^*$ and every n we have

$$\rho_A(h) \leqq \rho_{A_n + t_n U(X)}(h) \leqq \rho_{A_n}(h) + t_n \| h \| .$$

It follows that for any $g \in U(X^*)$ and $\varepsilon > 0$ and for any n such that $t_n < \varepsilon$ and j such that $\rho_{A_n}(g - f_{n,j}) < \varepsilon$ then

$$\rho_A(g - f_{n,j}) \leqq \varepsilon + t_n(\| f_{n,j} \| + \| g \|) \leqq \varepsilon + 2t_n < 3\varepsilon .$$

Thus the countable set $\bigcup_{n=1}^{\infty} W_n$ is a ρ_A-dense subset of X^* and hence D has the AP.

Here is the main result of this section.

THEOREM 5.3.5. Let $T: X \to Y$ be a continuous linear operator such that $T(U(X)) \subset Y$ has the AP. Then T factors through an Asplund space. That is, there are an Asplund space Z and continuous linear operators $T_1: X \to Z$ and $T_2: Z \to Y$ such that $T = T_2 \circ T_1$.

Proof. Let $K \equiv \overline{T(U(X))}$ and for each n let

$$K_n \equiv 2^n K + 2^{-n} U(Y) .$$

Let $\| \cdot \|_n$ be the Minkowski functional for K_n on Y and Y_n the Banach space $(Y, \| \cdot \|_n)$. Then Y and Y_n are isomorphic. Now let $Z \subset (\oplus \sum Y_n)_{c_0}$ be the subspace consisting of all constant sequences. That is, $(y_n) \in Z$ if and only if $y_j = y_1$ for $j = 1,2,\dots$ and $\lim_n \| y_1 \|_n = 0$. Define $T_1: X \to Z$ by

$$T_1(x) \equiv (T(x), T(x), \dots) \qquad \text{for } x \in X$$

and $T_2: Z \to Y$ by

$$T_2((y,y,\ldots)) \equiv y \quad \text{for} \quad (y,y,\ldots) \in Z .$$

This completes the construction and it remains to verify the various properties claimed for T_1, T_2 and Z.

Observe that T_1 really does have range in Z since for any $x \in X$, $\|T(x)\|_n \leqq 2^{-n}\|x\|$ by definition of K_n. Clearly both T_1 and T_2 are continuous and linear and $T = T_2 \circ T_1$ so it remains to prove that Z is an Asplund space. It is immediate that Z is a complete subspace of $(\oplus \sum Y_n)_{c_o}$ so it suffices to show that $U(Z) \subset Z$ has the AP, and by Lemma 5.3.1(b) it suffices to show that $U(Z)$ considered as a subset of $(\oplus \sum Y_n)_{c_o}$ has the AP. But by Lemma 5.3.2, $U(Z) \subset (\oplus \sum Y_n)_{c_o}$ has the AP if $P_n(U(Z))$ has the AP for each coordinate projection P_n of $(\oplus \sum Y_n)_{c_o}$ onto Y_n for $n = 1,2,\ldots$. Now

$$P_n(U(Z)) = \{y \in Y_n : \lim_j \|y\|_j = 0 \text{ and } \|y\|_j \leqq 1 \text{ for each } j\}$$

$$\subset \bigcap_{j=1}^{\infty} \{y \in Y: \|y\|_j \leqq 1 \text{ for each } j\} = \bigcap_{j=1}^{\infty} K_j .$$

The fact that $\bigcap_{j=1}^{\infty} K_j$ has the AP was established in Lemma 5.3.4 and hence the proof of Theorem 5.3.5 is complete.

Here are some consequences of obvious interest of the above factorization.

<u>THEOREM</u> 5.3.6. Let D be a bounded subset of a Banach space Y. Then D has the AP if and only if there are an Asplund space Z and a continuous linear operator $T_2: Z \to Y$ such that $D \subset T_2(U(Z))$.

Proof. If Z is an Asplund space and $T_2: Z \to Y$ is a continuous linear opera[tor] then $T_2(U(Z))$ has the AP by Lemma 5.3.1(a). If $D \subset T_2(U(Z))$ then D has the AP as well.

Conversely, suppose that $D \subset Y$ has the AP, let X be the Banach space $\ell_1(D)$ and let $T: X \to Y$ be the map defined at $(w_d) \in \ell_1(D)$ by

$$T((w_d)) \equiv \sum_{d \in D} w_d d .$$

Since $T(U(X)) \subset \overline{\text{aco}}(D)$, a set with the AP, it follows from Theorem 5.3.5 that ther[e] are an Asplund space Z and continuous linear operators $T_1: X \to Z$ and $T_2: Z \to Y$

with $T = T_2 \circ T_1$. By multiplying T_1 and T_2 by the appropriate scalars we may also assume that $\|T_1\| \leq 1$. Suppose now that $d \in D$. Then $d = T(w)$ for the element w of $U(X)$ which is 1 at d and 0 elsewhere. Hence $d = T_2 \circ T_1(w)$ and since $T_1(w) \in U(Z)$ it follows that $d \in T_2(U(Z))$. That is, $D \subset T_2(U(Z))$ as was to be shown.

<u>THEOREM</u> 5.3.7. Let K be a weak* compact convex subset of X^*. If K has the RNP then there are an Asplund space Z, a weak* compact convex subset C of Z^* and a continuous linear operator $T: X \to Z$ such that Range(T) is dense in Z and $K = T^*(C)$. Thus in particular, (K,weak*) and (C,weak*) are affinely homeomorphic.

Proof. Let $e: X \to C(K)$ be the evaluation map: that is,

$$e(x) \equiv \hat{x}|_K \quad \text{for } x \in X.$$

In order to show that $e(U(X))$ has the AP it suffices (by Theorem 5.2.11(g)) to prove that $e^*(U(C(K)^*))$ has the RNP. For $f \in K$ let $\delta_f \in U(C(K)^*)$ denote the point mass at f. Then $e^*(\delta_f) = f$. Recall that $ex(P_t(K)) = \{\delta_f\colon f \in K\}$. Since e^* is weak* continuous and K is weak* compact and convex, it follows that $e^*(\mu) \in K$ for each $\mu \in P_t(K)$. But $U(C(K)^*) = aco(P_t(K))$ and hence $e^*(U(C(K)^*)) \subset aco(K)$. In order to see that $e^*(U(C(K)^*))$ has the RNP, then, it suffices to check that $aco(K)$ has the RNP. Now Theorem 4.2.13 shows that $K|_Y$ is separable for each separable subspace Y of X. If $\{f_i|_Y\}$ is a countable dense subset of $K|_Y$ it is easy to see that the set of rational convex combinations of $\{\pm f_i|_Y\}$ is a countable dense subset of $aco(K)|_Y$. Hence by Theorem 4.2.13 again, $aco(K)$ has the RNP. Consequently $e^*(U(C(K)^*))$ has the RNP and hence $e(U(X))$ has the AP.

By Theorem 5.3.5 there are an Asplund space Y and continuous linear operators $T_1: X \to Y$ and $T_2: Y \to C(K)$ such that $e = T_2 \circ T_1$. Now $f = e^*(\delta_f) = T_1^* \circ T_2^*(\delta_f)$ for each $f \in K$. Thus T_1^* maps the set $E \equiv \{T_2^*(\delta_f)\colon f \in K\} \subset Y^*$ onto $K \subset X^*$. Let $Z \equiv \overline{T_1(X)} \subset Y$ and let $T: X \to Z$ be the map $T(x) \equiv T_1(x)$ for all $x \in X$. (Note: Range(T) $\subset Z$ while Range(T_1) $\subset Y$.) Since Y is an Asplund space so is Z. Since $\{\delta_f\colon f \in K\}$ is weak* compact, the set $E \subset Y^*$ is also weak* compact. Thus so is $C \equiv E|_Z \subset Z^*$. Furthermore, $T^*(C) = T_1^*(E) = K$. Finally, since Range(T) is dense in Z by definition of Z, T^* is 1-1 on Z^* and hence $C = T^{*-1}(K)$ is convex. The proof of Theorem 5.3.7 is thus complete.

§4. Weak metrizability.

In the statement of each reformulation of the AP there is a stringent condition placed on some set associated with the given one and stated in terms of the containing space or its dual. In the theorem below (due to Fitzpatrick [1980]) the condition is the weak metrizability of separable subsets.

THEOREM 5.4.1 (Fitzpatrick). Let $D \subset X$ be a bounded set. Then the following are equivalent.

(a) D has the AP.

(b) Each separable subset of $\overline{aco}(D)$ is weakly metrizable.

(c) If $A \subset D$ is separable then $\overline{aco}(A)$ is weakly first countable at 0.

Proof.

(a) $\Longrightarrow$ (b) If D has the AP then so does $\overline{aco}(D)$. Suppose that $A \subset \overline{aco}(D)$ is separable. Then A has the AP so there is a countable ρ_A-dense subset W of X^*. Let $Y \equiv \overline{span}(W) \subset X^*$ and define $j\colon X \to Y^*$ to be the restriction of the canonical injection of X into X^{**} to Y (so that $j(x)(f) \equiv f(x)$ for each $f \in Y$ and $x \in X$). Note that $j\colon (X,\text{weak}) \to (Y^*,\text{weak*})$ is continuous. We will demonstrate that $j\colon (A,\text{weak}) \to (j(A),\text{weak*})$ is a homeomorphism (and since Y is separable - which implies that bounded sets of Y^* like $j(A)$ are weak* metrizable - it will follow that (A,weak) is metrizable.) Let $i\colon Y \to X^*$ be the identity embedding. Then $j(x) = i^*(\hat{x})$. Since $x \to \hat{x}$ is a homeomorphism of (X,weak) into $(X^{**},\text{weak*})$ in order to show that $j\colon (A,\text{weak}) \to (j(A),\text{weak*})$ is a homeomorphism it is sufficient to show that $i^*|_{w^*cl(\hat{A})}$ is a homeomorphism (in the weak* topologies) and since $w^*cl(\hat{A})$ is weak* compact, it suffices to show that $i^*|_{w^*cl(\hat{A})}$ is 1-1. Now, if $f \in X^*$ then

$$\rho_A(f) = \sup\{|f(x)|: x \in A\} = \sup\{|\psi(f)|: \psi \in w^*cl(\hat{A})\}$$

and each $\psi \in w^*cl(\hat{A})$ is thus bounded on the ρ_A-unit ball. Consequently each

$\psi \in w^*cl(\hat{A})$ is ρ_A-continuous on X^*. Hence if two members of $w^*cl(\hat{A})$ agree on Y they must be equal since Y is ρ_A-dense in X^*. That is, i^* is 1-1 on $w^*cl(\hat{A})$ as was to be shown.

(b) ⟹ (c) is obvious.

(c) ⟹ (a) Let $K \subset \overline{aco}(D)$ be a closed bounded separable absolutely convex set. (It will be shown that the hypotheses imply that K has the AP and it follows from this that D has the AP as well.) Note that K is weakly first countable at 0 by hypothesis. Hence there is a sequence $\{F_n\}$ of finite subsets of X^* such that the sequence $\{(F_n)_\circ \cap K\}$ is a base of weak neighborhoods of 0. (See Lemma 5.4.2 below for the definitions of the polar notation used as well as for the needed polar calculations.) Pick any $g \in X^*$. Since $\{g\}_\circ$ is a weak neighborhood of 0 there is an n such that

$$\{g\}_\circ \supset (F_n)_\circ \cap K$$

and consequently

$$g \in \{g\}_\circ{}^\circ \subset ((F_n)_\circ \cap K)^\circ \tag{11}$$

Observe that since F_n is finite, $(F_n)_\circ{}^\circ = aco(F_n)$ is weak* compact. Let $F \equiv \bigcup_{n=1}^{\infty} aco(F_n)$ and record for future reference that F is a norm separable set since each $aco(F_n)$ is finite dimensional. Since $aco(F_n) + K^\circ$ is weak* closed (being the sum of a weak* compact and a weak* closed set) and absolutely convex and contains both $(F_n)_\circ{}^\circ$ and K° it follows from Lemma 5.4.2(d) below that

$$((F_n)_\circ \cap K)^\circ \subset aco(F_n) + K^\circ . \tag{12}$$

Thus given $g \in X^*$ pick n such that $g \in ((F_n)_\circ \cap K)^\circ$ (cf. (11)) and conclude that $g \in aco(F_n) + K^\circ$ from (12). That is, $X^* = F + K^\circ$. Thus for $g \in X^*$ and a positive integer m we have $mg = f + h$ for some $f \in F$ and $h \in K^\circ$. Consequently $\rho_K(g - m^{-1}f) = \rho_K(m^{-1}h) \leqq m^{-1}$. Since $m^{-1}f \in F$ it follows that F is ρ_K-dense in X^* and since F is separable (hence ρ_K-separable) X^* is ρ_K-separable. Thus K has the AP. From the remarks at the beginning of this part of the proof, we are done.

The well known material of the next lemma may be found, for example, in Kelley and Namioka [1963] Section 16.

LEMMA 5.4.2. Let $B \subset X$ and $W \subset X^*$ be arbitrary subsets. Then the polars of B and W are, respectively,

$$B^\circ \equiv \{f \in X^*: |f(x)| \leq 1 \text{ for each } x \in B\}$$

and

$$W_\circ \equiv \{x \in X: |g(x)| \leq 1 \text{ for each } g \in W\} .$$

Furthermore $(W_\circ)^\circ$ is abbreviated $W_\circ{}^\circ$ with similar obvious interpretations of $B^\circ{}_\circ$ and $W_\circ{}^\circ{}_\circ$. The following calculational aids hold.

(a) If $B_1 \subset B_2 \subset X$ then $B_2{}^\circ \subset B_1{}^\circ$.

(b) For $B \subset X$ the set B° is weak* closed and absolutely convex. Similarly, for $W \subset X^*$ the set $W_\circ$ is norm closed and absolutely convex.

(c) For $B \subset X$, $B^\circ{}_\circ = \overline{\text{aco}}(B)$. Similarly, for $W \subset X^*$, $W_\circ{}^\circ = \text{w*cl}(\text{aco}(W))$

(d) Let $F \subset X^*$ and suppose that $K \subset X$ is a closed bounded absolutely convex set. Then

$$(F_\circ \cap K)^\circ = \text{w*cl}(\text{co}(F_\circ{}^\circ \cup K^\circ)) .$$

Proof. (a) and (b) are immediate from the definitions.

(c) For any $W \subset X^*$, $W_\circ$ is a closed absolutely convex set. Since it is clear that $B \subset B^\circ{}_\circ$ it follows that $\overline{\text{aco}}(B) \subset B^\circ{}_\circ$. But if $x \notin \overline{\text{aco}}(B)$ there is an $f \in X^*$ such that

$$f(x) > 1 > \sup\{f(y): y \in \overline{\text{aco}}(B)\} = \sup\{|f(y)|: y \in B\} .$$

Thus $f \in B^\circ$ and hence $x \notin B^\circ{}_\circ$. That is, $B^\circ{}_\circ \subset \overline{\text{aco}}(B)$. The conclusion follows. The second statement may be proved analogously.

(d) Let A and B be absolutely convex and norm closed subsets of X. Then $A = A^\circ{}_\circ$ and $B = B^\circ{}_\circ$ by part (c) of this lemma. Observe that for any subsets C and D of X^*, $(C \cup D)_\circ = C_\circ \cap D_\circ$. Hence

$$A \cap B = A^\circ{}_\circ \cap B^\circ{}_\circ = (A^\circ \cup B^\circ)_\circ .$$

It follows that

(13) $$(A \cap B)^\circ = (A^\circ \cup B^\circ)_\circ{}^\circ = w^*cl(aco(A^\circ \cup B^\circ))$$

from (c). Using (13) and (b), for A and B absolutely convex norm closed subsets of X we have

(14) $$(A \cap B)^\circ = w^*cl(co(A^\circ \cup B^\circ)) .$$

Now the set $F_\circ$ of part (d) of the lemma is absolutely convex and norm closed by part (b). The desired conclusion is then an immediate consequence of (14) with $F_\circ$ and K in place of A and B, respectively.

The following corollary is immediate.

COROLLARY 5.4.3. X is an Asplund space if and only if each of its bounded separable subsets is weakly metrizable.

The final result of this section provides a nice nonlinear criterion for the AP. (See Example 5.4.5 for some cautionary remarks.)

COROLLARY 5.4.4. Let $K \subset X$ be a closed bounded absolutely convex set. If (K,weak) is homeomorphic to (D,weak) for some bounded set $D \subset Y$ with the AP then K has the AP.

Proof. Let $A \subset K$ be separable (and hence separable in the weak topology). Then its homeomorphic image B in D is weakly separable and hence norm separable. B is therefore weakly metrizable by Theorem 5.4.1 and hence so is A. Again by Theorem 5.4.1, then, K has the AP.

The results presented in §4, in keeping with those of most of these notes, are fairly delicate in the sense that even innocent appearing modifications of the hypotheses may well lead to drastic changes in the conclusions. Here is one such example, due to Fitzpatrick [1980].

EXAMPLE 5.4.5. There is a weakly metrizable closed bounded convex and separable set D which lacks the AP. (It is not absolutely convex, however. Compare with Theorem 5.4.1(b).) D is a subset of ℓ_1 and there is a continuous linear operator T on ℓ_1 such that T(D) is not weakly metrizable. Moreover, there is a convex continuous function F on ℓ_1 which is nowhere D-differentiable. The construction of D is elementary enough, as is the verification of the above statements. Indeed, if e_n denotes the nth unit vector in ℓ_1, let $D \equiv \overline{co}(\{0,e_1,e_2,\ldots\}) \subset \ell_1$. Then since ℓ_1^* is not separable, $U(\ell_1)$ fails to have the AP (Theorems 4.4.1 and 5.2.12) and hence D fails to have the AP by Theorem 5.2.11(d) since $\overline{aco}(D) = U(\ell_1)$. In order to see that D is weakly metrizable it suffices to check that the weak and norm topologies agree on D. For each n let f_n denote the point of $\ell_1^* = \ell_\infty$ with 0's in the first $n-1$ spots and 1's elsewhere. If $x \in D$ and $\varepsilon > 0$ pick n so large that $\sum_{j=n}^{\infty} x_j < \varepsilon$ (where $x \equiv (x_j)$). Then

$$W \equiv \{y \in D: f_n(y) < \varepsilon \text{ and } |y_j - x_j| < \varepsilon/n \text{ for } 1 \leq j \leq n\}$$

is a relative weak neighborhood of x. It is easy to check that $\|y - x\| \leq 3\varepsilon$ for each $y \in W$ and hence there are relative weak neighborhoods of arbitrarily small norm diameter about each point of D. Thus the topologies coincide, whence D is weakly metrizable as stated. Now let $T: \ell_1 \to \ell_1$ be the map

$$T(x_1,x_2,x_3,\ldots) \equiv (x_2 - x_1, x_4 - x_3, \ldots) \quad \text{for } (x_1,x_2,\ldots) \in \ell_1 .$$

Then $T(D) = U(\ell_1)$ so T(D) is not weakly metrizable even though D is. (If D had been absolutely convex in addition to being closed bounded convex and separable, then D would have had the AP. Hence T(D) would have had the AP as well by Lemma 5.3.1(a), and consequently T(D) would have been weakly metrizable by a combination of Theorems 5.2.11(d) and 5.4.1(b).) Finally, for $x \in \ell_1$ let $F(x) \equiv \|T(x)\|$. Then it is easy to check that F is convex and continuous on ℓ_1 but nowhere D-differentiable on ℓ_1.

§5. The GSP.

Equimeasurable subsets of $L^1(\mu)$ as defined below play an important role in the study of nuclear operators. See, for example, Grothendieck [1955], Chapter 1, §2, no. 2 and no. 3, especially pages 78–88. Roughly speaking, Stegall [1981] chose as his candidates for a local duality pairing with RNP sets, those bounded sets which are preimages of equimeasurable subsets of $L^\infty(\mu)$ under continuous linear operators. In Theorem 5.5.4 below, we show that such sets, called GSP sets, are exactly the sets with the Asplund Property, thus completing the cycle of equivalences of the AP, the SP, and the GSP.

DEFINITION 5.5.1. Let (Ω,F,μ) be a finite measure space. Then a subset $E \subset L^1(\Omega,F,\mu)$ is equimeasurable if for each $\varepsilon > 0$ there is a set $B \in F$ with $\mu(B) > \|\mu\| - \varepsilon$ such that $\{f\chi_B: f \in E\}$ is a relatively norm compact subset of $L^\infty(\Omega,F,\mu)$.

For any finite measure space (Ω,F,μ), by J we shall mean the natural inclusion operator $J: L^\infty(\mu) \to L^1(\mu)$.

DEFINITION 5.5.2. Let $D \subset X$ be a bounded set. Then D has the GSP if and only if whenever (Ω,F,μ) is a finite measure space and $T: X \to L^\infty(\mu)$ is a continuous linear operator then $J\circ T(D) \subset L^1(\mu)$ is equimeasurable.

The following result of Grothendieck [1955] (Proposition 9, p. 64) is at the heart of Stegall's approach. It provides the link between GSP sets and nuclear operators.

THEOREM 5.5.3 (Grothendieck). Let K be a closed bounded absolutely convex subset of X. Then K has the GSP if and only if for each Banach space Y and continuous linear operator $T: Y \to X$ with $T(U(Y)) \subset K$, and for each probability space (Ω,F,P) and continuous linear operator $S: X \to L^\infty(\Omega,F,P)$ the operator

$$J\circ S\circ T: Y \to L^1(\Omega,F,P)$$

is nuclear. That is, there are sequences $\{F_i: i = 1,2,\dots\}$ in $L^\infty(\Omega,F,P)$ and $\{f_i: i = 1,2,\dots\}$ in $L^1(\Omega,F,P)$ such that $\sum_{i=1}^\infty \|F_i\|\|f_i\|_1 < \infty$ and for each $y \in Y$, $J\circ S\circ T(y) = \sum_{i=1}^\infty F_i(y)f_i$.

We turn now to the central result of this section. Fitzpatrick [1980] established the equivalence of the SP, the AP and the GSP, though in a somewhat circuitous fashion. He and Uhl [1981] recently outlined the relatively direct proof below.

THEOREM 5.5.4. Let $K \subset X$ be a bounded set. Then K has the Asplund Property if and only if K has the GSP. Consequently, the AP, the SP and the GSP are equivalent properties for bounded subsets of Banach spaces.

Proof. The second statement follows from the first, coupled with Theorem 5.2.1. To begin with, let Y be an Asplund space and μ a finite measure on the measurable space (Ω, G). Let $T: Y \to L^\infty(\mu)$ be a bounded linear operator. We show that $T(U(Y))$ is equimeasurable. Define $m: G \to Y^*$ by

$$m(A) \equiv T^*(\chi_A) \qquad \text{for} \quad A \in G$$

(where $\chi_A \in L^1(\mu)$ and $\hat{\chi}_A \in (L^\infty(\mu))^*$ are identified). Then m is a vector measure for which $\|m(A)\| \leq \|T^*\| \, |\mu|(A)$ for each $A \in G$. Thus $m \ll |\mu|$ and in fact $AR(m) \subset \|T^*\| U(Y^*)$. Since Y^* has the RNP by Theorem 5.2.12, there is a $g \in L^1_{\|T^*\| U(Y^*)}(\mu)$ such that $m(A) = \int_A g d\mu$ for each $A \in G$. That is, $T^*(\chi_A) = \int_\Omega \chi_A g d\mu$ for $A \in G$, and by standard approximation methods, more generally,

$$T^*(f) = \int_\Omega fgd\mu \qquad \text{for} \quad f \in L^1(\mu) .$$

Now T may be described directly in terms of g. Indeed, if $f \in L^1(\mu)$ and $y \in Y$ then

$$\int_\Omega f(\omega)(\hat{y} \circ g)(\omega) d\mu(\omega) = T^*(f)(y) = f(T(y)) = \int_\Omega f(\omega) T(y)(\omega) d\mu(\omega)$$

whence

$$\hat{y} \circ g = T(y) \qquad [a.s.\mu] \qquad \text{for each} \quad y \in Y .$$

Let $\{s_n : n = 1,2,\ldots\}$ be a sequence of $\|T^*\| U(Y^*)$-valued simple functions which converge pointwise almost surely to g. By Egoroff's theorem, given $\varepsilon > 0$ there is an $A_\varepsilon \in G$ with $\mu(A_\varepsilon^c) < \varepsilon$ and $s_n \to g$ uniformly on A_ε. For each n let $T_{n,\varepsilon}$ and $T_{\infty,\varepsilon}$ be the linear operators from Y into $L^\infty(\mu)$ given by

$$T_{n,\varepsilon}(y) \equiv (\hat{y}\circ s_n)\chi_{A_\varepsilon} \quad \text{and} \quad T_{\infty,\varepsilon}(y) \equiv (\hat{y}\circ g)\chi_{A_\varepsilon} = T(y)\chi_{A_\varepsilon} .$$

Then the range of each $T_{n,\varepsilon}$ is finite dimensional and it is easy to see that $\| T_{n,\varepsilon} - T_{\infty,\varepsilon} \| \to 0$. Hence $T_{\infty,\varepsilon}$ is a compact operator. It follows that $\{T(y)\chi_{A_\varepsilon} : y \in U(Y)\} \subset L^\infty(\mu)$ is relatively compact. Hence $T(U(Y))$ is equimeasurable.

Suppose then that $K \subset X$ is a bounded set which has the Asplund Property. Let (Ω, F, μ) be a finite measure space. By Theorem 5.3.6 there are an Asplund space Y and bounded linear operator $T_2: Y \to X$ with $T_2(U(Y)) \supset K$. If $T_1: X \to L^\infty(\mu)$ is any bounded linear operator and $T \equiv T_1 \circ T_2: Y \to L^\infty(\mu)$ then $T(U(Y))$ is equimeasurable from above and thus $T_1(K) \subset T(U(Y))$ is equimeasurable as well. That is, K has the GSP.

In order to establish the converse, assume that $K \subset Y$ is a bounded set which does not have the Asplund Property. Let $Z \equiv \overline{\text{span}}(K)$ and let $X \equiv \ell_1(K)$. Define the operator $T: X \to Z$ at $(w_k) \in X$ by $T((w_k)) \equiv \sum_{k\in K} w_k k$. Since $\text{aco}(K) \subset T(U(X)) \subset \overline{\text{aco}}(K)$, the set $T(U(X)) \subset Z$ does not have the AP. By virtue of Theorem 5.2.3, $T^*(U(Z^*))$ lacks the RNP and by Theorem 4.2.13(d) there is a δ-tree $\{g_n : n = 1,2,\dots\}$ in $T^*(U(Z^*))$ for some $\delta > 0$. Since T has dense range, T^* is 1-1. Let $f_n \equiv T^{*-1}(g_n)$ and observe that $\{f_n : n = 1,2,\dots\}$ is a tree since its 1-1 image is, and that $f_n \in U(Z^*)$ for each n .

Let Δ denote the Cantor set, $\{\Delta_n : n = 1,2,\dots\}$ the usual Haar system of compact subsets of Δ and μ the probability measure on $\mathcal{B}(\Delta)$ such that $\mu(\Delta_n) = 2^{-k}$ for $2^k \leqq n < 2^{k+1}$. Determine a linear operator

$$S: \text{span}(\{\chi_{\Delta_n}\}_{n=1}^\infty) \subset L^1(\mu) \to Z^*$$

by the conditions $S(\chi_{\Delta_n}) \equiv \frac{1}{2^k} f_n$ for each k and n such that $2^k \leqq n < 2^{k+1}$. (The linearity of S causes no difficulty since $\chi_{\Delta_{2n}} + \chi_{\Delta_{2n+1}} = \chi_{\Delta_n}$ and $f_{2n} + f_{2n+1} = 2f_n$ for each n .) Observe that if $f \in \text{span}(\{\chi_{\Delta_n}\}_{n=1}^\infty)$ there are k and $\beta_n \in \mathbb{R}$, $2^k \leqq n < 2^{k+1}$, such that $f = \sum_{n=2^k}^{2^{k+1}-1} \beta_n \chi_{\Delta_n}$. Then

$$\|S(f)\| = \left\| \frac{1}{2^k} \sum_{n=2^k}^{2^{k+1}-1} \beta_n f_n \right\| \leqq \frac{1}{2^k} \sum_{n=2^k}^{2^{k+1}-1} |\beta_n| = \|f\|$$

whence S has a unique extension to a bounded linear operator, again called S , from $L^1(\mu)$ to Z^* . The main point to keep in mind is that $T^*\circ S(2^k \chi_{\Delta_n}) = g_n$ for each n and k , $2^k \leq n < 2^{k+1}$.

If $m_1: \mathcal{B}(\Delta) \to X^*$ is the vector measure

$$m_1(A) \equiv T^*\circ S(\chi_A) \qquad \text{for each } A \in \mathcal{B}(\Delta)$$

then clearly

$$\{g_n\}_{n=1}^{\infty} \subset AR(m_1) \subset T^*\circ S(U(L^1(\mu))) .$$

The argument that m_1 cannot have a Radon-Nikodým derivative with respect to μ go[es] as follows. Suppose temporarily that $F \equiv (dm_1/d\mu)$ exists in $L^1_{X^*}(\mu)$. Let F_n b[e] the finite σ-algebra generated by $\{\Delta_i: 1 \leq i < 2^{n+1}\}$ and observe that $\mathbb{E}[F|F_n]$ ha[s] constant value $\frac{1}{\mu(\Delta_i)} m_1(\Delta_i) = g_i$ on Δ_i for $2^n \leq i < 2^{n+1}$. Hence

$$\mathbb{E}[F|F_n] = \sum_{i=2^n}^{2^{n+1}-1} g_i \chi_{\Delta_i} .$$

For each n ,

$$\begin{aligned}\|\mathbb{E}[F|F_{n+1}] - \mathbb{E}[F|F_n]\| &= \Big\| \sum_{i=2^{n+1}}^{2^{n+2}-1} g_i\chi_{\Delta_i} - \sum_{i=2^n}^{2^{n+1}-1} g_i\chi_{\Delta_i} \Big\| \\ &= \Big\| \sum_{i=2^n}^{2^{n+1}-1} [g_{2i}\chi_{\Delta_{2i}} + g_{2i+1}\chi_{\Delta_{2i+1}} - g_i(\chi_{\Delta_{2i}} + \chi_{\Delta_{2i+1}}) \\ &= \sum_{i=2^n}^{2^{n+1}-1} [\|g_{2i} - g_i\| + \|g_{2i+1} - g_i\|]\frac{1}{2^n} \geq \delta .\end{aligned}$$

This is impossible by virtue of Lévy's Theorem 1.3.1, which states in part that $\{\mathbb{E}[F|F_n]: n = 1,2,\ldots\}$ must be Cauchy in $L^1_{X^*}(\mu)$. Thus $(dm_1/d\mu)$ fails to exist. Hence by Theorem 2.2.6 there is an $\varepsilon > 0$ such that whenever $A_\varepsilon \in \mathcal{B}(\Delta)$ and $\mu(A_\varepsilon)$ [>] $1 - \varepsilon$ then $AR(m_1|_{A_\varepsilon})$ is not relatively compact. See 2.2.6 for this notation. Fi[x] any $A_\varepsilon \in \mathcal{B}(\Delta)$ such that $\mu(A_\varepsilon) > 1 - \varepsilon$, and let $V: X \to L^\infty(\mu)$ be the operator de-fined at x by $V(x) \equiv S^*\circ T(x)\chi_{A_\varepsilon}$. Then for any $f \in L^1(\mu)$ and $x \in X$,

$$V^*(f)(x) = \int_\Delta (S^*\circ T)(x)(\omega)(f\chi_{A_\varepsilon})(\omega)\,d\mu(\omega) = (f\chi_{A_\varepsilon})[(S^*\circ T)(x)] = [(T^*\circ S)(f\chi_{A_\varepsilon})](x)\ .$$

It follows that $V^*(f) = (T^*\circ S)(f\chi_{A_\varepsilon})$ for each $f \in L^1(\mu)$ and hence

$$\begin{aligned} V^*(U(L^1(\mu))) &\supset \{T^*\circ S\left(\frac{\chi_A}{\mu(A)}\chi_{A_\varepsilon}\right) : A \in \mathcal{B}(\Delta),\ \mu(A) > 0\} \\ &\supset \{T^*\circ S\left(\frac{\chi_A}{\mu(A)}\right) : A \in \mathcal{B}(\Delta),\ \mu(A) > 0,\ A \subset A_\varepsilon\} \\ &= AR(m_1|_{A_\varepsilon})\ . \end{aligned}$$

Thus $V^*\colon L^1(\mu) \to X^*$ is not compact and hence V is not compact. That is, $V(U(X)) = \{S^*\circ T(x)\chi_{A_\varepsilon} : x \in U(X)\}$ is not relatively compact in $L^\infty(\mu)$. Since A_ε was any $\mathcal{B}(\Delta)$ element of measure exceeding $1 - \varepsilon$, $S^*\circ T(U(X))$ is not equimeasurable. Thus $T(U(X)) \subset Z$ does not have the GSP. We have just established that $K \subset Z$ lacks the GSP whenever $Z = \overline{\text{span}}(K)$ and K is a bounded set which does not have the AP.

In order to complete the proof we will need the fact that $L^\infty(\mu)$ is injective. We give an indication of the proof of this fact here, and refer to Lindenstrauss and Tzafriri [1973], Part II, Chapter 4, Sections d and n, and Theorem II.5.7(ii), p. 201 for much more powerful results in this same vein. Recall that if X is a Banach lattice then X^* is an order complete Banach lattice. That is, whenever $\{g_\alpha\}_{\alpha\in A}$ is a net in X^* which is bounded above, then $\sup_\alpha g_\alpha$ exists in X^*. See, for example, Kelley and Namioka [1963], pages 243 and 244 or Lindenstrauss and Tzafriri [1979], page 3. Let $T\colon Z \to L^\infty(\mu)$ be a bounded linear operator and suppose that $Z \subset Y$. Let $\rho\colon Y \to L^\infty(\mu)$ be the subadditive and positive homogeneous map $\rho(y) \equiv \|T\|\,\|x\| \cdot 1$ for $y \in Y$ where 1 denotes the function constantly 1 in $L^\infty(\mu)$. Evidently $T(z) \leq \rho(z)$ as elements of $L^\infty(\mu)$ for each $z \in Z$. The usual proof of the Hahn-Banach theorem may be mimicked exactly with T in place of the linear functional, using the order completeness of $L^\infty(\mu)$, to extend T one dimension at a time. The conclusion is that T has a norm preserving linear extension to $\hat{T}\colon Y \to L^\infty(\mu)$. Now suppose that $K \subset Y$ is bounded and has the GSP. Let $Z \equiv \overline{\text{span}}(K)$ and let $T\colon Z \to L^\infty(\Omega,\mathcal{F},\mu)$ be a bounded linear operator for some finite measure space $(\Omega,\mathcal{F},\mu)$.

Since T extends to a bounded linear operator $\hat{T}: Y \to L^{\infty}(\mu)$ from above, $J \circ \hat{T}(K) = J \circ T(K)$ is equimeasurable where $J: L^{\infty}(\mu) \to L^{1}(\mu)$ denotes the inclusion operator. It follows that K, considered as a subset of Z, also has the GSP. But then, by the results of the previous paragraph, $K \subset Z$ has the Asplund Property. Finally, to conclude that $K \subset Y$ has the Asplund Property, use Lemma 5.3.1(a). This completes the proof of Theorem 5.5.4.

We end this section with a brief historical comment. Our development of sets with the GSP has been independent of the first approaches to this subject. Asplund [1968], followed somewhat later by Namioka and Phelps [1975] and, for a finishing touch, Stegall [1978a], combined to prove Theorem 5.2.12 (in an entirely different way). Stegall's paper [1981] includes all but those theorems of the first four sections of this chapter explicitly attributed to others and, even in those results attributed to others, the techniques involved often stem from Stegall. The Stegall Property was singled out by Namioka and Fitzpatrick and formed the basis of the approach taken in Fitzpatrick [1980]. We have leaned heavily on that work of Fitzpatrick in these notes. Quite recently we learned of a series of papers by Reynov which precedes Stegall [1981] and which contains versions of much of the material covered by Stegall. See the addendum, Section 5.9, for specific references.

§6. Differentiable bump functions and Asplund spaces.

Let X be a Banach space with an equivalent norm λ which is Fréchet differentiable at each point other than 0 and suppose that $Y \subset X$ is a separable linear subspace. Let $U(Y)$ and $U(Y^*)$ denote the closed unit balls in the λ and λ-dual norm, λ^*, respectively, so that $\lambda = \lambda_{U(Y^*)}$ in the notation of Lemma 5.2.9. Let $\{y_n : n = 1,2,\ldots\}$ be a norm convergent sequence in Y with limit $y \neq 0$. If $f_n \in \partial\lambda(y_n)$ then $f_n(y_n) = \lambda(y_n)$ by Lemma 5.2.9. Since $f_n \in U(Y^*)$, a bounded set, we have

$$\lambda(y) = \lim_n \lambda(y_n) = \lim_n f_n(y_n) \leq \liminf_n \, (f_n(y) + \|f_n\| \|y_n - y\|)$$

$$= \liminf_n f_n(y) \leq \limsup_n f_n(y) \leq \sup\{g(y) : g \in U(Y^*)\} = \lambda(y) .$$

That is, $\lim_n f_n(y) = \lambda(y)$. Consequently, when $f \in \partial\lambda(y)$ -- so that $f(y) = \lambda(y)$ by Lemma 5.2.9 -- it follows that $\lim_n \| f_n - f \| = 0$ by Lemma 5.2.10(b). Denote the Fréchet derivative of λ at any nonzero $w \in Y$ by $\lambda'(w)$. Then for each $w \neq 0$, $\partial\lambda(w) = \{\lambda'(w)\} \subset Y^*$. Thus the above argument demonstrates that the map $y \to \lambda'(y)$ is a norm-norm continuous map on $Y \setminus \{0\}$ into Y^* and hence $\{\lambda'(y): y \in Y \setminus \{0\}\}$ is separable since Y is. We now show that this set of functionals is dense in the λ^*-unit sphere of Y^* , $S(Y^*)$. It is easy to see by direct calculation that $\lambda^*(\lambda'(y)) = 1$ for each $y \neq 0$. Suppose that $g \in Y^*$, $y \in Y$, and that $\lambda^*(g) = \lambda(y) = g(y) = 1$. Then clearly $g(z) - 1 \leq \lambda(z) - 1$ for each $z \in Y$ and hence $g(z) - g(y) \leq \lambda(z) - \lambda(y)$. That is, $g \in \partial\lambda(y)$, whence $g = \lambda'(y)$. The Bishop Phelps Theorem 3.3.1 guarantees that norm attaining functionals such as g above are dense in $S(Y^*)$ and hence $S(Y^*)$ is separable. Thus Y^* is separable. It follows that X^* has the RNP (Theorem 4.4.1) and that X is Asplund (Theorem 5.2.12). That is, we have just proved what we have called Corollary 5.6.8 below.

The main result of this section (Theorem 5.6.2) may turn out to be no more general than what has just been established above, but it remains an open question as to whether the existence of a differentiable function identically zero outside some bounded set on a Banach space X is sufficient to guarantee the existence of an equivalent norm on X which is Fréchet differentiable at each nonzero point of X . It is easy to establish the converse -- see the alternative proof of Corollary 5.6.8 provided after its statement, for example. Moreover, these two conditions are equivalent, and each is equivalent to the fact that X^* is separable, when X itself is separable. See Leach and Whitfield [1972]. Alternatively, observe that when X is separable, X^* is separable if and only if X^* has the RNP, if and only if X is Asplund. Thus if X has a differentiable function on it of bounded nonempty support, using Theorem 5.6.2 below, X^* is separable. The remaining implication -- if X^* is separable then X has an equivalent Fréchet differentiable norm -- goes back to Restrepo [1964]. It follows that Theorem 5.6.2 is at least as general as its first corollary! The proof below has been modified from the original one due to Ekeland and Lebourg [1976] by John Rainwater [unpublished notes]. Further information may be found in the survey by Ekeland [1979] and that of Sullivan [1981] where it is shown that Ekeland's original lemma actually characterizes complete metric spaces. (Lemma 5.6.6, due to Phelps [1964], is the special case of Ekeland's lemma needed in these notes.) We begin with a definition.

DEFINITION 5.6.1. (a) Let $b: X \to \mathbb{R}$. The support of b is the norm closure of the set $\{x \in X: b(x) \neq 0\}$.

(b) A differentiable bump function on X is a function $b: X \to \mathbb{R}$ which is Fréchet differentiable at each $x \in X$ and whose support is a bounded nonempty subset of X

THEOREM 5.6.2 (Ekeland-Lebourg). If there is a differentiable bump function on a Banach space X then X is an Asplund space.

The proof presented here is quite geometrical and rests on some of the ideas of Bishop and Phelps (see Section 3 of Chapter 3). Observe that if $F: X \to \mathbb{R}$ and $f \in X^*$ supports F at x (so that the graph of the affine function g given by $g(y) \equiv f(y - x) + F(x)$ for $y \in X$ lies everywhere below that of F, while both share the point $(x,F(x))$ -- i.e., f is a subdifferential of F), F may grow arbitrarily quickly compared to f near x. In order to study the differentiability characteristics of F near x, an upper bound for the growth of F (relative to that of f) near x is needed as well. In this regard, the "conical support condition" in the next definition is quite natural and leads to Lemma 5.6.6 very much as the discussion at the beginning of 3.3 leads to the proof of Theorem 3.3.1.

DEFINITION 5.6.3. Let $F: X \to \mathbb{R}$ and suppose that $\varepsilon > 0$. Then F is locally ε-supported by $f \in X^*$ at $x \in X$ if for some $\delta > 0$

(15) $$F(y) \leq F(x) + f(y-x) + \varepsilon\|y-x\| \qquad \text{if} \qquad \|y-x\| < \delta .$$

Denote by $S^*_\varepsilon F(x)$ the set $\{f \in X^*: f \text{ locally } \varepsilon\text{-supports } F \text{ at } x\}$ and by $S_\varepsilon F$ the set $\{x \in X: S^*_\varepsilon F(x) \neq \emptyset\}$.

The first lemma collects some straightforward but useful tools. In fact, 5.6.4(d) provides an analytic justification for the above definitions.

LEMMA 5.6.4. Let $F,G: X \to \mathbb{R}$ and $\varepsilon,\varepsilon' > 0$. Then

(a) $S_\varepsilon F \subset S_{\varepsilon'} F$ if $0 < \varepsilon < \varepsilon'$.

(b) $S^*_\varepsilon F(x) \subset S^*_{\varepsilon'} F(x)$ if $0 < \varepsilon < \varepsilon'$.

(c) $S^*_\varepsilon F(x) + S^*_{\varepsilon'} G(x) \subset S^*_{\varepsilon+\varepsilon'}(F+G)(x)$.

(d) If F is convex and $f \in S^*_\varepsilon F(x)$ then there is a $\delta > 0$ such that $|F(y) - F(x) - f(y-x)| < \varepsilon\|y-x\|$ for $\|y-x\| < \delta$.

(e) Suppose that $F: X \to \mathbb{R}$ is convex. If $f \in S^*_\varepsilon F(x)$ and $g \in \partial F(x)$ then $\|f-g\| \leq \varepsilon$.

Proof. (a), (b), and (c) are obvious.

(d) If $\delta > 0$ is chosen so that (15) holds for $\|y-x\| < \delta$ then since $\|(2x-y)-x\| < \delta$, (15) yields

$$F(2x-y) \leq F(x) + f(x-y) + \varepsilon\|y-x\| . \tag{16}$$

Since F is convex, $\frac{1}{2} F(2x-y) + \frac{1}{2} F(y) \geq F(x)$ so that $F(2x-y) \geq 2F(x) - F(y)$. Combined with (16) this latter inequality becomes

$$F(x) - F(y) \leq f(x-y) + \varepsilon\|y-x\|$$

which, combined with (15), proves this part of the lemma.

(e) Since $f \in S^*_\varepsilon F(x)$ there is a $\delta > 0$ such that (15) holds for $\|y-x\| < \delta$. Since $g(y-x) \leq F(y) - F(x)$ for all y , and in particular for y such that $\|y-x\| < \delta$, it follows that

$$(g-f)(y-x) \leq \varepsilon\|y-x\| \quad \text{for } \|y-x\| < \delta .$$

Consequently $\|g-f\| \leq \varepsilon$.

The next lemma might well be anticipated from Lemma 5.6.4, parts (d) and (e).

LEMMA 5.6.5. If $F: X \to \mathbb{R}$ is convex and continuous then F is Fréchet differentiable at x if and only if $x \in \bigcap_{\varepsilon>0} S_\varepsilon F$.

Proof. If F is differentiable at x with derivative $F'(x)$ then clearly $F'(x) \in S^*_\varepsilon F(x)$ for each $\varepsilon > 0$ whence $x \in \bigcap_{\varepsilon>0} S_\varepsilon F$. Conversely, suppose that

$x \in \bigcap_{\varepsilon>0} S_\varepsilon F$ and that $g \in \partial F(x)$. It follows from Lemma 5.6.4(e) that $\|g-f\| \leq \varepsilon$ whenever $f \in S_\varepsilon^* F(x)$ so that diameter $(S_\varepsilon^* F(x)) \leq 2\varepsilon$. Moreover, $S_\varepsilon^* F(x)$ is non-empty for each $\varepsilon > 0$ since $x \in S_\varepsilon F$. Hence there is an $f \in X^*$ such that $\{f\} = \bigcap_{\varepsilon>0} \overline{S_\varepsilon^* F(x)}$. In order to prove that $F'(x) = f$, fix $\varepsilon > 0$ and pick $f_1 \in S_{\varepsilon/2}^* F(x)$ such that $\| f-f_1 \| < \frac{\varepsilon}{2}$. Then there is a $\delta > 0$, $\delta < 1$, such that

$$F(y) \leq F(x) + f_1(y-x) + \frac{\varepsilon}{2} \|y-x\| \qquad \text{for } \|y-x\| < \delta$$

and it follows that

$$F(y) \leq F(x) + f(y-x) + \varepsilon\|y-x\| \qquad \text{for } \|y-x\| < \delta .$$

That is, $f \in S_\varepsilon^* F(x)$ for each $\varepsilon > 0$ which, by Lemma 5.6.4(d), forces $f = F'(x)$. This completes the proof.

The proof of the Ekeland-Lebourg theorem (5.6.2) thus reduces to showing that $\bigcap_{\varepsilon>0} S_\varepsilon F$ is a dense G_δ set whenever F is convex and continuous on a Banach space with a differentiable bump function. This will be done in two steps, the first of which is to show that $S_\varepsilon F$ is dense for each $\varepsilon > 0$. Here the idea is to show that given an open set V there is a partial ordering $\preccurlyeq$ on a subset of V such that any maximal element relative to $\preccurlyeq$ is in fact in $S_\varepsilon F$. The only place in the proof of Theorem 5.6.2 where the existence of a differentiable bump function is used is in forcing the maximal elements to belong to V (rather than to $\overline{V}$). The next lemma, due to Phelps [1964] is highly reminiscent of the material in §3.3.

<u>LEMMA</u> 5.6.6. Let $G: X \to \mathbb{R} \cup \{-\infty\}$ be upper semicontinuous. Suppose that dom G ($\equiv \{x \in X: G(x) \neq -\infty\}$) is nonempty and that G is bounded above on X. Then for any $r > 0$ there is an $x_0 \in \text{dom } G$ such that

$$G(x) \leq G(x_0) + r\|x-x_0\| \qquad \text{for all } x \in X .$$

(N.B. Not only does the conclusion of this lemma imply that $0 \in S_r^* G(x_0)$ but that <u>globally</u> the cone

$$(x_0, G(x_0)) + \{(x,t) \in X \times \mathbb{R}: t \geq r\|x\|\}$$

lies above the graph of G.)

Proof. For $x,y \in \operatorname{dom} G$ let $x \succcurlyeq y$ mean $G(x) - G(y) \geq r\|x-y\|$. Thus $\succcurlyeq$ is both transitive and antisymmetric. Given $z \in \operatorname{dom} G$ let $C \equiv \{x \in X\colon x \succcurlyeq z\}$. We will produce a maximal element x_o of X which belongs to C. (This will complete the proof since, for such an x_o and for any $x \neq x_o$, $x \not\succcurlyeq x_o$ so that $G(x) - G(x_o) < r\|x-x_o\|$.)

Note that C is closed since G is upper semicontinuous. Let $M \equiv \sup_{x\in C} G(x) < \infty$ and choose $x_1 \in C$ such that $G(x_1) > M - 2^{-1}$. Let $C_1 \equiv \{x\colon x \succcurlyeq x_1\}$ and $M_1 \equiv \sup_{x\in C_1} G(x)$. Suppose now that sets $C_1,\ldots,C_n$ and numbers $M_1,\ldots,M_n$ have already been defined. Let $x_{n+1} \in C_n$ be chosen so that $G(x_{n+1}) > M_n - 2^{-(n+1)}$, let $C_{n+1} \equiv \{x\colon x \succcurlyeq x_{n+1}\}$ and let $M_{n+1} \equiv \sup_{x\in C_{n+1}} G(x)$. Evidently $C_{n+1} \subset C_n \subset C$ by transitivity, C_{n+1} is closed, and $M_{n+1} \leqq M_n$ for each n. If $x \in C_n$ then $x \succcurlyeq x_n$ and

$$G(x) \leqq M_n \leqq M_{n-1} < G(x_n) + 2^{-n} .$$

Thus $r\|x-x_n\| \leqq G(x) - G(x_n) < 2^{-n}$ if $x \in C_n$. Hence $\lim_n \text{diameter } C_n = 0$ so there is a unique point x_o in the intersection of the C_n's. If $x \in X$ and $x \succcurlyeq x_o$ then $x \succcurlyeq x_n$ whence $x \in C_n$ for each n and thus $x = x_o$. That is, x_o is a maximal element of X, and clearly $x \in C$. The proof is thus complete.

It might seem initially that this lemma provides a little too much a little too easily. Indeed, given a convex continuous function $F\colon X \to \mathbb{R}$ and an open set V in X let $G(x) \equiv (-\infty)\chi_{\overline{V}^c}(x)$ for $x \in X$. Then G is upper semicontinuous and hence the sum $F+G$ is upper semicontinuous, bounded above (if V is small enough), and $\operatorname{dom}(F+G) = \overline{V}$. Given $\varepsilon > 0$, Lemma 5.6.6 guarantees that there is an $x_o \in \overline{V}$ such that $(F+G)(x) \leqq (F+G)(x_o) + \|x-x_o\|$ for all $x \in X$ and in particular for $x \in \overline{V}$, $F(x) \leqq F(x_o) + \varepsilon\|x-x_o\|$. If $x_o \in V$ then $x_o \in S_\varepsilon F$ as desired. However, it might well happen (as trivial examples show) that the only choice for x_o lies in $\overline{V}\backslash V$ and no conclusions may be drawn. The existence of a differentiable bump function on X, however, allows us to force $x_o \in V$ as the following proposition demonstrates.

PROPOSITION 5.6.7. Suppose that X has a differentiable bump function b. Then for each upper semicontinuous function $G\colon X \to \mathbb{R} \cup \{-\infty\}$ and for each $\varepsilon > 0$ the set $S_\varepsilon G$ is dense in interior $(\operatorname{dom} G)$.

Proof. There is nothing to prove if interior (dom G) $= \emptyset$ so assume that $z \in$ interior (dom G) and that V is an arbitrary neighborhood of z. By taking an even smaller neighborhood if necessary, assume both that G is bounded above on V (remember: G is upper semicontinuous) and that $V \subset$ interior (dom G). By appropriate translation and scalar multiplication we may assume that $b(z) \neq 0$, that the support of b is a subset of V and that $b \leq 0$ everywhere (by replacing b by $-b^2$ if necessary). Let $b_1 \equiv 1/b$, with $b_1(x) = -\infty$ when $b(x) = 0$. Then $b_1: X \to \mathbb{R} \cup \{-\infty\}$ is upper semicontinuous, bounded above by 0, and $z \in \operatorname{dom} b_1 \subset V$ It follows that $G_1 \equiv G + b_1$ is upper semicontinuous, bounded above, and that $z \in \operatorname{dom} G_1 = (\operatorname{dom} G) \cap (\operatorname{dom} b_1) \subset V$. Lemma 5.6.6 with $r = \varepsilon/2$ applies to G_1 so there is an $x_0 \in \operatorname{dom} G_1 \subset V$ such that

$$G_1(x) \leq G_1(x_0) + \frac{\varepsilon}{2} \|x-x_0\| \qquad \text{for all } x \in X .$$

Hence $0 \in S^*_{\varepsilon/2}\, G_1(x_0)$. Since $\operatorname{dom} G_1 \subset \operatorname{dom} b_1$ it follows that $b_1(x_0) \neq -\infty$ so that b_1 is differentiable at x_0. If $-b_1'(x_0)$ is denoted by f, then clearly $f \in S^*_{\varepsilon/2}(-b_1)(x_0)$. It follows from Lemma 5.6.4(c) that

$$f = f + 0 \in S^*_{\varepsilon/2}(-b_1)(x_0) + S^*_{\varepsilon/2}\, G_1(x_0) \subset S^*_{\varepsilon}G(x_0) .$$

Thus $x_0 \in S_\varepsilon(G) \cap V$, which completes the proof.

We are now in a position to prove the Ekeland-Lebourg theorem.

<u>Proof of 5.6.2.</u> Let $F: X \to \mathbb{R}$ be convex and continuous. For each $n \geq 1$, $\delta > 0$ and $x \in X$ let

$$K_{n,\delta}(x) \equiv \{f \in X^*: \text{for some } y \in X,\ \|y-x\| < \delta,\ y \in S_{1/n}F \text{ and } f \in S^*_{1/n}F(y)\} .$$

Now define V_n by

$$V_n \equiv \{x \in X: \text{diameter } K_{n,\delta}(x) < \frac{9}{n} \text{ for some } \delta > 0\} .$$

It is clear that V_n is open for each n. The proof of this theorem is broken into two steps.

<u>Step 1</u>. We show here that $S_{1/n}F \subset V_n$ for each n. Indeed, suppose that

$x_o \in S_{1/n}F$. Then for some $\delta > 0$ and any $f \in S^*_{1/n}F(x_o)$ it follows from Lemma 5.6.4(d) that

(17) $$F(x) \leqq F(x_o) + f(x-x_o) + \frac{1}{n}\|x-x_o\|$$

and

(18) $$F(x) \geqq F(x_o) + f(x-x_o) - \frac{1}{n}\|x-x_o\|$$

whenever $\|x-x_o\| < 2\delta$. Choose $y_o \in X$, $\|y_o-x_o\| < \delta$ and let $g \in \partial F(y_o)$. For any $x \in X$ such that $\|x-y_o\| = \|x_o-y_o\|$ evidently $\|x-x_o\| < 2\delta$ so that (17) holds for this x while (18) holds for y_o in place of x . Thus

$$\begin{aligned} g(x-y_o) &\leqq F(x) - F(y_o) = F(x) - F(x_o) + [F(x_o) - F(y_o)] \\ &\leqq f(x-y_o) + \frac{1}{n}\|x-x_o\| - f(y_o-x_o) + \frac{1}{n}\|y_o-x_o\| \\ &\leqq f(x-y_o) + \frac{3}{n}\|x_o-y_o\| . \end{aligned}$$

Consequently $(g-f)(x-y_o) \leqq \frac{3}{n}\|x-y_o\|$ whenever $\|x-y_o\| = \|x_o-y_o\|$ and hence $\|g-f\| \leqq \frac{3}{n}$.

Suppose now that $y_o \in S_{1/n}F$ and that $\|y_o-x_o\| < \delta$. Lemma 5.6.4(e) shows that for $g \in \partial F(y_o)$, $\|g-h\| \leqq \frac{1}{n}$ for each $h \in S^*_{1/n}F(y_o)$. From the previous paragraph it also follows that for any $f \in S^*_{1/n}F(x_o)$ and $g \in \partial F(y_o)$ then $\|f-g\| \leqq \frac{3}{n}$. Consequently diameter $K_{n,\delta}(x_o) \leqq 2\left(\frac{3}{n} + \frac{1}{n}\right) < \frac{9}{n}$. That is, $x_o \in V_n$.

<u>Step 2</u>. It was just established that $S_{1/n}F \subset V_n$ for each n and it was shown in Lemma 5.6.7 that $S_{1/n}F$ is dense. It follows that $W \equiv \bigcap_{n=1}^{\infty} V_n$ is a dense G_δ set. We show that $W \subset S_\varepsilon F$ for each $\varepsilon > 0$ and hence that W is exactly the set of points of Fréchet differentiability of F by Lemma 5.6.5. Let $x_o \in W$. Then for each n there is a δ_n such that diameter $K_{n,\delta_n}(x_o) < \frac{9}{n}$. Observe that $K_{n,\delta_n}(x_o) \neq \emptyset$ since $S_{1/n}F$ is dense. There is no loss in assuming that $\{\delta_n : n = 1, 2,\ldots\}$ is decreasing and it follows that $\{K_{n,\delta_n} : n = 1,2,\ldots\}$ is a nested sequence of sets in X . Since the diameters of these sets tend to 0 as $n \to \infty$ there is a unique $f \in X^*$ in the intersection of their closures. For any n , the ball $U_{1/n}(f)$ intersects $K_{n,\delta_n}(x_o)$ and hence

$$\|f-g\| \leqq \frac{1}{n} + \frac{9}{n} = \frac{10}{n} \quad \text{whenever } g \in S^*_{1/n}F(x) , \quad x \in S_{1/n}F , \quad \|x-x_o\| < \delta_n .$$

Now concentrate on x's as above -- i.e., assume $x \in S_{1/n}F$ and $\|x-x_o\| < \delta_n$. If $h \in \partial F(x)$ then by Lemma 5.6.4(e), $\|h-g\| \leqq \frac{1}{n}$ for $g \in S^*_{1/n}F(x)$ and hence $\|h-f\| \leqq \|h-g\| + \|g-f\| \leqq \frac{11}{n}$. Since $h \in \partial F(x)$, $h(x_o-x) \leqq F(x_o) - F(x)$ and thus

$$F(x) \leqq F(x_o) + h(x-x_o) \leqq F(x_o) + f(x-x_o) + \frac{11}{n}\|x-x_o\| .$$

That is,

$$(19) \qquad F(x) \leqq F(x_o) + f(x-x_o) + \frac{11}{n}\|x-x_o\| \quad \text{for each } x \in S_{1/n}F \text{ such that } \|x-x_o\| < \delta_n .$$

The inequality (19) thus maintains for a dense subset of the ball $U_{\delta_n}(x_o)$ (since $S_{1/n}F$ is dense) and since each of the functions in (19) is continuous, this same inequality maintains for each $x \in U_{\delta_n}(x_o)$. Thus $x_o \in S_{11/n}F$. We have just proved that $W \subset \bigcap_{n=1}^{\infty} S_{11/n}F = \bigcap_{\varepsilon>0} S_\varepsilon F$. The proof of Theorem 5.6.2 is thus complete.

<u>COROLLARY</u> 5.6.8. Let X be a Banach space with an equivalent norm which is Fréchet differentiable at each point other than 0. Then X is an Asplund space.

Proof. One proof was given in the first paragraph of this section. Here is another one. It suffices by Theorem 5.6.2 to show that the hypotheses of this corollary guarantee that there is a differentiable bump function on X. If λ is an equivalent norm on X differentiable away from 0 then λ^2 is everywhere Fréchet differentiable and $b \circ \lambda^2 : X \to \mathbb{R}$ is a differentiable bump function where b is any differentiable bump function on $\mathbb{R}$ which is nonzero at some $r > 0$. (For example $b(r) \equiv \exp((r^2-1)^{-1})$ if $|r| < 1$ and $b(r) \equiv 0$ for $|r| \geqq 1$ will do.)

There is an easy improvement of Corollary 5.6.8 available to us. Observe first that if λ denotes the norm on X so that $\lambda \equiv \lambda_{U(X^*)}$ in the terminology of 5.2.9 and if λ is Gateaux differentiable in a neighborhood of $x \neq 0$ (i.e., if $\partial\lambda$ is single valued in a neighborhood of x) then

$$(20) \qquad \lambda \text{ is Fréchet differentiable at } x \text{ if and only if } \partial\lambda \text{ is continuous at } x .$$

Indeed, Lemma 5.2.10(a) establishes one half of (20). For the other half, suppose that λ is Fréchet differentiable at $x \neq 0$, that $\lim_n \|x-x_n\| = 0$ and that both $\{f_n\} = \partial\lambda(x_n)$ for each n and $\{f\} = \partial\lambda(x)$. Because there is a subdifferential of λ at x_n in $U(X^*)$ by Lemma 5.2.9, the single valuedness of $\partial\lambda$ at x_n forces $\|f_n\| \leq 1$. Moreover, $f_n(x) = f_n(x_n) + f_n(x-x_n) \geq \|x_n\| - \|x-x_n\|$, again by 5.2.9. Thus $\lim_n f_n(x) = \|x\|$. It follows from 5.2.10(b) that $\lim_n \|f_n - f\| = 0$ as was to be shown. Thus "$\partial\lambda: X\setminus\{0\} \to X^*$ is norm-norm continuous for some equivalent norm λ on X" may be substituted with impunity for the hypotheses of Corollary 5.6.8. It is now clear that the next corollary is somewhat stronger.

<u>COROLLARY</u> 5.6.9. Let λ be an equivalent norm on X such that $\partial\lambda: X\setminus\{0\} \to X^*$ is everywhere singlevalued and norm-weak continuous. Then X is an Asplund space. (N.B. Such a norm is called <u>very smooth</u> by Diestel and Faires [1974].)

Proof. If Y is a separable subspace of X and if λ_1 denotes the restriction of λ to Y, then λ_1 is singlevalued on $Y\setminus\{0\}$ by the Hahn-Banach theorem. Let $\{y_n: n = 1,2,\ldots\}$ be a sequence in Y with $\lim_n \lambda_1(y_n - y) = 0$ for some $y \neq 0$ and suppose that both $\{f_n\} = \partial\lambda_1(y_n)$ for each n and $\{f\} = \partial\lambda_1(y)$. Let $\tilde{f}_n \in X^*$ be the unique norm-preserving extension of f_n and similarly let $\tilde{f}$ extend f without increasing its norm. Note that $\tilde{f}_n(y_n) = f_n(y_n) = \lambda(y_n)$ by Lemma 5.2.9 so that $\{\tilde{f}_n\} = \partial\lambda(y_n)$. Similarly, $\{\tilde{f}\} = \partial\lambda(y)$. By assumption $\tilde{f}_n \to \tilde{f}$ (weakly). It follows that $f_n = \tilde{f}_n|_Y \to \tilde{f}|_Y = f$ (weak convergence here too) and hence, since Y is separable, $\partial\lambda_1(Y\setminus\{0\})$ is weakly separable. Let λ_1^* denote the λ_1-dual norm on Y^*. As previously observed (see, for example, the first paragraph of this section), $\partial\lambda_1(Y\setminus\{0\}) = \{g \in Y^*: g(y) = \lambda_1^*(g) = \lambda_1(y) = 1 \text{ for some } y \in Y\}$, and this latter set is norm dense in $\{f \in Y^*: \lambda_1^*(f) = 1\}$ by the Bishop-Phelps theorem (3.3.1). Consequently, the λ_1^*-unit sphere is weakly separable, hence norm separable, and so is Y^*. It follows from Theorem 4.4.1 that X^* has the RNP and from Theorem 5.2.12 that X is Asplund.

The next corollary, stated somewhat differently, was first proved by Asplund [1968].

<u>COROLLARY</u> 5.6.10. Let X be a Banach space and suppose that $F: X \to \mathbb{R}$ is convex and continuous and that F is Fréchet differentiable on a dense subset of X. Then

the set of points at which F is Fréchet differentiable is a dense G_δ set. Thus X is an Asplund space if each convex continuous function on X is differentiable on some dense subset of X (depending on the function).

Proof. The proof of Theorem 5.6.2 only depends on the fact that whenever F is a convex continuous function on X then $S_{1/n}F$ is dense for each n. The hypotheses of this corollary guarantee somewhat more: $\bigcap_{n=1}^{\infty} S_{1/n}F = \{x: F'(x) \text{ exists}\}$ is dense. Hence the conclusion is that of Theorem 5.6.2: $\bigcap_{\varepsilon>0} S_\varepsilon F = \{x: F'(x) \text{ exists}\}$ is a dense G_δ set.

Below are two versions of a central problem remaining in this area.

<u>PROBLEM</u> 5.6.11. Let X be a Banach space. If X has a differentiable bump function must X have an equivalent norm which is Fréchet differentiable at each point other than 0?

<u>PROBLEM</u> 5.6.12. If X is an Asplund space does X have an equivalent norm which is Fréchet differentiable at each nonzero point of X ?

Of course Theorem 5.6.2 shows that a positive solution of 5.6.12 forces a positive solution of 5.6.11. Note that if 5.6.11 has a positive solution then any such X has a <u>continuously</u> differentiable bump function. Indeed, this depends on the observation made in the first paragraph of this section that the Fréchet derivative of a norm function is continuous wherever it exists.

The final result of this section is formally unrelated to the earlier material but complies with the spirit of identifying Asplund spaces in terms of differentiability behavior of special functions on the space.

<u>PROPOSITION</u> 5.6.13. Let X be a Banach space. Then the following are equivalent.

(a) X is an Asplund space.

(b) Each equivalent norm on X is Fréchet differentiable at some point of X

(c) Each equivalent norm on X is Fréchet differentiable on a dense subset of

Proof. Evidently (a) $\Longrightarrow$ (c) $\Longrightarrow$ (b).

not (a) $\Longrightarrow$ not (b) If X is not Asplund then X^* lacks the RNP and hence there is a bounded subset B of X^* which is not weak* dentable (Theorem 4.4.1). It is straightforward to demonstrate that each of the following sets in turn is not weak* dentable:

$$D \equiv w^*cl(B)\ ;\quad w^*cl(co(D \cup -D))\ ;\quad U(X^*) + w^*cl(co(D \cup -D))\ ;$$
$$U \equiv w^*cl[U(X^*) + w^*cl(co(D \cup -D))]\ .$$

Note that U is a weak* compact convex set symmetric about 0 and interior$(U) \neq \emptyset$. Hence it is the closed unit ball of a dual norm, say the dual of the norm λ on X. If λ were Fréchet differentiable at some $x \in X$ then clearly $x \neq 0$. Since $\lambda(x) = M(U,\hat{x})$, its derivative, $\lambda'(x)$, belongs to U (Lemma 5.2.9) and, from Lemma 5.2.10(b), $\lambda'(x)$ is a weak* denting point of U. Since this is impossible, λ is nowhere differentiable and the proof is complete.

§7. Duals of spaces with the RNP.

If X^* has the RNP then X is Asplund and we have seen in detail what may be expected in the way of smoothness for such spaces. Suppose, on the other hand, that X has the RNP. It is natural to inquire about the structure of X^*, the subject of the present section. Not surprisingly, X^* exhibits some nice smoothness characteristics (first discussed by Collier [1976]; stronger results have been given by Fitzpatrick [1978].) Collier's theorem follows as an easy consequence of the machinery at our disposal. Some of the results of Godefroy [1979] on the preduals and second duals of duals of RNP spaces round out the section.

The first results serve to emphasize the duality between convexity and differentiability in convenient terms.

PROPOSITION 5.7.1. (a) Let $C \subset X^*$ be a weak* compact convex set. Then λ_C is Fréchet differentiable at $x \in X$ with Fréchet derivative f if and only if $f \in C$ and x weak* strongly exposes C at f. (As usual, $\lambda_C(x) \equiv M(C,\hat{x})$ for $x \in X$.)

(b) Let D be a closed bounded convex subset of X. Here and elsewhere in this section let $\lambda_D^*(f) \equiv M(D,f)$ for $f \in X^*$. Then λ_D^* is Fréchet differentiable at g with derivative $G \in X^{**}$ if and only if $G = \hat{y}$ for some $y \in D$ and g strongly exposes D at y.

Proof. The proof is by and large a rewording of Lemmas 5.2.9 and 5.2.10.

(a) If λ_C is Fréchet differentiable at x with derivative f then $\partial\lambda_C(x) = \{f\}$ By Lemma 5.2.9 there is an element of C in $\partial\lambda_C(x)$. Hence $f \in C$ and $f(x) = \lambda_C(x)$. The fact that x weak* strongly exposes C at f is a restatement of par of 5.2.10(b). Conversely, if $f \in C$ and x weak* strongly exposes C at f then $f(x) = \lambda_C(x)$ and the fact that λ_C is Fréchet differentiable at x follows from 5.2.9 (which yields $f \in \partial\lambda_C(x)$) followed by 5.2.10(b).

(b) Let $\tilde{D} \subset X^{**}$ denote the set $w^*cl(\hat{D})$. Then $\lambda_D^*(g) = \lambda_{\tilde{D}}(g)$ for each $g \in X^*$ Thus by part (a), λ_D^* is Fréchet differentiable at g with derivative $G \in X^{**}$ if and only if $G \in \tilde{D}$ and g weak* strongly exposes $\tilde{D}$ at G. Suppose then that $G \in \tilde{D}$ and that g weak* strongly exposes $\tilde{D}$ at G. If $\{\hat{y}_\beta\}$ is a net in $\hat{D}$ converging weak* to G then $\hat{y}_\beta(g) \to G(g)$. Because g weak* strongly exposes $\tilde{D}$ at G it follows that $\lim_\beta \|\hat{y}_\beta - G\| = 0$ so that G is in the norm closure of $\hat{D}$. That is, $G \in \hat{D}$. If $G = \hat{y}$ for some $y \in D$ it is clear that g strongly exposes D at y. Conversely, if for some $y \in D$ and $g \in X^*$, g strongly exposes D a y it is a straightforward matter to show that g in fact weak* strongly exposes $\tilde{D}$ at $\hat{y}$, and it follows from (a) that λ_D^* is Fréchet differentiable at g.

A crude version of Collier's result appears next.

PROPOSITION 5.7.2. The following are equivalent.

(a) X has the RNP.

(b) λ_D^* is Fréchet differentiable on a dense G_δ subset of X^* for each closed bounded convex set $D \subset X$.

(c) λ_D^* has at least one point at which it is Fréchet differentiable whenever D is a closed bounded convex subset of X.

(d) If λ is an equivalent norm on X with dual norm λ^* on X^*, then λ^* is Fréchet differentiable at some point of X^*.

Proof.

(a) ⟹ (b) λ_D^* is Fréchet differentiable at each point of X^* which strongly exposes some point of D by 5.7.1(b). That is, λ_D^* is Fréchet differentiable at eac point of $SE(D)$, a dense G_δ subset of X^* by Theorem 3.5.4.

(b) $\Longrightarrow$ (c) and (c) $\Longrightarrow$ (d) are obvious.

(d) $\Longrightarrow$ (a) Let $U \equiv \{x: \lambda(x) \leq 1\}$ for an equivalent norm λ on X. Then there is a point $g \in X^*$ which strongly exposes U at x for some $x \in U$ by 5.7.1(b) since $\lambda^* \equiv \lambda^*_U$ in the notation of 5.2.9. Evidently x is a denting point of U, and hence U is dentable. That is, the unit ball of each equivalent norm on X is dentable under these hypotheses, and the desired conclusion follows from the last sentence of the paragraph following the proof of Lemma 2.3.5.

It should be observed that whenever D is a closed bounded convex subset of X, λ^*_D is a subadditive, positive homogeneous functional which is both norm continuous and weak* lower semicontinuous on X^*. Important from our perspective is the fact that if λ^* is any norm continuous weak* lower semicontinuous positive homogeneous and subadditive nonnegative functional on X^* then there is a closed bounded convex set $D \subset X$ such that $\lambda^* = \lambda^*_D$. (Indeed, let

$$D \equiv \{x \in X: g(x) \leq \lambda^*(g) \text{ for all } g \in X^*\} .$$

The details are not difficult.)

The next definition and theorem form the content (with different proofs) of Collier's work [1976]. The construction in the proof below was a standard feature in the early work on Asplund spaces.

DEFINITION 5.7.3. A dual Banach space X^* is called a weak* Asplund space if each norm continuous and weak* lower semicontinuous convex function on X^* is Fréchet differentiable on a dense G_δ subset of X^*.

THEOREM 5.7.4. The following are equivalent.

(a) X has the RNP.

(b) X^* is weak* Asplund.

(c) Each equivalent dual norm on X^* is Fréchet differentiable on a dense G_δ set.

(d) Each equivalent dual norm on X^* is Fréchet differentiable at some point.

Proof. Clearly (b) $\Longrightarrow$ (c) and (c) $\Longrightarrow$ (d). (d) $\Longrightarrow$ (a) was established in Proposition 5.7.2. It remains to prove that (a) $\Longrightarrow$ (b). Let $F: X^* \to \mathbb{R}$ be a continuous and weak* lower semicontinuous convex function and let $\mathcal{O}$ be any open set in X^*. In light of Corollary 5.6.10 it suffices to produce a point of $\mathcal{O}$ at which F is Fréchet differentiable. Pick $g \in \mathcal{O}$ and $\delta > 0$ such that $U_\delta[g] \subset \mathcal{O}$. By translation and appropriate addition of a constant to F there is no loss of generality in assuming that $g = 0$, that $F(0) = -1$, and that $F(f) \leq -\frac{1}{2}$ for $f \in U_\delta[0] \subset X^*$. Build a norm on $X^* \times \mathbb{R}$ by taking the gauge functional λ^* for the set

$$U \equiv \{(f,r) \in X^* \times \mathbb{R}\colon \ \|f\| \leq \delta \text{ and } F(f) \leq r \leq -F(-f)\}\ .$$

Since U is a symmetric bounded convex set in $X^* \times \mathbb{R}$ with nonempty interior, λ^* defines an equivalent norm on $X^* \times \mathbb{R}$. Moreover, since F is weak* lower semicontinuous, U is weak* compact and hence λ^* defines an equivalent dual norm on X^* $\mathbb{R}$. Thus by letting

$$D \equiv \{(x,r) \in X \times \mathbb{R}\colon \ g(x,r) \leq \lambda^*(g,r) \text{ for } (g,r) \in X^* \times \mathbb{R}\}$$

it follows that $\lambda^*_D = \lambda^*$ and λ_U defines an equivalent norm on $X \times \mathbb{R}$ whose dual norm is λ^*. Proposition 5.7.2(b) thus applies. (It is easy to check that $X \times \mathbb{R}$ has the RNP if X does. More generally, see Theorem 5.8.1.) It follows that λ^* is Fréchet differentiable on a dense G_δ subset of X^*. Of course each nonzero multiple of a point at which λ^* is differentiable is again a point of differentiability of λ^* so there must be a point $(f,r) \in X^* \times \mathbb{R}$ at which λ^* is Fréchet differentiable such that $F(f) = r$ and $\|f\| < \delta$. In particular, for this pair $(f,F(f))$ we have $\lambda^*(f,F(f)) = 1$. The proof is completed by demonstrating that F is Fréchet differentiable at f. According to Proposition 5.7.1(there is a point (x_o,t_o) in D such that $(x_o,t_o)^\wedge \in (X^* \times \mathbb{R})^*$ is the Fréchet derivative of λ^* at $(f,F(f))$ and $(f,F(f))$ strongly exposes D at (x_o,t_o). In particular, since $(0,0)$ is in the interior of D, $(x_o,t_o) \neq (0,0)$. Note that for $h \in X^*$ sufficiently close to 0 there is a constant M_1 such that $|F(f+h) - F(f)| \leq M_1\|h\|$ since F is locally Lipschitz by Lemma 5.2.6(d). Consequently there is a constant M_2 such that $\|(h,F(f+h) - F(f))\| \leq M_2\|h\|$ when h is sufficiently close to 0. Since $(x_o,t_o)^\wedge$ is the derivative of λ^* at $(f,F(f))$, given $h \in X^*$

such that $\|f + h\| < \delta$ we have

$$\begin{aligned} 0 &= \lambda^*[(f+h,F(f+h))] - \lambda^*[(f,F(f))] \\ &= \lambda^*[(f,F(f)) + (h,F(f+h) - F(f))] - \lambda^*[(f,F(f))] \\ &= (h,F(f+h) - F(f))(x_o,t_o) + o(\|(h,F(f+h)-F(f))\|) . \end{aligned}$$

That is

$$t_o[F(f+h) - F(f)] + h(x_o) = o(\|(h,F(f+h) - F(f))\|) = o(\|h\|) .$$

If $t_o = 0$ then $h(x_o) = o(\|h\|)$ which would force $x_o = 0$. But $(x_o,t_o) \neq (0,0)$ from above. Hence

$$F(f+h) - F(f) = \left(-\frac{1}{t_o}\hat{x}_o\right)(h) + o(\|h\|)$$

whence F is Fréchet differentiable at f with derivative $-\frac{1}{t_o}\hat{x}_o$. The proof is complete.

Let us turn attention now to a different aspect of duals of spaces with the RNP: namely, how such spaces behave with respect to their second duals and their preduals. The discussion here follows Godefroy [1979]. (Elementary examples demonstrate that the results which follow are not powerful enough to characterize weak* Asplund spaces. See Proposition 5.7.10 and Theorem 5.7.9 for elucidation of this comment.) Further results may be found in Godefroy [1981].

<u>DEFINITION</u> 5.7.5. (a) For any Banach space X let $i_X: X \to X^{**}$ denote the canonical injection map, so that $i_X(x) \equiv \hat{x}$ for $x \in X$.

(b) X is a <u>dualoid</u> if there is a norm one projection of X^{**} onto X.

(c) X <u>is</u> said to be <u>the unique normed predual</u> of X^* if whenever Y is a Banach space and $T: X^* \to Y^*$ is an isometric isomorphism of X^* onto Y^* then $T^*\circ i_Y: Y \to X^{**}$ is an isometric isomorphism of Y onto $\hat{X}$. (N.B. If X is the unique normed predual of X^* then Y is isometrically isomorphic to X whenever Y^* and X^* are isometrically isomorphic.)

It is clear that for any Banach space X, X^* is a dualoid since $i_{X^*}\circ(i_X)^*: X^{***} \to X^{***}$ is a norm one projection onto $(X^*)^\wedge$. Godefroy's main result (below) from our perspective singles out this projection when X^* is weak* Asplund.

THEOREM 5.7.6. Let X have the RNP. Then

(a) X is the unique normed predual of X^* ; and

(b) There is a unique norm one projection of X^{***} onto $(X^*)^\wedge$.

One of the main ingredients of the proof of this theorem is the following orthogonality result.

PROPOSITION 5.7.7. Let $F \in X^{**}$. Then the following are equivalent.

(a) $\|F + \hat{x}\| \geqq \|x\|$ for each $x \in X$.

(b) $F^{-1}(0) \cap U(X^*)$ is weak* dense in $U(X^*)$.

Proof.

(b) $\Longrightarrow$ (a) Given $\varepsilon > 0$ and $x \in X$ let

$$A_x \equiv \{f \in U(X^*): f(x) > \|x\| - \varepsilon\}.$$

Then A_x is a nonempty relatively weak* open subset of $U(X^*)$ so there is an $f_o \in A_x \cap F^{-1}(0)$. Consequently

$$(F + \hat{x})(f_o) = f_o(x) > \|x\| - \varepsilon$$

so that $\|F + \hat{x}\| > \|x\| - \varepsilon$. This half of the proposition follows now since $\varepsilon > 0$ was arbitrary.

not (b) $\Longrightarrow$ not (a) Pick any $F \in X^{**}$, $\|F\| = 1$ such that $F^{-1}(0) \cap U(X^*)$ is not weak* dense in $U(X^*)$. Since $w^*cl(F^{-1}(0) \cap U(X^*))$ is a symmetric weak* closed proper convex subset of $U(X^*)$ there is an $x_o \in X$, $\|x_o\| = 1$ and an $\alpha > 0$, $\alpha <$ such that

$$F^{-1}(0) \cap U(X^*) \subset \{f \in U(X^*): |f(x_o)| \leqq \alpha\} .$$

Let $B_{x_o} \equiv \{f \in U(X^*): f(x_o) > \alpha\}$. Since B_{x_o} is convex it is clear that the restriction of F to B_{x_o} is of constant sign, and by replacing x_o by $-x_o$ if nec essary, it is henceforth assumed that $F(f) > 0$ for $f \in B_{x_o}$.

Choose $\varepsilon > 0$ such that $(1-\varepsilon)^2 - \varepsilon > \alpha$. It will be shown that

(21) if $g \in U(X^*)$ and $g(x_0) > 1-\varepsilon$ then $F(g) \geq \varepsilon$.

Suppose not. Choose $f_0 \in U(X^*)$ such that $F(f_0) = -1+\varepsilon$ and $g_0 \in U(X^*)$ such that $g_0(x_0) > 1-\varepsilon$ and $F(g_0) < \varepsilon$. Then

$$F((1-\varepsilon)g_0 + \varepsilon f_0) < (1-\varepsilon)\varepsilon + \varepsilon(-1+\varepsilon) = 0$$

while

$$((1-\varepsilon)g_0 + \varepsilon f_0)(x_0) > (1-\varepsilon)(1-\varepsilon) + \varepsilon(-1) > \alpha .$$

Thus for $h \equiv (1-\varepsilon)g_0 + \varepsilon f_0$ we have $h \in B_{x_0}$ and $F(h) < 0$. This is impossible and thus (21) has been established. By symmetry it also follows that

(22) if $g \in U(X^*)$ and $g(x_0) < -1+\varepsilon$ then $F(g) \leq -\varepsilon$.

Now an element of X contradicting (a) of the proposition will be constructed. First note that for any $c > 0$ and $g \in U(X^*)$

(23) if $|g(x_0)| \leq 1-\varepsilon$ then $|(F - c\hat{x}_0)(g)| \leq 1 + c(1-\varepsilon)$.

Moreover, if $g \in U(X^*)$ then for any sufficiently large c it follows from (21) and (22) that

(24) if $|g(x_0)| > 1-\varepsilon$ then $|(F - c\hat{x}_0)(g)| \leq c - \varepsilon$.

Consequently, (23) and (24) may be combined to show that for c sufficiently large,

$$\|F - c\hat{x}_0\| \leq \max\{c-\varepsilon, 1+c(1-\varepsilon)\} < c = \|cx_0\|$$

which establishes not (a) (for the element $x \equiv cx_0$). Thus the proof of the proposition is complete.

The importance of the set of elements F of X^{**} "orthogonal" to X as in Proposition 5.7.7 will become apparent in what follows. This set and a useful subset of X^* are singled out first.

NOTATION 5.7.8. (a) Let $\text{Orth}(X) \equiv \{F \in X^{**}: \|F + \hat{x}\| \geq \|x\| \text{ for each } x \in X\}$.
(b) Let $C(X) \equiv \{f \in U(X^*)$: the identity map id: $(U(X^*),\text{weak*}) \to (U(X^*),\text{norm})$ is continuous at $f\}$.

THEOREM 5.7.9. Let X be a Banach space such that $\mathrm{Orth}(X)$ is a weak* closed line subspace of X^{**}. Then the following are equivalent.

(a) X is a dual space.

(b) $X^{**} = X \oplus \mathrm{Orth}(X)$.

(c) X is a dualoid.

Moreover, if X is as above and satisfies any -- hence all -- of the listed propert then there is a unique projection of norm one of X^{**} onto $\hat{X}$, and the normed pred of X is unique.

Proof. Clearly (a) $\Longrightarrow$ (c).

(c) $\Longrightarrow$ (b) Let $P: X^{**} \to \hat{X}$ be a surjective norm one projection. If $F \in P^{-1}(0)$ $x \in X$ then $\|F + \hat{x}\| \geqq \|P(F + \hat{x})\| = \|x\|$. Hence $P^{-1}(0) \subset \mathrm{Orth}(X)$. Suppose next t $F \in \mathrm{Orth}(X)$. Then $P(F) - F \in P^{-1}(0)$ so $P(F) - F \in \mathrm{Orth}(X)$. Since $\mathrm{Orth}(X)$ is linear space, $P(F)$ $(= P(F) - F + F)$ belongs to $\mathrm{Orth}(X)$ as well. Thus $\|P(F) + \hat{x}\| \geqq \|x\|$ for all $x \in X$. But $\mathrm{Range}(P) = \hat{X}$. Thus for $\hat{x} \equiv -P(F)$ we hav $\|P(F) - P(F)\| \geqq \|-P(F)\|$ and it follows that $P(F) = 0$. That is, $F \in P^{-1}(0)$ and hence $P^{-1}(0) = \mathrm{Orth}(X)$. Thus

$$X^{**} = \mathrm{Range}(P) \oplus \mathrm{Ker}(P) = \hat{X} \oplus \mathrm{Orth}(X) .$$

(b) $\Longrightarrow$ (a) For each $F \in X^{**}$ write $F = F_1 + F_2$ for $F_1 \in \hat{X}$ and $F_2 \in \mathrm{Orth}(X)$. Then $P: X^{**} \to \hat{X}$ given by $P(F) \equiv F_1$ is a projection of norm one by definition of $\mathrm{Orth}(X)$. Since $\mathrm{Orth}(X)$ is assumed to be a weak* closed subspace of X^{**}, the se

$$V \equiv \bigcap_{F \in \mathrm{Orth}(X)} F^{-1}(0) \qquad \text{(often written } (\mathrm{Orth}(X))_{\perp}\text{)}$$

is a closed subspace of X^* which has the property that

$$\mathrm{Orth}(X) = \bigcap_{f \in V} f^{-1}(0) \qquad \text{(often written } V^{\perp}\text{)}$$

Then V^* is easily seen to be isometrically isomorphic to $X^{**}/\mathrm{Orth}(X)$. (If $[F]$ $X^{**}/\mathrm{Orth}(X)$ and if $G \in [F]$ then associate $[F]$ with $G|_V$.) By hypothesis, $X^{**}/\mathrm{Orth}(X)$ is isometrically isomorphic to $\hat{X}$, hence to X. That is, X is isometrically isomorphic to the dual space V^*. Thus properties (a), (b), and (c) are equivalent.

If X satisfies (a) (or (b) or (c)) and $\mathrm{Orth}(X)$ is a weak* closed subspace of X^{**} then exactly as in the proof of (c) $\Longrightarrow$ (b), any norm one projection of X^{**} onto $\hat{X}$ must necessarily be the projection parallel to $\mathrm{Orth}(X)$. Thus the uniqueness of the projection is clear. We now verify that there is a unique normed predual of X under these hypotheses. Suppose that $V^* = X$ for a Banach space V and that for some Banach space Y, $T: X \to Y^*$ is a surjective isometric isomorphism. Here are two norm one projections of X^{**} onto $\hat{X}$:

$$i_X \circ (i_V)^* \qquad \text{and} \qquad P \equiv i_X \circ T^{-1} \circ (T^* \circ i_Y)^* .$$

Consequently by the uniqueness just established, $P = i_X \circ (i_V)^*$ and in particular, the kernels of these operators agree. That is,

$$\mathrm{Ker}((i_V)^*) = \mathrm{Ker}(i_X \circ (i_V)^*) = \mathrm{Ker}(P) = \mathrm{Ker}((T^* \circ i_Y)^*) .$$

But $\mathrm{Ker}((i_V)^*) = \{F \in X^{**}: F(\hat{v}) = 0 \text{ for each } \hat{v} \in \hat{V}\}$ $(\equiv (V)^{\perp})$ and $\mathrm{Ker}(T^* \circ i_Y)^*) = \{F \in X^{**}: F(f) = 0 \text{ for each } f \in (T^* \circ i_Y)(Y)\}$ $(\equiv [(T^* \circ i_Y)(Y)]^{\perp})$. Since both $\hat{V}$ and $T^* \circ i_Y(Y)$ are closed subspaces of X^* it follows that $\hat{V} = T^* \circ i_Y(Y)$, which was to be shown. The proof is complete.

The next proposition makes it clear that the hypotheses of Theorem 5.7.9 are met for many spaces of interest to us.

PROPOSITION 5.7.10. $\mathrm{Orth}(X)$ is a weak* closed linear subspace of X^{**} whenever X has an equivalent norm which is Fréchet differentiable on a dense set.

Proof. If $x \in X$, $\|x\| = 1$, is a point at which the norm is differentiable with derivative f, then $\|f\| = 1$ and $f(x) = 1$. Then x weak* strongly exposes $U(X^*)$ at f (Proposition 5.7.1(a)) and hence in particular, f is a point of $C(X)$ (see 5.7.8(b)). Note that $\mathrm{co}(C(X))$ must be weak* dense in $U(X^*)$. Otherwise, by the density of points of differentiability of the norm there would be an $x_0 \in X$ at which the norm is differentiable such that for some $\varepsilon > 0$, $|f(x_0)| \leq 1-\varepsilon$ for each $f \in C(X)$. If $f_0 \in U(X^*)$ denotes the derivative of the norm at x_0 then $f_0 \in C(X)$ from above and yet $f_0(x_0) = 1$. This is impossible. Hence $\mathrm{co}(C(X))$ is weak* dense in $U(X^*)$.

Now if $F \in \text{Orth}(X)$ then F is 0 on some weak* dense subset of $U(X^*)$ by Proposition 5.7.7. Hence if $f \in U(X^*)$ is a point at which $F: (U(X^*),\text{weak*}) \to \mathbb{R}$ is continuous then $F(f) = 0$. But observe that each point of $C(X)$ is such a point of relative weak* continuity of F and hence $F(f) = 0$ for each $f \in C(X)$. Consequently F is 0 on the weak* dense subset $\text{co}(C(X))$ of $U(X^*)$ when $F \in \text{Orth}(X)$. On the other hand if $G \in X^{**}$ is 0 on the weak* dense set $\text{co}(C(X)) \subset U(X^*)$ then automatically $G \in \text{Orth}(X)$ by Proposition 5.7.7. That is,

$$\text{Orth}(X) = \bigcap_{f \in \text{co}(C(X))} \hat{f}^{-1}(0) .$$

Thus $\text{Orth}(X)$ is a weak* closed subspace of X^{**} as was to be shown.

Proof of 5.7.6. If X has the RNP then X^* is weak* Asplund (Theorem 5.7.4) so each equivalent dual norm on X^* is Fréchet differentiable on a dense set. By Proposition 5.7.10 applied to X^*, Theorem 5.7.9 may be invoked with X^* in place of X in its statement. Since X^* is a dual space, the last part of Theorem 5.7.9 establishes Theorem 5.7.6.

The final result of this section is an immediate consequence of Proposition 5.7.10 and Theorem 5.7.9.

COROLLARY 5.7.11. If X is a dualoid Asplund space then X is a dual space.

§8. Stability.

A number of stability results have already been established for the sets and spaces of interest to us in various parts of these notes. (See, for example, Theorems 2.3.6, 4.1.13, and 5.2.11; Lemmas 5.3.1—5.3.4; and Corollaries 4.2.14 and 5.4.4 The following compendium contains just a sample of additional results known on this subject. Parts (b), (c), (g), and (h) appeared in Namioka and Phelps [1975] while (e) is due to Edgar [1977]. Part (f) comes from Stegall [1981] while (d) seems to be independently due to Stegall [1975] and to Namioka and Phelps [1975].

THEOREM 5.8.1. Let Y be a subspace of X and let Z be any Banach space.

(a) If $T: X \to Z$ is a continuous surjective linear operator and X is an

Asplund space then so is Z .

(b) If X is Asplund then X/Y is Asplund.

(c) If X is Asplund, then so is Y .

(d) If both Y and X/Y are Asplund then X is Asplund.

(e) If both Y and X/Y have the RNP then X has the RNP.

(f) If $T: Y \to Z$ is a surjective continuous linear operator and if X^* has the RNP then Z^* has the RNP.

(g) $\left(\oplus \sum_{\alpha \in A} X_\alpha\right)_{c_o}$ is an Asplund space if each X_α is Asplund.

(h) $\left(\oplus \sum_{\alpha \in A} X_\alpha\right)_{\ell_p}$ is an Asplund space if each X_α is Asplund and $1 < p < \infty$.

(i) $\left(\oplus \sum_{\alpha \in A} X_\alpha\right)_{\ell_p}$ has the RNP when $1 \leqq p < \infty$ if each X has the RNP.

Proof.

(a) Since $U(X) \subset X$ has the AP so does $T(U(X)) \subset Z$ (Lemma 5.3.1(a)) and hence so does $\overline{T(U(X))}$ (Theorem 5.2.11(d)). By the Baire category theorem this latter set has a nonempty interior since T is surjective. Hence $U(Z) \subset Z$ has the AP and Z is Asplund.

(b) follows directly from (a).

(c) Since $U(X) \subset X$ has the AP so does its subset $U(Y) \subset X$. Thus $U(Y) \subset Y$ has the AP by Lemma 5.3.1(b). In other words, Y is an Asplund space.

(d) Recall that $(X/Y)^*$ is isomorphic to $Y^\perp \subset X^*$ (where, as usual, $Y^\perp \equiv \{f \in X^*: f|_Y = 0\}$) and the weak* topology of $(X/Y)^*$ transfers under this isomorphism to the weak* topology of X^* restricted to Y^* . Moreover, Y^* is isomorphic to $X^*/Y^\perp$ and the quotient map $Q: X^* \to X^*/Y^\perp$ is actually the restriction map to Y . Note also that Q is weak*-weak* continuous. By using Theorem 5.2.12, the hypotheses thus imply that both $Y^\perp$ and $X^*/Y^\perp$ have the RNP. Once (e) of the present theorem is established, then, it will follow that X^* has the RNP, and hence that X is an Asplund space.

(e) Let $(\Omega, \mathcal{F}, P)$ be a probability space and $m: \mathcal{F} \to X$ a vector measure with average range in $U(X)$. Let $Q: X \to X/Y$ be the quotient map. Then $Q \circ m: \mathcal{F} \to X/Y$ is a vector measure with average range in $U(X/Y)$ so there is an $F: \Omega \to U(X/Y)$ which is Bochner integrable and $\int_A F dP = Q(m(A))$ for each $A \in \mathcal{F}$. Michael's selection theorem (see, for example, Parthasarathy [1972] or Michael [1956]) applies to this situation and guarantees that there is a norm continuous function $\psi: U(X/Y) \to X$ such that $Q\circ\psi$ is the identity map on $U(X/Y)$ and $\|\psi([x])\| \leq 2$ for each $[x] \in U(X/Y)$. Note that $\psi\circ F: \Omega \to X$ is Bochner integrable and the measure $m_1: \mathcal{F} \to X$ given by

$$m_1(A) \equiv \int_A \psi\circ F dP - m(A) \qquad \text{for} \quad A \in \mathcal{F}$$

has average range in $3U(X)$ by construction. Moreover,

$$Q(m_1(A)) = \int_A (Q\circ\psi)\circ F dP - Q(m(A)) = 0 \qquad \text{for} \quad A \in \mathcal{F}$$

and hence $m_1(A) \in Y$ for $A \in \mathcal{F}$. That is, the average range of m_1 is in $3U(Y)$ and consequently, by hypothesis, there is a Bochner integrable function $G: \Omega \to 3U(Y$ with $m_1(A) = \int_A G dP$ for all $A \in \mathcal{F}$. Thus

$$m(A) = \int_A \psi\circ F dP - m_1(A) = \int_A (\psi\circ F - G) dP$$

so the Radon-Nikodým derivative $\frac{dm}{dP}$ exists (and is $\psi\circ F - G$).

(f) Since X^* has the RNP, X is Asplund. Thus Y is Asplund by (c) and hence is Asplund by (a). It follows that Z^* has the RNP.

(g),(h) Once (i) is established, parts (g) and (h) follow from Theorem 5.2.12 and the facts that for any collection of Banach spaces $\{X_\alpha\}_{\alpha\in A}$,

$$[(\oplus \sum X_\alpha)_{c_0}]^* \text{ is isometrically isomorphic to } (\oplus \sum X_\alpha^*)_{\ell_1}$$

and

$$[(\oplus \sum X_\alpha)_{\ell_p}]^* \text{ is isometrically isomorphic to } (\oplus \sum X_\alpha^*)_{\ell_q}$$

for $1 < p < \infty$ and $\frac{1}{p} + \frac{1}{q} = 1$.

(i) Let K be a bounded convex subset of $(\oplus \sum_{\alpha\in A} X_\alpha)_{\ell_p}$ and for each subset $B \subset$ let $(\oplus \sum_{\alpha\in B} X_\alpha)_{\ell_p}$ be abbreviated X_B. Regard X_B as a subspace of X_A as well. If $\alpha \equiv \sup\{\|x\|^p: x \in K\}$ and $\varepsilon > 0$, pick $x \in K$ such that $\|x\|^p > \alpha - \varepsilon^p$. Then choose a finite subset F of A so that, if $P_F: X_A \to X_F$ denotes the natural projection, then $\|P_F(x)\| > \alpha - \varepsilon^p$. Now X_F has the RNP since F is finite (see the next paragraph for the straightforward verification of this fact) and hence $K_1 \equiv \overline{P_F(K)}$ has the RNP. Thus $K_1 = \overline{co}(\text{str exp}(K_1))$ (Theorem 3.5.4) and since $K_1 \cap \{y \in X : \|y\|^p > \alpha - \varepsilon^p\} \neq \emptyset$ there must be a $y \in \text{str exp}(K_1)$ with $\|y\|^p > \alpha - \varepsilon^p$. It follows that there is a slice S' of $P_F(K)$ (the restriction to $P_F(K)$ of a slice containing y in K_1) with diameter$(S') < \varepsilon$ and $\|z\|^p > \alpha - \varepsilon^p$ for each $z \in S'$. Let $S \equiv P_F^{-1}(S') \cap K$. Thus S is a slice of K. If $v \in S$ then

$$\|v\|^p = \|P_F(v)\|^p + \|v - P_F(v)\|^p \geq \|v - P_F(v)\|^p + \alpha - \varepsilon^p$$

so that $\|v - P_F(v)\| \leq \varepsilon$. Consequently, for any $v,w \in S$ we have

$$\|v - w\| \leq \|v - P_F(v)\| + \|P_F(v) - P_F(w)\| + \|P_F(w) - w\| < 3\varepsilon .$$

That is, diameter$(S) < 3\varepsilon$. It follows that X_A has the RNP since each bounded convex subset of X_A has slices of arbitrarily small diameter. (See Theorem 2.3.6 part (10) or part (11).)

In order to complete the proof it should be observed for the sake of thoroughness that a finite product of spaces with the RNP again has the RNP. Clearly, it suffices to show that $K_1 \times K_2 \subset X_1 \times X_2$ has the RNP whenever K_i has the RNP for $i = 1,2$. ($K_i \subset X_i$ is a closed bounded convex set.) Let $P_i\colon X_1 \times X_2 \to X_i$ be the natural projection. If (Ω,G,P) is a probability space and $m\colon G \to X_1 \times X_2$ is a measure with $m \ll \mu$ and $AR(m) \subset K_1 \times K_2$ then $P_i \circ m \ll \mu$ and $AR(P_i \circ m) \subset K_i$. Hence there is an $f_i \in L^1_{K_i}(P)$ with $d(P_i \circ m)/d\mu = f_i$. Evidently $(f_1,f_2) \in L^1_{K_1 \times K_2}(P)$ is the Radon-Nikodým derivative of $m = (P_1 \circ m , P_2 \circ m)$ with respect to μ.

§9. Addendum.

We recently learned of a series of papers by O. I. Reynov [1975, 1977, 1978a, 1978b, 1978c, 1979, 1981], dealing with some of the same topics as have appeared in this chapter, albeit from a slightly different point of view. Reynov calls a bounded set $B \subset X$ measurable in itself if for each $\mu \in P_t(U(X^*),\text{weak}^*)$, the set of functions $\{\hat{x}|_{U(X^*)} : x \in B\} \subset L^1(\mu)$ is equimeasurable. (See Definition 5.5.1.) If $B \subset X$ is bounded then B is measurable in itself if and only if it has the GSP. As Reynov's papers in which such sets are first discussed precede Stegall's work in this area, a revision of some of the historical remarks in this chapter is apparentl in order. The 1981 papers of Stegall and of Reynov both include detailed discussion of Banach spaces generated by a set with the GSP. (They are called GSG spaces.) Re nov [1981] also devotes considerable attention to an abstract formulation of weak* compact convex sets with the Radon-Nikodým Property: A compact Hausdorff space K is a compact of type RN if it is homeomorphic to a weak* compact subset D of a dua Banach space X^* such that whenever $\mu \in P_t(D,\text{weak}^*)$ and $\varepsilon > 0$ then there is a norm compact set $A_\varepsilon \subset D$ with $\mu(A_\varepsilon) > 1 - \varepsilon$. (N.B. By Theorem 4.3.11 and Lemma 4.3.10, a weak* compact convex set D in a dual space X^* has the RNP if and only if whenever $\mu \in P_t(D,\text{weak}^*)$ and $\varepsilon > 0$ then there is a norm compact set $A_\varepsilon \subset D$ with $\mu(A_\varepsilon) > 1 - \varepsilon$.) Reynov [1981] establishes that a compact Hausdorff space K is of type RN if and only if $C(K)$ is a GSG space and that, if K is of type RN there is a dense metrizable subset M of K each point of which is a G_δ in K.

S. Fitzpatrick (Separably related sets and the Radon-Nikodým Property; preprint has combined his earlier work [1980] with that of Stegall [1981] and Reynov [1981 in the following manner. Say that $K \subset X^*$ is separably related to $A \subset X$ provided that, in the notation of §5.2, (K,ρ_D) is separable for each countable bounded set $D \subset A$. Then weak* compact sets $K \subset X^*$ are RN sets in the sense of Reynov precise when they are separably related to X. Moreover, from our definition of the SP, a bounded set $A \subset X$ has the GSP - equivalently, is measurable in itself - precisely when X^* is separably related to A. The "separably related" language is convenien and natural for the study of duality between RN sets and GSP sets and much of the material in sections 1-4 of this chapter has been reformulated in this way in Fitzpa rick's paper. As he points out there, when dealing with nonconvex sets with a Radon Nikodým like property, the usual RNP definition is quite inadequate, while, as indic ted from the duality result mentioned above, weak* compact RN sets à la Reynov carry forward the desirable characteristics into the nonconvex realm nicely.

CHAPTER 6. INTEGRAL REPRESENTATIONS

The Krein-Milman theorem was the precursor of the sharper Choquet-Bishop-deLeeuw integral representation theory for compact convex sets (see Alfsen [1971], Choquet [1969] or Phelps [1966]). Since the theorem of Lindenstrauss [1966] that ℓ_1 has the KMP (Theorem 3.3.6) there has been considerable interest in developing an integral representation theory (by analogy) for closed bounded convex sets which have the KMP. The major difficulty in such a venture is that, while measures abound on compact sets by the Riesz representation theorem, the dual space of the space of continuous bounded real-valued functions on a non locally compact metric space consists of finitely additive measures - cf. Dunford and Schwartz [1958] - and an integral representation theory based on finitely additive measures is virtually useless. Nevertheless, Edgar [1975a] successfully used the MCP to provide a satisfactory analogue to the Choquet existence theorem (see Theorem 6.2.9) for closed bounded convex separable sets with the RNP. Since then a rather complete theory has emerged for closed bounded convex (possibly nonseparable) sets with the RNP. Indeed, the existence and uniqueness questions have been definitively settled; the natural orderings on appropriate sets have been successfully compared and an important new ordering introduced; and the limitations of the theory are reasonably well (though not completely) understood. Moreover a weak equivalence of the "integral representation property" and the RNP for closed bounded convex sets has been established. Each of these topics will be discussed. Just as integral representations have proved themselves valuable tools in the study of compact convex sets which arise in analysis it is to be hoped that as the subject matter of this chapter becomes more widely known, it too will see diverse application. (See §7.12 for some information about probabilistic applications.) Below is some notation which will be used without further explanation throughout this chapter.

NOTATIONAL CONVENTIONS FOR CHAPTER 6.

(1) K denotes a closed bounded convex subset of a Banach space X.

(2) Let ω_1 be the first uncountable ordinal and $\Omega \equiv \{0,1\}^{\omega_1}$ the uncountable product of two point spaces with the product topology. For each countable ordinal α

let $A_\alpha \equiv \{\omega \in \Omega: \omega(\alpha) = 0\}$ and denote by P the product measure on the σ-algebra generated by $\{A_\alpha: \alpha < \omega_1\}$ whose factor measures assign mass 1/2 to each of {0} and {1}. Let $\mathcal{F}$ be the P-completion of the above σ-algebra and $\mathcal{F}_\beta$ (β a countable ordinal) the P-completion of the σ-algebra generated by $\{A_\alpha: \alpha < \beta\}$. Thus $\mathcal{F}_0 = \{\phi,\Omega\}$ and $\mathcal{F} = \bigcup_{\alpha<\omega_1} \mathcal{F}_\alpha$.

(3) The countable product of two point spaces endowed with the product topology will be written $\tilde{\Omega}$. (Thus $\tilde{\Omega} \equiv \prod_{i=1}^{\infty} \{0,1\}$ is a compact metric space.) The natural product measure on $\mathcal{B}(\tilde{\Omega})$ (the Borel σ-algebra - cf. 4.3.1) whose factors assign mass 1/2 to both {0} and {1} will be written $\tilde{P}$. Thus $(\tilde{\Omega},\mathcal{B}(\tilde{\Omega}),\tilde{P})$ is a purely nonatomic probability space.

§1. The separable extremal ordering.

Before Mankiewicz's paper [1978] two partial orders (the "Choquet" and the "dilation" orders - see 6.3.7) were used in the study of integral representations on sets with the RNP. Each is analogous to a well known order in the compact case and each has severe (though manageable) technical drawbacks. A third ordering introduced by Edgar (cf. 6.3.7(c)) proved quite useful (see, for example, Edgar [1976] and Bourgin and Edgar [1976]) but is somewhat artificial. Mankiewicz's modification of Edgar's ordering provides an easy to use partial order on $L^1_K(\Omega,\mathcal{F},P)$ which emphasizes the separable nature of the Bochner integral (and hence the RNP).

<u>DEFINITION</u> 6.1.1. (Mankiewicz). Let $f,g \in L^1_K(\Omega,\mathcal{F},P)$. Write $f <_{SE} g$ if there is an $\alpha < \omega_1$ with $\mathbb{E}[g|\mathcal{F}_\alpha] = f$ [a.s.P]. This ordering is called the <u>separable extremal ordering</u>.

Observe that $<_{SE}$ really is a partial ordering by virtue of elementary properties of the conditional expectation operator (see for example Theorem 1.2.3). Justification for the name "separable extremal" appears in 6.2.8 and the proof of 6.2.9.

Roughly speaking the purpose of each of the partial orderings is to provide a means of comparing how close to ex(K) various probability measures on K "live". The measures on K naturally associated with $<_{SE}$ are the distributions of P by L^1_K - functions, that is, those of the form f(P) for $f \in L^1_K(P)$. (See 4.3.7 for this notation.) The first result characterizes this collection of probability measures. (It first appeared in slightly different form in Bourgin and Edgar [1976].) Recall that the collection of tight probability measures on K is denoted by $P_t(K)$ (cf. 4.3.2).

PROPOSITION 6.1.2. $\{f(P): f \in L^1_K(P)\} = \{f(\tilde{P}): f \in L^1_K(\tilde{P})\} = P_t(K)$.

Proof. If $f \in L^1_K(P)$ then f is almost separably valued by Theorem 1.1.7. Hence if $A \in F$ is chosen so that $P(A) = 0$ and $f(\Omega\backslash A)$ is separable then it is immediate that the f(P) measure of the complete separable metric space $\overline{f(\Omega\backslash A)}$ is one. But then f(P) considered as a measure restricted to $\overline{f(\Omega\backslash A)}$ is tight by Lemma 4.3.10 and hence f(P) on F is tight. That is, $\{f(P): f \in L^1_K(P)\} \subset P_t(K)$. Similarly, $\{f(\tilde{P}): f \in L^1_K(\tilde{P})\} \subset P_t(K)$.

Since $\tilde{\Omega}$ may be naturally embedded as a factor of Ω, each function $\tilde{f}$ in $L^1_K(\tilde{P})$ has an obvious extension to an element f of $L^1_K(P)$ (namely $f(\omega) \equiv \tilde{f}(\omega|_{\tilde{\Omega}})$ for $\omega \in \Omega$). Clearly $\tilde{f}(\tilde{P}) = f(P)$ and it follows that $\{f(P): f \in L^1_K(P)\} \supset \{\tilde{f}(\tilde{P}): \tilde{f} \in L^1_K(\tilde{P})\}$. Thus the proof will be completed by demonstrating that $\{\tilde{f}(\tilde{P}): \tilde{f} \in L^1_K(\tilde{P})\} \supset P_t(K)$.

As in Proposition 1.1.4 denote by [A] (for $A \in B(\tilde{\Omega})$) the equivalence class $\{A' \in B(\tilde{\Omega}): P(A \Delta A') = 0\}$. Let $[B(\tilde{\Omega})] \equiv \{[A]: A \in B(\tilde{\Omega})\}$ and note that $\tilde{P}$ determines a metric on $[B(\tilde{\Omega})]$ given by $\tilde{P}([A], [B]) \equiv \tilde{P}(A \Delta B)$. Since $\tilde{\Omega}$ is compact metric, $([B(\tilde{\Omega})],\tilde{P})$ is a separable metric space. Since $\tilde{P}$ is purely nonatomic, a theorem of Carathéodory (see, for example, Royden [1968] Theorem 2 p. 321) asserts that there is a σ-isomorphism between $([B(\tilde{\Omega})],\tilde{P})$ and $([G],\lambda)$ where λ denotes Lebesgue measure on [0,1], G is the collection of Lebesgue measurable sets in [0,1] and $([G],\lambda)$ is the metric space associated with $([0,1],G,\lambda)$ as above. Now suppose that $\mu \in P_t(K)$. Then there is a sequence of compact sets whose union has full μ-measure and hence there is a complete separable metric space $D \subset K$ with $\mu(D) = 1$. Then $([B(D)],\mu)$ is σ-isomorphic to a subalgebra of $([G],\lambda)$ by the Carathéodory result just used and hence there is a σ-isomorphism Φ between $([B(D)],\mu)$ and a subalgebra of $([B(\tilde{\Omega})],\tilde{P})$. If $Q: B(D) \to [B(D)]$ denotes the quotient map then $\Phi \circ Q$ is clearly a σ-homomorphism from $B(D)$ onto a subalgebra of $([B(\Omega)],P)$ and Royden [1968], Theorem 11, p. 329 asserts that $\Phi \circ Q$ is induced by a point mapping. That is, there is a set $A \in B(\tilde{\Omega})$ with

$\tilde{P}(A) = 1$ and a $\mathcal{B}(\tilde{\Omega}) - \mathcal{B}(D)$ measurable function $f: A \to D$ such that

$$f^{-1}(B) = \Phi \circ Q(B) \quad \text{for each} \quad B \in \mathcal{B}(D).$$

Fix $x_o \in D$ and let $f(\omega) \equiv x_o$ for $\omega \in \tilde{\Omega}\backslash A$. (In this way f may be considered a measurable map from $\tilde{\Omega}$ into D.) For each $B \in \mathcal{B}(D)$ we have

$$f(P)(B) = P(f^{-1}(B)) = P(\Phi \circ Q(B)) = P(\Phi([B])) = \mu(B)$$

by definition of "σ-isomorphism", whence $f(P) = \mu$. Now f is bounded, separably valued and measurable (hence weakly measurable). Thus $f \in L^1_D(\tilde{P}) \subset L^1_K(\tilde{P})$ by Theorem 1.1.7 as was to be shown.

If the geometry of K is to be reflected in a partial order on $P_t(K)$ (or some related set) the maximal measures - if there are any - should be those which "live" in some sense on ex(K). Without compactness to help, the existence of maximal measures is usually impossible to guarantee, and if such measures do exist, serious exertion is necessary to prove this is the case. (See, for example, Edgar [1976] and St. Raymond [1975].) Not so with the separable extremal ordering. The following straightforward result of Mankiewicz [1978] provides by contrast to the proofs of the corresponding results for the other orderings, ample justification for its early introduction in these notes.

THEOREM 6.1.3. Suppose that K has the RNP. If $f \in L^1_K(P)$ then there is a $g \in L^1_K(P)$ such that $f <_{SE} g$ and g is $<_{SE}$ - maximal.

Proof. Observe first that if $h \in L^1_K(P)$ then there is a $\gamma < \omega_1$ such that $h \in L^1_K(\Omega, F_\gamma, P)$. (In fact, since h is the almost sure limit of a sequence of simple functions each of which belongs to $L^1_K(\Omega, F_{\beta_i}, P)$ for some $\beta_i < \omega_1$, γ may be taken to be $\sup_i \beta_i$.)

Suppose there were a net $\{f_\alpha: \alpha < \omega_1\}$ of $L^1_K(P)$ functions satisfying the three conditions

(1) $f \in L^1_K(\Omega, F_\alpha, P)$ for each $\alpha < \omega_1$;

(2) $f_\alpha = \mathbb{E}[f_\beta | F_\alpha]$ whenever $\alpha \leq \beta < \omega_1$;

(3) For each $\alpha < \omega_1$ there is a $\beta > \alpha$, $\beta < \omega_1$ with $P_1(\{\omega \in \Omega: f_\alpha(\omega) \neq f_\beta(\omega)\}) > 0$.

If $A \in F$ then (since $F = \bigcup_{\alpha<\omega_1} F_\alpha$) it is possible to find $\alpha < \omega_1$ with $A \in F_\alpha$, and we let $m(A) \equiv \int_A f_\alpha dP$. Then m is well defined by (2) and it is clear that m is an X-valued measure on F such that $m << P$. Moreover, $AR(m) \subset K$. (Indeed, for any $F \in X^*$ and $A \in F_\alpha$, $F(m(A)) = \int_A F \circ f_\alpha dP \leqq (\sup\{F(x): x \in K\})(P(A))$ - recall: f_α is K valued - and the conclusion follows from the separation theorem.) But K has the RNP. Hence there is a $g \in L^1_K(P)$ such that $\int_A g dP = m(A)$ for $A \in F$ and it follows that $\mathbb{E}[g|F_\alpha] = f_\alpha$ for each $\alpha < \omega_1$. Since there is a $\gamma < \omega_1$ such that $g \in L^1_K(\Omega, F_\gamma, P)$ (see the first paragraph of this proof) it follows that for $\gamma < \alpha < \omega_1$

$$f_\alpha = \mathbb{E}[g|F_\alpha] = \mathbb{E}[\mathbb{E}[g|F_\gamma]|F_\alpha] = g \quad [\text{a.s.}P].$$

This violates condition (3) of the net $\{f_\alpha: \alpha < \omega_1\}$. Hence no such net exists.

If, on the other hand, for some $f \in L^1_K(P)$ the conclusion of Theorem 6.1.3 were false for f, a net satisfying (1), (2) and (3) above could be constructed. Indeed, assume the theorem were false for f. Since $f \in L^1_K(\Omega, F_{\alpha_o}, P)$ for some $\alpha_o < \omega_1$ set $f_{\alpha_o} \equiv f$ and for each $\beta < \alpha_o$ let $f_\beta \equiv \mathbb{E}[f_{\alpha_o}|F_\beta]$. Assume next that a net $\{f_\alpha: \alpha < \gamma\}$ has been given satisfying conditions (1) and (2) for some γ, $\alpha_o < \gamma < \omega_1$. The net may be extended at least to $\{f_\alpha: \alpha \leqq \gamma\}$ as follows. If $\gamma = \beta + 1$ for some β then $f <_{SE} f_\beta$ and f_β is not $<_{SE}$ - maximal by hypothesis. There is, then, an $h \in L^1_K(P)$ such that $f_\beta <_{SE} h$ and $P(\{\omega \in \Omega: h(\omega) \neq f_\beta(\omega)\}) > 0$. Then for some countable ordinal $\xi > \gamma$, $h \in L^1_K(\Omega, F_\xi, P)$ and hence by letting $f_\xi \equiv h$ and $f_\alpha \equiv \mathbb{E}[f_\xi|F_\alpha]$ for $\beta < \alpha < \xi$, conditions (1) and (2) continue to be met for the extended net $\{f_\alpha: \alpha \leqq \xi\}$. It remains then, to consider the case of γ a limit ordinal. In this case pick a fixed sequence $\alpha_1 < \alpha_2 < \ldots < \gamma$ with $\lim_i \alpha_i = \gamma$. Then $(f_{\alpha_i}, F_{\alpha_i})_{i=1}^\infty$ is a K-valued martingale so that by Theorem 2.3.6 part 4 there is an $f_\gamma \in L^1_K(\Omega, F_\gamma, P)$ such that $\lim_n f_{\alpha_n} = f_\gamma$ [a.s.P]. Because $\mathbb{E}[f_{\alpha_n}|F_{\alpha_k}] = f_{\alpha_k}$ [a.s.P] for each $k \leqq n$ it follows that $\mathbb{E}[f_\gamma | F_{\alpha_k}] = f_{\alpha_k}$ for each k. Thus if $\beta < \gamma$ then $\beta < \alpha_k$ for some k and hence

$$\mathbb{E}[f_\gamma \,|\, \mathcal{F}_\beta] = \mathbb{E}[\mathbb{E}[f_\gamma \,|\, \mathcal{F}_{\alpha_k}] \,|\, \mathcal{F}_\beta] = \mathbb{E}[f_{\alpha_k} \,|\, \mathcal{F}_\beta] = f_\beta \quad [\text{a.s.}P]$$

by inductive hypothesis (2). (Note that Lévy's Theorem 1.3.1 guarantees that in fact $\lim_{\beta<\gamma} f_\beta = f_\gamma$ [a.s.P].) The inductive step is thus complete. Hence there is a net $\{f_\alpha : \alpha < \omega_1\}$ satisfying (1), (2) and (3) under the hypothesis that the theorem being proved is false. Since this is impossible by what was established in the second paragraph, the proof is complete.

§2. Existence.

In this section a precise version of the following imprecise statement will be proved under appropriate hypotheses:

(I) If x_o is a point of a closed bounded convex set C there is a probability measure μ on C such that $\int_C x d\mu(x) = x_o$ and μ lives on ex(C).

There are some technical points which must be settled before stating and proving the main result (Theorem 6.2.9) and a natural starting point is with the notion of barycenter.

DEFINITION 6.2.1. Let $\mu \in P_t(K)$. Then $x_o \in K$ is the barycenter or center of mass of μ if $\int_K F(x) d\mu(x) = F(x_o)$ for each $F \in X^*$.

The existence and uniqueness assumptions about x_o in the above definition are justified by the first lemma (due to Bourgin [1971]).

Lemma 6.2.2. If $\mu \in P_t(K)$ then μ has a unique barycenter x_o. The point x_o belongs to K and $x_o = \int_K x d\mu(x)$ (as a Bochner integral).

Proof. Let $\{D_i\}$ be a (perhaps finite) sequence of pairwise disjoint compact subsets of K with $\mu(D_i) > 0$ for each i and $\mu(\cup D_i) = 1$. Let $\mu_i \in P_t(K)$ be the measure defined by the conditions

$$\mu_i(A) = \frac{1}{\mu(D_i)} \mu(A \cap D_i) \quad \text{for} \quad A \in \mathcal{B}(K).$$

Then $\mu_i(D_i) = 1$ for each i and hence from the classical result about barycenters of probability measures on compact convex sets (see for example Phelps [1966] Proposition 1.1, p. 5) there is a point $x_i \in \overline{co}(D_i)$ (since this latter set is compact and convex) which serves as barycenter for μ_i. Now $\bar{x} \equiv \Sigma\, \mu(D_i)x_i$ exists and belongs to K since $\Sigma\, \mu(D_i) = 1$ and K is a bounded set, and for each $f \in X^*$

$$f(x_o) = \sum \mu(D_i)f(x_i) = \sum \mu(D_i)\int_K f(x)d\mu_i(x) = \int_K f(x)d\mu(x).$$

Hence x_o is a barycenter of μ. But x_o is the only barycenter μ has since X* separates the points of X. Finally, since $\cup\, D_i$ is a separable set of μ-measure one, the identity map id: $(K,\mathcal{B}(K),\mu) \to K$ is Bochner integrable (Theorem 1.1.7) whence $\int_K id(x)d\mu(x) = \int_K xd\mu(x)$ exists as a Bochner integral. If $f \in X^*$ then by Theorem 1.1.8(e)

$$f\left(\int_K xd\mu(x)\right) = \int_K f(x)d\mu(x) = f(x_o).$$

Again since X* separates the points of X, it follows that $\int_K xd\mu(x) = x_o$ as was to be shown.

Technical problems are caused in the compact theory by the fact that the set of extreme points of a compact convex set need not be measurable with respect to a given probability measure. As indicated in (I) it is important to know enough about the set ex(C) to understand in what sense it could be expected that a probability measure on C live on ex(C). The next definition and theorem hold the key. (Compare 6.2.3 below to Definition 4.3.8, and see also the discussion following 4.3.8.)

DEFINITION 6.2.3. Let $(E,\mathcal{G})$ be a measurable space. If $\mathcal{M}_\mu$ denotes the μ-completion of $\mathcal{G}$ (cf. 4.3.8) for each probability measure μ on $\mathcal{G}$ then let

$$\mathcal{U}(\mathcal{G}) \equiv \{\mathcal{M}_\mu : \mu \text{ is a probability measure on } \mathcal{G}\}.$$

$\mathcal{U}(\mathcal{G})$ is known as the σ-algebra of universally measurable subsets of E relative to $\mathcal{G}$.

Note that as long as the universe M is completely metrizable and separable, there is no distinction between the tight universally measurable σ-algebra $\mathcal{U}_t(\mathcal{B}(M))$ and $\mathcal{U}(\mathcal{B}(M))$ by Lemma 4.3.10. The distinction between these σ-algebras will be important in later sections, however, when dealing with $\mathcal{U}(\mathcal{B}(K))$ for K not necessarily separable. The immediate importance of $\mathcal{U}(\mathcal{G})$ ($\mathcal{G}$ a σ-algebra) stems from the following theorem due to Lusin [1917] (cf. Diestel [1975] Appendix 1 for a proof).

THEOREM 6.2.4. If M_1 is a complete separable metric space, M_2 is metric and $\psi: M_1 \to M_2$ is continuous then $\psi(M_1) \in U(B(M_2))$.

Some basic information about universally measurable σ-algebras is gathered below. Recall first, though, that the metrizable continuous image of a complete separable metric space is said to be analytic. By Lusin's theorem, then, if A is an analytic subset of a metric space M it is μ-measurable for each probability measure μ on M. It is well known that each Borel subset of M is analytic when M is complete separable metric, and that if the complement of an analytic set is analytic, the set was Borel to begin with. (See, for example, Cohn [1980] Chapter 8 or Hoffmann-Jørgensen [1970] for these and other tidbits.) Moreover, even when $M = [0,1]$ the σ-algebra generated by the analytic subsets of $[0,1]$ is properly contained in $U(B([0,1]))$. (See Hoffmann-Jørgensen [1970], Theorem 18, p. 186 and Sierpinski and Szpilrajn [1936].)

LEMMA 6.2.5. Let (E_1,G_1) and (E_2,G_2) be measurable spaces.

(1) If $h: E_1 \to E_2$ is $G_1 - G_2$ measurable it is $U(G_1) - U(G_2)$ measurable as well

(2) Let μ be a probability measure on (E_1,G_1). Then $U(M_\mu) = M_\mu$ (where M_μ denotes the μ-completion of G_1).

(3) $U(U(G_1)) = U(G_1)$.

(4) Let K' be a closed bounded convex separable subset of a Banach space X. Let $f_1: K' \to K'$ be $U(B(K')) - U(B(K'))$ measurable and suppose that $g: \Omega \to K'$ is $F_{\alpha_0} - B(K')$ measurable for some $\alpha_0 < \omega_1$. Then $f_1 \circ g \in L^1_{K'}(\Omega, F_{\alpha_0}, P)$.

Proof. (1), (2) and (3) follow directly from the definitions. In order to verify (4) suppose that V is open in K'. Then $f_1^{-1}(V) \in U(B(K'))$. Since g is $F_{\alpha_0} - B(K')$ measurable it is $U(F_{\alpha_0}) - U(B(K'))$ measurable by (1) and since F_{α_0} is P - complete, $U(F_{\alpha_0}) = F_{\alpha_0}$ by (2). Consequently g is $F_{\alpha_0} - U(B(K'))$ measurable so that

$$(f_1 \circ g)^{-1}(V) = g^{-1}(f^{-1}(V)) \in F_{\alpha_o}$$

for each open V in K'. Consequently $f_1 \circ g$ is $F_{\alpha_o} - U(B(K'))$ measurable, hence weakly measurable. Since $f_1 \circ g$ is separably valued and has bounded range, Theorem 1.1.7 applies and $f_1 \circ g \in L^1_{K'}(\Omega, F_{\alpha_o}, P)$.

That no serious measurability question arise when K is separable is a consequence of the next result due to Bourgin [1971].

THEOREM 6.2.6. Let C be a closed convex separable subset of a Banach space. Then $ex(C) \in U(B(C))$.

Proof. Let $D \equiv \{(x,x): x \in C\} \subset C \times C$. Then $M_1 \equiv C \times C \setminus D$ is an open subset of the complete separable metric space $C \times C$ and is thus separable and completely metrizable (see, for example, Cohn [1980] Proposition 8.1.4, p. 254). Since the map $\psi: M_1 \to C$ given by $\psi(x,y) \equiv (1/2)x + (1/2)y$ is clearly continuous it follows that $\psi(M_1)$ is (an analytic set and hence) universally measurable (Theorem 6.2.4). But $\psi(M_1) = C \backslash ex(C)$. Thus $ex(C)$ $(= C \backslash (C \backslash ex(C)))$ must be universally measurable as well.

(N.B. The same proof establishes that $ex(C) \in U(B(C))$ whenever C is an analytic convex metrizable subset of a topological vector space.)

The following selection theorem, essentially due to Von Neumann [1949], will play an important role in the existence Theorem 6.2.9 when following tree branches to find extreme points. A proof may also be found in Diestel [1975] Appendix 2.

THEOREM 6.2.7. Let M_1 be a separable completely metrizable space and M_2 a metric space. If $\psi: M_1 \to M_2$ is continuous and surjective then there is an $F: M_2 \to M_1$ which is $U(B(M_2)) - B(M_1)$ measurable such that $\psi(F(x)) = x$ for each $x \in M_2$.

It is time to justify the name given the ordering $<_{SE}$. The definition below identifies measures which obviously deserve to be called separable extremal, and the theorem following it shows that such measures come from $<_{SE}$ - maximal elements.

DEFINITION 6.2.8. (Mankiewicz). Let $\mu \in P_t(K)$. Then μ is separable extremal if whenever H is a separable subspace of X such that $\mu(H \cap K) = 1$ then

$\mu(\mathrm{ex}(H \cap K)) = 1$.

Two brief observations about 6.2.8 are in order. First, $\mathrm{ex}(H \cap K) \in \mathcal{U}(\mathcal{B}(H \cap K)$ by 6.2.6 so that $\mathrm{ex}(H \cap K)$ is μ-measurable for $\mu \in P_t(K)$; and secondly, $\mathrm{ex}(H \cap K)$ may well properly contain $H \cap \mathrm{ex}(K)$. Theorem 6.2.9 below is due to Mankiewicz [1978 in its present form. As will be seen in §3, another result of Mankiewicz shows that it is equivalent to an earlier existence theorem of Edgar [1976] who used the dilation ordering.

THEOREM 6.2.9 (Edgar; Mankiewicz). Let K be a closed bounded convex subset of a Banach space X and suppose that K has the RNP. If $x_o \in K$ then there is a probability measure $\mu \in P_t(K)$ which is separable extremal and which has barycenter x_o.

Proof. Let $f \in L^1_K(P)$ be the constant function $f(\omega) \equiv x_o$ for $\omega \in \Omega$. By Theorem 6.1.3 there is a $g \in L^1_K(P)$ such that $f <_{SE} g$ and g is $<_{SE}$ - maximal. Let $\mu \equiv g(P)$. Then $\mu \in P_t(K)$ by Proposition 6.1.2. Moreover, since $\mathbb{E}[g|F_o] = f$ we have

$$\int_K x d\mu(x) = \int_\Omega g dP = \int_\Omega f dP = x_o .$$

Hence μ has barycenter x_o.

It remains to show that μ is separable extremal. If not there is a separable subspace H of X such that $\mu(K') = 1$ (where $K' \equiv H \cap K$) and $\mu(\mathrm{ex}(K')) < 1$. Thus there is an $A \in F$ with $P(A) > 0$ such that $g(\omega) \notin \mathrm{ex}(K')$ for each $\omega \in A$. Let $D \equiv \{(x,x): x \in K'\} \subset K' \times K'$ and define $\psi: K'\times K'\backslash D \to K'\backslash \mathrm{ex}(K')$ by $\psi(x,y) \equiv (1/2)x + (1/2)y$. Note that ψ is surjective and continuous. Since K' is an open subset of the complete separable metric space $K' \times K'$ it is completely metrizable (see Cohn [1980] Proposition 8.1.4, p. 254) and hence the selection Theorem 6.2.7 applies. Let $F: K'\backslash \mathrm{ex}(K') \to K'\times K'\backslash D$ be $\mathcal{U}(\mathcal{B}(K'\backslash \mathrm{ex}(K'))) - \mathcal{B}(K'\times K'\backslash D)$ measurable and satisfy $\psi(F(x)) = x$ for each $x \in K'\backslash \mathrm{ex}(K')$. Write F in the form (f_o, f_1) where $f_i: K'\backslash \mathrm{ex}(K') \to K'$ for $i = 0,1$ and observe that

$$(1/2)f_o(x) + (1/2)f_1(x) = x \quad \text{for each} \quad x \in K'\backslash \mathrm{ex}(K').$$

Moreover, $f_o(x) \neq f_1(x)$ for $x \in K'\backslash \mathrm{ex}(K')$ since Range (F) misses the diagonal D of

$K' \times K'$. Extend f_o and f_1 by letting

$$f_o(x) \equiv f_1(x) \equiv x \quad \text{for} \quad x \in ex(K').$$

It is not difficult to check that f_o and f_1 are $\mathcal{U}(\mathcal{B}(K')) - \mathcal{U}(\mathcal{B}(K'))$ measurable maps (cf. 6.2.5 part (3) and 6.2.6).

Since $g \in L^1_K(P)$ there is an $\alpha_o < \omega_1$ such that g is $\mathcal{F}_{\alpha_o}$-measurable (see the proof of 6.1.3). Thus $f_o \circ g$ and $f_1 \circ g$ are both in $L^1_{K'}(\Omega, \mathcal{F}_{\alpha_o}, P)$ by 6.2.5 part (4). Now let $h: \Omega \to K'$ be the function

$$h(\omega) \equiv (f_o \circ g)(\omega)\, \chi_{A_{\alpha_o+1}}(\omega) + (f_1 \circ g)(\omega)\, \chi_{A^c_{\alpha_o+1}}(\omega) \quad \text{for } \omega \in \Omega \;.$$

(Recall: A_β was defined in the conventions for Chapter 6 near the beginning of the chapter.) Then $h \in L^1_{K'}(\Omega, \mathcal{F}_{\alpha_o+1}, P)$ whence $h \in L^1_K(\Omega, \mathcal{F}, P)$. Moreover, it is straightforward to check that $g = \mathbb{E}[h|\mathcal{F}_{\alpha_o}]$ whence $g <_{SE} h$. But $A \subset \{\omega \in \Omega: g(\omega) \neq h(\omega)\}$ so that $P(\{\omega \in \Omega: g(\omega) \neq h(\omega)\}) > 0$. This contradicts the $<_{SE}$-maximality of g and hence μ must in fact be separable extremal.

<u>COROLLARY</u> 6.2.10. Let K be a closed bounded convex subset of a Banach space X and suppose that K has the RNP. If $\nu \in \mathcal{P}_t(K)$ there are a separable extremal measure $\mu \in \mathcal{P}_t(K)$ and functions $f, g \in L^1_K(P)$ such that $f(P) = \nu$, $g(P) = \mu$ and $f <_{SE} g$. (In particular, ν and μ have the same barycenter.)

Proof. If $\nu \in \mathcal{P}_t(K)$, apply Proposition 6.1.2 to produce $f \in L^1_K(P)$ with $f(P) = \nu$. Then mimic the proof of 6.2.9 to produce $g \in L^1_K(P)$ and $\mu = g(P)$ as desired.

§3. <u>Partial orderings.</u>

The separable extremal ordering of §1 is one of several partial orderings used to study integral representations of closed bounded convex sets with the RNP. Its main advantage lies in the ease with which it can be applied to prove the existence Theorem 6.2.9. It does not, unfortunately, appear to lend itself as well to the question of uniqueness of representation. On the other hand, at least one of the partial orders introduced here enters naturally in the uniqueness discussion. (Moreover, each of the orderings of this section is more natural than $<_{SE}$ at the very least in the sense that each is defined on the object of direct interest - $P_t(K)$ - rather than on the construct $L^1_K(P)$.) The representation of $P_t(K)$ in the dual sphere of two Banach spaces - one equipped with the weak* topology and the other with the norm topology - and their interconnections, occupies the first portion of this section. This is followed with the introduction of the partial orderings and the proof of their equivalence. The remainder of the section includes a geometric characterization of maximal elements as well as a key link with the separable extremal ordering. These results set the stage for a discussion of uniqueness which proceeds in §4.

One of the Banach spaces of interest here is $C_b(K)$ (cf. 4.3.3(a)). The other appears in 6.3.1(c).

<u>DEFINITION AND NOTATION</u> 6.3.1.

(a) Denote by $\|f\|_\infty$ the supremum norm for $f \in C_b(K)$.

(b) For $f \in C_b(K)$ let

$$\|f\|_L \equiv \sup \{\frac{|f(x) - f(y)|}{\|x - y\|} : x \neq y;\ x,y \in K\} \in [0,\infty].$$

(c) $\mathrm{Lip}(K) \equiv \{f \in C_b(K): \|f\|_L < \infty\}$.

When Lip(K) is equipped with the norm $\|\cdot\|_{BL}$ given by

$$\|f\|_{BL} \equiv \|f\|_L + \|f\|_\infty$$

then $(\mathrm{Lip}(K), \|\cdot\|_{BL})$ is a Banach space. It is called the space of <u>bounded Lipshitz functions on K</u>.

The following convention may at first seem troublesome, but the important theorem of Dudley [1966] (cf. 6.3.3) eliminates any apparent ambiguity.

<u>CONVENTION</u> 6.3.2. $P_t(K)$ may be considered as naturally embedded in either $C_b(K)^*$ or $\mathrm{Lip}(K)^*$. When $P_t(K)$ is considered a subset of $C_b(K)^*$ it will henceforth be assumed to be equipped with the relative weak*topology of $C_b(K)^*$ (so a net μ_α converges to μ - all in $P_t(K) \subset C_b(K)^*$ - if $\int g d\mu_\alpha \to \int g d\mu$ for each $g \in C_b(K)$). When, on the

other hand, $P_t(K)$ is considered as a subset of Lip(K)* its topology will be the relative norm topology as a subset of (Lip(K)*, $\|\cdot\|_{BL}^*$) (where $\|\cdot\|_{BL}^*$ denotes the dual $\|\cdot\|_{BL}$ norm).

Note that when K is not compact, $P_t(K) \subset C_b(K)^*$ is not closed in $C_b(K)^*$ even though $P_t(K) \subset \text{Lip}(K)^*$ is (according to the next theorem) a closed set. Nevertheless, there are very close ties between these two representations of $P_t(K)$ as is detailed below.

<u>THEOREM</u> 6.3.3 (Dudley). $P_t(K) \subset \text{Lip}(K)^*$ is a closed convex subset of S(Lip(K)*). Moreover, the identity map between $P_t(K) \subset C_b(K)^*$ and $P_t(K) \subset \text{Lip}(K)^*$ is a homeomorphism. (In particular, $P_t(K) \subset C_b(K)^*$ is completely metrizable.)

Some measure theoretic preliminaries precede the proof of Theorem 6.3.3.

<u>DEFINITION</u> 6.3.4. A subset $B \subset P_t(K)$ is <u>tight</u> if for each $\varepsilon > 0$ there is a compact subset D of K such that $\mu(D) > 1 - \varepsilon$ for each $\mu \in B$. (Compare with 4.3.2(a).)

The following well known results are included for the sake of thoroughness.

<u>PROPOSITION</u> 6.3.5. Suppose that μ_n n = 1,2,... and μ belong to $P_t(K)$.

(a) If $\mu_n \to \mu$ (weak*) in $P_t(K) \subset C_b(K)^*$ then $\{\mu_n\}_{n=1}^{\infty}$ is tight.

(b) The sequence $\{\mu_n: n = 1,2,\ldots\}$ converges (weak*) to μ in $P_t(K) \subset C_b(K)^*$ if and only if $\liminf_n \mu_n(0) \geqq \mu(0)$ for each open subset 0 of K.

(c) If $\{\mu_n\}_{n=1}^{\infty}$ is tight then there are $\nu \in P_t(K)$ and a subsequence $\{\mu_{n_i}: i = 1,2,\ldots\}$ of $\{\mu_n: n = 1,2,\ldots\}$ with $\mu_{n_i} \to \nu$ (weak*) in $P_t(K) \subset C_b(K)^*$.

Proof.

(a) Since each μ_n is tight there is a complete separable metric subset C of K with $\mu_n(C) = 1$ for each n. Let $\varepsilon > 0$. Temporarily fix a positive integer M, let $\{x_i\}_{i=1}^{\infty}$ be a countable dense subset of C and for each positive integer k let

$$C_k \equiv \{x \in C: \|x - x_k\| \leqq \tfrac{1}{M}(1 - \tfrac{1}{k}) \quad \text{for} \quad i = 1,\ldots,k\}.$$

Then C_k is closed and has a finite $\frac{1}{M}$ - net, interior $(C_{k+1}) \supset C_k$ and $\bigcup_{k=1}^{\infty} C_k = C$.

Choose $f_k \in C_b(K)$ so that $0 \leqq f_k \leqq 1$, $f_k|_{C_k} = 1$ and $f_k|_{K\setminus C_{k+1}} = 0$. Then

$$\chi_{C_k} \leqq f_k \leqq f_{k+1} \leqq \chi_{C_{k+2}} \quad \text{for each } k$$

so there is a k_o such that $\int f_{k_o} d\mu > 1 - \varepsilon/2^M$. Choose N so that $\int f_{k_o} d\mu_i > 1 - \varepsilon/2^M$ for $i \geqq N$ and k_i $(i = 1,\ldots,N-1)$ such that $\int f_{k_i} d\mu_i > 1 - \varepsilon/2^M$. Let $k \equiv \max\{k_o, k_1, \ldots, k_{N-1}\}$ and note that $\int f_k d\mu_i > 1 - \varepsilon/2^M$ for each $i = 1,2,\ldots$. It follows that $\mu_i(C_{k+1}) > 1 - \varepsilon/2^M$ for each i. Thus for each M there is a closed set D_M (called C_{k+1} until now) with a finite $1/M$ net such that $\mu_i(D_M) > 1 - \varepsilon/2^M$ for each i. Let $D \equiv \bigcap_{M=1}^{\infty} D_M$. Then D is compact and $\mu_i(D) \geqq 1 - \varepsilon$ for each i. That is, $\{\mu_n\}_{n=1}^{\infty}$ is tight.

(b) Fix $\varepsilon > 0$. If $\mu_n \to \mu$ (weak*) in $P_t(K) \subset C_b(K)^*$ and $D \subset K$ is closed, pick any open set $V \supset D$ with $\mu(V) < \mu(D) + \varepsilon$ and choose $f \in C_b(K)$ so that $f|_D = 1$, $f|_{K\setminus V} = 0$ and $0 \leqq f \leqq 1$. Since $\lim_n \int f d\mu_n = \int f d\mu$ by hypothesis and since $\mu_n(D) \leqq \int_K f d\mu_n$ it follows that

$$\limsup_n \mu_n(D) \leqq \lim \int f_n d\mu = \int f d\mu < \mu(D) + \varepsilon.$$

Since $\varepsilon > 0$ is arbitrary, $\liminf_n \mu_n(D^c) \geqq \mu(D^c)$ for each closed D.

Conversely, if $\liminf_n \mu_n(0) \geqq \mu(0)$ for each open $0 \subset K$ in order to prove that $\mu_n \to \mu$ (weak*) in $P_t(K) \subset C_b(K)^*$ it evidently suffices to show that $\lim_n \int f d\mu_n = \int f d\mu$ whenever $f \in C_b(K)$ and $0 < f < 1$. Fix $M > 0$ and for $i = 0,1,\ldots,M$ let

$$D_i \equiv \{x \in K: \tfrac{i}{M} \leqq f(x)\} \quad \text{and} \quad 0_i \equiv \{x \in K: \tfrac{i}{M} < f(x)\}.$$

A standard calculation shows that

$$\text{(II)} \qquad \frac{1}{M}\sum_{i=1}^{M} \mu(D_i) \leqq \int f d\mu \leqq \frac{1}{M} + \frac{1}{M}\sum_{i=1}^{M} \mu(D_i)$$

with similar inequalities when μ is replaced by μ_n and D_i is replaced by 0_i throughout. By assumption, $\limsup_k \mu_k(D_i) \leqq \mu(D_i)$ while $\liminf_k \mu_k(0_i) \geqq \mu(0_i)$ for each i. It follows from the versions of (II) with the D_i's that $\limsup_k \int f d\mu_k \leqq \frac{1}{M} + \int f d\mu$ and from the versions of (II) with the 0_i's, that $\int f d\mu \leqq \liminf_k \int f d\mu_k$. Since M is arbitrary, the result follows.

(c). Choose a sequence $\{D_k: k = 1,2,...\}$ of compact subsets of K with the properties $D_k \subset D_{k+1}$ and $\mu_i(D_k) > 1 - 1/2^k$ for $i,k = 1,2,...$. Then $\mu_i|_{D_k}$ (i.e. the measure whose value at $A \in \mathcal{B}(K)$ is $\mu_i(A \cap D_k)$) may naturally be considered an element of $C(D_k)^*$. Denote by $\mathcal{R}(D_k)$ the truncated positive cone

$$\mathcal{R}(D_k) \equiv \{t\mu: \mu \in \mathcal{P}_t(D_k) \text{ and } 0 \leqq t \leqq 1\}.$$

Since D_k is compact metric, $C(D_k)$ is separable whence bounded subsets of $C(D_k)^*$ (like $\mathcal{R}(D_k)$) are weak* metrizable. Thus $\mathcal{R}(D_k)$ is weak* compact and metrizable.

Let $\{\mu_{1,n}: n = 1,2,...\}$ be a subsequence of $\{\mu_n: n = 1,2,...\}$ such that $\{\mu_{1,n}|_{D_1}: n = 1,2,...\}$ converges in $\mathcal{R}(D_1)$ (weak* topology) to some $\nu_1 \in \mathcal{R}(D_1)$. Suppose now that $\{\mu_{k,n}: n = 1,2,...\}$ and $\nu_k \in \mathcal{R}(D_k)$ have been specified with $\mu_{k,n}|_{D_k} \underset{n}{\to} \nu_k$ (weak* topology of $C(D_k)^*$). The sequence of measures $\{\mu_{k,n}|_{D_{k+1}\setminus D_k}: n = 1,2,...\}$ in $\mathcal{R}(D_{k+1})$ has a weak* convergent subsequence, say $\{\mu_{k+1,n}|_{D_{k+1}\setminus D_k}: n = 1,2,...\}$ with limit $\gamma_{k+1} \in \mathcal{R}(D_{k+1})$ and it follows that $\{\mu_{k+1,n}|_{D_{k+1}}: n = 1,2,...\}$ converges (weak*) in $\mathcal{R}(D_{k+1})$ to $\nu_{k+1} \equiv \nu_k + \gamma_{k+1}$. (N.B. ν_k has been identified with its natural extension to a measure in $\mathcal{R}(D_{k+1})$.) Now each ν_k may be extended to a tight nonnegative measure on K (again written ν_k) and $\nu_k(K) \leqq 1$ for each k. Moreover, if $j < k$ then by part (b) of the proposition (using the fact that $D_k \setminus D_j$ is open in D_k)

$$\|\nu_j - \nu_k\| = \|\nu_j|_{D_j} - \nu_k|_{D_j}\| + \|\nu_k|_{D_k\setminus D_j}\| = \|\nu_k|_{D_k\setminus D_j}\|$$

$$= \nu_k(D_k\setminus D_j) \leqq \liminf_n\, (\mu_{k,n}|_{D_k})(D_k\setminus D_j)$$

$$\leqq 1 - (1 - 1/2^j) = 1/2^j .$$

Thus $\{\nu_k: k = 1,2,...\}$ is a norm Cauchy sequence in the norm closed set $\{t\mu: \mu \in \mathcal{P}_t(K),\ 0 \leqq t \leqq 1\} \subset C_b(K)^*$. Since $\nu_k(K) = \int 1 d\nu_k = \lim_n \int 1 d\mu_{k,n}|_{D_k} \geqq 1 - 1/2^k$ for each k it follows that $\lim_k \nu_k(K) = 1$. Hence if ν is the norm limit of $\{\nu_k: k = 1,2,...\}$ then $\nu \in \mathcal{P}_t(K)$. But for any open subset 0 of K and for any k

$$\liminf_n \mu_{n,n}(O) \geqq \liminf_n \mu_{n,n}(O \cap D_k) = \liminf_n \mu_{n,n}|_{D_k}(O) \geqq \nu_k(O)$$

by part (b) of this proposition (since $\mu_{n,n}|_{D_k} \underset{n}{\to} \nu_k$ (weak*)). It follows that $\liminf_n \mu_{n,n}(O) \geqq \nu(O)$ for each open $O \subset K$ and again by part (b), $\nu_{n,n} \to \nu$ (weak*). This completes the proof of the proposition.

In the course of the proof of Theorem 6.3.3 it will be established implicitly that $P_t(K) \subset C_b(K)^*$ is metrizable. The main step is isolated below.

LEMMA 6.3.6. The weak* topology on $P_t(K) \subset C_b(K)^*$ is first countable at each point.

Proof. Given $\mu \in P_t(K)$ let $\{D_n: n = 1,2,\ldots\}$ be an increasing sequence of compact subsets of K with $\lim_n \mu(D_n) = 1$. For each n let H_n be a countable dense collection of functions in $C(D_n)$. (Recall: D_n is compact metric so $C(D_n)$ is separable.) Identify each function in H_n with one of its extensions to an element of $C_b(K)$ which has the same sup norm as the original function. Next, for each pair of positive integers k and n let $h_{k,n} \in C_b(K)$ be any function for which $h_{k,n}|_{D_n} = 0$, $h_{k,n}|_{K\setminus(D_n+k^{-1}U(X))} = 1$ and $0 \leqq h_{k,n} \leqq 1$. Let

$$H \equiv \{h_{k,n}: k,n = 1,2,\ldots\} \cup \bigcup_{n=1}^{\infty} H_n.$$

Then H is a countable set and hence the collection

$$V \equiv \{V(G,k): G \subset H \text{ is finite, } k = 1,2,\ldots\}$$

is countable where, for each finite $G \subset H$ and positive integer k

$$V(G,k) \equiv \{\nu \in P_t(K): \left|\int gd(\mu - \nu)\right| < \frac{1}{k} \text{ for each } g \in G\}.$$

It remains to establish that V is a base for the weak* topology on $P_t(K)$ at μ, and since V is closed under finite intersections, it suffices to prove that V is a subbase at μ. In fact, another easy reduction shows that the proof will be complete once it is established that for $f \in C_b(K)$, $\|f\|_\infty = 1$ and for $\varepsilon > 0$, there is a $V \in V$ such that $V \subset \{\nu \in P_t(K): |\int fd(\mu - \nu)| < \varepsilon\}$. Thus suppose that $f \in C_b(K)$, $\|f\|_\infty = 1$ and that $\varepsilon > 0$. Choose n so that $\mu(K\setminus D_n) < \varepsilon/27$ and

$g \in H_n \subset H$ so that $\|(f - g)|_{D_n}\| < \varepsilon/8$. (In particular, then, $\|g\|_\infty < 2$.) Then for some positive integer $k > 27/\varepsilon$

$$|f(x) - g(x)| < \varepsilon/6 \text{ for each } x \in D_n + k^{-1}U(X).$$

Suppose that $\nu \in V(\{g\} \cup \{h_{k,n}\},k)$. Then

$$\nu(K\backslash(D_n + k^{-1}U(X)) \leqq \int h_{k,n}d\nu < \frac{1}{k} + \int h_{k,n}d\mu < 2\varepsilon/27.$$

Hence

$$\left|\int fd(\mu - \nu)\right| \leqq \left|\int gd(\mu - \nu)\right| + \left|\int (f - g)d(\mu - \nu)\right| < \frac{\varepsilon}{3} + \int |f - g|d(\mu + \nu)$$

$$\leqq \frac{\varepsilon}{3} + \int_{D_n+k^{-1}U(X)} (\varepsilon/6)d(\mu + \nu) +$$

$$\int_{K\backslash(D_n+k^{-1}U(X))} (\|f\|_\infty + \|g\|_\infty)d(\mu + \nu)$$

$$\leqq \frac{\varepsilon}{3} + \frac{\varepsilon}{3} + 3(\mu + \nu)(K\backslash(D_n + k^{-1}U(X))) \leqq \varepsilon$$

whence

$$V(\{g\} \cup \{h_{k,n}\},k) \subset \{\nu \in P_t(K): \left|\int fd(\mu - \nu)\right| < \varepsilon\}$$

as was to be shown. This completes the proof of 6.3.6.

We are now in a position to prove Dudley's Theorem 6.3.3. For the duration of its proof the notation $d(x,A)$ $(x \in X,\ A \in X)$ denotes the norm distance of x to the set A,

$$d(x,A) \equiv \inf\{\|x - y\|: y \in A\}.$$

<u>Proof of 6.3.3</u>. The proof that $P_t(K)$ is a closed subset of $S(\mathrm{Lip}(K)^*)$ will appear last.

Suppose that $\{\mu_n: n = 1,2,\ldots\}$ is a sequence in $P_t(K)$ and that for some $\mu \in P_t(K)$, $\lim_n \|\mu_n - \mu\|^*_{BL} = 0$. Fix an open subset O of K and for each n let $D_n \equiv \{x \in O: d(x,K\backslash O) \geqq \frac{1}{n}\}$. Then D_n is closed, $D_n \subset D_{n+1}$ for each n and $\bigcup_{n=1}^{\infty} D_n = O$. Let $f_n(x) \equiv \max\{0, 1-nd(x,D_n)\}$ for $x \in K$. Since the function $x \to 1 - nd(x,D_n)$

belongs to Lip(K), $f_n \in$ Lip(K) as well for each n. Observe that $0 \leqq f_n \leqq 1$, $f_n|_{D_n} = 1$ and $f_n|_{K\setminus O} = 0$. Given $\varepsilon > 0$ choose k so that $\mu(D_k) > \mu(O) - \varepsilon/2$ (and hence $\int f_k d\mu > \mu(O) - \varepsilon/2$). Next, choose N so that $|\int f_k d(\mu_n - \mu)| < \varepsilon/2$ for $n \geqq N$. It follows that for each $n \geqq N$

$$\mu_n(O) \geqq \int f_k d\mu_n \geqq \mu(O) - \varepsilon/2 - \varepsilon/2 = \mu(O) - \varepsilon$$

and since $\varepsilon > 0$ was arbitrary, $\liminf_n \mu_n(O) \geqq \mu(O)$. Thus $\mu_n \to \mu$ (weak* in $P_t(K) \subset C_b(K)^*$) by Proposition 6.3.5(b).

Conversely, suppose that $\{\mu_\alpha\}$ is a net in $P_t(K) \subset C_b(K)^*$ which converges weak* to $\mu \in P_t(K)$, and that $\{\mu_\alpha\}$ fails to converge in $\|\cdot\|_{BL}^*$ - norm to μ. Because the weak* topology of $P_t(K) \subset C_b(K)^*$ is first countable at μ (Lemma 6.3.6) it follows that there is also a sequence $\{\mu_n : n = 1,2,\ldots\}$ in $P_t(K)$ converging weak* to μ such that, for some $\varepsilon > 0$, $\|\mu_n - \mu\|_{BL}^* \geqq 2\varepsilon$ for each n, and consequently there is a sequence $\{f_n : n = 1,2,\ldots\}$ in U(Lip(K)) such that

$$\text{(III)} \qquad \left|\int f_n d(\mu_n - \mu)\right| \geqq \varepsilon \quad \text{for } n = 1,2,\ldots .$$

Let C be a complete separable subset of K such that $\mu_n(C) = 1$ for each n and suppose that $\{x_i\}_{i=1}^\infty$ is a countable dense subset of C. By the usual Cantor diagonalization procedure, by going to a subsequence if necessary we may assume that for each i, $\lim_n f_n(x_i)$ exists. Since $\|f_n\|_L \leqq 1$ for each n it follows that the function f, defined on $\{x_i\}_{i=1}^\infty$ by $f(x_i) \equiv \lim_n f_n(x_i)$ for each i, is uniformly continuous and thus has an extension at first to C and then to an element of $C_b(K)$ in such a way that the extension (denoted by f as well) satisfies $\|f\|_\infty \leqq 1$. Note that if $D \subset C$ is compact then $f_n|_D$ converges uniformly to $f|_D$ (again since $\|f_n\|_L \leqq 1$ for each n).

Suppose that $\delta > 0$. Then by Proposition 6.3.5(a) there is a compact set $D \subset C$ such that $\mu_n(D) > 1 - \delta/2$ for each n. Then

$$\text{(IV)} \qquad \left|\int_K f_n d\mu_n - \int_K f d\mu\right| \leqq \left|\int_K f d\mu_n - \int_K f d\mu\right| + \left|\int_D f_n d\mu_n - \int_D f d\mu_n\right| + \left|\int_{K\setminus D} f_n d\mu_n - \int_{K\setminus D} f d\mu_n\right| .$$

The first term on the right side of (IV) approaches 0 as $n \to \infty$ since $\mu_n \to \mu$ (weak*) and the second converges to 0 as $n \to \infty$ since $f_n \to f$ uniformly on D. Since $\|f\|_\infty + \|f_n\|_\infty \leqq 2$, the last term is dominated by δ and hence

(V) $$\lim_n \int_K f_n d\mu_n = \int_K f d\mu.$$

But $\mu(C) \geqq \limsup_n \mu_n(C) = 1$ by Proposition 6.3.5(b) whence $\mu(C) = \mu_n(C) = 1$ for each n. Hence (since $\lim_n f_n(x) = f(x)$ for $x \in C$)

$$\lim_n \int f_n d\mu = \int f d\mu.$$

that is, from (V),

$$\lim_n \left| \int f_n d\mu_n - \int f_n d\mu \right| = 0.$$

This contradicts (III) and shows that the identity map between $P_t(K) \subset \text{Lip}(K)^*$ and $P_t(K) \subset C_b(K)^*$ is indeed a homeomorphism.

It remains to show that $P_t(K) \subset S(\text{Lip}(K)^*)$ and that it is closed. Note that if $\mu \in P_t(K)$ then $\mu(1) = 1$ so $\|\mu\|^*_{BL} \geqq 1$, and if $f \in U(\text{Lip}(K))$ then $|\mu(f)| \leqq 1$ since $\|f\|_\infty \leqq 1$. Thus $P_t(K) \subset S(\text{Lip}(K)^*)$. Now suppose that $\{\mu_n: n = 1,2,\dots\}$ is a Cauchy sequence in $P_t(K) \subset \text{Lip}(K)^*$ and let C be a complete separable subset of K with $\mu_n(C) = 1$ for each n. It will be shown that $\{\mu_n: n = 1,2,\dots\}$ is tight. (This will enable us to apply Proposition 6.3.5(c) and produce a subsequence $\{\mu_{n_i}: i = 1,2,\dots\}$ and a $\nu \in P_t(K)$ with $\mu_{n_i} \to \nu$ (weak*). Then, applying what has earlier been established in the proof of this theorem, $\lim_i \|\mu_{n_i} - \nu\|^*_{BL} = 0$, and since $\{\mu_n: n = 1,2,\dots\}$ is a $\|\cdot\|^*_{BL}$ - Cauchy sequence, $\lim_n \|\mu_n - \nu\|^*_{BL} = 0$ as needs to be shown.)

In order to show that $\{\mu_n: n = 1,2,\dots\}$ is tight let $\{x_i\}_{i=1}^\infty$ be a countable dense subset of C, fix $\varepsilon > 0$ and for $j = 1,2,\dots$ let

$$f_j(x) \equiv \min\{1, \frac{2}{\varepsilon} d(x, \bigcup_{i=1}^{j} U_{\varepsilon/2}[x_i]\} \quad \text{for} \quad x \in K.$$

Then $f_j \in \text{Lip}(K)$, $\|f_j\|_\infty \leqq 1$, $\|f_j\|_L \leqq 2/\varepsilon$ and $0 \leqq f_{j+1} \leqq f_j$. Given $\delta > 0$ choose k_o such that whenever $n \geqq k$ then

$$\|\mu_n - \mu_{k_o}\|^*_{BL} = \frac{\delta\varepsilon}{3(2+\varepsilon)} .$$

Since $\lim_j f_j(x) = 0$ for each $x \in C$ it follows that $\lim_j \int f_j d\mu_{k_o} = 0$ (recall: $\mu_{k_o}(C) = 1$). Hence there is an integer i such that $\int f_i d\mu_{k_o} < \delta/3$ and hence $\int f_i d\mu_n < 2\delta/3$ for $n \geqq k_o$. If $1 \leqq m \leqq k - 1$ choose $i(m)$ so that $\int f_{i(m)} d\mu_m < \delta$ and let $j \equiv \max\{i(1),\ldots,i(m-1),i\}$. Then

(VI) $$\int f_j d\mu_n < \delta \quad \text{for} \quad n = 1,2,\ldots .$$

If $\bigcup_{i=1}^{j} U_\varepsilon[x_i] \equiv D(\varepsilon,\delta)$ then $D(\varepsilon,\delta)$ is a closed set with a finite ε-net and $f_j(x) = 1$ for $x \in K\backslash D(\varepsilon,\delta)$. From (VI) it follows that $\mu_n(D(\varepsilon,\delta)) \geqq 1 - \delta$ for each n. For any $\xi > 0$ let

$$D(\xi) \equiv \bigcap_{n=1}^{\infty} D(\frac{1}{n}, \frac{\xi}{2^{n+1}}).$$

Then $D(\xi)$ is compact and $\mu_n(D(\xi)) \geqq 1 - \xi$ for each n. That is, $\{\mu_n\}_{n=1}^{\infty}$ is tight, and from the preceding remarks, this completes the proof.

The orderings of interest in this section (and which will be used in the discussion of uniqueness in §4) appear below. The first two are obvious analogues of well known orderings in the compact case, (see, for example, Phelps [1966]) while the third is due to Edgar [1976].

<u>DEFINITION</u> 6.3.7. (a). If $\mu,\nu \in P_t(K)$ write $\mu <_C \nu$ if $\int_K f d\mu \leqq \int_K f d\nu$ for each conv $f \in C_b(K)$. $<_C$ is called the <u>Choquet ordering</u>.

(b). Suppose that $\mu,\nu \in P_t(K)$ and that $T: K \to P_t(K) \subset \mathrm{Lip}(K)^*$. Assume that T satisfies the following conditions:

(1) T is $U(B(K)) - B(\mathrm{Lip}(K)^*)$ measurable and T is separably valued. (Hence $T \in L^1_{P_t(K)}(K,U(B(K)),\mu)$.)

(2) $\int_K T d\mu = \nu$.

(3) There is a set $A \in B(K)$ with $\mu(A) = 1$ such that for each $x \in A$, $T(x)$ has barycenter x. (That is, $\int y dT(x)(y) = x$ [a.s.μ].) Then <u>T is called a μ-dilation</u> and ν is said to be a dilation of μ. Write $\mu <_D \nu$ if such a T exists for μ and ν.

(c). Let $\mu,\nu \in P_t(K)$. Suppose that for some probability space (Γ,G,Q) and σ-algebra $H \subset G$ there are functions $\pi_1,\pi_2 \in L^1_K(\Gamma,G,Q)$ such that $\pi_1(Q) = \mu$, $\pi_2(Q) = \nu$ and

$\pi_1 = \mathbb{E}[\pi_2 | \mathcal{H}]$. Then write $\mu <_E \nu$.

It is not at all clear apriori that $<_C$, $<_D$ or $<_E$ is a partial ordering nor is it evident that these orderings are equivalent. The relationship between these orderings and the separable extremal ordering is also far from transparent. Most of these topics will be dealt with in this section, and the rest will be covered in §7.9. In proving the orderings of 6.3.7 equivalent it will be necessary to produce a Radon-Nikodým derivative of a certain measure-valued measure on a probability space, and what is required is the interesting result below, an immediate consequence of work done independently by Edgar [1975b] and by Goldman [1977]. (Note that without parts of Theorem 6.3.3, the statement of Theorem 6.3.8 would not even make sense. For a broader view - at the cost of increased technique - see the remarks in §6.6 and see §7.12.)

<u>THEOREM</u> 6.3.8. $P_t(K) \subset \mathrm{Lip}(K)^*$ has the RNP.

(N.B. $P_t(K)$ as a norm closed subset of $C_b(K)^*$ rarely has the KMP let alone the RNP. For example, when $K = [0,1]$, Lebesgue measure does not belong to the norm closed convex hull of $\mathrm{ex}(P_t([0,1]))$.

<u>Proof of 6.3.8.</u> Let (Γ, G, Q) be a probability space and $m: G \to \mathrm{Lip}(K)^*$ a vector measure with $m \ll Q$ such that $AR(m) \subset P_t(K)$. Write m_A in place of $m(A)$ for $A \in G$. Since m_Γ is tight, given $\varepsilon > 0$ there are compact sets $C_1 \subset C_2 \subset \ldots \subset K$ with $m_\Gamma(C_n) \geqq 1 - 2^{-2n}\varepsilon^2$ for each n. Let $Q_n: G \to \mathbb{R}$ be the measure

$$Q_n(A) \equiv m_A(C_n) \quad \text{for} \quad A \in G.$$

Then $Q_n \geqq 0$ and when $A \in G$ satisfies $Q(A) > 0$ then, $\frac{1}{Q(A)} m_A \in P_t(K)$. Hence $Q_n(A) = m_A(C_n) \leqq Q(A)$ and it follows that $0 \leqq Q_n \leqq Q$ as measures. Thus by the classical Radon-Nikodým Theorem 1.0.1, there is an $f_n \in L^1(Q)$ such that $\int_A f_n dQ = m_A(C_n)$ for each $A \in G$ and since $m_A(C_n) \leqq Q(A)$ for each such A, $0 \leqq f_n \leqq 1$ [a.s.Q]. Let

$$A_n \equiv \{z \in \Gamma: f_n(z) \geqq 1 - 2^{-n}\varepsilon\}.$$

Then

$$1 - 2^{-2n}\varepsilon^2 \leqq m_\Gamma(C_n) = \int_\Gamma f_n dQ = \int_{A_n} f_n dQ + \int_{A_n^c} f_n dQ$$

$$\leqq Q(A_n) + (1 - 2^{-n}\varepsilon)(1 - Q(A_n)).$$

Consequently $1 - 2^{-n}\varepsilon \leqq Q(A_n)$. Let $A_\varepsilon \equiv \bigcap_{n=1}^{\infty} A_n$ so that $\mu(A_\varepsilon) \geqq 1 - \varepsilon$. If $A \in G$, $A \subset A_\varepsilon$ and $Q(A) > 0$ then

$$m_A(C_n) = \int_A f_n dQ \geqq (1 - 2^{-n}\varepsilon)Q(A)$$

so that $\frac{m_A(C_n)}{Q(A)} \geqq 1 - 2^{-n}\varepsilon$. Thus $AR(m|_{C_n})$ is a tight subset of $P_t(K) \subset C_b(K)^*$. (See Theorem 2.2.6 for this notation.) From Theorem 6.3.3, $P_t(K) \subset C_b(K)^*$ is metrizable and in conjunction with Proposition 6.3.6(c) it follows that tight subsets of $P_t(K) \subset C_b(K)^*$ are relatively weak* compact. Again by Theorem 6.3.3, then, $AR(m|_{C_n}) \subset P_t(K) \subset \mathrm{Lip}(K)^*$ is relatively norm compact. Thus the hypotheses of Theorem 2.2.6 part 2 are satisfied so that $\frac{dm}{dQ}$ exists. It follows that $P_t(K) \subset \mathrm{Lip}(K)^*$ has the RNP as was to be shown.

The proof of the next theorem is due to Edgar [1978a] based in part on a paper of Strassen [1965]. Earlier results of Bourgin and Edgar [1976], Edgar [1976] and St. Raymond [1975] originally established the various equivalences stated below. (Observe: at this stage of our development, $<_C$, $<_D$ and $<_E$ are not known to be partial orders.)

THEOREM 6.3.9. Let $\mu, \nu \in P_t(K)$. Then the following are equivalent:

(a) $\mu <_C \nu$. (b) $\mu <_D \nu$. (c) $\mu <_E \nu$.

Proof.

(b) ⇒ (c) Let T be a μ-dilation for which $\int T d\mu = \nu$. Let $\Gamma \equiv K \times K$, $G \equiv B(K \times K)$, $H \equiv B(K) \times \{\emptyset, K\}$ and π_1 and π_2 be, respectively, the first and second coordinate projections of $K \times K$ onto K. For $D \in G$ and $x \in K$ let $D_x \equiv \{y \in K: (x,y) \in D\}$. By the monotone class theorem (see, for example, Halmos [1950] p. 143, Theorem A) given $\gamma \in P_t(K)$ the map $x \to \gamma(D_x)$ is $B(K)$-measurable. Slightly more generally, the map $x \to (\Sigma \chi_{A_i} \gamma_i)(D_x)$ is $B(K)$-measurable whenever $\{A_i\}$ is a finite collection chosen from $\mathcal{B}(K)$ and $\gamma_i \in P_t(K)$ for each i. But since T is $\mathcal{U}(\mathcal{B}(K))$ - $\mathcal{B}(\mathrm{Lip}(K)^*)$ measurable and separably valued, it is strongly $\mathcal{U}(\mathcal{B}(K))$ measurable (cf. Theorem 1.1.7) and hence it is the almost sure limit of a sequence of simple functions. It follows that $x \to T(x)(D_x)$ is $\mathcal{U}(\mathcal{B}(K))$ measurable for each $D \in \mathcal{G}$. Consequently we may define a measure Q on $\mathcal{G}$ by

$$Q(D) \equiv \int_K T(x)(D_x)d\mu(x) \quad \text{for} \quad D \in \mathcal{G}$$

and standard arguments yield

$$\int_{K\times K} fdQ = \int_K\int_K f(x,y)dT(x)(y)d\mu(x)$$

for any bounded f in $L^1(Q)$.

In order to show that $\mu <_E \nu$ first note that when $A \in \mathcal{B}(K)$ then

$$\pi_1(Q)(A) = Q(\pi_1^{-1}(A)) = \int_K T(x)((A \times K)_x)d\mu(x) = \int_A T(x)(K)d\mu(x) = \mu(A)$$

while

$$\pi_2(Q)(A) = Q(\pi_2^{-1}(A)) = \int_K T(x)((K \times A)_x)d\mu(x) = \int_K T(x)(A)d\mu(x)$$

$$= [\int_K T(x)d\mu(x)](A) = \nu(A).$$

Hence $\pi_1(Q) = \mu$ and $\pi_2(Q) = \nu$. Finally, if $A \in \mathcal{B}(K)$ and $g \in X^*$ then

$$\text{(VII)} \quad \begin{aligned} \int_{A\times k} g\circ\pi_2 dQ &= \int_A\int_K g(y)dT(x)(y)d\mu(x) = \int_A g(x)d\mu(x) \\ &= \int_A\int_K g(x)dT(x)(y)d\mu(x) = \int_{A\times K} g\circ\pi_1 dQ. \end{aligned}$$

Since π_1 is clearly $\mathcal{H}$-measurable, it follows from (VII) that $\pi_1 = \mathbb{E}[\pi_2 | \mathcal{H}]$. That is, $\mu <_E \nu$.

<u>(c) ⇒ (a)</u> Suppose that $\pi_1 = \mathbb{E}[\pi_2 | \mathcal{H}]$, that $\pi_1(Q) = \mu$ and that $\pi_2(Q) = \nu$. Let $f \in C_b(k)$ be convex. By Jensen's inequality (see, for example, Meyer [1966] p. 29).

$$\int_K fd\mu = \int_K fd\pi_1(Q) = \int_{K\times K} f\circ\pi_1 dQ = \int_{K\times K} f\circ\mathbb{E}[\pi_2 | \mathcal{H}]dQ$$

$$\leqq \int_{K\times K} E[f\circ\pi_2 | \mathcal{H}]dQ = \int_{K\times K} f\circ\pi_2 dQ = \int_K fd\pi_2(Q) = \int_K fd\nu .$$

Thus $\mu <_C \nu$.

(a) $\Rightarrow$ (b) The proof of this part is somewhat involved. Briefly, we introduce the vector space V of measurable functions on K to $C_b(K)$ which assume only finitely many values, together with the sublinear functional ρ on V which μ-averages the upper envelopes of the elements of V. The constant functions in V form a subspace (which looks like $C_b(K)$) on which the linear functional which takes ν-averages (henceforth called L) is dominated by ρ . Once the Hahn-Banach theorem is invoked, a $C_b(K)^*$-valued measure m may be defined at each $A \in \mathcal{U}(\mathcal{B}(K))$ to be that linear functional on $C_b(K)$ whose value at $f \in C_b(K)$ is (the extended version of) L evaluated at the element of V which is f on A and 0 elsewhere. It turns out that $AR(m) \subset \mathcal{P}_t(K)$ and since $\mathcal{P}_t(K) \subset \text{Lip}(K)^*$ has the RNP, $\frac{dm}{d\mu} \equiv T$ exists. Then T will (almost) be the sought for μ-dilation. The details follow.

For $f \in C_b(K)$ denote by $\hat{f}$ the <u>upper envelope</u> of f. Thus

$$\hat{f}(x) \equiv \inf \{g(x): g \in C_b(K), \text{ g affine on K}, g \geqq f\}.$$

The following easily verified facts will be needed about the upper envelope function. Let f, $g \in C_b(K)$ and $t > 0$. Then

(1) $\hat{f}$ is concave and upper semicontinuous.

(2) $f \leqq \hat{f}$.

(3) $\widehat{f+g} \leqq \hat{f} + \hat{g}$.

(4) $|\hat{f}(x) - \hat{g}(x)| \leqq \|f - g\|$ for each $x \in K$.

(5) $\widehat{tf} = t\hat{f}$.

(6) $\hat{f}(x) = \inf \{h(x): h \in C_b(K), \text{ h concave}, h \geqq f\}$.

(7) $f = \hat{f}$ if f is concave.

(Properties (6) and (7) follow from the separation theorem in $X \times \mathbb{R}$. For example, to establish (6) let $\tilde{f}(x) \equiv \inf \{h(x): h \in C_b(K), \text{ h concave}, h \geqq f\}$ and suppose $\tilde{f}(x_0) < \hat{f}(x_0)$ for some $x_0 \in K$. Since $\tilde{f}$ is upper semicontinuous it is possible to strictly separate $(x_0, \hat{f}(x_0))$ from $\{(x,t): x \in K, t \leqq \tilde{f}(x)\}$ - a closed convex set in $X \times \mathbb{R}$ - by a hyperplane Z in $X \times \mathbb{R}$. The function $g: X \to \mathbb{R}$ defined at x to be the unique real number g(x) such that $(x, g(x)) \in Z$ is easily seen to be affine and continuous. It also has the properties $g \geqq f$ and $g(x_0) < \hat{f}(x_0)$, which is impossible

Thus $\tilde{f} \geqq \hat{f}$ on K and since the reverse inequality is immediate, $\tilde{f} = \hat{f}$ on K.)

Let V be the vector space over $\mathbb{R}$

$$V \equiv \{\psi\colon K \to C_b(K)\colon \psi \text{ is } \mathcal{B}(K) - \mathcal{B}(C_b(K)) \text{ measurable and } \psi \text{ is a simple function}\}$$

and let $\rho\colon V \to \mathbb{R}$ be defined by

$$\rho(\psi) \equiv \int_K \widehat{\psi(x)}(x)\,d\mu(x) \quad \text{for} \quad \psi \in V.$$

(Note that if $\psi \in V$ takes value f_j on A_j $(j = 1,\ldots, n)$ where $f_j \in C_b(K)$ for each j and $\{A_j\}_{j=1}^n$ is a partition of K chosen from B(K) then for any $x \in K$ and for the unique i for which $x \in A_i$

$$\widehat{\psi(x)}(x) = \hat{f}_i(x) = \sum_{j=1}^n \hat{f}_j(x)\chi_{A_j}(x).$$

Hence $x \to \psi(x)(x)$ is a bounded $\mathcal{B}(K)$ measurable function so the definition of ρ makes sense.) Moreover, from the listed properties of the upper envelope function, ρ is a sublinear functional on V.

For $f \in C_b(K)$ and $A \in \mathcal{B}(K)$ let $\chi_A \otimes f \in V$ be the function

$$\chi_A \otimes f(x) \equiv \begin{cases} f & \text{if} \quad x \in A \\ 0 & \text{otherwise} \end{cases}$$

and let $W \equiv \{\chi_K \otimes f\colon f \in C_b(K)\}$. Note that W is a linear subspace of V. Now define L on W by

$$L(\chi_K \otimes f) \equiv \int_K f\,d\nu \quad \text{for} \quad \chi_K \otimes f \in W.$$

We will show that $L \leqq \rho$ on W. For this purpose fix $f \in C_b(K)$ and let $H \equiv \{h \in C_b(K)\colon$ h is concave, $h \geqq f\}$. Observe that the pointwise infimum of any two elements of H again belongs to H. Thus given a compact subset D of K, since C(D) is separable there is a sequence $\{h_n\colon n = 1,2,\ldots\}$ in H such that $h_1 \geqq h_2 \geqq \cdots \geqq \hat{f}$ and $\inf_n h_n(x) = \hat{f}(x)$ for each $x \in D$. Consequently $\lim_n \int_D h_n d\mu = \int_D \hat{f} d\mu$ and since μ is tight

$$\text{(VIII)} \qquad \int_K \hat{f}\,d\mu = \inf\left\{\int_K h\,d\mu\colon h \in H\right\}.$$

But for $h \in H$, $\int f d\nu \leqq \int h d\nu$ since $f \leqq h$, and since $-h$ is convex, $\int h d\nu \leqq \int h d\mu$ by hypothesis. That is, using (VIII) and property (6) of upper envelopes,

$$\int f d\nu \leqq \inf \{\int_K h d\mu : h \in H\} = \int_K \hat{f} d\mu .$$

In other words, $L(\chi_K \otimes f) \leqq \rho(\chi_K \otimes f)$ as predicted. Thus the Hahn-Banach theorem guarantees an extension of L (again called L) to a linear functional on V such that $L(\psi) \leqq \rho(\psi)$ for each $\psi \in V$.

For each $A \in \mathcal{U}(\mathcal{B}(K))$ let m_A be the linear functional on $C_b(K)$

$$m_A(f) \equiv L(\chi_A \otimes f) \quad \text{for} \quad f \in C_b(K).$$

Now for $f \in C_b(K)$, $f \geqq 0$, we have $\widehat{(-f)} \leqq 0$ and hence

$$m_A(-f) = L(\chi_A \otimes (-f)) \leqq \rho(\chi_A \otimes (-f)) = \int_A \widehat{(-f)} d\mu \leqq 0$$

so that

(IX) $$m_A \geqq 0 \quad \text{on} \quad C_b(K).$$

Moreover, for any affine $g \in C_b(K)$, $\hat{g} = g$ so that

$$m_A(g) = L(\chi_A \otimes g) \leqq \rho(\chi_A \otimes g) = \int_A \hat{g} d\mu = \int_A g d\mu$$

and similarly (since $-g$ is also affine) $m_A(-g) \leqq \int_A -g d\mu$. That is

(X) $$m_A(g) = \int_A g d\mu \text{ for each affine } g \in C_b(K).$$

In particular, then,

(XI) $$m_A(1) = \mu(A).$$

Define m by

$$m(A) \equiv m_A \quad \text{for each} \quad A \in \mathcal{U}(\mathcal{B}(K)).$$

We will establish that m is a countably additive $C_b(K)^*$-valued measure with $AR(m) \subset \mathcal{P}_t(K)$. First note that (by going back to the definition of m_A) it is easily checked that m is finitely additive. Next, for $B \in \mathcal{U}(\mathcal{B}(K))$ and for $f \geqq 0$, $f \in C_b(K)$ it follows from (IX) that

(XII) $$0 \leq m_B(f) = m_K(f) - m_{K\setminus B}(f) \leq m_K(f) = L(\chi_K \otimes f) = \int_K f\,d\nu.$$

In particular, then, if $\{f_n : n = 1,2,...\}$ is a decreasing sequence in $C_b(K)$ with $\lim_n f_n(x) = 0$ for each $x \in K$ then

$$0 \leq \lim_n m_B(f_n) \leq \lim_n \int f_n d\nu = 0.$$

Hence (by the Daniell criterion for countable additivity- see, for example, Royden [1968] Proposition 21, p. 299) m_B corresponds to a countably additive measure on $\mathcal{U}(\mathcal{B}(K))$. (Another easy consequence of (XII) is that $m_B \in C_b(K)^*$ for each $B \in \mathcal{U}(\mathcal{B}(K))$.) Now that m_B is countably additive, though, it follows from (XII) that as measures, $0 \leq m_B \leq \nu$ on $\mathcal{U}(\mathcal{B}(K))$ and since ν is tight, so is m_B for each $B \in \mathcal{U}(\mathcal{B}(K))$. Hence by (XI), $\frac{m_A}{\mu(A)} \in \mathcal{P}_t(K)$ whenever $A \in \mathcal{U}(\mathcal{B}(K))$ and $\mu(A) > 0$. That is, $AR(m) \subset \mathcal{P}_t(K)$. Finally, recall that m is finitely additive and for each A, $m_A \geq 0$. Thus from (XI), for $\{A_n : n = 1,2,...\}$ a decreasing sequence in $\mathcal{U}(\mathcal{B}(K))$ with empty intersection, we have

$$\lim_n \|m(A_n)\| = \lim_n m_{A_n}(1) = \lim_n \mu(A_n) = 0$$

and hence m is countably additive.

Since $\mathcal{P}_t(K) \subset \mathrm{Lip}(K)^*$ has the RNP by Theorem 6.3.8, there is a $T \in L^1_{\mathcal{P}_t(K)}(K,\mathcal{U}(\mathcal{B}(K)),\ )$ such that

(XIII) $$\int_A T d\mu = m_A \quad \text{for each} \quad A \in \mathcal{U}(\mathcal{B}(K)).$$

(We may also assume that T is separably valued by modifying T on a set of μ-measure zero if necessary.) In particular, then, $\int_K T d\mu = m_K$. But for each $f \in C_b(K)$

$$\int f dm_K = L(\chi_K \otimes f) = \int f d\nu$$

and hence $m_K = \nu$. That is, $\int_K T d\mu = \nu$. In order to show that T is a μ-dilation and thus complete the proof of Theorem 6.3.9, choose any $g \in X^*$ and $A \in \mathcal{U}(\mathcal{B}(K))$. Then $\int_A T d\mu = m_A$ and since "evaluation at g" is a continuous linear functional on $\mathrm{Lip}(K)^*$ it follows from (X) that

(XIV) $$\int_A \int_K g(y)dT(x)(y)d\mu(x) = \int_A T(x)(g)d\mu(x) = m_A(g) = \int_A g d\mu .$$

Since $A \in \mathcal{U}(\mathcal{B}(K))$ is arbitrary in (XIV), $\int_K g(y)dT(x)(y) = g(x)$ for each x in a set $K\backslash Z(g)$ of μ-measure one, possibly depending on g. The elementary Lemma 6.3.10 below establishes that there is a closed convex separable subset C of K such that $T(x)(C) = 1$ for each $x \in K$ since T is separably valued. If $\{g_n\}_{n=1}^{\infty} \subset X^*$ separates the points of C and if $x \in K\backslash \bigcup_{n=1}^{\infty} Z(g_n)$ (a set of μ-measure one) then $\int g_n dT(x) = g_n(x)$ for each n. Since the barycenter of $T(x)$ lies in C by Lemma 6.2.2 it follows that $T(x)$ has barycenter x. Thus $\mu <_D \nu$ as was to be shown.

<u>LEMMA</u> 6.3.10. Let $T: K \to P_t(K) \subset \mathrm{Lip}(K)^*$ be separably valued. Then there is a closed convex separable subset C of K such that $T(x)(C) = 1$ for each $x \in K$.

Proof. If $\{\xi_n\}_{n=1}^{\infty}$ is a countable dense subset of $T(K) \subset P_t(K)$, for each n let C_n be a σ-compact subset of K such that $\xi_n(C_n) = 1$ and let $C \equiv \overline{\mathrm{co}}(\bigcup_{n=1}^{\infty} C_n)$. Then C is separable, closed bounded and convex. If $x \in K$ then let $\{\gamma_n : n = 1,2,\ldots\}$ be a sequence chosen from $\{\xi_n\}_{n=1}^{\infty}$ with $\lim_n \|\gamma_n - T(x)\|_{BL}^* = 0$. By Theorem 6.3.3, $\gamma_n \to T(x)$ (weak*) in $P_t(K) \subset C_b(K)^*$ and hence by Proposition 6.3.5(b), $\limsup_n \gamma_n(C) \leqq T(x)(C)$ But $\gamma_n(C) = 1$ for each n whence $T(x)(C) = 1$ as well.

Now that $<_C$, $<_D$ and $<_E$ are equivalent, in order to show they are partial orderings it suffices to show that one of them is. The first part of the following result appeared in Bourgin and Edgar [1976] although it apparently is much older. (See LeCam [1957].)

<u>THEOREM</u> 6.3.11. $<_C$ is a partial ordering. Hence $<_D$ and $<_E$ are partial orderings as well.

Proof. Since $<_C$ is clearly both reflexive and transitive we need only show that if $\mu <_C \nu$ and $\nu <_C \mu$ then $\mu = \nu$ for $\mu,\nu \in P_t(K)$. Suppose first that $D \subset K$ is compact and convex and for each n let $f_n(x) \equiv \inf \{n\|x - y\| : y \in D\}$ for $x \in K$. Let $g_n(x) \equiv \max \{1, f_n(x)\}$ $(x \in K)$. Then f_n and g_n belong to $C_b(K)$ and $\lim_n (g_n(x) - f_n(x)) = \chi_D(x)$ for $x \in K$. Since μ and ν agree on convex functions in $C_b(K)$ and since $|g_n(x) - f_n(x)| \leqq 1$ for $x \in K$ it follows that

$$\mu(D) = \lim_n \int (g_n - f_n)d\mu = \lim_n \int (g_n - f_n)d\nu = \nu(D).$$

Hence μ and ν agree on compact convex subsets of K.

Suppose next that B is a compact (but not necessarily convex) subset of K and that O is a relatively open subset of K with $B \subset O$. Construct $D_1,\ldots,D_n$ compact convex subsets of K such that $B \subset \bigcup_{i=1}^{k} D_i \subset O$ and observe that since μ and ν agree on all compact convex subsets of K (and in particular on all intersections chosen from $\{D_1,\ldots,D_n\}$) we have $\mu(\bigcup_{i=1}^{k} D_i) = \nu(\bigcup_{i=1}^{k} D_i)$. Thus $\mu(B) = \nu(B)$ for all compact $B \subset K$ whence $\mu = \nu$.

For the purposes of application to integral representations, it is mainly the collection of separable extremal (as well as, potentially at least at this stage, the collection of $<_C$- or $<_D$- or $<_E$- maximal) measures which is of interest since in some sense each measure in such a collection has its support in ex(K). In the remainder of this section these sets of measures will be compared and a geometrically pleasing and easily applied criterion for maximality described. These results will then be applied in §4. The next definition and theorem are due to Edgar [1976].

<u>DEFINITION</u> 6.3.12. Let $\mu \in P_t(K)$. A subset D of K is <u>μ-movable</u> if $D \in \mathcal{U}(\mathcal{B}(K))$ and if there is a μ-dilation T such that $T(x) \neq \varepsilon_x$ for each $x \in D$. (Here ε_x denotes the point mass at x. Thus $\varepsilon_x(A) = \chi_A(x)$ for $A \in \mathcal{U}(\mathcal{B}(K))$.)

<u>THEOREM 6.3.13</u>. μ is $<_D$-maximal (or $<_C$-maximal or $<_E$-maximal) if and only if $\mu(D) = 0$ for each μ-movable subset D of K.

Proof. Let $\mu \in P_t(K)$, $D \in \mathcal{U}(\mathcal{B}(K))$, $\mu(D) > 0$ and suppose that D is μ-movable by a μ-dilation T. Then $\int T d\mu = \nu$ for some $\nu \in P_t(K)$ by Theorem 6.3.3. Of course $\mu <_D \nu$ and for this half of the proof it remains to show that $\mu \neq \nu$. By Lemma 6.3.10 there is a separable closed convex subset C of K such that $T(x)(C) = 1$ for each $x \in K$. If $\{g_n : n = 1,2,\ldots\}$ is a sequence in X* which separates the points of C with $\|g_n\| = 1$ for each n, let $f \equiv \sum_{n=1}^{\infty} 2^{-n} g_n^2$. Then $f \in C_b(K)$ is a convex function and $f|_C$ is strictly convex. Let $\gamma \in P_t(K)$ and suppose that $\gamma(C) = 1$. If γ has barycenter x_0 (hence $x_0 \in C$ by Lemma 6.2.2) and $\gamma \neq \varepsilon_{x_0}$ then there is a compact subset A of C with

$\gamma(A) > 0$ and $\gamma(C\backslash A) > 0$. Now $\gamma_1 \equiv \frac{1}{\gamma(A)}\gamma|_A$ and $\gamma_2 \equiv \frac{1}{1-\gamma(A)}\gamma|_{C\backslash A}$ are in $P_t(K)$, they have barycenters x_1 and x_2 respectively in C, and for an appropriate choice of A, $x_1 \neq x_0$. (The intersection of C with some closed ball not containing x_0 has positive measure and inside that ball there is a compact A of positive measure.) It is easy to check that $\int_K f d\gamma_1 \geqq f(x_1)$ and $\int_K f d\gamma_2 \geqq f(x_2)$. Consequently $\int_K f d\gamma = \gamma(A) \int_K f d\gamma_1 + (1 - \gamma(A)) \int_K f d\gamma_2 \geqq \gamma(A) f(x_1) + (1 - \gamma(A)) f(x_2) > f(x_0)$. Thus for μ-almost every x in D, T(x) has barycenter x and $\int_K f(y) dT(x)(y) > f(x)$. It follows that

$$\begin{aligned}\int_K f d\nu &= \int_K\int_K f(y)dT(x)(y)d\mu(x)\\ &= \int_D\int_K f(y)dT(x)(y)d\mu(x) + \int_{K\backslash D}\int_K f(y)dT(x)(y)d\mu(x)\\ &> \int_D f(x)d\mu(x) + \int_{K\backslash D} f(x)d\mu(x) = \int f d\mu.\end{aligned}$$

Hence $\mu \neq \nu$ and μ is not $<_D$-maximal.

Conversely, suppose that $\mu \in P_t(K)$ and that $\mu(D) = 0$ for each μ-movable set D. Let T be a μ-dilation and let C be a closed convex separable set in K with $\mu(C) = 1$. Then

$$D \equiv \{x \in C: T(x) \neq \varepsilon_x\} \in U(B(K))$$

and clearly D is μ-movable. Hence $\mu(D) = 0$. That is, $T(x) = \varepsilon_x$ [a.s.μ]. It follow[s] that $\int_K T d\mu = \mu$ and thus μ is $<_D$-maximal.

The link between $<_D$-maximal measures and separable extremal ones could not be nicer. It was discovered by Mankiewicz [1978] and allows us to speak glibly of maximal measures without regard to which ordering among $<_C$, $<_D$, $<_E$ or $<_{SE}$ is being referred to, without fear of reprisal. (By virtue of this next result, the existenc[e] Theorem 6.2.9 may be invoked while struggling with different orderings in the quest for uniqueness.) For a more detailed description of the relationships between $<_D$ and $<_{SE}$, see §7.9.)

<u>THEOREM</u> 6.3.14. If $\mu \in P_t(K)$ then μ is $<_D$-maximal (or $<_C$-maximal or $<_E$-maximal) if and only if μ is separable extremal.

Proof. If $\mu \in P_t(K)$ is not $<_D$-maximal then there is a set $D \in U(\mathcal{B}(K))$ with $\mu(D) > 0$ which is μ-movable (Theorem 6.3.13). Let T be a μ-dilation with $T(x) \neq \varepsilon_x$ for $x \in D$ and let C be a closed convex separable subset of K with $T(x)(C) = 1$ for each $x \in K$ (Lemma 6.3.10). For each $x \in D$ there is a closed convex $B_x \subset C$ with $T(x)(B_x) > 0$ and $x \notin B_x$. The measure

$$\frac{1}{T(x)(B_x)} T(x)|_{B_x}$$

is tight and has barycenter x_1 in B_x. Thus $x_1 \neq x$ and $T(x)(B_x) < 1$. If x_2 denotes the barycenter of

$$\frac{1}{1 - T(x)(B_x)} T(x)|_{K \setminus B_x}$$

then $x_2 \in C$ and $x = T(x)(B_x)x_1 + (1 - T(x)(B_x))x_2$, a nontrivial convex combination of points of C. It follows that $x \in C\setminus ex(C)$ for each $x \in D$. Let $H \equiv \overline{span}(C)$ and observe that H is separable and that $C \subset K \cap H$. Thus $x \notin ex(K \cap H)$ since $x \notin ex(C)$. That is, $\mu(D) > 0$ and $D \cap ex(K \cap H) = \emptyset$. Thus μ is not separable extremal.

Suppose, on the other hand, that μ is not separable extremal. Then there is a separable subspace H of X such that $\mu(H \cap K) = 1$ and $\mu(ex(H \cap K)) < 1$. Let $K_1 \equiv K \cap H$, let $D \equiv \{(x,x): x \in K_1\} \subset K_1 \times K_1$ and define $\psi: K_1 \times K_1 \setminus D \to K_1\setminus ex(K_1)$ by $\psi(x,y) \equiv (1/2)x + (1/2)y$. (See the proof of Theorem 6.2.9 for more details in what follows.) Apply Theorem 6.2.7 to produce a $U(\mathcal{B}(K_1\setminus ex(K_1))) - \mathcal{B}(K_1 \times K_1 \setminus D)$ measurable $F: K_1\setminus ex(K_1) \to K_1 \times K_1 \setminus D$ such that $\psi(F(x)) = x$ for all x in $K_1\setminus ex(K_1)$. Extend F to a map from K_1 into $K_1 \times K_1$ by letting $F(x) \equiv (x,x)$ for $x \in ex(K_1)$, write $F = (f_0, f_1)$ where each f_i has domain and range in K_1 and observe that each f_i is $U(\mathcal{B}(K_1))$ - $U(\mathcal{B}(K_1))$ measurable. For any $x_0 \in K$ define $T: K \to P_t(K)$ as follows:

$$T(x) \equiv \begin{cases} \frac{1}{2}\varepsilon_{f_0(x)} + \frac{1}{2}\varepsilon_{f_1(x)} & \text{if } x \in K_1 \\ \varepsilon_{x_0} & \text{if } x \in K\setminus K_1. \end{cases}$$

Then T is separably valued and it is routine to establish that T is actually a μ-dilation. (N.B. The measurability of the f_i's above determined the measurability restriction imposed on μ-dilations in Definition 6.3.7(b).) Let $\nu \equiv \int T d\mu$ so that

$\nu \in P_t(K)$ and $\mu <_D \nu$. Exactly as in the proof of Theorem 6.3.13, since $T(x) \neq \varepsilon_x$ on a set of positive μ-measure (since $\mu(ex(K_1)) < 1 = \mu(K_1)$) it follows that $\mu \neq \nu$. That is, μ is not $<_D$-maximal, and the proof of Theorem 6.3.14 is thus complete.

The following corollary (which incorporates Edgar's original and historically important contribution [1975a] to integral representations on sets with the RNP) is now obvious.

COROLLARY 6.3.15. Suppose that K is a separable closed bounded convex subset of a Banach space.

(a) If $\mu \in P_t(K)$ is separable extremal (or $<_C$- , $<_D$- or $<_E$- maximal) then $\mu(ex(K)) = 1$.

(b) If K also has the RNP and if $x_0 \in K$ then there is a $\nu \in P_t(K)$ with barycenter x_0 such that $\nu(ex(K)) = 1$.

§4. Uniqueness.

A finite dimensional simplex S is usually defined to be the convex hull of finitely many affinely independent points. It has the property that each point of S is a convex combination of the vertices of S in a unique way. (Or, in the language of probability measures, for each point x_0 of S there is a unique probability measure μ on $\mathcal{B}(S)$ with $\mu(ex(S)) = 1$ whose barycenter is x_0). In extending this result to the case of infinite dimensional simplices, the first job is to provide an adequate definition for a possibly infinite dimensional simplex. The conventional wisdom is to abandon any attempt at a direct generalization of the above definition and to use instead an algebraic characterization of finite dimensional simplices which requires no modification in order to serve as a general definition. This will be our approach as well.

Say that a subset S of a vector space V over $\mathbb{R}$ is in general position if S lies in a hyperplane of V which misses the origin. (Note that for any subset S of V, $S \times \{1\} \subset V \times \mathbb{R}$ is in general position.) If S is a convex subset of V and S is in general position let

$$C(S) \equiv \{ts: t \geqq 0, s \in S\}$$

denote the cone over S with vertex 0. Define a partial ordering on span(S) by $x \leqq y$ if and only if $y - x \in C(S)$.

DEFINITION 6.4.1. Let S be a convex subset of a vector space V over $\mathbb{R}$ and assume that S is in general position. (If it is not, replace S by $S \times \{1\}$ and V by $V \times \mathbb{R}$.) Then S is a simplex if and only if the partial ordering $\leqq$ on span(S) with positive cone $C(S)$ is a lattice ordering on span(S). (That is, S is a simplex if and only if each two elements of span(S) have a least upper bound - equivalently, each two elements of span(S) have a greatest lower bound - in span(S).)

It is not difficult to prove that finite dimensional simplices may in fact be characterized as in 6.4.1. (See, for example, Phelps [1966], Proposition 9.11, p. 75.) The main theorem of this section may now be stated. It is due to Bourgin and Edgar [1976]. (Independently, St. Raymond [1975] established a version of Theorem 6.4.2 when K is separable.)

THEOREM 6.4.2. (Bourgin, Edgar). Let K be a closed bounded convex subset of a Banach space X and assume that K has the RNP. Then K is a simplex if and only if for each $x_0 \in K$ there is a unique separable extremal measure $\mu \in P_t(K)$ with barycenter x_0.

(N.B. Theorem 6.4.2 may of course be equivalently formulated in terms of $<_C$ -, $<_D$ -, or $<_E$ - maximal measures. See Theorems 6.3.9 and 6.3.14.)

The proof of Theorem 6.4.2 requires two preliminary results, the first of which is often called the Riesz decomposition lemma.

LEMMA 6.4.3. Let S be a simplex in V and assume that S is already in general position. Let $a_1,\dots,a_n$ and $b_1,\dots,b_k$ be points of $C(S)$ such that $\sum_{i=1}^{n} a_i = \sum_{j=1}^{k} b_j$. Then there are points $z_{i,j}$ $(i = 1,\dots,n,\ j = 1,\dots,k)$ in $C(S)$ such that

$$a_i = \sum_{j=1}^{k} z_{i,j} \quad \text{for each } i \quad \text{and} \quad b_j = \sum_{i=1}^{n} z_{i,j} \quad \text{for each } j.$$

This will be displayed in tabular form:

$$\begin{array}{cccccc} & b_1 & b_2 & \cdot\ \cdot\ \cdot & b_k \\ a_1 & z_{1,1} & z_{1,2} & \cdot\ \cdot\ \cdot & z_{1,k} \\ a_2 & z_{2,1} & z_{2,2} & \cdot\ \cdot\ \cdot & z_{2,k} \\ \cdot & \cdot & \cdot & & \cdot \\ \cdot & \cdot & \cdot & & \cdot \\ \cdot & \cdot & \cdot & & \cdot \\ a_n & z_{n,1} & z_{n,2} & \cdot\ \cdot\ \cdot & z_{n,k} \end{array}$$

Proof. The proof is by induction on k and n and easily reduces to the case $n = k = 2$. Thus suppose $a_i, b_i \in C(S)$ $i = 1,2$ and that $a_1 + a_2 = b_1 + b_2$. Let

$$z_{1,1} \equiv a_1 \wedge b_1; \quad z_{1,2} \equiv a_1 - z_{1,1}; \quad z_{2,1} \equiv b_1 - z_{1,1}; \quad z_{2,2} \equiv a_2 - z_{2,1}.$$

Then clearly $0 \leqq z_{1,1}$, $0 \leqq z_{1,2}$ and $0 \leqq z_{2,1}$. Moreover,

$$z_{1,2} + a_2 = a_1 + a_2 - z_{1,1} = b_1 + b_2 - z_{1,1} = z_{2,1} + b_2$$

whence $z_{2,2} = a_2 - z_{2,1} = b_2 - z_{1,2}$. It follows that $z_{1,j} + z_{2,j} = b_j$ for $j = 1,2$ and $z_{i,1} + z_{i,2} = a_i$ for $i = 1,2$. The proof will be completed by showing that $z_{2,2} \geqq 0$. Note first that

$$z_{1,2} \wedge z_{2,1} = (a_1 - z_{1,1}) \wedge (b_1 - z_{1,1}) = a_1 \wedge b_1 - z_{1,1} = 0.$$

But $z_{2,1} \leqq z_{2,1} + b_2 = z_{1,2} + a_2$ so that

$$z_{2,1} = z_{2,1} \wedge (z_{1,2} + a_2) \leqq z_{2,1} \wedge z_{1,2} + z_{2,1} \wedge a_2 = z_{2,1} \wedge a_2.$$

That is, $z_{2,1} \leqq a_2$ whence $0 \leqq a_2 - z_{2,1} = z_{2,2}$.

The other prerequisite to the proof of Theorem 6.4.2 is another algebraic result which takes care of the substantive part of the "easy" direction of the proof of that theorem: namely, if each point of K has a unique separable extremal representing measure then K is a simplex.

LEMMA 6.4.4. Let K be a closed bounded convex subset of a Banach space and denote by M the collection of separable extremal measures in $P_t(K)$. Then $M \subset C_b(K)^*$ is a simplex.

Proof. In order to check that M is convex, suppose that $t \in (0,1)$ and that $\mu, \nu \in M$. If $D \in U(B(K))$ is $t\mu + (1 - t)\nu$-movable then there is a $t\mu + (1 - t)\nu$-dilation T such that $T(x) \neq \varepsilon_x$ for each $x \in D$. Since $0 < t < 1$ it follows that T is both a μ-dilation and a ν-dilation as well, and hence (since μ and ν are $<_D$- maximal by Theorem 6.3.14) $\mu(D) = \nu(D) = 0$ by Theorem 6.3.13. It follows that $(t\mu + (1 - t)\nu)(D) = 0$ and thus $t\mu + (1 - t)\nu$ is $<_D$-maximal (Theorem 6.3.13), hence separable extremal (Theorem 6.3.14). That is, $t\mu + (1 - t)\nu \in M$. Now $M \subset P_t(K)$ and $P_t(K) \subset C_b(K)^*$ is already in general position. The proof that M is a simplex proceeds in two elementary steps: the first establishes that $P_t(K)$ is a simplex while in the second, the location of M in $P_t(K)$ is described (indirectly) as that of an extremal subset (whose cone induces an hereditary subordering of that induced by $P_t(K)$.)

If $\mu_1, \mu_2 \in C(P_t(K))$ then let

$$\mu_1 \vee \mu_2 \equiv (\mu_1 - \mu_2)^+ + \mu_2$$

where $(\mu_1 - \mu_2)^+$ denotes the positive part of $\mu_1 - \mu_2$ in its Hahn decomposition.) (Equivalently,

$$\mu_1 \vee \mu_2 = \left(\frac{d\mu_1}{d(\mu_1 + \mu_2)} \vee \frac{d\mu_2}{d(\mu_1 + \mu_2)}\right) (\mu_1 + \mu_2).)$$

in general, for $\mu, \nu \in \text{span}(P_t(K))$ write $\mu \equiv \mu_1 - \mu_2$ and $\nu \equiv \nu_1 - \nu_2$ where $\mu_i, \nu_i \in C(P_t(K))$ for $i = 1,2$. Then let

$$\mu \vee \nu \equiv (\mu_1 + \nu_2) \vee (\nu_1 + \mu_2) - (\mu_2 + \nu_2).$$

It is easily checked that $\text{span}(P_t(K))$ with the ordering induced by $C(P_t(K))$ is a lattice with $\mu \vee \nu$ the least upper bound for μ and ν. Hence $P_t(K)$ is a simplex.

Temporarily denote by $\leqq_P$ the partial ordering induced by $C(P_t(K))$ on $\text{span}(P_t(K))$ (and by $\wedge_P$ the greatest lower bound in this ordering) and let $\leqq_M$ denote the partial ordering on $\text{span}(M)$ induced by $C(M)$. Observe that

(XV) if $\gamma \in C(P_t(K))$, $\mu \in C(M)$ and $\gamma \leqq_P \mu$ then $\gamma \leqq_M \mu$.

Indeed, suppose that $D \in U(B(K))$ were γ-movable by a γ-dilation T. By modifying T if necessary on a set of γ-measure zero we may assume that $\{x \in K: T(x)$ has barycenter $x\}$ has μ-measure one as well as γ-measure one and hence T may be considered both a γ-dilation and a μ-dilation. Then $\mu(D) = 0$ (Theorems 6.3.13 and 6.3.14) and since $\gamma \leqq_P \mu$, $\gamma(D) = 0$ as well. That is, $\gamma \in C(M)$ and $\gamma \leqq_M \mu$ (Theorems 6.3.13 and 6.3.14 again). Condition (XV) is often expressed as saying that $C(M)$ is an hereditary subcone of $C(P_t(K))$.

If $\mu,\nu \in C(M)$ then $\mu,\nu \in C(P_t(K))$ and hence they have a greatest lower bound in the $\leqq_P$ ordering since $P_t(K)$ is a simplex. Let $\xi \equiv \mu \wedge_P \nu$. (Note that $0 \leqq_P \mu$ and $0 \leqq_P \nu$ together imply that $0 \leqq_P \xi$.) Since $0 \leqq_P \xi \leqq_P \mu$ it follows from (XV) that $\xi \in C(M)$ and that $\xi \leqq_M \mu$. Similarly, $\xi \leqq_M \nu$. In order to show that ξ is the $\leqq_M$ greatest lower bound of μ and ν, then, it clearly suffices to establish that whenever $\alpha \in C(M)$ satisfies $\alpha \leqq_M \mu$ and $\alpha \leqq_M \nu$ then $\alpha \leqq_M \xi$. But $0 \leqq_M \alpha \leqq_M \mu$ implies that $0 \leqq_P \alpha \leqq_P \mu$, and $\alpha \leqq_M \nu$ implies $\alpha \leqq_P \nu$. Since $\xi = \mu \wedge_P \nu$ it follows that $\alpha \leqq_P \xi$. Since $\xi \in C(M)$ from above, it follows from (XV) that $\alpha \leqq_M \xi$. That is, ξ is the $\leqq_M$ greatest lower bound of μ and ν. Finally, suppose that $\mu,\nu \in \mathrm{span}(M)$ (rather than the previous, more restrictive assumption that $\mu,\nu \in C(M)$). An easy calculation shows that (for $\mu \equiv \mu_1 - \mu_2$, $\nu \equiv \nu_1 - \nu_2$, $\mu_i,\nu_i \in C(M)$ for $i = 1,2$)

$$\mu \wedge_M \nu \equiv (\mu_1 + \nu_2) \wedge_M (\mu_2 + \nu_1) - (\mu_2 + \nu_2).$$

That is, $\leqq_M$ is a lattice ordering on $\mathrm{span}(M)$ and the proof is complete.

The proof of 6.4.2 need be delayed no longer.

Proof of 6.4.2. Suppose first that for each $x_0 \in K$ there is a unique separable extremal measure $\mu \in P_t(K)$ with barycenter x_0. If $M \equiv \{\mu \in P_t(K): \mu$ is separable extremal$\}$ the barycenter map $r: M \to K$ which associates $\mu \in M$ with its barycenter in K is by assumption one-one and surjective. Since r is always affine and since M is a simplex (Lemma 6.4.4) it follows that $r(M) = K$ is also a simplex.

Conversely, suppose that K is a simplex and that $x_0 \in K$. By Theorem 6.2.9 there is at least one $\mu \in M$ with barycenter x_0 and hence the proof of the theorem reduces to showing that when K is a simplex and $x_0 \in K$ there is at most one $\mu \in M$ with barycenter x_0. In fact, we will show that

(XVI) Given $\mu_1,\mu_2 \in P_t(K)$ both with barycenter x_0, there is a $\mu_3 \in P_t(K)$ such that both $\mu_1 <_E \mu_3$ and $\mu_2 <_E \mu_3$.

(Then, assuming (XVI), if $\mu_1,\mu_2 \in M$ both have barycenter x_0, pick μ_3 as in (XVI). Since $\mu \in M$ if and only if μ is $<_E$-maximal by Theorem 6.3.14, it follows that $\mu_1 = \mu_3 = \mu_2$ and hence uniqueness is guaranteed.)

Thus suppose $\mu_1,\mu_2 \in P_t(K)$ and that both measures have barycenter x_0. The notational conventions outlined at the beginning of this chapter will be used in the remainder of this section, and in addition, for each n = 1,2,..., G_n will denote the finite σ-algebra of subsets of $\tilde{\Omega}$

$$G_n \equiv \{A \times \prod_{i=n+1}^{\infty} \{0,1\}: A \subset \prod_{i=1}^{n} \{0,1\} \text{ is arbitrary}\}.$$

According to Proposition 6.1.2 there are functions f and g in $L^1_K(\tilde{\Omega},\mathcal{B}(\tilde{\Omega}),\tilde{P})$ such that $f(\tilde{P}) = \mu_1$ and $g(\tilde{P}) = \mu_2$. Let $f_n \equiv \mathbb{E}[f|G_n]$ and $g_n \equiv \mathbb{E}[g|G_n]$. Then (f_n,G_n) and (g_n,G_n) are K-valued martingales on $(\tilde{\Omega},\mathcal{B}(\tilde{\Omega}),\tilde{P})$. By Lévy's Theorem 1.3.1, then, $\lim_n f_n = f$ [a.s.$\tilde{P}$] and $\lim_n g_n = g$ [a.s.$\tilde{P}$].

We will construct a sequence $\{Q_n: n = 1,2,...\}$ of probability measures on $G_n \times G_n$ together with a sequence $\{h_n: n = 1,2,...\}$ of functions from $\tilde{\Omega} \times \tilde{\Omega}$ to K such that for each n

(1) $Q_{n+1}|_{G_n \times G_n} = Q_n$.

(2) h_n is $G_n \times G_n - \mathcal{B}(K)$ measurable.

(3) $\int_{A_1 \times A_2} h_{n+1} dQ_{n+1} = \int_{A_1 \times A_2} h_n dQ_n$ for $A_1, A_2 \in G_n$.

(4) $Q_n(A \times \tilde{\Omega}) = Q_n(\tilde{\Omega} \times A) = \tilde{P}(A)$ for $A \in G_n$.

(5) $\int_{\tilde{\Omega} \times A} h_n dQ_n = \int_A f d\tilde{P}$ for $A \in G_n$.

(6) $\int_{A \times \tilde{\Omega}} h_n dQ_n = \int_A g d\tilde{P}$ for $A \in G_n$.

Assume for the time being that such sequences $\{Q_n : n = 1,2,...\}$ and $\{h_n : n = 1,2,...$ exist. For each A in the algebra $\mathcal{A} \equiv \bigcup_{n=1}^{\infty} G_n \times G_n$ set $\bar{Q}(A) \equiv Q_k(A)$ for any k for which $A \in G_k \times G_k$ (and observe that $\bar{Q}$ is well defined by (1)). If $\{A_i : i = 1,2,...\}$ is a pairwise disjoint sequence in $\mathcal{A}$ whose union $\bigcup_{i=1}^{\infty} A_i \equiv A$ also belongs to $\mathcal{A}$ then A belongs to some $G_k \times G_k$ and is, therefore, compact. But each A_i also belongs to some $G_j \times G_j$ and is consequently open. It follows that finitely many of the A_i's cover A and that there is an n such that $A_i = \emptyset$ for $i > n$. Since $\bar{Q}$ is finitely additive on $\mathcal{A}$,

$$\bar{Q}(\bigcup_{i=1}^{\infty} A_i) = \bar{Q}(\bigcup_{i=1}^{n} A_i) = \sum_{i=1}^{n} \bar{Q}(A_i) = \sum_{i=1}^{\infty} \bar{Q}(A_i) .$$

Thus $\bar{Q}$ is countably additive on $\mathcal{A}$. By the Carathéodory extension theorem (see, for example, Royden [1968] Theorem 8, p. 257) there is a unique probability measure Q on $\mathcal{B}(\tilde{\Omega} \times \tilde{\Omega})$, the σ-algebra generated by $\mathcal{A}$, such that $Q(A) = \bar{Q}(A)$ for $A \in \mathcal{A}$. That is, $Q(A) = Q_m(A)$ for $A \in G_m \times G_m$ for each m. Conditions (2) and (3), rephrased, state that $(h_n, G_n \times G_n)$ is a K-valued martingale based on $(\tilde{\Omega} \times \tilde{\Omega}, \mathcal{B}(\tilde{\Omega} \times \tilde{\Omega}), Q)$. Thus by Theorem 2.3.6 part 4, $h \equiv \lim_n h_n$ exists [a.s.Q]. Since Q is the unique extension of $\bar{Q}$ it follows from (4) that $Q(\tilde{\Omega} \times A) = \tilde{P}(A)$ for each $A \in \mathcal{B}(\tilde{\Omega})$. Define $\bar{f}: \tilde{\Omega} \times \tilde{\Omega} \to K$ by $\bar{f}(\xi,\omega) \equiv f(\omega)$ for $(\xi,\omega) \in \tilde{\Omega} \times \tilde{\Omega}$. Then $\bar{f}$ is $\{\emptyset, \tilde{\Omega}\} \times \mathcal{B}(\tilde{\Omega}) - \mathcal{B}(K)$ measurable and $\bar{f}(Q) = f(P) = \mu_1$. Moreover, from (5), if $A \in G_n$ then

$$\int_{\tilde{\Omega}\times A} h dQ = \int_{\tilde{\Omega}\times A} h_n dQ = \int_{\tilde{\Omega}\times A} h_n dQ_n = \int_A f d\tilde{P}$$

and hence, at first for each $A \in \bigcup_{n=1}^{\infty} G_n$ and then more generally for each $A \in \mathcal{B}(\tilde{\Omega})$

$$\text{(XVII)} \qquad \int_{\tilde{\Omega}\times A} h dQ = \int_{\tilde{\Omega}\times A} \bar{f} dQ.$$

Thus $\mathbb{E}[h|\{\emptyset, \tilde{\Omega}\} \times \mathcal{B}(\tilde{\Omega})] = \bar{f}$. A similar modification of g yields a function $\bar{g}: \tilde{\Omega} \times \tilde{\Omega} \to K$ such that $\mathbb{E}[h|\mathcal{B}(\tilde{\Omega}) \times \{\emptyset, \tilde{\Omega}\}] = \bar{g}$ and $\bar{g}(Q) = g(P) = \mu_2$. Let $\mu_3 \equiv h(Q)$. Then according to the definition of the $<_E$ -ordering, we have $\mu_1 <_E \mu_3$ and $\mu_2 <_E \mu_3$. That is, modulo the construction of sequences $\{Q_n : n = 1,2,...\}$ and $\{h_n : n = 1,2,...$ which satisfy conditions (1) through (6) above, (XVI) has been established.

It thus remains to construct sequences $\{Q_n : n = 1,2,...\}$ and $\{h_n : n = 1,2,...\}$ which satisfy (1) - (6). For each n = 1,2,... let π_n denote the canonical projection of $\tilde{\Omega}$ onto $\prod_{i=1}^{n} \{0,1\}$. Thus the atoms of G_n are the sets $\pi_n^{-1}(i_1,\ldots,i_n)$ where each $i_k = 0,1$ and the atoms of $G_n \times G_n$ are the sets

$$\pi_n^{-1}(i_1,\ldots,i_n) \times \pi_n^{-1}(j_1,\ldots,j_n) \quad \text{where} \quad i_k, j_k = 0,1 .$$

Without loss of generality assume that $K \subset F^{-1}(1)$ for some nonzero $F \in X^*$ (since $K \times \{1\} \subset X \times \mathbb{R}$ has the RNP if K does).

Since f_1 and g_1 are G_1-measurable they depend only on the first coordinate of their arguments. Note that

$$\tfrac{1}{2}f_1(0,\text{---}) + \tfrac{1}{2}f_1(1,\text{---}) = x_0 = \tfrac{1}{2}g_1(0,\text{---}) + \tfrac{1}{2}g_1(1,\text{---})$$

(where '---' indicates that any choice of 0's and 1's may be entered). The Riesz decomposition Lemma 6.4.3 asserts the existence of elements $z(i;j)$ $(i,j = 0,1)$ each in $C(K)$ such that, in tabular form

	$\tfrac{1}{2}f_1(0,\text{---})$	$\tfrac{1}{2}f_1(1,\text{---})$
$\tfrac{1}{2}g_1(0,\text{---})$	$z(0;0)$	$z(0;1)$
$\tfrac{1}{2}g_1(1,\text{---})$	$z(1;0)$	$z(1;1)$

Define Q_1 on the atoms of $G_1 \times G_1$ by

$$Q_1(\pi_1^{-1}(i) \times \pi_1^{-1}(j)) \equiv F(z(i;j))$$

and extend it to a measure on $G_1 \times G_1$ (also denoted by Q_1). Because $\Sigma\,\{z(i;j): i,j = 0,1\} = x_0 \in K$ and $F(x_0) = 1$ it follows that Q_1 is a probability measure. Let $h_1\colon \tilde{\Omega} \times \tilde{\Omega} \to K$ be the function depending on the first coordinate of each $\tilde{\Omega}$-factor defined by

$$h_1((i_1,\text{---}) \times (j_1,\text{---})) \equiv \begin{cases} [F(z(i;j))]^{-1} z(i;j) & \text{if } z(i;j) \neq 0 \\ x_0 \quad \text{if } z(i;j) = 0 . \end{cases}$$

Properties (2), (4), (5) and (6) are easy to check. For example, to verify (5) when $A \equiv \pi_1^{-1}(j) \in G_1$ write

$$\int_{\tilde{\Omega}\times A} h\,dQ_1 = h_1((0,---) \times (j,---))Q_1(\pi_1^{-1}(0) \times \pi_1^{-1}(j))$$

$$+ h_1((1,---) \times (j,---))Q_1(\pi_1^{-1}(1) \times \pi_1^{-1}(j)) = z(0;j) + z(1;j)$$

$$= \frac{1}{2}f_1(j,---) = \int_A f_1 d\tilde{P} = \int_A f d\tilde{P}.$$

Suppose next that $Q_1,\ldots,Q_n$ and $h_1,\ldots,h_n$ have been constructed and that they satisfy (1) through (6). Let

$$z(i_1,\ldots,i_n;j_1,\ldots,j_n)$$

$$\equiv h_n((i_1,\ldots,i_n,---) \times (j_1,\ldots,j_n,---))Q_n(\pi_n^{-1}(i_n,\ldots,i_n) \times \pi_n^{-1}(j_1,\ldots,j$$

for each $i_k, j_k = 0,1$. Then (5) becomes

$$\sum \{z(i_1,\ldots,i_n;j_1,\ldots,j_n) : i_k = 0,1\} = 2^{-n}f_n(j_1,\ldots,j_n,---).$$

Also $\mathbb{E}[f_{n+1}|G_n] = f_n$ so that

$$2^{-n}f_n(j_1,\ldots,j_n,---) =$$

$$2^{-(n+1)}f_{n+1}(j_1,\ldots,j_n,0,---) + 2^{-(n+1)}f_{n+1}(j_1,\ldots,j_n,1,---).$$

The Riesz decomposition Lemma 6.4.3 yields

	$2^{-(n+1)}f_{n+1}(j_1,\ldots,j_n,0,---)$	$2^{-(n+1)}f_{n+1}(j_1,\ldots,j_n,1,---$
$z(0,\ldots,0;j_1,\ldots,j_n)$	$w(0,\ldots,0;j_1,\ldots,j_n,0)$	$w(0,\ldots,0;j_1,\ldots,j_n,1)$
.	.	.
.	.	.
.	.	.
$z(1,\ldots,1;j_1,\ldots,j_n)$	$w(1,\ldots,1;j_1,\ldots,j_n,0)$	$w(1,\ldots,1;j_1,\ldots,j_n,1)$

But then

$$2^{-n}g_n(i_1,\ldots,i_n,---) = \sum \{z(i_1,\ldots,i_n;j_1,\ldots,j_n) : j_k = 0,1\}$$

$$= \sum \{w(i_1,\ldots,i_n;j_1,\ldots,j_n,j_{n+1}) : j_k = 0,1\}.$$

On the other hand,

$$2^{-n}g_n(i_1,\ldots,i_n,---) =$$

$$2^{-(n+1)}g_{n+1}(i_1,\ldots,i_n,0,---) + 2^{-(n+1)}g_{n+1}(i_1,\ldots,i_n,1,---).$$

Hence Lemma 6.4.3 applies again and gives

	$2^{-(n+1)}g_{n+1}(i_1,\ldots,i_n,0,---)$	$2^{-(n+1)}g_{n+1}(i_1,\ldots,i_n,1,---)$
$w(i_1,\ldots,i_n;0,\ldots,0)$	$z(i_1,\ldots,i_n,0;0,\ldots,0)$	$z(i_1,\ldots,i_n,1;0,\ldots,0)$
.	.	.
.	.	.
.	.	.
$w(i_1,\ldots,i_n;1,\ldots,1)$	$z(i_1,\ldots,i_n,0;1,\ldots,1)$	$z(i_1,\ldots,i_n,1;1,\ldots,1)$

Thus

$$\sum \{z(i_1,\ldots,i_{n+1};j_1,\ldots,j_{n+1}): j_k = 0,1\} = 2^{-(n+1)}g_{n+1}(i_1,\ldots,i_{n+1},---)$$

and

$$\sum \{z(i_1,\ldots,i_{n+1};j_1,\ldots,j_{n+1}): i_k = 0,1\} = 2^{-(n+1)}f_{n+1}(j_1,\ldots,j_{n+1},---).$$

Define Q_{n+1} on the atoms of $G_{n+1} \times G_{n+1}$ by

$$Q_{n+1}(\pi_{n+1}{}^{-1}(i_1,\ldots,i_{n+1}) \times \pi_{n+1}{}^{-1}(j_1,\ldots,j_{n+1})) \equiv F(z(i_1,\ldots,i_{n+1};j_1,\ldots,j_{n+1}))$$

and extend Q_{n+1} to a measure on $G_{n+1} \times G_{n+1}$ again denoted by Q_{n+1}. Define $h_{n+1}: \tilde{\Omega} \times \tilde{\Omega} \to K$ to be that function depending on the first $n + 1$ coordinates of each $\tilde{\Omega}$ factor given by

$$h_{n+1}((i_1,\ldots,i_{n+1},---) \times (j_1,\ldots,j_{n+1},---))$$

$$\equiv \begin{cases} a \quad \text{if} \quad z(i_1,\ldots,i_{n+1};j_1,\ldots,j_{n+1}) = 0 \\ [F(z(i_1,\ldots,i_{n+1};j_1,\ldots,j_{n+1}))]^{-1}z(i_1,\ldots,i_{n+1};j_1,\ldots,j_{n+1}) \text{ otherwise.} \end{cases}$$

Properties (1) - (6) may be easily verified. As a typical example we check (1). It suffices to show that

$$Q_{n+1}(\pi_n^{-1}(i_1,\ldots,i_n) \times \pi_n^{-1}(j_1,\ldots,j_n))$$
$$= Q_n(\pi_n^{-1}(i_1,\ldots,i_n) \times \pi_n^{-1}(j_1,\ldots,j_n)).$$

But

$$Q_{n+1}(\pi_n^{-1}(i_1,\ldots,i_n) \times \pi_n^{-1}(j_1,\ldots,j_n))$$
$$= \sum \{Q_{n+1}(\pi_{n+1}^{-1}(i_1,\ldots,i_{n+1}) \times \pi_{n+1}^{-1}(j_1,\ldots,j_{n+1})): i_{n+1},j_{n+1} = 0,1\}$$
$$= \sum \{F(z(i_1,\ldots,i_{n+1};j_1,\ldots,j_{n+1})): i_{n+1},j_{n+1} = 0,1\}$$
$$= F(w(i_1,\ldots,i_n;j_1,\ldots,j_n,0)) + F(w(i_1,\ldots,i_n;j_1,\ldots,j_n,1))$$
$$= F(z(i_1,\ldots,i_n;j_1,\ldots,j_n))$$
$$= Q_n(\pi_n^{-1}(i_1,\ldots,i_n) \times \pi_n^{-1}(j_1,\ldots,j_n)).$$

Thus the proof of Theorem 6.4.2 is complete.

§5. Examples.

Although the existence Theorem 6.2.9 is as sharp as possible when K is a closed bounded separable convex set with the RNP (Corollary 6.3.15(b)) there are, possibly, improvements when the separability assumption on K is dropped. Many reasonable conjectures may be stated. For example, it might seem plausible that ex(K) is always a G_δ-set (or a Borel set, or in $\mathcal{U}(\mathcal{B}(K))$ or in $\mathcal{U}_t(\mathcal{B}(K))$ at least) and that for each $x_0 \in K$ there is a $\mu \in \mathcal{P}_t(K)$ with barycenter x_0 such that $\mu(ex(K)) = 1$. Correspondingly stronger results might then be expected when K is also a simplex. The following examples, however, put several such thoughts to rest. The first one was pointed out to me by G.A. Edgar, and it depends in large part on the work of Lindenstrauss [1972].

EXAMPLE 6.5.1. There is a closed bounded convex subset K of a Banach space such that K has the RNP and $ex(K) \notin \mathcal{U}_t(\mathcal{B}(K))$. (Hence, in particular, $ex(K) \notin \mathcal{U}(\mathcal{B}(K))$.) In fact, K may be chosen to be weakly compact.

Construction. The by now classical examples of Bishop and deLeeuw [1959] of compact convex sets whose extremal subsets are arbitrarily bad are subsets of $(C(D)^*,\text{weak}^*)$ for suitably chosen compact Hausdorff spaces D. Lindenstrauss [1972], p. 258, noted that the Bishop-deLeeuw examples may in fact be constructed with D an Eberlein compact (i.e. with D homeomorphic to a weakly compact subset of some Banach space). See Lindenstrauss [1972] for information about such sets. It follows from Theorems 3.2 and 3.3 of this same source that when D is an Eberlein compact, $(U(C(D)^*), \text{weak}^*)$ is affinely homeomorphic to a weakly compact convex subset of $(c_o(\Gamma),\text{weak})$ for some set Γ, and consequently the examples of Bishop and deLeeuw may be constructed as weakly compact convex subsets of some $c_o(\Gamma)$. Thus there is a weakly compact convex set K in $c_o(\Gamma)$ for some Γ such that $ex(K) \notin \mathcal{U}_t(\mathcal{B}(K,\text{weak}))$. But then, by Theorem 4.4.4, $ex(K) \notin \mathcal{U}_t(\mathcal{B}(K,\text{norm}))$, and since clearly $\mathcal{U}_t(\mathcal{B}(K,\text{norm})) \supset \mathcal{U}(\mathcal{B}(K,\text{norm})$, we conclude $ex(K) \notin \mathcal{U}(\mathcal{B}(K,\text{norm}))$ as well. Finally, since weakly compact convex subsets of Banach spaces have the RNP by Theorem 3.6.1, the construction is complete.

It follows from Theorem 6.2.6 that any closed bounded convex subset of a Banach space for which $ex(K) \notin \mathcal{U}(\mathcal{B}(K))$ must be non-separable. Even in the separable case, though, there is some question about the Borelean nature of ex(K). This question (first raised in Bourgin [1971]) has been settled by example in Jayne and Rogers [1977] (Example 3, p. 270) and the interested reader is referred to that source for details. In short, they demonstrate the following.

EXAMPLE 6.5.2. There is a closed bounded convex subset K of ℓ_1 such that ex(K) is not a Borel set.

(N.B. Ex(K) is not analytic, either, since if it were, using the fact that $K\setminus ex(K)$ is analytic, both would be Borel.) Jayne and Rogers [1977] contains a great deal of additional interesting information about other pathological sets of extreme points.

The next example is, in some sense, more critical. It dashes all hope of having tight representing measures live on ex(K) even when ex(K) is a nice set. (It is due to Davis, Edgar and Johnson, and appears in Edgar [1976].)

EXAMPLE 6.5.3. There is a closed bounded convex subset K of a Hilbert space H and a point x_0 of K with the following properties:

(1) ex(K) is a closed, discrete set.

(2) If $\mu \in P_t(K)$ has barycenter x_0 then $\mu(\text{ex}(K)) = 0$.

(3) If μ is a separable extremal measure in $P_t(K)$ then $\mu(A) = 1$ for each $A \in \text{Baire}(K,\text{weak})$ such that $A \supset \text{ex}(K)$. (See Notation 4.3.1(b).)

Construction. Let λ denote Lebesgue measure on [0,1] and let $\tilde{K}_1 \equiv P_t([0,1]) \subset C[0,1]^*$. Then $\tilde{K}_1$ is a weak* compact convex metrizable simplex (cf. the second paragraph of the proof of Lemma 6.4.4). Since [0,1] is homeomorphic to $\{\varepsilon_x : x \in [0,1]\} = \text{ex}(\tilde{K}_1)$ under the map h which associates x with ε_x, it is easy to check directly that the only measure $\nu \in P_t(\tilde{K}_1)$ with barycenter λ for which $\nu(\text{ex}(\tilde{K}_1)) = 1$ is $\nu = h(\lambda)$. Because $\tilde{K}_1$ is compact and convex and metrizable, it is affinely homeomorphic to a compact convex subset K_1 of separable Hilbert space H_1. If $x_0 \in K_1$ is the image under this affine homeomorphism of $\lambda \in \tilde{K}_1$ then K_1 and x_0 have the property that whenever $\mu \in P_t(K_1)$ has barycenter x_0 and $\mu(\text{ex}(K_1)) = 1$ then μ has no discrete part. Now let H_2 be a (nonseparable) Hilbert space with orthonormal basis indexed by $\text{ex}(K_1)$: say $\{e_y : y \in \text{ex}(K_1)\}$ is an orthonormal basis of H_2. Let $H \equiv H_1 \oplus H_2$, let

$$D \equiv \{y + e_y : y \in \text{ex}(K_1)\} \cup \{y - e_y : y \in \text{ex}(K_1)\}$$

and let $K \equiv \overline{\text{co}}(D)$.

In order to check that K has the desired properties it will first be established that ex(K) = D. Indeed, if $\pi: H_1 \oplus H_2 \to H_1$ is the natural projection onto H_1 then $\pi(D) = \text{ex}(K_1)$ and thus $\pi(K) = K_1$. Now, the only weak accumulation point of $\{e_y : y \in \text{ex}(K_1)\}$ is 0 and hence weak closure(D) $\subset D \cup K_1$. By the Milman Theorem 3.2.2, then, (since K is weakly compact and convex) ex(K) $\subset$ weak closure(D) $\subset D \cup K_1$ and since $y = \frac{1}{2}(y + e_y) + \frac{1}{2}(y - e_y)$ for each $y \in \text{ex}(K_1)$, ex(K) $\subset$ D. On the other hand it may be verified that the functional $e_y \in H^*$ ($y \in \text{ex}(K_1)$) exposes $y + e_y \in K$ and hence $y + e_y \in \text{ex}(K)$ for $y \in \text{ex}(K_1)$. A similar argument shows that, in fact, $D \subset \text{ex}(K)$ whence, from above, D = ex(K).

Observe that D is a norm closed subset of K and that each two points of D are of distance at least $\sqrt{2}$ from one another. Thus K satisfies the first of its claimed properties. Next, suppose that $\nu \in P_t(K)$ has barycenter x_0 and that $\nu(ex(K)) > 0$. Then $\pi(\nu) \in P_t(K_1)$ has barycenter $\pi(x_0) = x_0$ and $\pi(\nu)(ex(K_1)) > 0$. But since ν is tight and since $ex(K) = D$ is discrete and closed (whence compact subsets are finite) ν must assign positive mass to some point of D and thus $\pi(\nu)$ must assign positive mass to some point of $\pi(D) = ex(K_1)$. Let T denote a $\pi(\nu)$-dilation such that $T(\pi(\nu)) \equiv \mu$ is $<_D$-maximal on K_1. Since K_1 is compact and convex (as well as metrizable), $\mu(ex(K_1)) = 1$. Of course T may not alter $\pi(\nu)$ on $ex(K_1)$ and hence μ must also assign positive mass to some point of $ex(K_1)$ just as $\pi(\nu)$ did. Since this contradicts the choice of K_1 (as explained in the first paragraph of this construction) it follows that $\nu(ex(K)) = 0$ as desired. The third property listed for Example 6.5.3 is an immediate consequence of the following result (which appears in Edgar [1976], Theorem 6.1).

PROPOSITION 6.5.4. Let K be a weakly compact convex subset of a Banach space. If $A \in$ Baire(K,weak) satisfies $ex(K) \subset A$ then $\nu(A) = 1$ for each separable extremal $\nu \in P_t(K)$.

Proof. Let $\tilde{\mu}$ and $\tilde{\nu}$ be any two probability measures on Baire(K,weak). Theorems 4.5 and 4.6 of Bourgin and Edgar [1976] establish that

(a) $\tilde{\nu}$ has a unique extension to a measure in $P_t(K)$. (This extension will be denoted by ν and the extension of $\tilde{\mu}$ by μ.)

(b) $\mu <_D \nu$ if and only if $\int f d\tilde{\mu} \leqq \int f d\tilde{\nu}$ for each weakly continuous convex bounded function f on K.

If $\nu \in P_t(K)$ is any separable extremal measure then ν is $<_D$- maximal (Theorem 6.3.14) and hence its restriction to Baire(K,weak), $\tilde{\nu}$, is maximal in the sense of the Choquet ordering for the Choquet-Bishop-deLeeuw theorem about compact convex sets in locally convex spaces. (See, for example, Phelps [1966], Definition, p. 24.) It follows from that theorem (ibidem, p. 24) that such a $\tilde{\nu}$ has the property that $\tilde{\nu}(A) = 1$ for any $A \in$ Baire(K,weak) such that $A \supset ex(K)$. Hence $\nu(A) = 1$ for such sets A, which completes the proof of the proposition.

In fact, Edgar [1976], p.159 has asked whether the above Proposition might not be a special case of a much more general phenomenon in the sense that perhaps whenever K is a closed bounded convex subset of a Banach space, K has the RNP, $\mu \in P_t(K)$ is separable extremal and $A \supset ex(K)$ is a weak Baire subset of K then it follows that $\mu(A) = 1$. Talagrand [1983a] has very recently established that the answer is "no", so the question of "support" of maximal measures on RNP sets remains open. As the final example of this section points out, the deterioration of the the of integral representations outside the context of sets with the RNP is dramatic. T example is due to Talagrand [1983a] and represents a significant improvement over an earlier example in the same spirit due to Edgar [1978b].

EXAMPLE 6.5.5. There is a closed bounded convex subset K of $c_0(\mathbb{R})$ such that $ex(K) = \emptyset$ and yet each point x_0 of K is the barycenter of a separable extremal measure $\mu \in P_t(K)$.

The construction of K is by transfinite induction up to the first uncountable ordinal ω_1. At the $0^{\underline{th}}$ stage let $C \subset \mathbb{R}^2$ (where we consider $\mathbb{R}^2$ as $c_0(T)$ for T the two point space $\{1,2\}$) be the unit disc:

$$C \equiv \{(x,y): x^2 + y^2 \leqq 1\}.$$

Let λ denote normalized Lebesgue measure on $ex(C)$ $(= \{(x,y): x^2 + y^2 = 1\})$. Thus $\lambda \in P_t(K)$ has barycenter $0 \in c_0(T)$ and $\lambda(ex(C)) = 1$. Let $\Gamma_0 \equiv \{1,2\}$ and let $K_0 \equiv C$.

Suppose now that for some countable ordinal $\xi > 0$, sets Γ_α and K_α have been defined for each $\alpha < \xi$ subject to the conditions

(a) $\Gamma_\beta \subset \Gamma_\alpha$ if $\beta < \alpha < \xi$;

(b) $|\Gamma_\alpha| = |\mathbb{R}|$ for $0 < \alpha < \xi$; ($|\cdot|$ denotes cardinality)

(c) K_α is a closed bounded convex subset of $U(c_0(\Gamma_\alpha))$ for $\alpha < \xi$.

In the obvious manner identify $c_0(\Gamma_\beta)$ with a subspace of $c_0(\Gamma_\alpha)$ for $\beta < \alpha < \xi$, and assume that, with this identification,

(d) $K_\beta \subset K_\alpha$ for $\beta < \alpha < \xi$.

If ξ is a limit ordinal then let $\Gamma_\xi \equiv \bigcup_{\alpha<\xi} \Gamma_\alpha$ and $K_\xi \equiv \overline{\bigcup_{\alpha<\xi} K_\alpha}$. If, on the othe hand, $\xi = \alpha + 1$ for some countable ordinal α then both Γ_α and K_α are known and we

let

$$\Gamma_\xi \equiv \Gamma_\alpha \cup (T \times K_\alpha).$$

For each $x \in c_o(\Gamma_\alpha)$ and $p \in c_o(T)$ let $x \cdot p \in c_o(\Gamma_\xi)$ be the point defined coordinatewise by

$$x \cdot p(\gamma) \equiv x(\gamma) \quad \text{for} \quad \gamma \in \Gamma_\alpha \quad \text{and}$$

$$x \cdot p(n,y) \equiv \begin{cases} p_n & \text{if } y = p \\ 0 & \text{if } y \neq p \end{cases} \quad \text{for} \quad (n,y) \in T \times K_\alpha$$

(where of course $p \equiv (p_1,p_2))$. Then let

$$K_\xi \equiv \overline{co}(\{x \cdot c\colon x \in K_\alpha\ ,\ c \in C\}).$$

Since $C \subset U(c_o(T))$ and $K_\alpha \subset U(c_o(\Gamma_\alpha)$ by hypothesis, $x \cdot c \in U(c_o(\Gamma_\xi))$ for $x \in K_\alpha$ and $c \in C$, whence K_ξ is a closed bounded convex subset of $U(c_o(\Gamma_\xi))$. Evidently $|\Gamma_\xi| = |\Gamma_\alpha| + |K_\alpha| = |\mathbb{R}|$. With the identification of $c_o(\Gamma_\alpha)$ as a subspace of $c_o(\Gamma_\xi)$ then $x \cdot 0 = x$ for $x \in c_o(\Gamma_\alpha)$. Consequently $K_\alpha \subset K_\xi$.

Let $K \equiv \bigcup_{\alpha<\omega_1} K_\alpha$ and $\Gamma \equiv \bigcup_{\alpha<\omega_1} \Gamma_\alpha$. Note that $|\Gamma| = |\mathbb{R}|$ and that any sequence in K actually belongs to K_α for some $\alpha < \omega_1$. Hence K is a closed bounded convex subset of $U(c_o(\mathbb{R}))$. The following lemmas provide the body of work necessary to demonstrate that K has the properties listed at the beginning of the example.

<u>LEMMA</u> 6.5.6. Fix $\alpha < \omega_1$ and let $\mu \in P_t(K_\alpha)$ be a separable extremal measure on K_α. Suppose that $\mu(\{x\}) = 0$ for each $x \in K_\alpha$. (Such a measure is called <u>diffuse</u>.) Then μ is separable extremal when considered as a measure on $K_{\alpha+1}$.

Proof. Let $\nu \in P_t(K_{\alpha+1})$ and assume that $\mu <_D \nu$. We will show that $\mu = \nu$, and because μ is $<_D$-maximal (cf. Theorem 6.3.14) on K_α, it suffices to show that $\nu \in P_t(K_\alpha)$ - i.e., that $\nu(K_\alpha) = 1$. Let $\pi\colon c_o(\Gamma_{\alpha+1}) \to c_o(\Gamma_\alpha)$ denote the natural projection. Then $\pi(\mu) <_D \pi(\nu)$. (It is easy to check that $\mu <_C \nu$ forces $\pi(\mu) <_C \pi(\nu)$. Now use Theorem 6.3.9.) Hence, since $\pi(\mu) = \mu$ is $<_D$-maximal, $\mu = \pi(\nu)$. If $\nu(K_\alpha) < 1$ then for some point $x \in K_{\alpha+1} \setminus K_\alpha$, each neighborhood of $x \in K_{\alpha+1}$ has positive ν-

measure. Note that $\pi(K_{\alpha+1}) = K_\alpha$. Hence if $x \in K_{\alpha+1} \backslash K_\alpha$ there is an index $\gamma \in \Gamma_{\alpha+1} \backslash \Gamma_\alpha$ such that $x(\gamma) \neq 0$ and without loss of generality we assume $x(\gamma) > 0$. Now γ is of the form (n_0, y_0) for $n_0 = 1$ or 2 and $y_0 \in K$ and thus for some $\varepsilon > 0$, $\varepsilon < 1$, $\nu(V) > 0$ for $V \equiv \{w \in K_{\alpha+1}: w(n_0, y_0) > \varepsilon\}$. Let $\nu_1 \equiv \frac{1}{\nu(V)} \nu|_V \in P_t(K_{\alpha+1})$ and let $B \equiv \varepsilon y_0 + (1 - \varepsilon) K_\alpha$.

We will prove that $\pi(\nu_1)(B) = 1$ by showing that $\pi(V) \subset B$. Suppose that $\sum_{i=1}^n t_i x_i \cdot c_i \in V$ for some choice of $t_i \geq 0$, $\Sigma t_i = 1$, $x_i \in K_\alpha$, $c_i \in C$. Then $\varepsilon' \equiv \sum_{x_i = y_0} t_i \geq (\Sigma t_i x_i \cdot c_i)(n_0, y_0) > \varepsilon$. (Here and elsewhere terms like $\sum_{x_i = y_0}$ mean summation over those indices i such that the formula below the Σ - in this case $x_i = y_0$ - is true.) But

$$\pi(\sum t_i x_i \cdot c_i) = \sum t_i x_i = (\sum_{x_i = y_0} t_i) y_0 + (\sum_{x_i \neq y_0} t_i x_i) = \varepsilon' y_0 + (1 - \varepsilon') b$$

where $b \equiv (1 - \varepsilon')^{-1} \sum_{x_i \neq y_0} t_i x_i$ if $\varepsilon' < 1$ (and $b \equiv y_0$ if $\varepsilon' = 1$) is an element of K_α. Thus

$$\pi(\sum t_i x_i \cdot c_i) = \varepsilon y_0 + (1 - \varepsilon) \frac{(\varepsilon' - \varepsilon) y_0 + (1 - \varepsilon') b}{1 - \varepsilon} \in B.$$

Since the set of points of V of the form $\Sigma t_i x_i \cdot c_i$ as above is dense in V, $\pi(V) \subset B$ follows by continuity. Consequently $\pi(\nu_1)(B) = 1$. Since $\pi(\nu_1) << \mu$, $\pi(\nu_1)$ is diffuse and in particular, $\pi(\nu_1)(\{y_0\}) = 0$. Hence there is a compact set $D \subset B \backslash \{y_0\}$ with $\pi(\nu_1)(D) > 0$. Observe that $\pi(\nu_1)$ is $<_D$-maximal in K_α by Theorem 6.3.13 (since it is absolutely continuous with respect to the $<_D$- maximal measure μ). A contradiction will be obtained by showing that D is a $\pi(\nu_1)$-movable set and hence must have $\pi(\nu_1)$-measure 0 by Theorem 6.3.13. A picture helps make the following argument transparent. Let $r \equiv \inf_{d \in D} \|d - y_0\|$. Then $r > 0$. For each $d \in D$ define points d_1 and d_2 as follows:

$$d_1 \equiv y_0 + (1 + \frac{\varepsilon r}{\|y_0 - d\|})(d - y_0);$$

$$d_2 \equiv y_0 + (1 - \frac{\varepsilon r}{\|y_0 - d\|})(d - y_0).$$

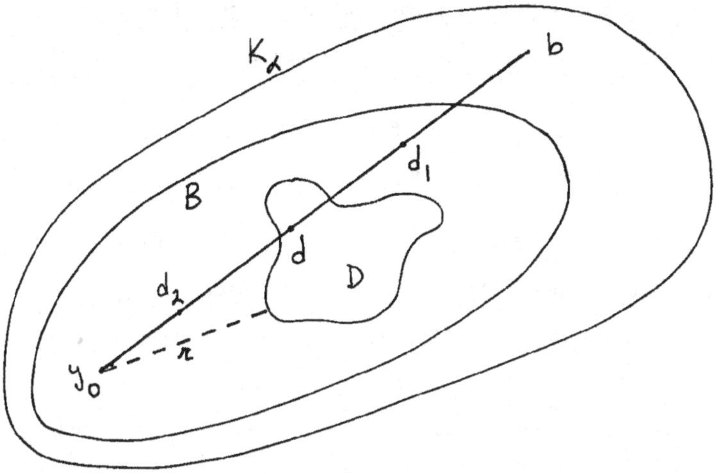

Note the following obvious facts about d_1 and d_2 .

(a) Both points lie on the line determined by y_0 and d;

(b) d_2 lies on the line segment joining y_0 to d and hence $d_2 \in B$;

(c) $d_1 \in K_\alpha$. (Indeed, since $d \in B$, there is a $b \in K_\alpha$ such that $d = \varepsilon y_0 + (1 - \varepsilon)b$. Then d_1 may be written in the form $d_1 = y_0 + t(b - y_0)$ where $t \equiv (\frac{\varepsilon r}{\|y_0 - d\|} + 1)(1 - \varepsilon) < (\varepsilon + 1)(1 - \varepsilon) < 1$. Thus d_1 lies on the line segment joining d and b.)

(d) $\frac{1}{2}d_1 + \frac{1}{2}d_2 = d$ and $d_1 \neq d \neq d_2$;

(e) d_1 and d_2 vary continuously with $d \in D$.

If A denotes a closed separable subset of K_α with $\pi(\nu_1)(A) = 1$ then the function $T: K_\alpha \to P_t(K)$ given by

$$T(x) \equiv \begin{cases} \frac{1}{2}\varepsilon_{d_1} + \frac{1}{2}\varepsilon_{d_2} & \text{if } x = d \in D; \\ \varepsilon_x \quad \text{if } x \in A \backslash D; \\ \varepsilon_{y_0} \quad \text{if } x \in K_\alpha \backslash (A \cup D) \end{cases}$$

is certainly a $\pi(\nu_1)$-dilation which moves each point of D. That is, D is $\pi(\nu_1)$-movable, and from the above discussion this is impossible. Hence ν_1 does not exist. It follows that $\mu = \nu$ as was to be shown.

The next lemma is given a proof similar to that of 6.5.6 by Talagrand [1983a]. The proof below perhaps elucidates some of the geometry in a different manner.

LEMMA 6.5.7. Fix $\alpha < \omega_1$ and let $x \in K_\alpha$. Then there is a $\mu \in P_t(K_{\alpha+1})$ which is separable extremal, which has barycenter x, and which is diffuse.

Proof. For $x \in K_\alpha$ fixed, let $f_x: C \to K_{\alpha+1}$ be the map $f_x(c) \equiv x \cdot c$. If μ denotes the image measure $f_x(\lambda)$ (where λ is the measure on $ex(C)$ with barycenter $0 \in C$ as described in the first few lines of this example) then it is easy to check that μ is diffuse, that μ has barycenter x and that $\mu(\{x \cdot c: c \in ex(C)\}) = 1$. The proof that μ is separable extremal will be accomplished by showing that $x \cdot c \in ex(K_{\alpha+1})$ for each $c \in ex(C)$ and hence that each μ-movable set in $K_{\alpha+1}$ has μ-measure 0. (Thus μ will be $<_D$-maximal by Theorem 6.3.13, which, by virtue of Theorem 6.3.14, guarantees that μ is indeed separable extremal on $K_{\alpha+1}$.)

Fix $c \in ex(C)$ and write $x \cdot c = \frac{1}{2}y + \frac{1}{2}z$ for points $y, z \in K_{\alpha+1}$. Given $\varepsilon > 0$ choose finite sets of scalars $s_i > 0$ and $t_j > 0$, $\Sigma s_i = \Sigma t_j = 1$, points $y_i, z_j \in K_\alpha$ and points $d_i, e_j \in C$ such that

$$\text{(XVIII)} \qquad \|y - \sum s_i y_i \cdot d_i\| < \varepsilon \text{ and } \|z - \sum t_j z_j \cdot e_j\| < \varepsilon.$$

Since coordinatewise y and z are within ε of their respective sums in (XVIII), evaluation of the above inequalities at the points $(1,x)$ and $(2,x)$ (in $T \times K_\alpha \subset \Gamma_{\alpha+1}$) yields

$$\text{(XIX)} \qquad \|c - [\frac{1}{2}\sum_{y_i=x} s_i d_i + \frac{1}{2}\sum_{z_j=x} t_j e_j]\| < \varepsilon.$$

Now

$$\frac{1}{2}\sum_{y_i=x} s_i d_i + \frac{1}{2}\sum_{z_j=x} t_j e_j$$

is a convex combination of points of C (including possibly 0). Since c lies on the unit circle a straightforward calculation in the plane $c_0(T)$ leads from (XIX) to

$$\frac{1}{2}\sum_{y_i=x} s_i + \frac{1}{2}\sum_{z_j=x} t_j > 1 - \sqrt{2\varepsilon}$$

and hence both

$$\sum_{y_i=x} s_i > 1 - \sqrt{2}\, 2\varepsilon \quad \text{and} \quad \sum_{z_j=x} t_j > 1 - \sqrt{2}\, 2\varepsilon \ .$$

But then

$$\text{(XX)} \qquad \|\sum s_i y_i \cdot d_i - \sum s_i x \cdot d_i\| = \|\sum_{y_i \neq x} s_i (y_i \cdot d_i - x \cdot d_i)\|$$

$$\leqq 2 \sum_{y_i \neq x} s_i < 4\sqrt{2}\varepsilon < 6\varepsilon \ .$$

Similarly, $\|\Sigma\, t_j z_j \cdot e_j - \Sigma\, t_j x \cdot e_j\| < 6\varepsilon$. Note that $\Sigma\, s_i x \cdot d_i = x \cdot (\Sigma\, s_i d_i)$ since $\Sigma\, s_i = 1$ and similarly, $\Sigma\, t_j x \cdot e_j = x \cdot (\Sigma\, t_j e_j)$. Thus (XX) and (XVIII) yield

$$\text{(XXI)} \qquad \|y - x \cdot (\sum s_i d_i)\| < 7\varepsilon \quad \text{and} \quad \|z - x \cdot (\sum t_j e_j)\| < 7\varepsilon \ .$$

But by evaluating (XXI) at (1,x) and (2,x) we obtain a more useful version of (XIX):

$$\text{(XXII)} \qquad \|c - [\tfrac{1}{2}\sum s_i d_i + \tfrac{1}{2}\sum t_j e_j]\| < 7\varepsilon \ .$$

Now either by calculating again in the plane $c_0(T)$ or by observing that c is a strongly exposed point of C so that $\|c - \Sigma\, s_i d_i\| \to 0$ as $\varepsilon \to 0$ and $\|c - \Sigma\, t_j e_j\| \to 0$ as $\varepsilon \to 0$, we conclude from (XXI) that

$$\|y - x \cdot c\| \leqq \|y - x \cdot (\sum s_i d_i)\| + \|x \cdot (\sum s_i d_i) - x \cdot c\| \leqq 7\varepsilon + \eta(\varepsilon)$$

where $\eta(\varepsilon) \to 0$ as $\varepsilon \to 0$. Since $\varepsilon > 0$ was arbitrary, $y = x \cdot c$. Hence also $z = x \cdot c$, and thus $x \cdot c \in ex(K_{\alpha+1})$ whenever $x \in K_\alpha$ and $c \in ex(C)$. From what has been discussed above, this completes the proof of Lemma 6.5.7.

Now in order to prove that each $x \in K$ is the barycenter of a separable extremal measure on K, we argue as follows. Pick $x \in K$, find $\alpha < \omega_1$ such that $x \in K_\alpha$ and choose $\mu \in P_t(K_{\alpha+1})$ which has barycenter x, is diffuse, and is separable extremal by Lemma 6.5.7. Suppose that $\nu \in P_t(K)$ and $\mu <_D \nu$. For each $\beta > \alpha$, $\beta < \omega_1$, let $\pi_\beta\colon c_0(\Gamma) \to c_0(\Gamma_\beta)$ be the natural projection. Then $\pi_{\alpha+1}(\nu) = \mu$ according to Lemma 6.5.6. In fact, according to that lemma, in order to show that $\pi_\beta(\nu) = \mu$ for each $\beta < \omega_1$ it suffices to prove that whenever ξ is a limit ordinal, $\alpha < \xi < \omega_1$ such that $\pi_\beta(\nu) = \mu$ for $\beta < \xi$, then $\pi_\xi(\nu) = \mu$ as well. But for $\omega \in K$,

$\lim_{\beta<\xi} \|\pi_\xi(\omega) - \pi_\beta(\omega)\| = 0$ and consequently the net $\{\pi_\beta(\nu): \beta < \xi\}$ converges weak* to $\pi_\xi(\nu)$ in $P_t(K) \subset C_b(K)^*$. Of course, since $\pi_\beta(\nu) = \mu$ for $\beta < \xi$, this forces $\pi_\xi(\nu) = \mu$ as well. Thus each point of K is the barycenter of a separable extremal probability measure on K. Either by noting that the measure constructed in Lemma 6.5.7 is diffuse or by the elementary observation that each point $x \in K_\alpha$ is the midpoint of the points $x \cdot c$ and $x \cdot (-c)$ for any $c \in ex(C)$, it is clear that $ex(K) = \emptyset$ This completes the example 6.5.5.

§6. The Integral Representation Property.

Let us agree that a closed convex set $S \subset X$ has the Integral Representation Property (IRP) if whenever $C \subset S$ is a closed bounded convex subset and $x \in C$ then there is a $\mu \in P_t(D)$ which is separable extremal with barycenter x. Accordi to Theorem 6.2.9, then, each S with the RNP has the IRP. On the other hand, the converse remains open in general. Recall that if S has the RNP then it has the KMP (Theorem 3.3.6) and since separable extremal measures intuitively "live", in some sense, on the set of extreme points, it is enticing to conjecture that if S has the IRP then S has the KMP as well. Nevertheless, this too remains an open question in general. When C is a separable closed bounded convex set it is true that $\mu(ex(C)) = 1$ for each separable extremal μ in $P_t(C)$ and hence $ex(C) \neq \emptyset$. In particular, then, if S is a separable closed convex set with the IRP then each of its closed bounded convex subsets has an extreme point and hence, by Proposition 3.1.1, S has the KMP. On the other hand, as was demonstrated in Example 6.5.5, there are closed bounded convex (nonseparable!) sets K for which $ex(K) = \emptyset$ in spite of the fact that each x in K is the barycenter of a separable extremal measure, so the relationship between the IRP and the existence of extreme points is somewhat elusive. We shall study here some weak forms of a converse to the implication RNP implies IRP. (See Theorem 6.6.1, Corollary 6.6.9 and Theorem 6.6.10.)

The first result of this section will establish that a closed bounded convex set $K \subset X$ has the RNP if and only if the countable product of K with itself has the IRP. In order to keep within the confines of Banach spaces, for the countable product of K with itself we shall mean the set $K^\sim$ defined by

(XXIII)
$$K^{\sim} \equiv \{(x_1,x_2,\ldots) \in (\oplus \sum_{i=1}^{\infty} X_i)_{\ell_2} : x_n \in K \text{ for each } n\}$$
where $(X_i, \|\cdot\|_i)$ is the space X equipped with norm $\|x\|_i \equiv 2^{-i}\|x\|$, $x \in X$.

As the discussion unfolds it will become evident that there is considerable latitude in the choice of interpretation of how to represent a countable product of K with itself. The choice of $K^{\sim}$ in (XXIII) is just one among many.

The exact form of the theorem we shall prove is as follows.

THEOREM 6.6.1. Let $K \subset X$ be a closed bounded convex set. Then K has the RNP if and only if $K^{\sim}$ (as in (XXIII) above) has the IRP.

E.G.F. Thomas [1979] proved a version of 6.6.1 as part of a systematic study of integral representations in a conuclear cone setting. More recently Blizzard [1980] constructed a closed bounded convex set of integral operators which is the closed convex hull of its extreme points but which contains a point (an integral, nonuclear operator in fact) which is not the barycenter of any separable extremal measure on the set. We have used the Blizzard example as the basis of a general geometric construction which yields Theorem 6.6.1. One of the central ingredients in all these efforts is a disintegration of measures theorem, the one stated below being a special case of a far more general result due to Edgar [1975b].

Let $(\Delta, \mathcal{B}(\Delta))$ be a topological space with its Borel σ-algebra and suppose that E is a metric space. For each $R \subset \Delta \times E$ let $R_{(t)}$ $(t \in \Delta)$ be the slice of R above t. Thus

$$R_{(t)} \equiv \{x \in E: (t,x) \in R\}.$$

Let $\pi_1: \Delta \times E \to E$ denote the first coordinate projection.

THEOREM 6.6.2. Let ν be a tight probability measure on $\mathcal{B}(\Delta)\times\mathcal{B}(E)$. (We use the notation described just above.) Denote $\pi_1(\nu)$ by γ. Then there is a family $\{\nu_t: t \in \Delta\}$ of probability measures on $\mathcal{B}(E)$ such that for each $R \in \mathcal{B}(\Delta)\times\mathcal{B}(E)$ both

(a) the function $t \to \nu_t(R_{(t)})$ is γ-measurable; and

(b) $\nu(R) = \int_\Omega \nu_t(R_{(t)})d\gamma(t)$.

A collection $\{\nu_t\colon t \in \Delta\}$ satisfying these properties is called a <u>strict disintegration of ν with respect to π_1</u>.

If E is a complete metric space the proof amounts to applying Theorem 6.3.8 to the vector measure $m\colon \mathcal{B}(\Delta) \to \mathrm{Lip}(E)^*$ given by

$$m(A)(B) \equiv \nu(A\times B) \quad \text{for} \quad A \in \mathcal{B}(\Delta),\ B \in \mathcal{B}(E) \tag{XXIV}$$

whose average range with respect to γ is a subset of $P_t(E)$. If $\frac{dm}{d\gamma} \equiv F$ and $F(t) \equiv \nu_t$ for each $t \in \Delta$ then $\{\nu_t\colon t \in \Delta\}$ may be shown by standard arguments to be a strict disintegration of ν with respect to π_1. In fact, if E is a Borel subset of a complete metric space M then by working in M in place of E and, at the end noting that $\nu_t(E) = 1$ for t in a set of γ-measure one, we may successfu repeat the above outline. Thus for Borel subsets E of complete metric spaces, Theorem 6.6.2 is a consequence of Theorem 6.3.8. In the case of interest to us here however, E will be of the form $\mathrm{ex}(K)$ for some separable closed bounded convex subset K of a Banach space, and the examples of Jayne and Rogers [1977] (see Examp 6.5.2) show that E need not be Borel in K. (Of course, by the proof of Theorem 6.2.6, $\mathrm{ex}(K)$ is always the complement of an analytic set, and is hence universally Borel measurable.) A theorem more general than 6.3.8 is needed, and though the idea of the proof of Theorem 6.6.2 remains to establish that $\frac{dm}{d\gamma}$ exists in a suitable sense and that its range is the sought for disintegration, the details require somewhat more care. They may be found in Edgar [1975b].

Observe that, as in the proof of Theorem 5.8.1(i), whenever $K \subset X$ is a closed bounded convex set with the RNP then the set $K^{\sim}$ of (XXIII) also has the RNP. Henc $K^{\sim}$ has the IRP and one direction of Theorem 6.6.1 has just been established. In th next definition and subsequent paragraph we discuss the conventions, notation and terminology which will then be used in the proof of the more substantive half of Theorem 6.6.1: if K lacks the RNP then $K^{\sim}$ lacks the IRP. Let us start with the formal definition of an ε-bush.

<u>DEFINITION</u> 6.6.3. <u>An ε-bush B in X</u> is a sequence of pairs

(XXV) $$B \equiv \{(b_{n,i}, r_{n,i}) : (n,i) \in A\}$$

where $b_{n,i}$, $r_{n,i}$ and A satisfy the following conditions:

(a) A is a subset of $\mathbb{N}\times\mathbb{N}$ (called the <u>bush indices</u> of B).

(b) $(0,1)$ is the only element of A with first coordinate 0.

(c) For each $n \geq 1$ there are only finitely many integers i such that $(n,i) \in A$ and the set of such pairs (n,i) is denoted by A_n. Intuitively A_n denotes the set of indices of the <u>n^{th} stage</u> of the bush B. Thus, with this terminology, the bush originates at the point $b_{0,1}$, the sole member of stage 0.

(d) $r_{0,1} = 1$ and for each $(n,i) \in A$, $0 < r_{n,i} \leq 1$.

(e) For each $(n,i) \in A$ there is an associated subset of A_{n+1} consisting of at least two elements and denoted henceforth by $A(n,i)$ such that

$$\sum \{r_{n+1,j} : (n+1,j) \in A(n,i)\} = 1 \qquad \text{and}$$

$$\sum \{r_{n+1,j} b_{n+1,j} : (n+1,j) \in A(n,i)\} = b_{n,i}.$$

The elements of $A(n,i)$ are called the <u>immediate successors</u> of (n,i). For each $n \geq 1$ and $(n,i) \in A$ there is a unique $(n-1,k) \in A$ such that $(n,i) \in A(n-1,k)$.

(f) For each $(n,i) \in A$ and $(n+1,j) \in A(n,i)$ we have $\|b_{n+1,j} - b_{n,i}\| \geq \varepsilon$.

There are a generalized Cantor set $\Delta_{0,1}$ and probability measure μ on $\mathcal{B}(\Delta_{0,1})$ naturally associated with an ε-bush B as will be described next. For any set A let $|A|$ denote its cardinality. Given an ε-bush as in (XXV) define a sequence $\{I_{n,k} : (n,i) \in A\}$ of closed subintervals of $[0,1]$ inductively, beginning with $I_{0,1} \equiv [0,1]$. Suppose that $I_{n,i}$ has been defined for each $(n,i) \in A$ with $n < k$. Let $(n-1,j) \in A$. Divide $I_{k-1,j}$ into $2|A_{(k-1,j)}| - 1$ equal closed subintervals which overlap only at the endpoints and discard the interiors of the even ones (moving left to right). Attach as subscripts to the remaining $|A(k-1,j)|$ closed subintervals the elements of $A(k-1,j)$. Evidently

(XXVI) $$\Delta_{0,1} \equiv \bigcap_{n=1}^{\infty} \bigcup_{(n,i)\in A_n} I_{n,i}$$

is a generalized Cantor set in $[0,1]$. Let

(XXVII) $$\Delta_{n,i} \equiv \Delta_{0,1} \cap I_{n,i} \quad \text{for each} \quad (n,i) \in A.$$

An alternative description of $\Delta_{0,1}$ will prove useful as well (which corresponds to the usual ternary description of the points of the Cantor set). Let ψ be a 1-1 correspondence between A and $\mathbb{N}^+$ and let $X_{n,i}$ be the Banach space $X_{\psi(n,i)}$ defined in (XXIII). With this identification we may write

$$(\oplus \sum_{(n,i)\in A} X_{n,i})_{\ell_2} \quad \text{in place of} \quad (\oplus \sum_{i=1}^{\infty} X_i)_{\ell_2}$$

and, as it is convenient, we shall condider $K^{\sim}$ a subset of the doubly indexed ℓ_2 sum above. Now observe that given $t \in \Delta_{0,1}$ there is a unique sequence $\{(n,i(n): n = 0,1,\ldots\}$ in A such that $t \in \bigcap_{n=0}^{\infty} \Delta_{n,i(n)}$. Let $\bar{\bar{t}}$ be the sequence of 0's and 1's (indexed by the elements of A) given by

(XXVIII) $$\bar{\bar{t}}_{m,j} \equiv \begin{cases} 1 & \text{if} \quad (m,j) = (n,i(n)) \quad \text{for some} \quad n = 0,1,\ldots \\ 0 & \text{otherwise for} \quad (m,j) \in A \end{cases}$$

and let

(XXIX) $$\bar{\bar{\Delta}}_{0,1} \equiv \{\bar{\bar{t}}: t \in \Delta_{0,1}\}.$$

Provide $\bar{\bar{\Delta}}_{0,1}$ the topology of coordinatewise convergence - i.e. the product topolog and note that $\Delta_{0,1}$ and $\bar{\bar{\Delta}}_{0,1}$ are then homeomorphic. Finally, let μ be the prob ability measure on $\mathcal{B}(\Delta_{0,1})$ which satisfies the conditions

(XXX) $$\mu(\Delta_{0,1}) = 1 \quad \text{and} \quad \mu(\Delta_{n+1,j}) = r_{n+1,j}\mu(\Delta_{n,i})$$
$$\text{whenever} \quad (n,i) \in A \quad \text{and} \quad (n+1,j) \in A(n,i).$$

(A formal argument establishing the existence of μ may be given exactly along the lines of that presented in 4.2.11. Uniqueness is direct.) We are now in a position to proceed with the main construction.

Proof of 6.6.1. We shall rely on the notation of Definition 6.6.3 and the subsequent discussion. If $K \subset X$ is a closed bounded convex set which lacks the RNP then K contains a separable closed bounded convex subset K_1 which also lacks the RNP. Since $K_1^{\sim} \subset K^{\sim}$, if it can be demonstrated that $K_1^{\sim}$ lacks the IRP then $K^{\sim}$ will lack the IRP as well. Thus from now on assume without loss of generality that K itself is a separable closed bounded convex subset of X and that K lacks the RNP. It will also be convenient to assume that $0 \in K \setminus ex(K)$. Let ρ be the Minkowski functional for K. (Thus, $\rho(x) = 1$ for each $x \in ex(K)$ if, indeed, there are any extreme points of K.) Since K lacks the RNP there is an ε-bush $B \equiv \{(b_{n,i}, r_{n,i}): (n,i) \in A\}$ (as defined in (XXV)) in K for some $\varepsilon > 0$ by Theorem 2.3.6. The set $C \subset K$ of central interest may now be defined:

$$C \equiv \{(x_{n,i}) \in K^{\sim}:$$

(XXXI)
$$\text{(a)} \quad x_{n,i} = \sum\{x_{n+1,j}: (n+1,j) \in A(n,i)\} \quad \text{for each } (n,i) \in A; \text{ and}$$

$$\text{(b)} \quad \sum_{(n,i)\in A_n} \rho(x_{n,i}) \leq 1 \quad \text{for each } n = 0,1,\ldots\}.$$

In the sequel references to the defining conditions (a) and (b) of the set C in (XXXI) will be abbreviated (XXXIa) and (XXXIb). Furthermore, sums as in (XXXIa) and (XXXIb) will be written more succinctly

$$x_{n,i} = \sum_{A(n,i)} x_{n+1,j} \quad \text{and} \quad \sum_{A_n} \rho(x_{n,i}) \leq 1.$$

The proof of this theorem is broken into four steps presented in Lemmas 6.6.4, 6.6.5, 6.6.6 and 6.6.7. Evidently the completion of such a program will establish Theorem 6.6.1.

LEMMA 6.6.4. C is a closed bounded convex separable subset of $(\oplus \sum_A X_{n,i})_{\ell_2}$.

Proof. C is separable since K is and C is convex since ρ is subadditive and positive homogeneous. Note that for any sequence $\{y_k: k = 1,2,\ldots\}$ in X and point $y \in X$, if $\lim_k \|y_k - y\| = 0$ then $\liminf_k \rho(y_k) \geqq \rho(y)$ since K is closed. In order to check that C is closed observe that if a sequence in C converges to some point of $(\oplus \sum_A X_{n,i})_{\ell_2}$ then it converges coordinatewise and from the above inequality on ρ, condition (XXXIb) is satisfied for each limit point of C. Sinc (XXXIa) is met by limit points of C as well, Lemma 6.6.4 has been established.

<u>LEMMA</u> 6.6.5. $\mathrm{Ex}(C) = \{x\bar{\bar{t}}: x \in \mathrm{ex}(K) \text{ and } \bar{\bar{t}} \in \bar{\bar{\Delta}}_{0,1}\}$.

Proof. Suppose first that $x \in \mathrm{ex}(K)$ and $\bar{\bar{t}} \in \bar{\bar{\Delta}}_{0,1}$. Since $0 \in K$ by assum tion, the point $x\bar{\bar{t}}$ belongs to $K^{\sim}$. Let $\{(n,i(n)): n = 0,1,\ldots\}$ be the unique sequence in A such that $t \in \bigcap_{n=0}^{\infty} \Delta_{n,i(n)}$. For any n, among the indices $(n+1,j)$ of $A(n,i(n))$ there is exactly one - namely $(n+1,i(n+1))$ - such that $\bar{\bar{t}}_{n+1,j} = 1$, while $\bar{\bar{t}}_{n+1,j} = 0$ for $(n+1,j) \in A(n,i(n))$, $j \neq i(n+1)$. Hence (XXXIa) is satisfi for the indices $(n,i(n))$. If, on the other hand, for a given n the index (n,i) belongs to A and $i \neq i(n)$ then $\bar{\bar{t}}_{n,i} = 0$ and $\bar{\bar{t}}_{n+1,j} = 0$ for each $(n+1,j) \in A(n,i)$. Thus for each $(n,i) \in A$, whether or not $i = i(n)$ it follows that $x\bar{\bar{t}}$ satisfies (XXXIa). As for (XXXIb), observe that $(x\bar{\bar{t}})_{n,i} = 0$ unless $i = i(n)$ ar hence the sum in (XXXIb) reduces to the single term $\rho(x\bar{\bar{t}}_{n,i(n)}) = \rho(x) = 1$ since $x \in \mathrm{ex}(K)$. We have just shown that $x\bar{\bar{t}} \in C$ whenever $x \in \mathrm{ex}(K)$ and $\bar{\bar{t}} \in \bar{\bar{\Delta}}_{0,1}$. Next suppose that for x and $\bar{\bar{t}}$ as above there are points $(y_{n,i})$ and $(z_{n,i})$ in C with $\frac{1}{2} y_{n,i} + \frac{1}{2} z_{n,i} = x\bar{\bar{t}}_{n,i}$ for each $(n,i) \in A$. Then

$$\frac{1}{2} y_{n,i(n)} + \frac{1}{2} z_{n,i(n)} = x\bar{\bar{t}}_{n,i(n)} = x \quad \text{for each } n$$

and since $x \in \mathrm{ex}(K)$ and $y,z \in C \subset K^{\sim}$ (whence $y_{n,i(n)}$ and $z_{n,i(n)}$ belong to K) we have $y_{n,i(n)} = z_{n,i(n)} = x$. But since $x \in \mathrm{ex}(K)$, $\rho(x) = 1$ and hence $\rho(y_{n,i(n)}) = \rho(z_{n,i(n)}) = 1$ for each n. Since $(y_{n,i}) \in C$, by (XXXIb) we have $\sum_{A_n} \rho(y_{n,i}) \leqq 1$ and since one of the terms in this sum is $\rho(y_{n,i(n)})$ it follows that $\rho(y_{n,i}) = 0$ for each $(n,i) \in A$ such that $i \neq i(n)$. That is,

$$y_{n,i} = \begin{cases} 0 & \text{if } i \neq i(n) \\ x & \text{if } i = i(n) \end{cases} = \begin{cases} x\bar{\bar{t}}_{n,i} & \text{if } i \neq i(n) \\ x\bar{\bar{t}}_{n,i} & \text{if } i = i(n) \end{cases} \quad \text{for each } (n,i) \in A.$$

That is, $(y_{n,i}) = x\bar{\bar{t}}$. Similarly, $(z_{n,i}) = x\bar{\bar{t}}$. Thus $\{x\bar{\bar{t}}: x \in ex(K),\ \bar{\bar{t}} \in \Delta_{0,1}\} \subset ex(K)$.

Finally, suppose that $(x_{n,i})$ is an extreme point of C. For any $(n,i) \in A$ let $A^1(n,i) \equiv A(n,i)$ and inductively, for $j \geqq 1$ let

$$A^{j+1}(n,i) \equiv \cup A(m,k) \quad \text{where the union extends over all } (m,k) \in A^j(n,i).$$

Thus $A^j(n,i)$ denotes the set of "$j^{\underline{th}}$ level" successors of (n,i). Denote the

$$S(n,i) \equiv \{(n,i)\} \cup \bigcup_{j=1}^{\infty} A^j(n,i).$$

Assume for now that there is an $m \geqq 1$ such that for some k and j, $k \neq j$, both (m,k) and (m,j) belong to A_m and $x_{m,k} \neq 0 \neq x_{m,j}$. In order to show that $(x_{n,i}) \notin ex(C)$ under such an assumption we create two points $(u_{n,i})$ and $(v_{n,i})$ in C distinct from $(x_{n,i})$ with $(x_{n,i})$ on the open line segment joining them. For $n \geqq m$ let

(XXXII) $$u_{n,i} \equiv \begin{cases} x_{n,i} & \text{if } (n,i) \in S(m,k) \\ 0 & \text{if } n \geqq m \text{ and } (n,i) \in A \setminus S(m,k). \end{cases}$$

Determine the coordinates $u_{n,i}$ for $n < m$ and $(n,i) \in A$ by using (XXXIa). Thus, first determine $u_{m-1,p}$ for $(m-1,p) \in A_{m-1}$ by (XXXIa), then $u_{m-2,q}$ for $(m-2,q) \in A_{m-2}$ and so on. In a similar vein let

$$v_{n,i} \equiv \begin{cases} 0 & \text{if } (n,i) \in S(m,k) \\ x_{n,i} & \text{if } n \geqq m \text{ and } (n,i) \in A \setminus S(m,k) \end{cases} \quad \text{for } n \geqq m$$

and determine $v_{n,i}$ for $n < m$, $(n,i) \in A$ by condition (XXXIa) as was done above for $u_{n,i}$. Observe that

(XXXIII) $$\sum_{A_n} \rho(u_{n,i}) + \sum_{A_n} \rho(v_{n,i}) = \sum_{A_n} \rho(x_{n,i}) \leq 1 \quad \text{for each} \quad n \geq m$$

from (XXXIb) since $(x_{n,i}) \in C$. Consequently $(u_{n,i})$ and $(v_{n,i})$ also satisfy (XXXIb) when $n \geq m$. Of course, for any $(n,i) \in A$ the point $u_{n,i}$ is the sum of its $p^{\underline{th}}$ level successors (indexed by $A^p(n,i)$) for any $p \geq 0$ and hence

(XXXIV) $$\rho(u_{n,i}) \leq \sum_{A^p(n,i)} \rho(u_{n+p,q}).$$

Using (XXXIII), induction and (XXXIV) it follows that

(XXXV) $$\sum_{A_n} \rho(u_{n,i}) \leq \sum_{A_m} \rho(u_{m,i}) \leq 1 \quad \text{for each} \quad n < m$$

and hence by (XXXIII), (XXXV) and the definition of $(u_{n,i})$ (see (XXXII)) we conclude $(u_{n,i}) \in C$. Similarly, $(v_{n,i}) \in C$. Since

$$\sum_{A_n} \rho(v_{n,i}) \leq \sum_{A_{n+1}} \rho(v_{n+1,i}) \leq 1$$

for each n by (XXXIII) and (XXXIV), it is possible to define $r \in \mathbb{R}$ by

$$r \equiv \sup_n \sum_{A_n} \rho(u_{n,i}) = \lim_n \sum_{A_n} \rho(u_{n,i}).$$

Note that $r \geq \rho(u_{m,k}) > 0$ and that, by (XXXIII),

$$1 - r \geq \sup_n \sum_{A_n} \rho(v_{n,i}) \geq \sum_{A_m} \rho(v_{m,i}) \geq \rho(v_{m,j}) > 0.$$

Hence the points $(\frac{1}{r} u_{n,i})$ and $(\frac{1}{1-r} v_{n,i})$ both belong to C and we have

$$(x_{n,i}) = r(\frac{1}{r} u_{n,i}) + (1-r)(\frac{1}{1-r} v_{n,i}),$$

a proper convex combination of distinct points of C. It follows that for each $n \geq 1$ there is at most one index $i(n)$ with $x_{n,i(n)} \neq 0$. On the other hand, the point of C all of whose coordinates are 0 is not an extreme point of C since

$0 \in K \setminus ex(K)$. Hence for some n there is an $(n,i(n)) \in A$ such that $x_{n,i(n)} \neq 0$ and then by (XXXIa), $x_{n,i(n)} = x_{p,i(p)}$ for each choice of $n,p \geqq 0$. Denote this common value by x. Then $x \in K$ and again by repeated use of (XXXIa) there is a $\bar{\bar{t}} \in \bar{\Delta}_{0,1}$ such that $(x_{n,i}) = x\bar{\bar{t}}$. Since x must clearly be an extreme point of K in order that $x\bar{\bar{t}}$ be an extreme point of C, the proof of Lemma 6.6.5 is complete.

LEMMA 6.6.6. The point $(\mu(\Delta_{n,i})b_{n,i})$ belongs to C.

Proof. Since $0 < \mu(\Delta_{n,i}) \leqq 1$ and both 0 and $b_{n,i}$ belong to K for each $(n,i) \in A$, $\mu(\Delta_{n,i})b_{n,i} \in K$ and hence $(\mu(\Delta_{n,i})b_{n,i}) \in K^{\sim}$. From (XXX) and the critical condition 6.6.3(e) for bushes, for any $(n,i) \in A$,

$$\sum_{A(n,i)} \mu(\Delta_{n+1,j})b_{n+1,j} = \mu(\Delta_{n,i}) \sum_{A(n,i)} r_{n+1,j}b_{n+1,j} = \mu(\Delta_{n,i})b_{n,i}.$$

Hence $(\mu(\Delta_{n,i})b_{n,i})$ satisfies (XXXIa) for $(n,i) \in A$. Finally, for any $n \geqq 0$,

$$\sum_{A_n} \rho(\mu(\Delta_{n,i})b_{n,i}) \leqq \sum_{A_n} \mu(\Delta_{n,i}) = \mu(\Delta_{0,1}) = 1$$

so that (XXXIb) is satisfied as well by this point. The lemma is proved.

LEMMA 6.6.7. $(\mu(\Delta_{n,i})b_{n,i})$ is not the barycenter of any separable extremal probability measure $\nu \in P_t(C)$.

The proof of 6.6.7 rests on the following general fact.

PROPOSITION 6.6.8. Let $\varepsilon > 0$ and suppose that $B \equiv \{(r_{n,i},b_{n,i}): (n,i) \in A\}$ is a bounded ε-bush in a Banach space X. Let $\Delta_{0,1}$ denote the generalized Cantor set associated with B and μ the natural probability measure on $\mathcal{B}(\Delta_{0,1})$ associated with B. (See Definition 6.6.3 and the discussion following it, through (XXX) for the notation used here.) Suppose that $\gamma \in P_t(\Delta_{0,1})$ and that M_γ denotes the γ-completion of $\mathcal{B}(\Delta_{0,1})$. Then there is no function $F \in L^1_X(\Delta_{0,1},M_\gamma,\gamma)$ such that

$$\int_{\Delta_{n,i}} F d\gamma = \mu(\Delta_{n,i})b_{n,i} \quad \text{for each } (n,i) \in A.$$

The proof of 6.6.8 will be given after we complete that of 6.6.7. Let $\nu \in P_t($ be separable extremal and suppose that ν has barycenter $(\mu(\Delta_{n,i})b_{n,i})$. Since C is separable, $\nu(\mathrm{ex}(C)) = 1$, and hence ν may be considered as a probability measu on the product space $\bar{\bar{\Delta}}_{0,1} \times \mathrm{ex}(K)$ by Lemma 6.6.5. Let $\pi_1 : \bar{\bar{\Delta}}_{0,1} \times \mathrm{ex}(K) \to \bar{\bar{\Delta}}_{0,1}$ be th first coordinate projection. According to the disintegration of measures Theorem 6.6.2 (and by identifying the homeomorphic spaces $\Delta_{0,1}$ and $\bar{\bar{\Delta}}_{0,1}$) there is a fami $\{\nu_t : t \in \Delta_{0,1}\}$ of probability measures on $\mathcal{B}(\mathrm{ex}(K))$ with the following properties: if γ denotes the probability measure $\pi_1(\nu)$ on $\mathcal{B}(\Delta_{0,1})$ then for each $R \in \mathcal{B}(\Delta_{0,1}) \times (\mathrm{ex}(K))$

(XXXVI) the function $t \to \nu_t(R_{(t)})$ from $\Delta_{0,1}$ to $\mathbb{R}$ is γ-measurable; and

(XXXVII) $\nu(R) = \int_{\Delta_{0,1}} \nu_t(R_{(t)})d\gamma(t).$

For each $t \in \Delta_{0,1}$ let $F(t)$ denote the barycenter of ν_t. Thus

(XXXVIII) $$F(t) \equiv \int_{\mathrm{ex}(K)} x d\nu_t(x).$$

We will show that $F \in L^1_X(\Delta_{0,1}, M_\gamma, \gamma)$ where M_γ denotes the γ-completion of $\mathcal{B}(\Delta_0,$ Observe that F is separably valued and bounded (since K is separable and bounded Let $\psi \in X^*$ be arbitrary. Then by (XXXVIII)

(XXXIX) $$\psi \circ F(t) = \int_{\mathrm{ex}(K)} \psi(x) d\nu_t(x).$$

Note that if $D \in \mathcal{B}(\mathrm{ex}(K))$ then the function $t \to \nu_t(D)$ $(= \nu_t((\Delta_{0,1} \times D)_{(t)}))$ is γ-measurable by (XXXVI) and hence for each simple function s on $(\mathrm{ex}(K), \mathcal{B}(\mathrm{ex}(K)))$ it follows that $t \to \int_{\mathrm{ex}(K)} s(x)d\nu_t(x)$ is γ-measurable as well. Thus, by virtue of (XXXIX), since ψ is continuous, $t \to \psi \circ F(t)$ is γ-measurable. Thus F is weakly measurable and from above (using Theorem 1.1.7), $F \in L^1_X(\Delta_{0,1}, M_\gamma, \gamma)$ as predicted.

Recall that ν has barycenter $(\mu(\Delta_{n,i})b_{n,i})$. Hence

$$(\mu(\Delta_{n,i})b_{n,i}) = \int_{\bar{\bar{\Delta}}_{0,1}\times ex(K)} z\,d\nu(z) = \int_{\bar{\bar{\Delta}}_{0,1}\times ex(K)} x\bar{\bar{t}}\,d\nu(x\bar{\bar{t}})$$

and in particular, for any fixed $(n,i) \in A$, by projection onto the $(n,i)^{th}$ co-ordinate,

$$\mu(\Delta_{n,i})b_{n,i} = \int_{\bar{\bar{\Delta}}_{0,1}\times ex(K)} x\bar{\bar{t}}_{n,i}\,d\nu(x\bar{\bar{t}}) = \int_{\bar{\bar{\Delta}}_{0,1}\times ex(K)} x\,d\nu(x\bar{\bar{t}})$$

(XL)

$$= \int_{\Delta_{n,i}} \left[\int_{ex(K)} x\,d\nu_t(x)\right]d\gamma(t)$$

by (XXXVII). But from (XXXVIII), the form of (XL) may be rewritten as

(XLI) $$\mu(\Delta_{n,i})b_{n,i} = \int_{\Delta_{n,i}} F(t)\,d\gamma(t) \quad \text{for each } (n,i) \in A.$$

However, (XLI) directly contradicts the conclusion of Proposition 6.6.8. Hence Lemma 6.6.7 has been established except for the proof of 6.6.8.

<u>Proof of 6.6.8</u>. Suppose to the contrary that a $\gamma \in P_t(\Delta_{0,1})$ and $F \in L^1_X(\Delta_{0,1},M_\gamma,\gamma)$ exist such that $\int_{\Delta_{n,i}} F\,d\gamma = \mu(\Delta_{n,i})b_{n,i}$ for each $(n,i) \in A$. Let $F_1 \equiv \mathbb{E}[F\,|\,\mathcal{B}(\Delta_{0,1})]$. Then

(XLII) $$\int_{\Delta_{n,i}} F_1\,d\gamma = \mu(\Delta_{n,i})b_{n,i} \quad \text{for each } (n,i) \in A$$

as well. Let $M\colon \mathcal{B}(\Delta_{0,1}) \to X$ be the vector measure

(XLIII) $$M(A) \equiv \int_A F_1\,d\gamma \quad \text{for } A \in \mathcal{B}(\Delta_{0,1}).$$

The following facts are not difficult to check.

(XLIV) $|M|$, the total variation of M, is given by the formula

$$|M|(A) = \int_A \|F_1(t)\|\,d\gamma(t) \quad \text{for each } A \in \mathcal{B}(\Delta_{0,1}); \text{ and}$$

(XLV) $\frac{dM}{d|M|}$ exists as a Bochner integrable function in $L^1_X(\Delta_{0,1},\mathcal{B}(\Delta_{0,1}),|M|)$.

Indeed, to verify (XLV), let $G\colon \Delta_{0,1} \to X$ be the function which is 0 whenever $F_1(t) = 0$ and which has value $\frac{F_1(t)}{\|F_1(t)\|}$ for all other $t \in \Delta_{0,1}$. It is easy to che[ck] that G is Bochner integrable - use Theorem 1.1.7 - and that $G = \frac{dM}{d|M|}$.

For each n let $f_n\colon \Delta_{0,1} \to X$ be the simple function

(XLVI) $$f_n \equiv \sum_{A_n} b_{n,i}\, \chi_{\Delta_{n,i}}.$$

Assume for convenience that $B \subset U(X)$ and let $\mathcal{F}_n$ be the finite σ-algebra generat[ed] by the atoms $\Delta_{n,i}$ for $(n,i) \in A_n$. Then by definition of μ in (XXX), $(f_n, \mathcal{F}_n)$ is a martingale with respect to the probability space $(\Delta_{0,1}, \mathcal{B}(\Delta_{0,1}), \mu)$ and we let $m\colon \mathcal{B}(\Delta_{0,1}) \to X$ be the vector measure associated with this martingale as is detailed next. For each k let $m_k\colon \mathcal{B}(\Delta_{0,1}) \to X$ be the vector measure

$$m_k(A) \equiv \int_A f_k\, d\mu \quad \text{for each } A \in \mathcal{B}(\Delta_{0,1})$$

and observe that $\lim_k m_k(A)$ exists for each $A \in \mathcal{B}(\Delta_{0,1})$. (Indeed, the sequence $\{m_k(A)\colon k = 1,2,\ldots\}$ is eventually constant if $A \in \mathcal{F}_n$ for some n. In general, given $\delta > 0$ and $A \in \mathcal{B}(\Delta_{0,1})$ choose N and $A' \in \mathcal{F}_N$ with $\mu(A \,\Delta\, A') < \frac{\delta}{2}$. Then if $k,n \geq N$ it is easy to check that $\|m_n(A) - m_k(A)\| < \delta$ since $B \subset U(X)$.) By the Hahn-Vitali-Saks Theorem 1.1.5, then,

$$m(A) \equiv \lim_k m_k(A) \quad \text{for } A \in \mathcal{B}(\Delta_{0,1})$$

defines a vector measure. Observe that

(XLVII) $$m(\Delta_{n,i}) = \mu(\Delta_{n,i})b_{n,i} \quad \text{for each } (n,i) \in A$$

since

$$m(\Delta_{n,k}) = \lim_k m_k(\Delta_{n,i}) = \lim_k \int_{\Delta_{n,i}} f_k\, d\mu = \int_{\Delta_{n,i}} f_n\, d\mu = \mu(\Delta_{n,i})b_{n,i}.$$

Moreover, as will be demonstrated next,

(XLVIII) $\frac{dm}{d\mu}$ does not exist as an element of $L^1_X(\Delta_{0,1},\mathcal{B}(\Delta_{0,1}),\mu)$.

To verify (XLVIII), suppose, to the contrary, that there is a function g in $L^1_X(\Delta_{0,1},\mathcal{B}(\Delta_{0,1}),\mu)$ such that $\int_A g d\mu = m(A)$ for each $A \in \mathcal{B}(\Delta_{0,1})$. If $A \in F_n$ then

$$\int_A g d\mu = m(A) = \lim_k m_k(A) = m_n(A) = \int_A f_n d\mu$$

whence $\mathbb{E}[g|F_n] = f_n$ for each n. Then by Lévy's Theorem 1.3.1, $\lim_n \|g(t) - f_n(t)\| = 0$ [a.s.μ]. But this is impossible since, for each $t \in \Delta_{0,1}$ and each n, if $f_n(t) = b_{n,i}$ and $f_{n+1}(t) = b_{n+1,j}$ then $(n+1,j)$ is an immediate successor of (n,i) and hence

$$\|f_{n+1}(t) - f_n(t)\| = \|b_{n+1,j} - b_{n,i}\| \geqq \varepsilon.$$

Thus (XLVIII) holds. Observe next that $|m|(A) \leqq \mu(A)$ for each $A \in \mathcal{B}(\Delta_{0,1})$ since $B \subset U(X)$ and hence $\frac{d|m|}{d\mu} \in L^1(\Delta_{0,1},\mathcal{B}(\Delta_{0,1}),\mu)$ and it is μ-almost surely bounded between 0 and 1. If $\frac{dm}{d|m|}$ existed as an element of $L^1_X(\Delta_{0,1},\mathcal{B}(\Delta_{0,1}),|m|)$ then $\frac{dm}{d\mu} = \frac{|dm|}{d|m|}\frac{d|m|}{d\mu}$ would exist and belong to $L^1_X(\Delta_{0,1},\mathcal{B}(\Delta_{0,1}),\mu)$. Since this contradicts (XLVIII), $\frac{dm}{d\mu}$ does not exist in $L^1_X(\Delta_{0,1},\mathcal{B}(\Delta_{0,1}),\mu)$.

In order to complete the proof of the proposition observe that by (XLII), (XLIII) and (XLVII),

(XLIX) $m(\Delta_{n,i}) = \mu(\Delta_{n,i})b_{n,i} = M(\Delta_{n,i})$ for each $(n,i) \in A$.

Consequently, m and M agree on each open set in $\Delta_{0,1}$ and hence on each Borel set in $\Delta_{0,1}$. We have just ascertained that $\frac{dm}{d|m|}$ fails to exist as an element of $L^1_X(\Delta_{0,1}),\mathcal{B}(\Delta_{0,1}),\mu)$ in the previous paragraph, and yet, since $m = M$, according to (XLV), $\frac{dm}{d|m|}$ does exist. This contradiction completes the proof of 6.6.8, hence of 6.6.7 and simultaneously of 6.6.1.

There are some natural situations in which the space $(\oplus \sum X_i)_{\ell_2}$ in Theorem 6.6.1 may be replaced by X. W. Schackermayer [1980] has noted the following Corollary of 6.6.1.

COROLLARY 6.6.9. Let X be a Banach space which is isomorphic with its square $X \times X$. Then X has the RNP if and only if X has the IRP.

Proof. One direction is immediate from Theorem 6.2.9. For the converse, suppose that $K \subset U(X)$ lacks the RNP for some closed bounded convex subset K of the unit ball. Then from Theorem 6.6.1, $K^{\sim} \subset (\oplus \sum_{i=1}^{\infty} X_i)_{\ell_2}$ lacks the IRP. The proof will be completed by showing that $K^{\sim}$ is affinely homeomorphic to a closed bounded convex subset K' of X (and hence K' lacks the IRP as needs to be shown).

Inductively define sequences $\{P_n: n = 1,2,\ldots\}$ and $\{R_n: n = 1,2,\ldots\}$ of bounded linear projections on X as follows. Let $P_0 \equiv R_0 \equiv$ identity of X. Assume that P_n and R_n have been chosen for some $n \geq 0$ and that $R_n(X) \equiv Y$ is isomorphic to X. Since X is isomorphic to its square so is Y and we let $I_Y: Y \to Y \times Y$ denote one such surjective isomorphism. Let $P_Y \equiv I_Y^{-1} \circ \pi_1 \circ I_Y$ and $R_Y \equiv I_Y^{-1} \circ \pi_2 \circ I_Y$ where $\pi_i: Y \times Y \to Y$ is the $i^{\underline{th}}$ coordinate projection for $i = 1,2$. Finally, let $P_{n+1} \equiv P_Y \circ R_n$ and $R_{n+1} \equiv R_Y \circ R_n$. Eliminate P_0 from further considerations and note that $\{P_n: n = 1,2,\ldots\}$ is then a sequence of bounded linear projections on X such that both $P_n(X)$ is isomorphic to X and $P_n \circ P_k = 0$ for each $n \neq k$, $n,k \geq 1$. For each n let $i_n: X \to P_n$ denote a surjective isomorphism for which $\|i_n\| = 2^{-n}$ and define $\psi: K^{\sim} \to X$ by

$$\psi((x_n)) \equiv \sum_{n=1}^{\infty} i_n(x_n).$$

Note that for $(x_n) \in K^{\sim}$ and $z \equiv \psi((x_n))$

$$\text{(L)} \qquad i_k(x_k) = P_k(z) \text{ for each } k \text{ and hence } z = \sum_{k=1}^{\infty} P_k(z).$$

Evidently ψ is affine and continuous on $K^{\sim}$. Let $K' \equiv \psi(K^{\sim})$. It is immediate from (L) that ψ is injective. Recall that $K \subset U(X)$ by assumption. Hence a sequence in $K^{\sim}$ converges if and only if it converges coordinatewise since, for any point (x_n) of $K^{\sim}$, $\|x_n\|_n \equiv 2^{-n}\|x_n\| \leq 2^{-n}$ by definition of $\|\cdot\|_n$ in (XXIII). That is, the topology on $K^{\sim}$ is the product topology. Similarly, using (L) it follows that a sequence $\{z_j: j = 1,2,\ldots\}$ in K' is Cauchy if and only if each of the sequences $\{P_k(z_j): j = 1,2,\ldots\}$ is Cauchy for $k = 1,2,\ldots$ since $\|P_k(z)\| \leq 2^{-k}$ for each and each $z \in K'$. Hence $\{z_j: j = 1,2,...\}$ in K' is Cauchy if and only if each of the sequences $\{i_k^{-1} \circ P_k(z_j): j = 1,2,...\}$ is Cauchy for $k = 1,2,...$ which, from above, is equivalent to the statement that the sequence of points

$$\psi^{-1}(z_1) = (i_k^{-1} \circ P_k(z_1)), \quad \psi^{-1}(z_2) = (i_k^{-1} \circ P_k(z_2)), \quad \psi^{-1}(z_3), \ ...$$

in $K^{\sim}$ is Cauchy. Thus ψ^{-1} is uniformly continuous. Consequently K' is closed since $K^{\sim}$ is closed, and ψ is indeed an affine homeomorphism between $K^{\sim}$ and the closed bounded convex subset K' of X. The proof is complete.

The last result of this section again deals with special conditions under which the IRP implies the RNP. It is nevertheless of quite a different nature than either Theorem 6.6.1 or Corollary 6.6.9. Indeed, rather than requiring that all closed bounded convex subsets of a given set satisfy the condition that each point be the barycenter of a separable extremal measure, the hypotheses below restrict attention to just the set itself. A more general result appears in Edgar [1978a], Corollary 2.7.

THEOREM 6.6.10. Let $C \subset X$ be a separable closed bounded convex simplex. Assume that for each $x \in C$ there is a unique separable extremal $\mu \in P_t(C)$ with barycenter x. Then C has the RNP.

The key to the proof of this theorem is a more general version of our Theorem 6.3.8 which says that (correctly interpreted), for any completely regular space T, $P_t(T)$ has the RNP. Notice that since C in the statement of 6.6.10 is separable, separable extremal measures actually live on $ex(C)$ and hence the hypotheses of 6.6.10 imply that the barycenter map $r: P_t(ex(C)) \to C$ is 1-1 and surjective. It is always continuous and affine. The argument then reduces to showing that the closed continuous 1-1 affine image of an RNP set has the RNP, a straightforward exercise. It is interesting to note that, in order to obtain the above result about the closed set C we were forced out of the domain of closed sets in order to deal with the set $ex(C)$. The subject of possibly nonclosed convex sets with the RNP is dealt with by several authors. See, for example, Edgar [1978a] and Rosenthal [1979].

CHAPTER 7. SELECTED TOPICS

In the early and mid 1970s most of the ongoing work concerning the Radon-Nikodým Property focused on the internal structure of spaces and sets with that property. The paper of Chatterji [1968] and especially that of Rieffel [1967] set the stage initially and, within seven or so years of that seminal work, some form of the geometric results of Chapters 2, 3 and 6, the duality theory (non localized) of Chapter 5 and the dual space theory of Chapter 4 was in print. The refinements of such results (presented in the previous chapters) are representative of the second stage of development after the basic theory had been outlined and a review of the initial broad strokes was undertaken. The past several years have seen the emergence of the RNP as a standard functional analytic tool, perhaps the ultimate compliment (or demise) for a successful theory. This represents a distinct third stage of development and an attempt will be made in this chapter to indicate some of the ways in which the RNP has recently been used. Because the body of material which could justifiably be included in a chapter such as this is so vast (and is expanding so rapidly) there is a certain arbitrariness to the choice of topics actually included, and several subjects we had originally intended to discuss have been eliminated for reasons of time and space.

The topics which appear below have been chosen for a variety of reasons. Several sections include updates of material touched on in some of the earlier chapters. This is especially true of §7.5 (the Kunen-Rosenthal quasi martingale approach to the Bourgain-Phelps theorem), §7.8 (on measurability) and §§7.9 and 7.12 (which overlap with some of the integral representation material of Chapter 6). Some of the deepest results in these notes appear in the examples. The separable infinite dimensional $\mathcal{L}_\infty$ spaces which have the RNP discussed by Bourgain and Delbaen [1980] (§7.2), the non-RNP subspaces of $L^1[0,1]$ which do not contain certain types of bush structures (§7.3) due to Bourgain and Rosenthal [1980a], the long James space studied by Edgar [1980] (§7.6) and the McCartney-O'Brien example [1980] of separable RNP spaces which do not embed in separable duals (§7.1) are each presented in considerable detail. Recent Banach lattice breakthroughs of Bourgain and Talagrand [1981] and especially of Talagrand [1983b] have received attention in §7.13 and a brief survey (without proofs) of progress on questions concerning Pettis integration and the weak Radon-Nikodým Property is given in §7.4. Two sections on bush considerations (§§7.6 and 7.11) and one on norm attaining operators (§7.10) round out the chapter. Most sections may be read independently of one another.

§1. A separable RNP space not isomorphic to a subspace of a separable dual.

For several years it was an open question whether the study of separable Banach spaces with the RNP was, in disguised form, that of subspaces of separable duals. (More precisely, the question went as follows: Let X be a separable Banach space and suppose that X has the RNP. Are there a separable dual space Y^* and isomorphism T from X into Y^*?) The answer is "no" and two (independently produced) examples appeared at approximately the same time. Bourgain and Delbaen [1980] have an example which is a $\mathcal{L}_\infty$ space while McCartney and O'Brien [1980] took an ℓ_1 sum of spaces discussed previously by McCartney [1980], each one of which is a subspace of $c_0 \oplus_{\ell_1} Z^*$ for a fixed separable dual space Z^*. Both examples are quite interesting. That of McCartney and O'Brien is discussed here (it is the more readily described of the two) and the Bourgain-Dalbaen example occupies §7.2. (More recently, Johnson and Lindenstrauss [1980] have produced a class of $\mathcal{L}_1$ spaces which have the Schur property and the RNP but do not embed in separable duals. See §7.2 for definitions.)

The idea behind McCartney's examples is straightforward. If ℓ_∞^n denotes n-dimensional ℓ_∞ then $Z^* \equiv (\oplus \sum_{n=1}^{\infty} \ell_\infty^n)_{\ell_1}$. (Note that $((\oplus \sum_{n=1}^{\infty} \ell_1^n)_{c_0})^* = Z^*$ so that Z^* is indeed a separable dual.) Let $Q\colon c_0 \oplus_{\ell_1} Z^* \to Z^*$ be the natural projection and for each n let F_n be the n-dimensional subspace of c_0 consisting of those points all of whose k^{th} coordinates are 0 for $k > n$. Thus, the "restriction to the first n coordinates" map R_n is an isometry from F_n onto ℓ_∞^n for each n. Fix δ, $0 < \delta < 1/4$. Then for each $x \in F_n$ the point

$$y \equiv ((1-\delta)x, 0, \ldots, 0, \delta R_n(x), 0, \ldots) \in c_0 \oplus_{\ell_1} Z^*$$

which is nonzero only in the first (i.e. the c_0^{th}) and the $(n+1)^{\text{st}}$ (i.e. the

ℓ_∞^n $\underline{\text{th}}$) coordinates has the property

$$\delta\|y\| = \delta((1-\delta)\|x\| + \delta\|R_n(x)\|) = \delta\|R_n(x)\| = \|Q(y)\|.$$

More generally, let Y_δ be the normed linear space

$$Y_\delta \equiv \text{span}(\{((1-\delta)x,0,\ldots,0,\delta R_n(x),0,\ldots): x \in F_n, \quad n = 1,2,\ldots\})$$
$$\subset c_o \oplus_{\ell_1} Z^*.$$

If $y \in Y_\delta$ then $y = ((1-\delta)\sum_{i=1}^k x_i, \delta R_1(x_1),\ldots,\delta R_k(x_k),0,\ldots)$

for some positive integer k and points $x_i \in F_i$ for $i = 1,\ldots,k$. Thus

$$\delta\|y\| \leqq \delta[(1-\delta)\sum_{i=1}^k \|x_i\| + \delta \sum_{i=1}^k \|R_i(x_i)\|]$$
$$= \delta \sum_{i=1}^k \|R_i(x_i)\| = \|Q(y)\|.$$

By continuity of the norm, then,

(1) $$\delta\|y\| \leqq \|Q(y)\| \text{ for each } y \in \overline{Y}_\delta \subset c_o \oplus_{\ell_1} Z^*.$$

Once consequence of this projection inequality (i.e. (1)) is that $\overline{Y}_\delta$ - and hence each of its closed subspaces - has the RNP. (Indeed, if not, then $U(\overline{Y}_\delta)$ contains an ε-bush B for some $\varepsilon > 0$ by Theorem 2.3.6 part 8. See also the beginning of §2.4. Now if $y_1,y_2 \in B$ and $\|y_1-y_2\| \geqq \varepsilon$ then $\|Q(y_1-y_2)\| \geqq \delta\|y_1-y_2\| \geqq \delta\varepsilon$ and it follows that $Q(B)$ is a $\delta\varepsilon$ bush in $U(\overline{Q(Y_\delta)}) \subset U(Z^*)$. Thus Z^* lacks the RNP which contradicts the conclusion of the Dunford-Pettis Theorem 4.1.3. Consequently $\overline{Y}_\delta$ has the RNP as was to be shown.)

For each $j = 1,2,\ldots$ let δ_j be the point of c_o which has a 1 in the $j^{\underline{\text{th}}}$ coordinate and 0's elsewhere, and inductively define a doubly indexed sequence $\{x_{n,i}: n = 0,1,\ldots; 1 \leqq i \leqq 2^n\}$ on $U(c_o)$ as follows:

$x_{0,1} \equiv \delta_1$. If $x_{k,i}$ is known for some $k \geq 1$ and some i,

(2) $1 \leq i \leq 2^k$ then let $x_{k+1,2i-1} \equiv x_{k,i} - \delta_{k+2}$ and

$x_{k+1,2i} \equiv x_{k,i} + \delta_{k+2}$.

(Thus $x_{1,1} = \delta_1 - \delta_2$, $x_{1,2} = \delta_1 + \delta_2$, $x_{2,1} = \delta_1 - \delta_2 - \delta_3$, $x_{2,2} = \delta_1 - \delta_2 + \delta_3$, etc.) Now observe that $x_{n,i} \in F_k$ whenever $n \leq k$ and $1 \leq i \leq 2^n$, and hence

(3)
$$X_\delta \equiv \overline{\text{span}}(\{((1-\delta)x_{n,i},0,\ldots,0,\delta R_k(x_{n,i}),0,\ldots): 1 \leq i \leq 2^n;$$
$$0 \leq n \leq k;\ k = 1,2,\ldots\}) \subset \overline{Y}_\delta$$

has the RNP from the discussion in the previous paragraph.

We will show that X_δ is not isometric with a separable dual space and that $X \equiv (\oplus \sum_{n=5}^{\infty} X_{1/n})_{\ell_1}$ is a separable RNP space not isomorphic to a separable dual. This can be done from scratch without too much trouble but we introduce some terminology of McCartney [1980] which will make the ensuing arguments even more transparent. Let

$$I \equiv \{(n,i) \in \mathbb{N} \times \mathbb{N}: n = 0,1,\ldots,;\ 1 \leq i \leq 2^n\}$$

and for $k = 1,2,\ldots$ let

$$I_k \equiv \{(n,i) \in I: n \leq k\}.$$

Using this notation, a <u>tree</u> (see §2.4) may then be considered as a doubly indexed sequence $\{y_{n,i}: (n,i) \in I\}$ such that

(4) $$(1/2)(y_{n+1,2i-1} + y_{n+1,2i}) = y_{n,i} \quad \text{for each} \quad (n,i) \in I,$$

and an <u>ε-tree</u> is a tree subject to the additional constraints

(5) $$\|y_{n+1,2i-1} - y_{n,i}\| = \|y_{n+1,2i} - y_{n,i}\| \geq \varepsilon \quad \text{for each} \quad (n,i) \in I.$$

Say that a finite sequence $\{y_{n,i}: (n,i) \in I_k\}$ is a <u>finite (k,ε)-tree</u> if (4) and (5) hold for each $(n,i) \in I_{k-1}$. (Hence a Banach space has the finite tree property

- see §2.4 - if for each ε, $0 < \varepsilon < 1$ and each $k = 2,3,\dots$ its unit ball contai a finite (k,ε)-tree.) Now suppose that for some ε, $0 < \varepsilon < 1$ and for some Banac space W, $U(W)$ contains a finite (k,ε)-tree for each $k = 2,3,\dots$. If these (k,ε)-trees are properly "lined up" with respect to one another there is hope of building an ε' tree from them for some $\varepsilon' > 0$. One possible meaning for "lined up" is described next.

DEFINITION 7.1.1. Let K be a closed bounded convex subset of a Banach space W.

(a) Then K has the Neighborly Tree Property for (δ,ε) if $0 < \delta < \varepsilon/4$ and there are finite (k,ε)-trees $\{w^k_{n,i}: (n,i) \in I_k\}$ for $k = 2,3,\dots$ in K such that for each $(n,i) \in I$ there is a closed δ ball in W which contains $w^k_{n,i}$ for each $k \geq n$. (That is, consider the (n,i)th point of each of the (k,ε)-trees for k sufficiently large that such a point is defined. Then all such points lie in some δ-ball.)

(b) K has the Neighborly Tree Property (NTP) if it has the Neighborly Tree Property for some (δ,ε).

(c) The Banach space W has the NTP if $U(W)$ has the NTP.

Returning to the X_δ spaces (see (3)) observe that $U(X_\delta)$ has the NTP for $(2\delta,2)$ since the sequence

$$\{((1-\delta)x_{n,i},0,\dots,0,\delta R_k(x_{n,i}),0,\dots): (n,i) \in I_k\}$$

is easily seen to be a finite $(k,2)$-tree in $U(X_\delta)$ for each k; and for a fixed $(n,i) \in I$, if $k \geq n$ then

$$(6)\quad \|((1-\delta)x_{n,i},0,\dots,0,\delta R_n(x_{n,i}),0,\dots) - ((1-\delta)x_{n,i},0,\dots,0,\delta R_k(x_{n,i}),0,\dots)\|$$

$$= \delta(\|R_n(x_{n,i})\| + \|R_k(x_{n,i})\|) = 2\delta.$$

(N.B. With $\varepsilon = 2$, $2\delta < 1/2 = \varepsilon/4$ in this case as is required by the definition.) Since X_δ has the RNP from above, it contains no bounded ε'-trees for any $\varepsilon' > 0$. Thus there are spaces with the NTP which fail to have bounded ε'-trees. As the following lemma demonstrates, X_δ cannot, for these same reasons, be (isometric with) a dual space.

LEMMA 7.1.2. Let K be a weak* compact convex subset of a dual Banach space W^*. If K has the NTP then K contains an ε-tree for some $\varepsilon > 0$. (In particular, a dual space with the NTP lacks the RNP.)

Proof. The last conclusion follows from the first since a Banach space with a bounded ε-tree lacks the RNP. Now suppose that K has the NTP for (δ,ε) and that $\{w^k_{n,i}: (n,i) \in I_k\}$ $k = 2,3,\ldots$ is a sequence of finite (k,ε)-trees in K as in Definition 7.1.1(a). For each $(n,i) \in I$ let $U_{n,i}$ denote the intersection with K of a closed ball of radius δ which contains $w^k_{n,i}$ for each $k \geqq n$. In the product space

$$P \equiv \prod(U_{n,i}, \text{weak*}): (n,i) \in I\}$$

consider the sequence $\{f_k: k = 1,2,\ldots\}$ defined by the conditions

$$f_k(n,i) \equiv \begin{cases} w^n_{n,i} & \text{if } k < n \\ & \\ w^k_{n,i} & \text{if } k \geqq n \end{cases} \qquad (n,i) \in I.$$

(Thus for a given $(n,i) \in I$ the sequence $\{f_k(n,i): k = 1,2,\ldots\}$ is eventually the (n,i)-th element of the k-th finite tree.) Since P is compact, $\{f_k: k = 1,2,\ldots\}$ has a cluster point, say f, in P. Observe that when $(n,i) \in I$ then

$$(1/2)f_k(n+1,2i-1) + (1/2)f_k(n+1,2i) = f_k(n,i) \qquad \text{when } k > n$$

and hence

$$f(n+1,2i-1) + f(n+1,2i) = f(n,i) \quad \text{for each } (n,i) \in I.$$

Also $f(n+1,2i-1) \in U_{n+1,2i-1}$ and $f(n,i) \in U_{n,i}$, subsets of K of diameter at most 2δ. Thus for any k,

$$\begin{aligned} &\|f(n+1,2i-1) - f(n,i)\| \\ &\geqq \|f_k(n+1,2i-1) - f_k(n,i)\| - \|f(n+1,2i-1) - f_k(n+1,2i-1)\| - \|f(n,i) - f_k(n,i)\| \\ &\geqq \varepsilon - 4\delta > 0 \end{aligned}$$

and similarly, $\|f(n+1,2i) - f(n,i)\| \geqq \varepsilon - 4\delta$. Thus with $w_{n,i} \equiv f(n,i)$ for $(n,i) \in I$ it follows that $\{w_{n,i}: (n,i) \in I\}$ is a K-valued $\varepsilon - 4\delta$ tree.

Suppose that $T: X \to Y^*$ (Y^* a separable dual) were a norm one isomorphism. Let

$$w^k_{n,i} \equiv ((1-\delta)x_{n,i},0,\ldots,0,\delta R_k(x_{n,i}),0,\ldots) \quad \text{for} \quad (n,i) \in I_k$$

for each $k = 1,2,\ldots$. Then $\{T(w^k_{n,i}): (n,i) \in I_k\}$ defines a finite tree in $U(Y^*$ for each k. It follows from (6) that for each $(n,i) \in I$ the closed ball centered at $w^n_{n,i}$ of radius 2δ, here denoted by $U_{n,i}$, contains $w^k_{n,i}$ for each $k \geqq n$. The image of $U_{n,i}$ under T is contained in a 2δ ball centered at $T(w^n_{n,i})$ and hence each point $T(w^k_{n,i})$ for $k \geqq n$ is contained in the 2δ-ball in Y^* centered at $T(w^n_{n,i})$. But Y^* has the RNP and hence contains no bounded ε' trees for any $\varepsilon' > 0$. According to Lemma 7.1.2, Y^* does not have the NTP either. In particular then, for any $\varepsilon' > 0$, $\{T(w^k_{n,i}): (n,i) \in I_k\}$ $k = 2,3,\ldots$ is NOT a sequence of (k,ε')-finite trees which satisfy the conditions of Definition 7.1.1 for the pair $(2\delta,\varepsilon')$. But the only things which can go awry are that $\varepsilon'/4$ may not exceed 2δ or, failing that (i.e. if $\varepsilon' > 8\delta$) then for some $(n,i) \in I$ and some $k > n$

$$\|T(w^k_{n,i}) - T(w^k_{n+1,2i-1})\| < \varepsilon'.$$

Consequently there is an $(n,i) \in I$ and a $k > n$ such that

$$\|T(w^k_{n,i}) - T(w^k_{n+1,2i-1})\| < 9\delta.$$

For this same choice of (n,i) and k,

$$2 = \|w^k_{n,i} - w^k_{n+1,2i-1}\| = \|T^{-1}(T(w^k_{n,i}) - T(w^k_{n+1,2i-1}))\|$$

$$\leq \|T^{-1}\|\|T(w^k_{n,i}) - T(w^k_{n+1,2i-1})\| < \|T^{-1}\|9\delta$$

whence

(7) $$\|T^{-1}\| > \frac{2}{9\delta}.$$

The main result of this section is now within easy reach.

THEOREM 7.1.3 (McCartney, O'Brien). Let $X \equiv (\oplus \sum X_{1/n})_{\ell_1}$ (where $X_{1/n}$ is defined in (3) above). Then X is a separable Banach space with the RNP and yet X is not isomorphic to a subspace of a separable dual Banach space.

Proof. Since each X_δ has the RNP, so does X by Theorem 5.8.1(i). If Y^* were a separable dual space and $T: X \to Y^*$ were a norm one isomorphism then the normalized restriction

$$T_n \equiv \|T|_{X_{1/n}}\|^{-1} T|_{X_{1/n}}$$

of T to $X_{1/n}$ is a norm one isomorphism of $X_{1/n}$ into Y^* which, by (7) satisfies the condition $\|T_n^{-1}\| > 2n/9$. Hence

$$\|T|_{X_{1/n}}\| \|(T|_{X_{1/n}})^{-1}\| > 2n/9 \quad \text{for each} \quad n$$

and since $\|T|_{X_{1/n}}\| \leq 1$ and $\|T^{-1}\| \geq \|(T|_{X_{1/n}})^{-1}\|$ it follows that $\|T^{-1}\| > 2n/9$ for each n, an absurdity which establishes the theorem.

We close this section with a few comments.

(I) The NTP is a truly neighborly property in the following senses:

(a) If Y is a subspace of X which contains a bounded ε-tree and if Z is (another) subspace of X such that for some $\delta > 0$, $\delta < \varepsilon/6$, each element of the ε-tree in Y is within δ of Z, then Z has the NTP.

(b) Suppose that X is separable Banach space which has the NTP for (δ,ε). Then X may be embedded in a Banach space containing a bounded ε' tree for some $\varepsilon' > 0$ in such a way that each element of T is within δ of some point of X.

(II) Most of what appears in §7.1 can be redone without conceptual complication for bushes in place of trees. In particular, the NBP (Neighborly Bush Property) may be defined analogously to the NTP of Definition 7.1.1 and then Lemma 7.1.2 and the two results listed in (I) above have direct analogues.

(III) As might be predicted, a dual Banach space X^* has the NTP (NBP) if and only if X^* contains a bounded ε-tree (bush) for some $\varepsilon > 0$, which happens if and only if X has a separable subspace Y such that Y^* has a bounded ε-tree (bush) for some $\varepsilon > 0$. See McCartney [1980] for details.

§2. A separable $\mathcal{L}_\infty$ space which has the RNP.

It is known (Proposition 7.2.5) that an infinite dimensional $\mathcal{L}_\infty$ space (see Definition 7.2.1) is not isomorphic to a subspace of a separable dual and hence a separable infinite dimensional $\mathcal{L}_\infty$ space with the RNP provides a second example of a separable RNP space not isomorphic to a subspace of a separable dual. (Another example appears in §7.1) This example, however, does far more. Indeed, its power as a counterexample to a wide variety of conjectures in the structure theory of Banach spaces is indicated in part in Theorem 7.2.6 below, and still other surprising aspects of this space may be found in the well written paper by Bourgain and Delbaen [1980] in which it originally appeared. In the first part of this section we outline the Bourgain-Delbaen construction without going into too much detail. The detailed argument is then presented after the proof of Lemma 7.2.2.

A sequence of positive integers $\{d_n: n = 1,2,\ldots\}$ with $\lim_n d_n = \infty$ will be defined inductively. (The first few are $d_1 = d_2 = 1$, $d_3 = 5$, $d_4 = 45$, $d_5 = 1305$.) For each n let F_n denote the finite dimensional subspace of ℓ_∞ of dimension d_n

(1) $$F_n \equiv \{(x_k) \in \ell_\infty: x_k = 0 \text{ for each } k > d_n\}$$

and let $\pi_n: \ell_\infty \to F_n$ denote the natural projection. The key to the construction lies in choosing a special isomorphic embedding of F_n into F_{n+1} which may be roughly described as follows. For $\lambda > 1$ (independent of n) let $0 < \delta < 1$ satisfy $1 + 2\delta\lambda \leqq \lambda$ and suppose that linear functionals $f_1,\ldots,f_{d_{n+1}-d_n}$ in F_n^* have already been described. Then let $i_{n,n+1}: F_n \to F_{n+1}$ be the map

(2) $$i_{n,n+1}(x) \equiv (x_1,\ldots,x_{d_n},f_1(x),\ldots,f_{d_{n+1}-d_n}(x),0,\ldots) \quad \text{for}$$
$$x \equiv (x_1,\ldots,x_{d_n},0,\ldots) \in F_n.$$

With proper care in the choice of $f_1,\dots,f_{d_{n+1}-d_n}$ it will follow that $\|i_{n,n+1}\| \leqq \lambda$ and that (cf. Lemma 7.2.3, part 2)

(3) $$\|i_{n,n+1}(x)\| \leqq \|\pi_m(x)\| + \delta\|x - i_{m,n}\circ\pi_m(x)\| \quad \text{whenever} \quad x \in F_n, \quad m < n$$

where $i_{m,n}: F_m \to F_n$ denotes the composition

(4) $$i_{m,n} \equiv i_{n-1,n} \circ \cdots \circ i_{m+1,m+2} \circ i_{m,m+1} \qquad m < n.$$

Since $i_{n,n+1}$ is a linear map which does not alter any of the first d_n coordinates of the points of F_n, $i_{n,n+1}$ is an isomorphism of F_n into ℓ_∞ and $\pi_n \circ i_{n,n+1}$ is the identity on F_n. Now suppose that $x \in F_m$ for some m. Then the sequence of $j^{\underline{th}}$ coordinates of the points

(5) $$x, i_{m,m+1}(x), i_{m,m+2}(x), \dots, i_{m,n}(x), \dots$$

is eventually constant for any positive integer j and hence it is possible to formally define the limit of the sequence appearing in (5), say $i_m(x)$, coordinate by coordinate:

(6) $$(i_m(x))_j \equiv \lim_n (i_{m,n}(x))_j.$$

Again by the careful choice of the functionals $f_1,\dots,f_{d_{n+1}-d_n}$, for each n it will turn out that $i_n: F_n \to \ell_\infty$ is an isomorphism into ℓ_∞ with $\|i_n\| \leqq \lambda$. Then the Banach space sought for is

(7) $$X \equiv \overline{\bigcup_{n=1}^{\infty} i_n(F_n)} \subset \ell_\infty.$$

The inequality (3) may be parlayed into

(8) $$\|x\| \geqq \|\pi_m(x)\| + \delta\|x - i_m\circ\pi_m(x)\| \quad \text{for each} \quad x \in X \quad \text{and each} \quad m$$

and as is shown in Lemma 7.2.7, the inequality (8) is the key to showing that X has the RNP and the Schur property (Lemma 7.2.8).

DEFINITION 7.2.1. (a) Let Y and Z be Banach spaces. Then the Banach-Mazur

<u>distance</u> between Y and Z is

$$d(Y,Z) \equiv \inf\{\|T\|\|T^{-1}\|: T: Y \to Z \text{ is a surjective isomorphism}\}.$$

(b) Given $\lambda \geqq 1$ and $1 \leqq p \leqq \infty$, a Banach space Y is said to be a $\mathcal{L}_{p,\lambda}$ space if whenever E is a finite dimensional subspace of Y there is a finite dimensional subspace F of Y with $E \subset F$ and $d(F, \ell_p^{\dim F}) \leqq \lambda$. (Here $\dim F$ means the dimension of F and $\ell_p^{\dim F}$ is the dim F-dimensional ℓ_p space.) In general, Y is a $\mathcal{L}_p$ space if Y is a $\mathcal{L}_{p,\lambda}$ space for some $\lambda \geqq 1$.

(c) The Banach space <u>Y has the Schur property</u> if each weakly compact subset of Y is compact in the norm topology.

The following plausible lemma will be used to establish that the space X in (7) is in fact a $\mathcal{L}_\infty$ space.

<u>LEMMA</u> 7.2.2. Let Y be a Banach space and suppose that $\lambda \geqq 1$. Let $\{Y_n: n=1,2,\ldots\}$ be an increasing sequence of finite dimensional subspaces of Y such that both

(a) $Y = \overline{\bigcup_{n=1}^{\infty} Y_n}$; and

(b) $d(Y_n, \ell_\infty^{\dim Y_n}) \leqq \lambda$ for each n.

Then for each $\varepsilon > 0$, Y is a $\mathcal{L}_{\infty,\lambda+\varepsilon}$ space.

Proof. Let E be a finite dimensional subspace of Y with basis $\{x_1,\ldots,x_k\}$ in $S(E)$. Since all norms on finite dimensional spaces are equivalent, there is a constant $M \geqq 1$ such that

$$\frac{1}{M}\max\{|t_i| : 1 \leqq i \leqq k\} \leqq \left\|\sum_{i=1}^{k} t_i x_i\right\| \leqq M\max\{|t_i| : 1 \leqq i \leqq k\}$$

for all choices $t_1,\ldots,t_k$ in $\mathbb{R}$. Given $0 < \varepsilon < 1$, fix $0 < \eta < \frac{1}{2kM}$ such that $\lambda\frac{1+2Mk\eta}{1-2Mk\eta} < \lambda+\varepsilon$ and choose N so large that $\|x_i - y_i\| < \eta$ for $i = 1,\ldots,k$ for some choice of points $y_1,\ldots,y_k \in Y_N$. Note that if $t_1,\ldots,t_k \in \mathbb{R}$ then

$$\frac{1}{2M}\max\{|t_i| : 1 \leq i \leq k\} \leq \left\| \sum_{i=1}^{k} t_i y_i \right\| \leq 2M \max\{|t_i| : 1 \leq i \leq k\}.$$

It follows from the Hahn-Banach theorem that there are linear functionals $h_1,\ldots,h_k$ in Y^* such that $\|h_i\| \leq 2M$ and

$$h_i(y_j) = \begin{cases} 1 & \text{if } i = j \\ 0 & \text{if } i \neq j \end{cases} \qquad \text{for } 1 \leq i,j \leq k.$$

Now define $T: Y_N \to Y$ by

(9) $$T(y) \equiv y + \sum_{i=1}^{k} h_i(y)(x_i - y_i) \quad \text{for } y \in Y_N$$

and observe that $T(y_i) = x_i$ for $i = 1,\ldots,k$ so that $E \subset T(Y_N)$. Moreover, from (9),

$$(1-2Mk\eta)\|y\| \leq \|T(y)\| \leq (1+2Mk\eta)\|y\| \quad \text{for each } y \in Y_N$$

and hence $d(Y_N, T(Y_N)) \leq \|T\|\|T^{-1}\| \leq \frac{1+2Mk\eta}{1-2Mk\eta}$. But since $d(Y_N, \ell_\infty^{\dim Y_N}) \leq \lambda$, it follows that $d(T(Y_N), \ell_\infty^{\dim Y_N}) \leq \lambda \frac{1+2Mk\eta}{1-2Mk\eta} \leq \lambda + \varepsilon$ as was to be shown.

The detailed construction of X is presented below. Fix $\lambda > 1$ and choose $0 < \delta < 1$ so small that $1+2\delta\lambda \leq \lambda$. Let $d_1 \equiv 1$. The construction of the integers d_n and associated linear functionals on F_n is by induction on n. By the end of the $(n-1)^{\underline{st}}$ stage of induction the integers $d_1, d_2, \ldots, d_{n-1}$ <u>and</u> d_n will already have been defined. The cardinality of the following set of quintuples of integers gives the size of the index set for the set $\mathcal{F}_n$ of linear functionals in F_n^* chosen at the $n^{\underline{th}}$ stage. The set of quintuples is:

(10) $$J_n \equiv \{(m,k,j,\alpha,\beta): 1 \leq m < n,\ 1 \leq k \leq d_m,\ 1 \leq j \leq d_n,\ \alpha = \pm 1,\ \beta = \pm 1\}.$$

(Thus the elements of $\mathcal{F}_n$ are elements of F_n^* which may be indexed by the integers between 1 and $|J_n|$ where $|J_n|$ denotes the cardinality of the set J_n.) The rule for choosing d_{n+1} from d_n is

(11) $$d_{n+1} \equiv d_n + |\mathcal{F}_n|$$

and if $f_1,\ldots,f_{d_{n+1}-d_n}$ is a listing of the functions in $\mathcal{F}_n$, the embedding of F_n into F_{n+1} is accomplished by the map $i_{n,n+1}\colon F_n \to F_{n+1}$ defined in (2) above:

$$i_{n,n+1}(x) \equiv (x_1,\ldots,x_{d_n},f_1(x),\ldots,f_{d_{n+1}-d_n}(x),0,\ldots)$$

for $x \in F_n$. In order to describe the functions in $\mathcal{F}_n$ observe that at each previo stage m of the induction an embedding of F_m into F_{m+1} will have been defined a hence there is a natural embedding $i_{m,n}\colon F_m \to F_n$ given by composition as in (4) above:

$$i_{m,n}(x) \equiv i_{n-1,n} \circ \cdots \circ i_{m+1,m+2} \circ i_{m,m+1}(x) \quad \text{for} \quad x \in F_m, \quad m < n.$$

The elements of $\mathcal{F}_n$ may then be described as follows: given $(m,k,j,\alpha,\beta) \in J_n$ and $x \equiv (x_1,\ldots,x_{d_n},0,\ldots) \in F_n$ let

$$f_{m,k,j,\alpha,\beta}(x) \equiv \alpha x_k + \beta\delta(x_j - (i_{m,n} \circ \pi_m(x))_j). \tag{12}$$

(N.B. Each $f_{m,k,j,\alpha,\beta}$ clearly depends also on the index n, though this dependen is suppressed.) We describe in complete detail the first few steps of this inducti procedure. As indicated above, in order to get started let $d_1 \equiv 1$.

<u>n = 1</u>. Since $d_1 = 1$, $F_1 = \{(x_1,0,\ldots) \in \ell_\infty\}$ (cf. (1)) and from (10), $J_1 = \emptyset$. Since the indexing set for $\mathcal{F}_1$ is empty, $\mathcal{F}_1 = \emptyset$ as well, and hence, from (11), $d_2 = 1 + 0 = 1$. Thus $F_2 = \{(x_1,0,\ldots) \in \ell_\infty\} = F_1$ and $i_{1,2}\colon F_1 \to F_2$ as determin by (2), is the identity map on F_1.

<u>n = 2</u>. From (10), $J_2 \equiv \{(1,1,1,\pm1,\pm1)\}$ is a set of cardinality 4. The elements of $\mathcal{F}_2$ may then be calculated from (12):

$$f_{1,1,1,+1,+1}(x_1,0,\ldots) \equiv x_1 + \delta(x_1-(i_{1,2}\circ\pi_1(x))_1) = x_1 + \delta(x_1-x_1) = x_1;$$

$$f_{1,1,1,+1,-1}(x_1,0,\ldots) \equiv x_1 - \delta(x_1-x_1) = x_1;$$

$$f_{1,1,1,-1,+1}(x_1,0,\ldots) \equiv -x_1 + \delta(x_1-x_1) = -x_1;$$

$$f_{1,1,1,-1,-1}(x_1,0,\ldots) \equiv -x_1 - \delta(x_1-x_1) = -x_1.$$

Now $d_3 = d_2 + |F_2| = 1 + 4 = 5$ and hence $F_3 \equiv \{(x_1,x_2,x_3,x_4,x_5,0,\ldots) \in \ell_\infty\}$. $i_{2,3}: F_2 \to F_3$ is the map

$$i_{2,3}(x_1,0,\ldots) \equiv (x_1,x_1,x_1,-x_1,-x_1,0,\ldots)$$

and

$$i_{1,3}: F_1 \to F_3 \text{ is } i_{1,3} \equiv i_{2,3} \circ i_{1,2} = i_{2,3}.$$

<u>$n = 3$.</u> The elements (m,k,j,α,β) of J_3 are as follows: $m = 1$ or 2 and since $d_1 = d_2 = 1$ and $1 \leq k \leq d_m$, we have $k = 1$. j may be $1,2,3,4$ or 5 and each of α and β is ± 1. Hence $J_3 = 2\cdot 1\cdot 5\cdot 2\cdot 2 = 40$ and thus $d_4 \equiv d_3 + |F_3| = 5 + |J_3| = 45$.

Here is the calculation of a typical element of F_3:

$$f_{2,1,4,-1,+1}(x_1,x_2,x_3,x_4,x_5,0,\ldots) = -x_1 + \delta(x_4 - (i_{2,3}\circ\pi_2(x))_4)$$

$$= -x_1 + \delta(x_4 - (x_1,x_1,x_1,-x_1,-x_1,0,\ldots)_4) = -x_1 + \delta(x_4 + x_1)$$

$$= (-1 + \delta)x_1 + \delta x_4.$$

$i_{3,4}: F_3 \to F_4$ is the embedding

$$i_{3,4}(x) \equiv (x_1,x_2,x_3,x_4,x_5,f_{1,1,1,+1,+1}(x),\ldots,f_{2,1,5,-1,-1}(x),0,\ldots)$$

while $i_{2,4}(x) = i_{3,4}\circ i_{2,3}$ and $i_{1,4} = i_{2,4}\circ i_{1,2} = i_{2,4}$.

<u>$n = 4$.</u> Note that in J_4, m may be 1, 2 or 3. If m is 1 or 2 then $k = 1$. If $m = 3$ then k may be 1, 2, 3, 4 or 5. j ranges between 1 and 45 and each of α and β is ± 1. Thus $|F_4| = |J_4| = 2\cdot 1\cdot 45\cdot 2\cdot 2 + 1\cdot 5\cdot 45\cdot 2\cdot 2 = 1260$. And so it goes.

A few immediate consequences of these inductive definitions are listed below.

(13) $$\pi_m \circ i_{m,n} = \text{identity on } F_m \text{ for } m < n;$$

(14) $$i_{m,n}\circ i_{k,m} = i_{k,n} \text{ for } k < m < n;$$

(15) $$(i_{n,k}(x))_j = x_j \quad \text{for } j \leqq d_n, \quad n < k \text{ and } x \in F_n;$$

(16) $$d(F_n, \ell_\infty^{d_n}) = 1 \quad \text{for each } n.$$

Two more consequences of the construction, not quite so immediate as the above, appear next. The first, together with Lemma 7.2.2, will show that X is a $\mathcal{L}_\infty$ space while the second is the main step on the way to the inequality (8).

LEMMA 7.2.3. (a) $\|i_{m,n}\| \leqq \lambda$ if $m < n$; and

(b) For each $x \in F_n$ and for each $m < n$

$$\|i_{n,n+1}(x)\| \geqq \|\pi_m(x)\| + \delta\|x - i_{m,n}\circ\pi_m(x)\|.$$

Proof. (a) The proof is by induction. Since $i_{1,2}$ is the identity on F_1, $\|i_{1,2}\| = 1 \leqq \lambda$. Now suppose that for some $n \geqq 2$, $\|i_{m,n}\| \leqq \lambda$ for each $m < n$. The proof that $\|i_{m,n+1}\| \leqq \lambda$ for each $m < n+1$ is broken into two steps:

Step 1. $m = n$. For $x \in F_n$

$$\|i_{m,n+1}(x)\| = \max(\{|x_i| : 1 \leqq i \leqq d_n\}, \{|f_{m,k,j,\alpha,\beta}(x)| : (m,k,j,\alpha,\beta) \in J_n\})$$

by definition. But for any $(m,k,j,\alpha,\beta) \in J_n$, using the obvious inequality $\|\pi_m\| \leqq 1$ as well as the inductive hypothesis,

$$\begin{aligned} |f_{m,k,j,\alpha,\beta}(x)| &\leqq |x_k| + \delta\|x - i_{m,n}\circ\pi_m(x)\| \\ &\leqq \|x\| + \delta(\|x\| + \lambda\|x\|) \leqq (1+2\delta\lambda)\|x\| \leqq \lambda\|x\| \end{aligned}$$

whence $\|i_{n,n+1}\| \leqq \lambda$.

Step 2. $m < n$. For $x \in F_m$

$$\|i_{m,n+1}(x)\| = \|i_{n,n+1}(i_{m,n}(x))\| = \max(\|i_{m,n}(x)\|, \{|f\circ i_{m,n}(x)| : f \in F_n\})$$

so that

(17) $$\|i_{m,n+1}\| \leqq \max(\|i_{m,n}\|, \{|f\circ i_{m,n}| : f \in F_n\})$$

and since $\|i_{m,n}\| \leqq \lambda$ by inductive hypothesis, it suffices to establish that

$\|f\circ i_{m,n}\| \leqq \lambda$ for each $f \in F_n$. Thus, pick $(m',k,j,\alpha,\beta) \in J_n$ and let $f \equiv f_{m',k,j,\alpha,\beta}$. For $x \in F_n$

$$(18) \qquad f\circ i_{m,n}(x) = \alpha(i_{m,n}(x))_j + \beta\delta((i_{m,n}(x))_j - (i_{m',n}\circ\pi_{m'}\circ i_{m,n}(x))_j).$$

Now $(i_{m,n}(x))_k = (\pi_{m'}\circ i_{m,n}(x))_k$ since $k \leqq d_{m'}$. By going back to the definition of f, such a substitution in the first term on the right hand side of (18) yields

$$(19) \qquad \|f\circ i_{m,n}\| \leqq \|\pi_{m'}\circ i_{m,n}\| + \delta\|i_{m,n} - i_{m',n}\circ\pi_{m'}\circ i_{m,n}\|.$$

If $m' \leqq m$ then $\pi_{m'}\circ i_{m,n} = \pi_{m'}$ and hence by inductive hypothesis, (19) becomes

$$\|f\circ i_{m,n}\| \leqq \|\pi_{m'}\| + \delta(\|i_{m,n}\| + \|i_{m',n}\circ\pi_{m'}\|) \leqq 1 + \delta(\lambda+\lambda) \leqq \lambda$$

while, if $m' > n$ then $i_{m',n}\circ\pi_{m'}\circ i_{m',n} = i_{m',n}\circ i_{m,m'} = i_{m,n}$ by (13) and (14) applied judiciously, whence (19) becomes

$$\|f\circ i_{m,n}\| \leqq \|\pi_{m'}\circ i_{m,n}\| + \delta(\|i_{m,n} - i_{m,n}\|) \leqq \lambda.$$

It follows from (17) that $\|i_{m,n+1}\| \leqq \lambda$ and the induction is complete.

<u>(b)</u> If $m < n$ then choose k and $\alpha = \pm 1$ such that $x_k = \|\pi_m(x)\|$ and choose $\beta = \pm 1$ and j such that $\beta\delta(x_j - (i_{m,n}\circ\pi_m(x))_j) = \delta\|x - i_{m,n}\circ\pi_m(x)\|$. Then

$$\|i_{n,n+1}(x)\| \geqq |f_{m,k,j,\alpha,\beta}(x)| = \|\pi_m(x)\| + \delta\|x - i_{m,n}\circ\pi_m(x)\|.$$

This completes the proof of Lemma 7.2.3.

Recall that $i_{k,k+1}$ only alters the entries between the $(d_k+1)^{\underline{st}}$ and $d_{k+1}^{\underline{st}}$ coordinates of a point of F_k. Thus starting with any n and $x \in F_n$ the successive points

$$x,\ i_{n,n+1}(x),\ i_{n,n+2}(x) \quad (= i_{n+1,n+2}(i_{n,n+1}(x))),\ldots,i_{n,k}(x),\ldots$$

fix more and more coordinates as k increases without bound. It follows that for each j and each n

$$(i_n(x))_j \equiv \lim_k (i_{n,k}(x))_j \quad \text{exists for each} \quad x \in F_n.$$

Then $i_n(x)$, the sequence of real numbers whose $j^{\underline{th}}$ coordinate is $(i_n(x))_j$ as defined above, has the following properties.

LEMMA 7.2.4. (a) $i_n: F_n \to \ell_\infty$ is a linear map for each n;

(b) $\|i_n\| \leqq \lambda$ for each n;

(c) $\pi_n \circ i_n$ = identity on F_n for each n;

(d) $d(F_n, i_n(F_n)) \leqq \lambda$ for each n;

(e) $i_n \circ i_{m,n} = i_m$ for each $m < n$;

(f) $i_m(F_m) \subset i_n(F_n)$ for each $m < n$;

(g) $\|x\| \geqq \|\pi_m(x)\| + \delta\|x - i_m \circ \pi_m(x)\|$ whenever $x \in i_n(F_n)$ and $m < n$.

Proof. (a) and (b) For each j and each $x \in F_n$, by Lemma 7.2.3(a)

(20) $$|(i_n(x))_j| = \lim_k |(i_{n,k}(x))_j| \leqq \lim_k \|i_{n,k}\| \|x\| \leqq \lambda\|x\|.$$

Clearly $i_n: F_n \to \ell_\infty$ is linear, and from (20), $\|i_n\| \leqq \lambda$.

(c) For any $k > n$, $\pi_n \circ i_{n,k}$ = identity on F_n by (13) from which (c) follows directly.

(d) $d(F_n, i_n(F_n)) \leqq \|\pi_n\|\|i_n\| \leqq \lambda$ from parts (b) and (c) just established.

(e) If $m < n$ and $x \in F_m$ then

$$(i_n \circ i_{m,n}(x))_j = \lim_k (i_{n,k}(i_{m,n}(x)))_j = \lim_k (i_{m,k}(x))_j = (i_m(x))_j$$

from which (e) follows.

(f) $i_m(F_m) = i_n \circ i_{m,n}(F_m) \subset i_n(F_n)$ for $m < n$ by (e) above.

(g) Suppose that $x \in i_n(F_n)$. For each $k > n$ it is easy to see that

$$\|i_{n,k} \circ \pi_n(x)\| \leqq \|i_{n,k+1} \circ \pi_n(x)\|$$

and from the definition of i_n coupled with the fact that $i_n \circ \pi_n$ = identity on $i_n(F_n)$ (cf. 7.2.4(c)) we have

$$\lim_k \| i_{n,k} \circ \pi_n(x) \| = \| i_n \circ \pi_n(x) \| = \|x\|.$$

Now suppose that $m < n \leqq k$ and that $x \in i_n(F_n)$. Then

(21) $$\|x\| \geqq \| i_{n,k+1} \circ \pi_n(x) \| = \| i_{k,k+1}(i_{n,k} \circ \pi_n(x)) \|.$$

Apply Lemma 7.2.3(b) with x there replaced by $i_{n,k} \circ \pi_n(x)$ to obtain

(22) $$\| i_{k,k+1}(i_{n,k} \circ \pi_n(x)) \| \geqq \| \pi_m \circ i_{n,k} \circ \pi_n(x) \| + \delta \| i_{n,k} \circ \pi_n(x) - i_{m,k} \circ \pi_m \circ i_{n,k} \circ \pi_n(x) \|.$$

Observe that $\pi_m \circ i_{n,k} \circ \pi_n = \pi_m$ and $i_{m,k} \circ \pi_m \circ i_{n,k} \circ \pi_n = i_{m,k} \circ \pi_m$ since $m < n \leqq k$. Moreover, using the fact that $\pi_n \circ i_n$ is the identity on F_n it is clear that $i_{n,k} \circ \pi_n(x) = \pi_k(x)$ since $x \in i_n(F_n)$. Thus (21) and (22) may be combined to yield

(23) $$\|x\| \geqq \| \pi_m(x) \| + \delta \| \pi_k(x) - i_{m,k} \circ \pi_m(x) \|.$$

Passing to the limit as $k \to \infty$ in (23) yields the desired conclusion. Thus the proof of Lemma 7.2.4 is complete.

As stated in (7), let $X \equiv \overline{\bigcup_{n=1}^{\infty} i_n(F_n)}$. Now it is obvious that $d(F_n, \ell_\infty^{d_n}) = 1$ and from Lemma 7.2.4(d), $d(F_n(i_n(F_n)) \leqq \lambda$. Thus $d(F_n, \ell_\infty^{d_n}) \leqq \lambda$ and it follows from Lemma 7.2.2 with Y_n replaced by $i_n(F_n)$ that X is an $\mathcal{L}_{\infty,\lambda+\varepsilon}$ space for each $\varepsilon > 0$. X is obviously infinite dimensional. As an immediate corollary of the next result, then, X is not isomorphic to a subspace of a separable dual.

<u>PROPOSITION</u> 7.2.5. Let Z be a separable infinite dimensional $\mathcal{L}_\infty$ space. Then Z does not semi-embed in a separable dual space. (Semi-embeddings were defined and briefly discussed in Section 4.1. See 4.1.11.)

This proposition is due to Bourgain and Rosenthal [1983].

<u>Proof of 7.2.5.</u> Some facts about $\mathcal{L}_\infty$ spaces will be used. They are gathered here (with references for further details).

(A) If Z is a $\mathcal{L}_\infty$ space then Z^{**} is a $\mathcal{L}_\infty$ space. (In fact, Z is a $\mathcal{L}_p$ space if and only if Z^* is a $\mathcal{L}_q$ space $1 \leqq p,q \leqq \infty$, $\frac{1}{p} + \frac{1}{q} = 1$. See

Lindenstrauss and Tzafriri [1973], Proposition II.5.8(ii), p. 203, or Lindenstrauss and Rosenthal [1969].)

(B) Y is isomorphic to a $\mathcal{L}_1$ space if and only if Y^* is injective. (See Lindenstrauss and Tzafriri [1973], Theorem II.5.7(ii), p. 201. Recall: a Banach space W is <u>injective</u> if there is an $\alpha < \infty$ such that for each Banach space $W' \supset W$ there is a continuous linear projection π from W' onto W with $\|\pi\| \leq \alpha$. (See Lindenstrauss and Tzafriri [1973], Part II, Chapter 4, sections d and n, or Lindenstrauss and Rosenthal [1969].)

(C) Let W be an injective Banach space and $S: W \to Y$ a continuous linear operator into the Banach space Y. If S is not weakly compact there is a subspace V of W such that V is isomorphic to ℓ_∞ and the restriction of S to V, $S|_V$ is an isomorphism of V into Y. (See Rosenthal [1970], Corollary 1.4.) In particular each infinite dimensional injective Banach space contains an isomorph of ℓ_∞.

Given this background, the proof of the proposition is not difficult. Let Z be a separable infinite dimensional $\mathcal{L}_\infty$ space, suppose that Y^* is a separable dual space and that $T: Z \to Y^*$ is a semi-embedding. Denote by π the natural projection of Y^{***} onto Y^*. Then $\pi \circ T^{**}: Z^{**} \to Y^*$. From (A), Z^{**} is a $\mathcal{L}_\infty$ space and from (B) each dual $\mathcal{L}_\infty$ space is injective. Hence Z^{**} is injective. If T is not a weakly compact operator then neither is $\hat{T}: \hat{Z} \to Y^*$ (where $\hat{T}$ is defined by $\hat{T}(\hat{z}) \equiv T(z)$ for $z \in Z$) and hence neither is T^{**} (since the restriction of T^{**} to $\hat{Z}$ is $\hat{T}$). Thus, if T is not weakly compact there is a subspace V of Z^{**} isomorphic to ℓ_∞ on which $\pi \circ T^{**}$ is an isomorphism by (C). In particular, then, $\pi \circ T^{**}(V) \subset Y^*$ is not separable. Since this is impossible, T must be a weakly compact operator. Since T is a semi-embedding $T(U(Z))$ must then be a weakly compact set. Let $\mathcal{T}$ denote the weak topology on Z induced by the set of functionals $\{F \circ T: F \in Y^{**}\} \subset Z^*$. Then $\mathcal{T}$ is a locally convex Hausdorff topology for Z and $(U(Z), \mathcal{T})$ is compact. Note that $\mathcal{T}$ is at least as weak as the norm topology on Z. A calculation shows that Z is isometrically isomorphic to the dual of the Banach space $A(U(Z), \mathcal{T})$, the space of affine continuous functions in the $\mathcal{T}$ topology on $U(Z)$ which take 0 to 0, in the supremum norm. That is, Z is a dual space under these hypotheses. But since Z is an infinite dimensional dual $\mathcal{L}_\infty$ space it is injective by (A) and (B) and hence by (C) contains an isomorphic copy of ℓ_∞. Thus Z cannot be separable. The proof is complete.

The main theorem is stated below. Some parts have already been established, and the others will be proved in the remainder of this section. Still other properties of X of significant interest may be found in Bourgain and Delbaen [1980] (where other classes of $\mathcal{L}_\infty$ spaces with considerably different properties but constructed in very much the same way as is X are also described).

THEOREM 7.2.6. (Bourgain, Delbaen). Given $\lambda > 1$ there is a separable infinite dimensional $\mathcal{L}_{\infty,\lambda}$ space X with the following properties:

(a) X is not isomorphic to a subspace of a separable dual Banach space.

(b) X has the RNP.

(c) Weakly compact subsets of X are norm compact (i.e. X has the Schur property).

(d) X has the weak compact extension property. That is, whenever $Y_1 \subset Y_2$ are Banach spaces and $T: Y_1 \to X$ is a weakly compact linear operator on Y_1 then there is a weakly compact linear operator $T_2: Y_2 \to X$ which extends T_1. (Thus, $T_2|_{Y_1} = T_1$.)

(e) X* is isomorphic to C[0,1]*.

(f) X contains no isomorphic copy of c_0.

(g) Each infinite dimensional subspace of X contains an isomorph of ℓ_1.

Proof. The fact that the space X constructed above is a $\mathcal{L}_{\infty,\lambda+\varepsilon}$ space for each $\varepsilon > 0$ was a consequence of Lemma 7.2.2. But we could have begun with any $\lambda > 1$ and $\varepsilon > 0$ and thus, for a prescribed $\lambda > 1$ we may construct X as above with λ replaced by $(1/2)(1+\lambda)$ and $\varepsilon \equiv (1/2)(\lambda-1)$ in Lemma 7.2.2. Part (a) of 7.2.6 was established in Proposition 7.2.5, as was indicated in the discussion preceding its statement. Part (b) is the content of Lemma 7.2.7 below, and parts (c) and (g) are to be found in Lemma 7.2.8. It is known that a Banach space Y is a $\mathcal{L}_\infty$ space if and only if it has the compact extension property. (That is, whenever $Y_1 \subset Y_2$ are Banach spaces and $T: Y_1 \to X$ is a compact linear operator there is a compact linear extension T_2 of T_1, $T_2: Y_2 \to X$. See Lindenstrauss and Tzafriri

[1973] p. 219 for details.) But by part (c), weakly compact operators arriving in X are really compact operators. Thus (d) follows. Lewis and Stegall [1973] proved that a separable infinite dimensional $\mathcal{L}_\infty$ space Y has dual isomorphic either to ℓ_1 or to $C[0,1]^*$, and that Y^* is isomorphic to $C[0,1]^*$ precisely when Y contains isomorphic copies of ℓ_1. Since X contains many isomorphic copies of ℓ_1 (see Lemma 7.2.8) X^* is isomorphic to $C[0,1]^*$. If X contains an isomorphic cop of c_o then, according to Lemma 7.2.8, c_o would contain an isomorphic copy of ℓ_1 (which is absurd since there are uncountably many continuous linear functionals on ℓ_1, all of distance at least 1 from one another, and if ℓ_1 could be embedded in c_o, each could be extended by the Hahn-Banach theorem to an element of c_o^*, making this space nonseparable). Thus the proof of Theorem 7.2.6 rests on the proofs of Lemmas 7.2.7 and 7.2.8.

LEMMA 7.2.7. X has the RNP.

Proof. First observe that the formula (8) holds. (That is, for each $x \in X$ and each m,

$$\|x\| \geqq \|\pi_m(x)\| + \delta\|x - i_m \circ \pi_m(x)\|.)$$

Indeed, if $x \in i_j(F_j)$ for some j then this inequality is a restatement of Lemma 7.2.4(g) when $m < j$, and if $m \geqq j$ then (8) is easy to check since then $x = i_m \circ \pi_m(x)$. Thus (8) holds for $x \in \bigcup_{j=1}^{\infty} i_j(F_j)$, a dense subset of X, and the inequality for general $x \in X$ follows immediately by continuity.

Let (Ω, F, P) be a probability space and $m: F \to X$ a measure absolutely continuous with respect to P for which $AR(m) \subset U(X)$. For each $A \in F$ and each n, by (8)

$$(24) \qquad \|m(A)\| \geqq \|\pi_n(m(A))\| + \delta\|m(A) - i_n \circ \pi_n(m(A))\|.$$

Take the supremum over all finite partitions of Ω chosen from F to obtain (from (24))

$$(25) \qquad |||m||| \geqq |||\pi_n \circ m||| + \delta |||m - i_n \circ \pi_n \circ m|||$$

(where $|||\cdot|||$ denotes the total variation norm). Given $\varepsilon > 0$ choose a finite partition $\{A_1,\ldots,A_k\}$ of Ω from F such that

$$|||m||| \leqq \sum_{i=1}^{k} \|m(A_i)\| + \varepsilon/2$$

and choose N so large that for each $n \geqq N$

(26) $$\sum_{i=1}^{k} \|\pi_n(m(A_i))\| \geqq \sum_{i=1}^{k} \|m(A_i)\| - \varepsilon/2 \geqq |||m||| - \varepsilon.$$

Note that $|||\pi_n \circ m||| \geqq \sum_{i=1}^{k} \|\pi_n(m(A_i))\|$. Hence from (26), $|||\pi_n \circ m||| \geqq |||m||| - \varepsilon$ and it follows from (25) that

(27) $$|||m - i_n \circ \pi_n \circ m||| \leqq \varepsilon/\delta \quad \text{for each } n \geqq N.$$

Of course $i_n \circ \pi_n \circ m << P$ and $AR(i_n \circ \pi_n \circ m) \subset \lambda U(X)$ since $\|i_n\| \leqq \lambda$ for each n. Moreover, $i_n \circ \pi_n \circ m$ is a measure with finite dimensional range (and in particular, with range in a space with the RNP). Hence there is a function $f_n \in L^1_{\lambda U(X)}(P)$ such that

$$\int_A f_n dP = i_n \circ \pi_n \circ m(A) \quad \text{for each } A \in F.$$

Observe that (27) may be rephrased to say that $\{i_n \circ \pi_n \circ m : n = 1,2,\ldots\}$ is Cauchy in $|||\cdot|||$ with limit m. Now if $h \in L^1_X(P)$ and $m_1 : F \to X$ is defined by $m_1(A) \equiv \int_A h dP$ then $|||m_1||| = \|h\|_1$. (Indeed, if h happens to be a simple function this is obvious, and for general h, approximation by simple functions routinely leads to this conclusion.) Hence

$$|||i_n \circ \pi_n \circ m - i_j \circ \pi_j \circ m||| = \|f_n - f_j\|_1$$

for each n and j, and it follows that $\{f_n : n = 1,2,\ldots\}$ is Cauchy in $L^1_{\lambda U(X)}(P)$. If f denotes its limit then by the bounded convergence theorem (since some subsequence of the f_n's converges almost surely to f)

$$\int_A f dP = \lim_n \int_A f_n dP = \lim_n i_n \circ \pi_n \circ m(A) = m(A).$$

Thus $\frac{dm}{dP}$ exists and X has the RNP as was to be shown.

LEMMA 7.2.8. X has the Schur property. Moreover, each infinite dimensional subspace of X contains a subspace isomorphic to ℓ_1.

Proof. Both statements are consequences of the following result (whose demonstration occupies the rest of this paragraph): Whenever $\{x^{(k)}: k = 1,2,\ldots\}$ is a sequence in S(X) with $\lim_k (x^{(k)})_j = 0$ for each j then there is a subsequence $\{y^{(n)}: n = 1,2,\ldots\}$ of $\{x^{(k)}: k = 1,2,\ldots\}$ which is equivalent to the usual basis in ℓ_1. Thus suppose $\{x^{(k)}: k = 1,2,\ldots\}$ is a sequence of norm one elements which converges coordinatewise to 0. Pick numbers $\varepsilon_1 > \varepsilon_2 > \cdots > 0$ such that both

$$\prod_{j=1}^{\infty} (1-\varepsilon_j) > 1/2 \quad \text{and} \quad \sum_{j=1}^{\infty} \varepsilon_j \leqq \frac{1}{4(1+\delta)} \tag{28}$$

(δ as in the definition of X). Let $n_1 \equiv 1$. A straightforward inductive argumen will produce integers $s_1, n_2, s_2, n_3, \ldots$ (in this order) which satisfy each of the following properties:

$$1 \equiv n_1 < n_2 < \cdots \quad \text{and} \quad s_1 < s_2 < \cdots ; \tag{29}$$

$$\| i_{s_j} \circ \pi_{s_j} (x^{(n_k)}) \| \leqq \varepsilon_k \quad \text{for } j < k; \tag{30}$$

$$\| \pi_{s_k}(x) \| \geqq (1-\varepsilon_{k+1}) \| x \| \quad \text{for } x \in \text{span}(\{x^{(1)}, \ldots, x^{(n_k)}\}); \tag{31}$$

$$\| x^{(n_j)} - i_{s_k} \circ \pi_{s_k} (x^{(n_j)}) \| \leqq \varepsilon_k \quad \text{for } j \leqq k. \tag{32}$$

(Note that (29), (31) and (32) are used to choose s_k once n_k is known, while (29 and (30) are used to choose n_k once s_j is known for $j < k$.) Let $y^{(k)} \equiv x^{(n_k)}$ for $k = 1,2,\ldots$. We must show that $\{y^{(k)}: k = 1,2,\ldots\}$ is equivalent to the usu ℓ_1 basis. For any $k > 1$ and $t_1,\ldots,t_k \in \mathbb{R}$ substitute $\sum_{i=1}^k t_i y^{(i)}$ for x in (8) to obtain the first inequality below:

$$\left\| \sum_{i=1}^{k} t_i y^{(i)} \right\|$$

$$\geq \|\pi_{s_{k-1}} (\sum_{i=1}^{k} t_i y^{(i)})\| + \delta \left\| \sum_{i=1}^{k} t_i y^{(i)} - i_{s_{k-1}} \circ \pi_{s_{k-1}} (\sum_{i=1}^{k} t_i y^{(i)}) \right\|$$

(33)

$$\geq \|\pi_{s_{k-1}} \sum_{i=1}^{k-1} t_i y^{(i)})\| - |t_k| \|\pi_{s_{k-1}} (y^{(k)})\| + \delta |t_k| \|y^{(k)} - i_{s_{k-1}} \circ \pi_{s_{k-1}} (y^{(k)})\|$$

$$- \delta \sum_{i=1}^{k-1} |t_i| \|y^{(i)} - i_{s_{k-1}} \circ \pi_{s_{k-1}} (y^{(i)})\| .$$

Now estimate each of the terms on the right side of (33) as follows:

(34) $$\|\pi_{s_{k-1}} (\sum_{i=1}^{k-1} t_i y^{(i)})\| \geq (1-\varepsilon_k) \left\| \sum_{i=1}^{k-1} t_i y^{(i)} \right\| \quad \text{by (31);}$$

(35) $$|t_k| \|\pi_{s_{k-1}} (y^{(k)})\| \leq |t_k| \|i_{s_{k-1}} \circ \pi_{s_{k-1}} (y^{(k)})\| \leq |t_k| \varepsilon_k \quad \text{by (30);}$$

$$\delta |t_k| \|y^{(k)} - i_{s_{k-1}} \circ \pi_{s_{k-1}} (y^{(k)})\|$$

(36)

$$\geq \delta |t_k| [\|y^{(k)}\| - \|i_{s_{k-1}} \circ \pi_{s_{k-1}} (y^{(k)})\|] \geq \delta |t_k| (1-\varepsilon_k) \quad \text{by (30);}$$

(37) $$\delta \sum_{i=1}^{k-1} |t_i| \|y^{(i)} - i_{s_{k-1}} \circ \pi_{s_{k-1}} (y^{(i)})\| \leq \delta \sum_{i=1}^{k-1} |t_i| \varepsilon_{k-1} \quad \text{by (32).}$$

Thus (33) through (37) combine to give

$$\left\| \sum_{i=1}^{k} t_i y^{(i)} \right\| \geq (1-\varepsilon_k) \left\| \sum_{i=1}^{k-1} t_i y^{(i)} \right\| - |t_k| \varepsilon_k + \delta |t_k| (1-\varepsilon_k) - \delta \varepsilon_{k-1} \sum_{i=1}^{k-1} |t_i|$$

$$\geq (1-\varepsilon_k) \left\| \sum_{i=1}^{k-1} t_i y^{(i)} \right\| + \delta |t_k| (1-\varepsilon_k) - (1+\delta) \varepsilon_{k-1} \sum_{i=1}^{k} |t_i|$$

since $\varepsilon_{k-1} > \varepsilon_k$. By induction, then

$$\left\|\sum_{i=1}^{k} t_i y^{(i)}\right\| \geqq (1-\varepsilon_k)\left\|\sum_{i=1}^{k-1} t_i y^{(i)}\right\| + \delta|t_k|(1-\varepsilon_k) - (1+\delta)\varepsilon_{k-1}\sum_{i=1}^{k}|t_i|$$

$$\geqq (1-\varepsilon_k)[(1-\varepsilon_{k-1})\left\|\sum_{i=1}^{k-2} t_i y^{(i)}\right\| + \delta|t_{k-1}|(1-\varepsilon_{k-1}) - (1+\delta)\varepsilon_{k-2}\sum_{i=1}^{k-1}|t_i|$$

$$+ \delta|t_k|(1-\varepsilon_k) - (1-\delta)\varepsilon_{k-1}\sum_{i=1}^{k}|t_i|$$

(38)
$$\geqq (1-\varepsilon_k)(1-\varepsilon_{k-1})\left\|\sum_{i=1}^{k-2} t_i y^{(i)}\right\| + \delta(|t_{k-1}| + |t_k|)(1-\varepsilon_k)(1-\varepsilon_{k-1})$$

$$- (1+\delta)(\varepsilon_{k-2} + \varepsilon_{k-1})(\sum_{i=1}^{k}|t_i|)$$

$$\geqq \cdots \geqq \delta(\sum_{i=1}^{k}|t_i|)(\prod_{i=1}^{k}(1-\varepsilon_i)) - (1+\delta)(\sum_{i=1}^{k-1}\varepsilon_i)(\sum_{i=1}^{k}|t_i|)$$

$$\geqq [\tfrac{1}{2}\delta - (1+\delta)\frac{\delta}{4(1+\delta)}][\sum_{i=1}^{k}|t_i|] = \frac{1}{4}\delta\sum_{i=1}^{k}|t_i| .$$

Consequently, for any $(t_i) \in \ell_1$, (38) gives

$$\frac{1}{4}\delta\|(t_i)\|_{\ell_1} \leqq \left\|\sum_{i=1}^{\infty} t_i y^{(i)}\right\| \leqq \sum_{i=1}^{\infty}|t_i| = \|(t_i)\|_{\ell_1}$$

and thus $\{y^{(k)}: k = 1,2,\ldots\}$ is equivalent to the usual ℓ_1 basis.

In order to prove Lemma 7.2.8, suppose first that $D \subset X$ is weakly compact but not compact in the norm topology. Using the Eberlein-Smulian theorem it is possible to choose $\varepsilon > 0$, a point $z \in D$ and a sequence $\{z^{(n)}: n = 1,2,\ldots\}$ in D such that $\|z^{(n)} - z^{(m)}\| \geqq \varepsilon$ for $n \neq m$ and yet $z^{(n)} \to z$ weakly. There is no harm in assuming $\|z^{(n)} - z\| \geqq \varepsilon/2$ for each n as well. For each n let $x^{(n)} \equiv \|z^{(n)} - z\|^{-1}(z^{(n)} - z)$ and note that both $\|x^{(n)}\| = 1$ and $x^{(n)} \to 0$ weakly. In particular, for each j, $\lim_n (x^{(n)})_j = 0$ and thus from what was established in the first paragraph of the proof of this lemma, there is a subsequence $\{y^{(k)}: k = 1,2,\ldots$ of $\{x^{(n)}: n = 1,2,\ldots\}$ which is equivalent to the usual ℓ_1 basis. Then $y^{(k)} \to$

weakly (since $x^{(k)} \to 0$ weakly) and yet $y^{(k)} \not\to 0$ weakly since the usual basis elements of ℓ_1 do not converge weakly to 0. Thus D is norm compact. That is, X has the Schur property.

Finally, suppose that $Z \subset X$ is an infinite dimensional subspace of X. Then choose a sequence in U(Z), say $\{z^{(n)}: n = 1,2,\ldots\}$ such that for some $\varepsilon > 0$, $\|z^{(n)} - z^{(m)}\| > \varepsilon$ whenever $m \neq n$. By an elementary diagonalization argument we may assume that for each j, $\lim_n (z^{(n)})_j$ exists. Now let $x^{(n)} \equiv \|z^{(2n+1)} - z^{(2n)}\|^{-1} \cdot (z^{(2n+1)} - z^{(2n)})$ for $n = 1,2,\ldots$ and observe that $\|x^{(n)}\| = 1$ and $\lim_n (x^{(n)})_j = 0$ for each j. From the first paragraph of the proof of this lemma there is a subsequence $\{y^{(k)}: k = 1,2,\ldots\}$ of $\{x^{(n)}: n = 1,2,\ldots\}$ equivalent to the usual ℓ_1 basis. Evidently, $W \equiv \overline{\text{span}}(\{y^{(k)}: k = 1,2,\ldots\})$ is a subspace of Z isomorphic to ℓ_1.

§3. A Banach space which both lacks the RNP and has no bounded δ-trees.

For several years it was unknown whether a Banach space had bounded δ-trees precisely when it lacked the RNP. (One implication is easy: if X has a bounded δ-tree it is not subset dentable and hence it lacks the RNP.) In 1979 Bourgain [1979] outlined a counterexample, and the details presented below come in the main from a version of this counterexample due to Bourgain and Rosenthal [1980a]. (The fact that the space we construct also lacks the KMP is due independently to Bourgain and to Elton.) In order to exploit the full power of the construction it will be convenient to think of trees in terms of the range of a certain type of martingale. This is made explicit below. (See also Definition 7.5.2 and the discussion preceeding it for a more general view.)

For each $n = 1,2,\ldots$ let $\mathcal{D}_n$ denote the σ-algebra of subsets of [0,1] generated by $\{[\frac{j-1}{2^n}, \frac{j}{2^n}): 1 \leqq j \leqq 2^n\}$ and let λ denote Lebesgue measure on $\mathcal{B}([0,1])$ (the Borel σ-algebra on [0,1]). Then a <u>dyadic martingale</u> is one based on $([0,1],\mathcal{B}([0,1]),\lambda)$ in which the $n^{\underline{th}}$ σ-algebra is $\mathcal{D}_n$. (That is, it is a martingale of the form $(F_n,\mathcal{D}_n)$.) Note that if X is a Banach space and $(F_n,\mathcal{D}_n)$ is an X-valued dyadic martingale then F_{n+1} is constant on each atom of $\mathcal{D}_{n+1}$ and hence the (constant) value of F_n on $[\frac{j-1}{2^n},\frac{j}{2^n})$ is the average of the value of F_{n+1} on $[\frac{2j-2}{2^{n+1}},\frac{2j-1}{2^{n+1}})$ and the value of F_{n+1} on $[\frac{2j-1}{2^{n+1}},\frac{2j}{2^{n+1}})$. Thus the range of a dyadic martingale may be viewed as a tree and evidently each tree in X arises from some dyadic martingale in this manner. Certainly a dyadic martingale $(F_n,\mathcal{D}_n)$ giving rise to a bounded δ-tree satisfies $\|F_{n+1}(t) - F_n(t)\| \geqq \delta$ for each n and each $t \in [0,1)$ and hence it satisfies the (strictly weaker) inequality

$$\int_{[0,1]} \|F_{n+1}(t) - F_n(t)\| d\lambda(t) \geqq \delta \quad \text{for each } n.$$

More generally, let $\{\mathcal{F}_n: n = 1,2,\ldots\}$ be an increasing sequence of finite sub σ-algebras of $\mathcal{B}([0,1])$. Then the range of an X-valued martingale $(F_n,\mathcal{F}_n)$ is a bush (and each bush in a Banach space appears as the range of some martingale $(F_n,\mathcal{F}_n)$ for an appropriate choice of finite σ-algebras $\mathcal{F}_n$ $n = 1,2,\ldots$). Thus in part (c) of the main Theorem (7.3.2 below) the special case $\mathcal{F}_n = \mathcal{D}_n$ for each n gives a stronger conclusion than that X has no bounded δ-trees for any $\delta > 0$. (The added generality obtained by using an arbitrary increasing sequence $\{\mathcal{F}_n: n = 1,2,\ldots\}$ of finite sub σ-algebras of $\mathcal{B}([0,1])$ rather than $\{\mathcal{D}_n: n = 1,2,\ldots\}$ will be useful in §7.6.)

<u>NOTATION AND TERMINOLOGY</u> 7.3.1.

(a) For any X-valued martingale $(f_n,\mathcal{F}_n)$ the <u>difference sequence associated with</u> $(F_n,\mathcal{F}_n)$ is the sequence $\{D_n: n = 1,2,\ldots\}$ of functions where $D_1 \equiv F_1$ and for each $n > 1$, $D_n \equiv F_n - F_{n-1}$.

(b) The norm on $L^1_X(\lambda)$ will be denoted by $\|\cdot\|_1$ to distinguish it from the norm in $L^1(\lambda)$, denoted by $|\cdot|_1$. Similarly, for essentially bounded functions, the essential sup norm will be written $\|\cdot\|_\infty$ for functions in $L^\infty_X(\lambda)$ while that for $L^\infty(\lambda)$ is written $|\cdot|_\infty$. The elements of $L^1_X(\lambda)$ and $L^\infty_X(\lambda)$ will generally be written as capital letters F, G, H, etc. while the elements of $L^1(\lambda)$ and $L^\infty(\lambda)$ will be lower case letters f, g, h, etc.

(c) Dummy variables will often be suppressed. Thus, for example, if $g \in L^1(\lambda)$ then $(g > \alpha) \equiv \{t \in [0,1]: g(t) > \alpha\}$.

(d) For $f,g \in L^1(\lambda)$ let

$$d(f,g) \equiv \inf\{\varepsilon > 0: \lambda(|f-g| \geqq \varepsilon) \leqq \varepsilon\}.$$

Thus d is a metric on $L^1(\lambda)$ which gives rise to the topology of convergence in measure and makes $L^1(\lambda)$ into a topological vector space. If $f \in L^1(\lambda)$ and $A \subset L^1(\lambda)$ then $d(f,A) \equiv \inf\{d(f,g): g \in A\}$.

(e) Let $\{q_n: n = 1,2,\dots\}$ be an increasing sequence of positive integers and let $\{F_n: n = 1,2,\dots\}$ be an increasing sequence of sub σ-algebras of $\mathcal{B}([0,1])$ for which

$$|F_n| \leqq q_n \quad \text{for} \quad n = 1,2,\dots .$$

($|F_n|$ denotes the cardinality of F_n.) Then we shall say that $\{F_n: n = 1,2,\dots\}$ __is dominated by__ $\{q_n: n = 1,2,\dots\}$.

__THEOREM__ 7.3.2 (Bourgain, Rosenthal). Let $\{q_n: n = 1,2,\dots\}$ be an increasing sequence of positive integers. Then there is a subspace X of $L^1(\lambda)$ such that whenever $\{F_n: n = 1,2,\dots\}$ is an increasing sequence of finite sub σ-algebras of $\mathcal{B}([0,1])$ which is dominated by $\{q_n: n = 1,2,\dots\}$ then

(a) X does not have the RNP.

(b) $U(X)$ is relatively compact in the topology of convergence in λ-measure;

(c) $\lim_n \inf \|D_n\|_1 = 0$ whenever (F_n, F_n) is a bounded X-valued martingale with associated difference sequence $\{D_n: 1,2,\dots\}$.

(d) X has the strong Schur property (cf. 7.3.11). In particular, each weakly convergent sequence converges in norm.

(e) (Bourgain; Elton) X lacks the KMP.

Our starting point in the construction of X as in the statement of Theorem 7.3.2 will be a technical result of Bourgain and Rosenthal which asserts the existence of a very special $L^1(\lambda)$ function (see Theorem 7.3.4 below). Before stating this result we make a convention which will remain in force throughout this section.

__CONVENTION__ 7.3.3. The Borel equivalence of $([0,1]^{\mathbb{N}}, \mathcal{B}([0,1]^{\mathbb{N}}), \lambda^{\mathbb{N}})$ and $([0,1], \mathcal{B}([0,1]), \lambda)$ induces an isometry between $L^1(\lambda)$ and $L^1(\lambda^{\mathbb{N}})$ which is also an isometry between these spaces equipped with the convergence in measure metrics.

(The convergence in measure metric d' on $L^1(\lambda^{\mathbb{N}})$ is defined as is d in 7.3.1(d) with $\lambda^{\mathbb{N}}$ in place of λ. Of course by $\lambda^{\mathbb{N}}$ we mean the product measure on the Bore sets of $[0,1]^{\mathbb{N}}$ each of whose factors is λ.) (See, for example, Royden [1968] Theorem 9, p. 327.) For the remainder of §7.3 the spaces $L^1(\lambda)$ and $L^1(\lambda^{\mathbb{N}})$ will be identified by this isometry and we shall work in whichever space is most convenient for the task at hand.

Recall that an indexed collection $\{f_i\}_{i\in I}$ of distinct functions in $L^1(\lambda)$ is <u>independent</u> if for each finite set of distinct indices $\{i_1,\ldots,i_k\}$ and corresponding set of Borel subsets $\{B_{i_1},\ldots,B_{i_k}\}$ of $\mathbb{R}$ it follows that

$$\lambda\Big(\bigcap_{j=1}^{k} f_{i_j}^{-1}(B_{i_j})\Big) = \prod_{j=1}^{k} \lambda(f_{i_j}^{-1}(B_{i_j})).$$

For any subset A of $L^1(\lambda)$ <u>the σ-algebra induced by A</u> is the smallest one containing all the sets of the form $f^{-1}(0)$ for 0 open in $\mathbb{R}$ and $f \in A$. The following facts may be verified directly (or see, for example, Breiman [1968] pages 36-39 and page 74).

(a) Suppose that $A \subset L^1(\lambda)$, that $g \in L^1(\lambda)$ and that $\mathcal{A}$ denotes the σ-algebra induced by A. If $\lambda(B \cap B_1) = \lambda(B)\lambda(B_1)$ for each $B \in \mathcal{A}$ and B_1 in th σ-algebra induced by $\{g\}$ then g is said to be independent of A and in this cas $E[g|\mathcal{A}]$ is almost surely the constant function $E[g]$.

(b) If f_1 and f_2 are independent then $E[f_1f_2] = E[f_1]E[f_2]$.

(c) If $A \subset L^1(\lambda)$, if $\mathcal{A}$ is the σ-algebra induced by A and if $f \in A$ then $E[f|\mathcal{A}] = f$ [a.s.λ].

Finally, recall that two functions <u>f and g</u> in $L^1(\lambda)$ <u>are identically distributed</u> if

$$\lambda(f \in B) = \lambda(g \in B) \quad \text{for each} \quad B \in \mathcal{B}([0,1]).$$

Theorem 7.3.4 below, the starting point of the Bourgain-Rosenthal construction, guarantees the existence of an infinite dimensional subspace of $L^1(\lambda)$ (generated b 1 and independent copies of a single function) which behaves almost like a one dimensional subspace as far as the convergence in measure metric d is concerned (part (b) of 7.3.4) while being as "spread out" as possible in the L^1-metric.

THEOREM 7.3.4. Let $0 < \varepsilon < 1$. There is an $L^1(\lambda)$ function h with the following properties:

(a) h is strictly positive, $|h|_1 = 1$, $|h-1|_1 > 2 - \varepsilon$ and $d(h,0) < \varepsilon$.

(b) Let $\{h_1,h_2,\ldots\}$ be an independent sequence of $L^1(\lambda)$ functions each with the same distribution as h. For each $g \in \overline{\operatorname{span}(\{1,h_1,h_2,\ldots\})}$, $|g|_1 \leqq 1$, there is a constant c with $|c| \leqq 1$ such that $\lambda(|g-c| \geqq \varepsilon) \leqq \varepsilon$. Hence $d(g,U(\operatorname{span}(\{1\})))$ $\leqq \varepsilon$.

(c) If $\{h_i: i = 1,2,\ldots\}$ is as in (b) above then

$$\lim_n \left|\frac{1}{n} \sum_{n=1}^{n} h_i - 1\right|_1 = 0.$$

(Note that from (a), h must "spike" large on a set of measure less than ε and otherwise remain positive but close to 0.) The proof of this theorem is not included. (It depends on an analysis of the functional analytic properties of a sequence of independent random variables. See Bourgain and Rosenthal [1980a] Theorem 1.2, p. 56 and Lemma 1.8(b), p. 60 for details.) Except for this omission, however, the details presented below are comprehensive. As the actual construction of X and verification of its properties is somewhat lengthy, an outline below precedes the detailed proof.

PROOF OUTLINE FOR THEOREM 7.3.2.

(I) Theorem 7.3.4 is generalized as follows: given a certain type of finite dimensional subspace X_o of $L^1(\lambda^{\mathbb{N}})$ (generalizing span({1}) of 7.3.4) and a finite subset W of $S(X_o)$ (generalizing {1} of 7.3.4) a finite dimensional subspace Y, $X_o \subset Y \subset L^1(\lambda^{\mathbb{N}})$ is constructed in 7.3.5 which simultaneously behaves almost one dimensionally insofar as the d-metric is concerned (as in 7.3.4(b)) while having bush like branches emanating from each element of W (as in 7.3.4(c)).

(II) Let (F_i,F_i) be a $U(L^1(\lambda))$-valued martingale with difference sequence $\{D_i: i = 1,2,\ldots\}$ for which $\inf_i \|D_i\|_1 \geqq \varepsilon$. Then there is a positive integer constant

$\beta(m,\varepsilon)$ for each m such that whenever Z is a finite dimensional subspace of $L^1(\lambda)$ of dimension m and k is an index with $k \geqq \beta(m,\varepsilon)$ then the average value of the L^1-distance from $F_k(t)$ to Z is at least $\varepsilon/3$. That is,

$$\int \inf\{\,|F_k(t) - f|_1 : f \in Z\}d\lambda(t) \geqq \varepsilon/3.$$

(III) For each finite dimensional subspace Z of $L^1(\lambda)$ and each ε, $0 < \varepsilon < 1$ let

$$\gamma(Z,\varepsilon) \equiv \max\{\delta: 0 < \delta \leqq \varepsilon \text{ and } \int_A |f|d\lambda \leqq \varepsilon|f|_1 \text{ for all } f \in Z \text{ and Borel sets } A \subset [0,1] \text{ with } \lambda(A) \leqq \delta\}.$$

Thus $\gamma(Z,\varepsilon)$ is a measure of uniform integrability of $U(Z)$. A constant $\delta(\varepsilon,Z)$ will be defined which depends on the integers q_n (of the statement of 7.3.2), on the constant $\gamma(Z,\varepsilon)$ and on the constant $\beta(\dim Z,\varepsilon)$ (as in (II)) where $\dim Z$ denotes the dimension of Z. (See formula (7).) $\delta(\varepsilon,Z)$ turns out to be "monotone" in the sense that when $0 < \varepsilon_1 \leqq \varepsilon$ and Z and Z_1 are finite dimensional subspace of $L^1(\lambda)$ with $Z \subset Z_1$ then $\delta(\varepsilon_1,Z_1) \leqq \delta(\varepsilon,Z)$. An increasing sequence $\{X_n: n = 1,2,\ldots\}$ of finite dimensional subspaces of $L^1(\lambda)$ is inductively constructed usin (I) with a resulting "approximate 1-bush" in the positive cone of the unit ball of $X \equiv L^1(\lambda)$-closure of $\bigcup_{n=1}^{\infty} X_n$. Such an approximate bush may easily be modified to produce a 1/2-bush in the positive cone of the unit sphere of X and thus X lacks the RNP. Moreover, from the "almost one dimensionality" of X_n in the d-metric as described in (I), the X_n's may also be chosen so that for each n and for each $f \in U(X)$ we have $d(f,X_n) < \delta(\frac{1}{n},X_n)$ from which it follows that $U(X)$ is relativel compact in the topology of convergence in λ-measure.

(IV) The space X described in (III) lacks the RNP and has d-relatively compact unit ball. The subsequence splitting lemma of Rosenthal together with the relative compactness of $(U(X),d)$ combine to provide a direct proof that X has the strong Schur property. The construction of Elton [1982] of a closed bounded convex set in X with no extreme points is easy enough: it is just the $L^1(\lambda)$-closure of the convex hull of the 1/2-bush constructed in (III) above. Finally, suppose that $\{F_n : n = 1,2,\ldots\}$ is dominated by $\{q_n : n = 1,2,\ldots\}$, that $(F_n,\mathcal{F}_n)$ is a U(X)-valued martingale with associated difference sequence $\{D_n : n = 1,2,\ldots\}$ and that for some $\varepsilon > 0$, $\|D_n\|_1 \geqq \varepsilon$ for each n. By an elementary perturbation we may also assume that Range $F_n \subset \bigcup_{j=1}^{\infty} X_j$ for each n as well. Fix indices j_o and n. (Some of the conditions on these indices will appear in the remainder of this paragraph.) Assume that F_{j_o} is X_n-valued and choose a $U(X_n)$-valued function G which is $\mathcal{F}_k$-measurable (where $k \equiv \beta(\dim X_n,\varepsilon)$) and which satisfies

$$d(G(t),F_k(t)) \leqq \delta(\tfrac{1}{n},X_n) \quad \text{for} \quad t \in [0,1].$$

(The existence of such a G is guaranteed by (III) above.) The fact that the martingale is tied to the sequence of σ-algebras $\{\mathcal{F}_n : n = 1,2,\ldots\}$ (which is itself dominated by the fixed sequence $\{q_n : n = 1,2,\ldots\}$) allows us to bound the d-distance between F_i and $\mathbb{E}[G|\mathcal{F}_i] \equiv G_i$ for $1 \leqq i \leqq k$ and in particular it may be shown that

$$d(G_{j_o}(t),F_{j_o}(t)) \leqq \gamma(X_n,\tfrac{1}{n}) \quad \text{for each} \quad t \in [0,1].$$

(This is the only place where the hypothesis that the martingale is tied to this sequence of σ-algebras is used.) Since both G and F_{j_o} take all their values in U(X) the above estimate on the d-distance between $G_{j_o}(t)$ and $F_{j_o}(t)$ may be transformed into an estimate on the size of $|\|G\|_1 - \|F_{j_o}\|_1|$. By choosing j_o and n large enough this difference may be made as small as desired by making both arbitrarily close to $\sup_i \|F_i\|_1$. Now for each t choose a set $A_t \in \mathcal{B}([0,1])$ of λ-measure close to one such that $F_k(t)$ and $G(t)$ are uniformly close on A_t. In order to obtain a contradiction observe first that if $\lambda(A_t)$ is sufficiently close to one for each t then $||F_k(t)\chi_{A_t}|_1 - |F_k(t)|_1|$ can be made as small as desired, and in fact it is possible to guarantee that $|\int |F_k(t)\chi_{A_t}|_1 d\lambda(t) - \|F_k\|_1|$ can be made as small as desired. On the other hand, since G is X_n-valued and by definition of $k \equiv \beta(\dim X_n, \varepsilon)$, $\|F_k - G\|_1 \geqq \varepsilon/3$, and since $F_k(t)$ and $G(t)$ are uniformly close on A_t, most of the mass of $\|F_k - G\|_1$ must occur on the sets A_t^c. But again because G is $U(X_n)$-valued and $\gamma(X_n,\frac{1}{n})$ measures the uniform integrability of $U(X_n)$, $\| G(t)\chi_{A_t^c}\|_1$ is very small if $\lambda(A_t^c)$ is sufficiently small, uniformly in t. Thus $\int \| F_k(t)\chi_{A_t^c}\|_1 d\lambda(t)$ must include most of the mass of $\|F_k - G\|_1$. Now write $\|F_k\|_1$ in the form

$$\|F_k\|_1 = \int \|F_k(t)\chi_{A_t^c}\|_1 d\lambda(t) + \int \|F_k(t)\chi_{A_t}\|_1 d\lambda(t).$$

The first term is approximately $\|F_k - G\|_1$ and hence can be made close to $\varepsilon/3$, while the second term is as good an approximation of $\|F_k\|_1$ as desired. This leads to the contradiction $\|F_k\|_1 > \sup_i \|F_i\|_1$.

The precise statement of the result alluded to in (I) above follows. Note first that a subset $\underline{A}$ of $L^1(\lambda^{\mathbb{N}})$ <u>is said to depend on only finitely many coordinates</u> if there is a positive integer j such that $f(x) = f(y)$ whenever $x,y \in [0,1]^{\mathbb{N}}$ and $x_i = y_i$ for $1 \leqq i \leqq j$.

<u>THEOREM</u> 7.3.5. Let X_o be a finite dimensional subspace of $L^1(\lambda^{\mathbb{N}})$ which depends on only finitely many coordinates and suppose that $0 < \varepsilon < 1$. For each finite subset W of the norm-one elements of X_o there exists a finite dimensional subspace Y of $L^1(\lambda^{\mathbb{N}})$ with the following properties:

(a) $X_o \subset Y$ and Y depends on only finitely many coordinates;

(b) $d(g,U(X_o)) < \varepsilon$ for each $g \in U(Y)$;

(c) there is an n such that for each $\phi \in W$ there is a finite subset $W'(\phi) \equiv \{g_1,\ldots,g_n\} \subset S(Y)$ with $\|g_i - \phi\|_1 > 2-\varepsilon$ for $1 \leqq i \leqq n$ and $\|\frac{1}{n}\sum_{i=1}^n g_i - \phi\|_1 < \varepsilon$.

The first step in the proof of 7.3.5 is the following elementary lemma.

<u>LEMMA</u> 7.3.6. Let X_o be a finite dimensional subspace of $L^1(\lambda^{\mathbb{N}})$ and $\varepsilon > 0$. Then there is a constant $\eta(X_o,\varepsilon)$, abbreviated η, such that $\eta > 0$ and for all $g \in U(L^1(\lambda^{\mathbb{N}}))$, if $d(g,X_o) < \eta$ then $d(g,U(X_o)) < \varepsilon$.

Proof. If the conclusion were false there would be sequences $\{g_n : n = 1,2,\ldots$ in $U(L^1(\lambda^{\mathbb{N}}))$ and $\{f_n : n = 1,2,\ldots\}$ in X_o such that both

$$d(g_n,f_n) < \frac{1}{n} \quad \text{and} \quad d(g_n,U(X_o)) \geqq \varepsilon \quad \text{for each } n.$$

Note that $|f_n|_1 > 1$ when $\frac{1}{n} < \varepsilon$ (since otherwise $d(g_n,U(X_o)) \leqq d(g_n,f_n) < \varepsilon$) and, by going back to the definition of d it follows that $d(\frac{g_n}{|f_n|_1},\frac{f_n}{|f_n|_1}) \leqq d(g_n,f_n)$.

Because X_o is finite dimensional there is a subsequence of $\{\frac{f_n}{|f_n|_1}: n = 1,2,\ldots\}$ (denoted with no loss of generality by $\{\frac{f_n}{|f_n|_1}: n = 1,2,\ldots\}$ again) and some $f \in X_o$ for which

$$\lim_n \left|\frac{f_n}{|f_n|_1} - f\right|_1 = 0.$$

Then $|f|_1 = 1$ and since an L^1-convergent sequence necessarily converges in measure,

$$\lim_n d(\frac{f_n}{|f_n|_1},f) = 0$$

as well. Consequently $\lim_n d(\frac{g_n}{|f_n|_1},f) = 0$ and again using the definition of d,

$$\lim_n d(\frac{|g_n|}{|f_n|_1}, |f|) = 0.$$

By the convergence in measure version of Fatou's lemma, then,

$$1 \leqq |f|_1 \leqq \liminf_n \frac{|g_n|_1}{|f_n|_1} \leqq \liminf_n \frac{1}{|f_n|_1}.$$

Since $|f_i|_1 > 1$ for all large i we conclude that $\lim_n |f_n|_1 = 1$ whence $\lim_n |f_n - f|_1 = 0$ and consequently $\lim_n d(f_n,f) = 0$. That is, we have demonstrated that

$$\begin{aligned}\varepsilon &\leqq \limsup_n d(g_n,U(X_o)) \leqq \limsup_n d(g_n,f) \\ &\leqq \limsup_n (d(g_n,f_n) + d(f_n,f)) = 0\end{aligned}$$

a contradiction which proves the lemma.

Proof of 7.3.5. For X_o, W and ε as in the statement of 7.3.5 let $\eta \equiv \eta(X_o,\varepsilon)$ be chosen as in Lemma 7.3.6 subject to the one additional constraint that $\eta \leqq \varepsilon$. Denote the elements of W by $\{\phi_1,\ldots,\phi_k\}$ and let

$$\xi \equiv \frac{\eta^2}{4k}.$$

For this ξ construct a function $h \in L^1(\lambda)$ with the properties guaranteed by Theorem 7.3.4. (Thus, h is strictly positive, $h \in S(L^1(\lambda))$, $\|h-1\|_1 > 2-\xi$, $d(h,0) < \xi$ and if $\{h_i: i = 1,2,\ldots\}$ is an independent sequence in $L^1(\lambda)$ where each h_i has the same distribution as h, then $d(g,U(\mathrm{span}(\{1\}))) \leqq \xi$ whenever $g \in U(\overline{\mathrm{span}}(\{1,h_1,h_2,\ldots\}))$.) Next choose an integer m so that X_o depends on at most the first m coordinates of $[0,1]^{\mathbb{N}}$ and define $L^1(\lambda^{\mathbb{N}})$ functions $h_1,h_2,\ldots$ by

$$h_i(x) \equiv h(x_{m+i}) \quad \text{for each } i \text{ and each } x \in [0,1]^{\mathbb{N}}.$$

Then $\{h_i: i = 1,2,\ldots\}$ is an independent sequence and each h_i has the same distribution as does h. Finally, choose n so that

$$\left\| \frac{1}{n}\sum_{i=1}^{n} h_i - 1\right\|_1 < \xi$$

(as guaranteed by 7.3.4(c)). For $\phi_i \in W$ let

$$h_{\phi_i j} \equiv h_{ni+j} \qquad (1 \leqq i \leqq k;\; 1 \leqq j \leqq n)$$

and let

$$W'(\phi) \equiv \{h_{\phi j}\phi: 1 \leqq j \leqq n\} \quad \text{for each } \phi \in W.$$

Define the subspace Y by

$$Y \equiv X_o \oplus \mathrm{span}(\{W'(\phi): \phi \in W\}).$$

Since X_o, $h_{\phi j}$ and $\phi \in W \subset X_o$ each depend on only finitely many coordinates, the same is true of Y.

In order to verify (b) of 7.3.5 pick any $g \in U(Y)$ and write

$$g = f_1 + \sum_{i,j} \alpha_{ij} h_{\phi_i j}\phi_i \quad \text{for some} \quad f_1 \in X_0 \quad \text{and constants} \quad \alpha_{ij}.$$

Let $f \equiv f_1 + \sum_{i,j} \alpha_{ij}\phi_i$ and $g_i \equiv \sum_j \alpha_{ij}(h_{\phi_i j} - 1)$ for $1 \leq i \leq k$. Then

$$g = f + \sum_i g_i\phi_i \quad \text{where} \quad f \in X_0 \quad \text{and} \quad g_i \in \operatorname{span}(\{h_{\phi_i j} : 1 \leq j \leq n\})$$

for each i. (Note: $\int g_i d\lambda^{\mathbb{N}} = 0$ since $\int h_{\phi_i j} d\lambda^{\mathbb{N}} = 1$ for all i,j.) As is shown next,

(1) $$|g_i|_1 \leq 2 \quad \text{for each} \quad i = 1,\ldots,k.$$

Indeed, fix i and let A_i be the σ-algebra induced by $\{f,\phi_1,\ldots,\phi_k,g_1,\ldots,g_{i-1},g_{i+1},\ldots,g_k\}$. Since g_i is independent of this collection we have $E[g_i|A_i] = E[g_i] = 0$. Consequently $E[g|A_i] = f + \sum_{j\neq i} g_j\phi_j$ and hence

$$|g_i|_1 = |g_i|_1|\phi_i|_1 = |g_i\phi_i|_1 \quad \text{(since } g_i \text{ and } \phi_i \text{ are independent)}$$

$$= |g - E[g|A_i]|_1$$

$$\leq |g|_1 + |E[g|A_i]|_1 \leq 2|g|_1 \leq 2.$$

Thus (1) holds. Theorem 7.3.4(b) applied to $\frac{1}{2}g_i$ states that there is a number c_i with $|c_i| \leq 2$ such that

(2) $$\lambda^{\mathbb{N}}(|g_i - c_i| \geq 2\xi) \leq \xi.$$

Since g_i is practically the constant c_i as far as the d-metric is concerned, a good approximation of g in X_0 is the function

$$\bar{g} \equiv f + \sum_{i=1}^{k} c_i\phi_i.$$

The verification of 7.3.5(b) will be completed by showing that $d(g,\bar{g}) < \eta$ (from which it follows that $d(g,U(X_0)) < \varepsilon$ by definition of η).

Let $\phi \equiv 2\xi \sum_{i=1}^k |\phi_i|$. Note that $g - \bar{g} = \sum_{i=1}^k (g_i - c_i)\phi_i$ so that

$$\lambda^{\mathbb{N}}(|g-\bar{g}| > \phi) \leqq \lambda^{\mathbb{N}}\left(\bigcup_{i=1}^{k} (|g_i - c_i||\phi_i| > 2\xi|\phi_i|)\right) \leqq k\xi$$

from (2). Now

$$(|g-\bar{g}| > \eta) \subset (|g-\bar{g}| > \phi) \cup (\phi > \eta)$$

so that

$$\begin{aligned}
\lambda^{\mathbb{N}}(|g-\bar{g}| > \eta) &\leqq \lambda^{\mathbb{N}}(|g-\bar{g}| > \phi) + \lambda^{\mathbb{N}}(\phi > \eta) \\
&\leqq k\xi + \int \chi_{(\phi>\eta)} d\lambda^{\mathbb{N}} \\
&\leqq k\xi + \int_{(\phi>\eta)} \frac{1}{\eta}|\phi| d\lambda^{\mathbb{N}} \\
&\leqq k\xi + \frac{1}{\eta}|\phi|_1 \leqq k\xi + \frac{1}{\eta}2k\xi \quad \text{(by definition of } \phi) \\
&< \eta \quad \text{(by definition of } \xi \text{ and the fact that} \\
&\qquad \eta \leqq \xi < 1.)
\end{aligned}$$

Thus 7.3.5(b) holds.

In order to check condition 7.3.5(c) observe that

$$\begin{aligned}
&\left|\frac{1}{n}\sum_{j=1}^{n} h_{\phi_i j}\phi_i - \phi_i\right|_1 \\
&= \int \left|\frac{1}{n}\sum_{j=1}^{n} h_{\phi_i j} - 1\right||\phi_i| d\lambda^{\mathbb{N}} \\
&= E\left[\left|\frac{1}{n}\sum_{j=1}^{n} h_{\phi_i j} - 1\right|\right] E[|\phi_i|] \quad \text{(since } \{h_{\phi_i 1},\ldots,h_{\phi_i n},\phi\} \text{ is an independent set)} \\
&= E\left[\left|\frac{1}{n}\sum_{j=1}^{n} h_j - 1\right|\right] \quad \text{(since } \left|\frac{1}{n}\sum_{j=1}^{n} h_{\phi_i j} - 1\right| \text{ and } \left|\frac{1}{n}\sum_{j=1}^{n} h_j -\right. \\
&\qquad \text{have the same distribution)} \\
&< \xi \leqq \varepsilon .
\end{aligned}$$

Also note that

$$|h_{\phi_i j}\phi_i - \phi_i|_1 = |h_{\phi_i j} - 1|_1 |\phi_i|_1 \quad \text{(by independence)}$$

$$= |h - 1|_1 \quad \text{(since } h_{\phi_i j} - 1 \text{ and } h - 1 \text{ have the same distribution)}$$

$$> 2 - \xi \geqq 2 - \varepsilon.$$

Thus (a), (b) and (c) of 7.3.5 have been verified and the proof is complete.

Step (II) of the proof outline of 7.3.2 asserts that any $U(L^1(\lambda))$-valued martingale $(F_i, \mathcal{F}_i)$ with difference sequence $\{D_i: i = 1,2,...\}$ such that $\|D_i\|_1 \geqq \varepsilon$ for all i must "spread out" and on average become uniformly far from each finite dimensional subspace of $L^1(\lambda)$ of a given dimension. (This may be viewed as a quantitative measure of the fact that finite dimensional subspaces of $L^1(\lambda)$ have the RNP, hence the MCP.) More precisely, we show the following.

PROPOSITION 7.3.7. Let $\varepsilon > 0$ and for each positive integer m let $\beta(m,\varepsilon) \equiv [\frac{36m}{\varepsilon^2} + 1]$. (Here only, $[\alpha]$ denotes the greatest integer less than or equal to α.) Suppose that Z is an m-dimensional subspace of $L^1(\lambda)$ and that $(F_i, \mathcal{F}_i)$ is an $L^1(\lambda)$-valued martingale such that both $\|F_i\|_\infty \leqq 1$ and $\|D_i\|_1 \geqq \varepsilon$ for all i. ($\{D_i: i = 1,2,...\}$ is the difference sequence for $(F_i, \mathcal{F}_i)$ and $\{\mathcal{F}_i: i = 1,2,...\}$ is as in the statement of 7.3.2.) If $n \geqq \beta(m,\varepsilon)$ then

$$\int \text{dist}_1(F_n(t), Z)\, d\lambda(t) \geqq \varepsilon/3$$

where $\text{dist}_1(F_n(t), Z)$ denotes $\inf\{|F_n(t) - g|_1 : g \in Z\}$.

Proof. If the conclusion were false then for some $n \leqq \beta(m,\varepsilon)$ we would have

$$\int \text{dist}_1(F_n(t), Z)\, d\lambda(t) \leqq \varepsilon/3.$$

Since F_n is constant on each atom of $\mathcal{F}_n$ there is an $\mathcal{F}_n$-measurable Z-valued function G_n such that

$$|G_n(t) - F_n(t)|_1 = \text{dist}_1(F_n(t), Z) \quad \text{for each } t \in [0,1].$$

Then $\|G_n\|_\infty \leqq 2$ since $\|F_n\|_\infty \leqq 1$ and $\text{dist}_1(F_n(t), Z) \leqq |F_n(t) - 0|_1 \leqq 1$. For each

$j < n$ let $G_j \equiv \mathbb{E}[G_n | F_j]$. Then

$$\|G_j\|_\infty \leqq \|G_n\|_\infty \leqq 2 \quad \text{for each } j \leqq n,$$

and

$$\|G_j - F_j\|_1 = \|\mathbb{E}[G_n - F_n | F_j]\|_1 \leqq \|G_n - F_n\|_1$$

$$= \int \text{dist}_1(F_n(t), Z)\, d\lambda(t) \leqq \varepsilon/3.$$

Consequently,

$$\|G_{j+1} - G_j\|_1 \geqq \|D_{j+1}\|_1 - \|F_j - G_j\|_1 - \|F_{j+1} - G_{j+1}\|_1 \geqq \varepsilon/3$$

for $1 \leqq j \leqq n-1$ and

$$\|G_1\|_1 \geqq \|F_1\|_1 - \varepsilon/3 = \|D_1\|_1 - \varepsilon/3 \geqq 2\varepsilon/3 > \varepsilon/3.$$

It follows that

(3) $$\left(\|G_1\|_1^2 + \sum_{j=1}^{n-1} \|G_{j+1} - G_j\|_1^2\right)^{1/2} \geqq \frac{\sqrt{n}}{3}\varepsilon.$$

The proof of this proposition will be completed by the next lemma which, applied to the "finite" martingale $(G_i, F_i)_{i=1}^n$ above shows that the left side of (3) is dominated by $2\sqrt{m}$. (This is impossible, however, since $2\sqrt{m} \geqq \frac{\sqrt{n}}{3}\varepsilon$ forces $n < \beta(m,\varepsilon)$ in contradiction to the choice of n.)

LEMMA 7.3.8. Let Z be an m-dimensional normed linear space and (F_i, F_i) a Z-valued martingale with $\|F_i\|_\infty \leqq 1$ for all i. ($\{F_n : n = 1,2,\dots\}$ is as in the statement of 7.3.2.) Then

$$\left(\sum \|D_i\|_1^2\right)^{1/2} \leqq \sqrt{m}$$

where $\{D_i : i = 1,2,\dots\}$ is the difference sequence for (F_i, F_i).

Proof. Suppose first that Z is Hilbert space with inner product $(\cdot,\cdot)$. Let

$L_Z^2(\lambda) \equiv \{F \in L_Z^1(\lambda): \int (F(t),F(t))d\lambda(t) < \infty\}$ be the linear space equipped with norm $\|F\|_2 \equiv (\int (F(t),F(t))d\lambda(t))^{1/2}$. Note that the inner product $<\cdot,\cdot>$ on $L_Z^2(\lambda)$ given by

$$<F,G> \equiv \int (F(t),G(t))d\lambda(t)$$

induces $\|\cdot\|_2$. Next observe that for $1 \leq i < j$,

(4) $$<D_i,D_j> = 0.$$

(Though (4) is true in general for difference sequences of Hilbert space valued martingales, we argue directly for this case.) Indeed, since D_i is $\mathcal{F}_i$-measurable,

$$D_i = \sum_{k\in I_i} z_k \chi_{A_k} \quad \text{for } z_k \in Z,\ \{A_k: k \in I_i\} \text{ the set of atoms of } \mathcal{F}_i.$$

Thus

$$<D_i,D_j> = \int (\sum_{k\in I_i} z_k\chi_{A_k}(t),D_j(t))d\lambda(t)$$

$$= \sum_{k\in I_i} \int_{A_k} (z_k,D_j(t))d\lambda(t)$$ ($(z_k\chi_{A_k}(t),D_j(t)) = (z_k,D_j(t))$ or 0 depending on whether or not $t \in A_k$.)

$$= \sum_{k\in I_i} \int (z_k,\chi_{A_k}(t)D_j(t))d\lambda(t)$$

$$= \sum_{k\in I_i} (z_k, \int_{A_k} D_j(t)d\lambda(t))$$ (since $(z_k,\cdot)$ is a continuous linear functional)

$$= \sum_{k\in I_i} (z_k,0) = 0$$ (since $A_k \in \mathcal{F}_i$ and $\mathbb{E}[D_j|\mathcal{F}_i] = 0$)

which proves (4). Thus $\{D_1,D_2,\ldots\}$ is orthogonal in $L_Z^2(\lambda)$ whence $\|F_n\|_2^2 = \sum_{j=1}^n \|D_j\|_2^2$ for each n. It follows that

$$\sum_{j=1}^{\infty} \|D_j\|_2^{\ 2} = \lim_n \|F_n\|_2^{\ 2} \leqq \limsup_n \|F_n\|_\infty^{\ 2} \leqq 1. \tag{5}$$

Now suppose that Z is an arbitrary m-dimensional Banach space. Then there is an isomorphism $T: Z \to \ell_m^2$ (m-dimensional Hilbert space) with $\|T^{-1}\|\|T\| \leqq \sqrt{m}$. (See John [1948].) Then $(T(F_n), F_n)$ is a martingale with $\|T(F_n)\|_\infty \leqq \|T\|$. Hence

$$\begin{aligned} \sum \|T(D_j)\|_1^{\ 2} &\leqq \sum \|T(D_j)\|_1^{\ 2} \qquad \text{(by the Schwarz inequality)} \\ &\leqq \|T\|^2 \qquad \text{(by (5))} \end{aligned}$$

whence

$$\sum \|D_j\|_1^{\ 2} \leqq \|T^{-1}\|^2 \sum \|T(D_j)\|_1^{\ 2} \leqq \|T^{-1}\|^2 \|T\|^2 \leqq m$$

as was to be shown.

The beginning of step (III) of the proof outline of 7.3.2 is verbatim as in the outline. Define for each finite dimensional subspace Z of $L^1(\lambda)$ and for each ε $0 < \varepsilon \leqq 1$

$$\gamma(Z,\varepsilon) \equiv \max\{\delta: 0 < \delta \leqq \varepsilon \text{ and } \int_A |f|\, d\lambda \leqq \varepsilon |f|_1 \text{ for each } f \in Z \text{ and each Borel set } A \subset [0,1] \text{ with } \lambda(A) \leqq \delta\}. \tag{6}$$

$\gamma(Z,\varepsilon)$ thus provides a measure of the uniform integrability of the finite dimensional ball $U(Z)$. Let $\dim Z$ denote the dimension of Z, recall that $|B|$ denotes the cardinality of a set B, and define $\delta(\varepsilon,Z)$ by

$$\delta(\varepsilon,Z) \equiv (q_{\beta(\dim Z,\varepsilon)})^{-\beta(\dim Z,\varepsilon)}\gamma(Z,\varepsilon) \tag{7}$$

where $\{q_n: n = 1,2,\ldots\}$ is given in the statement of Theorem 7.3.2 and β is the constant of Proposition 7.3.7. Since β increases as $\dim Z$ increases and as ε decreases, and since $q_i \leqq q_{i+1}$ for each i while γ decreases as Z increases and ε decreases, it follows that when $Z \subset Z_1 \subset L^1(\lambda)$ with $\dim Z_1 < \infty$, and

$0 < \varepsilon_1 \leqq \varepsilon$ then $\delta(\varepsilon_1, Z_1) \leqq \delta(\varepsilon, Z)$. This is sufficient background for the next result (where the Banach space X of 7.3.2 is finally constructed).

THEOREM 7.3.9. There is an increasing sequence $\{X_m: m = 1,2,\ldots\}$ of finite dimensional subspaces of $L^1(\lambda)$ such that the Banach space $X \equiv L^1(\lambda)$-closure of $\bigcup_{m=1}^{\infty} X_m$ has the following properties:

(a) X fails to have the RNP;

(b) for each n and each $f \in U(X)$, $d(f, U(X_n)) < \delta(\frac{1}{n}, x_n)$.

In particular, X has relatively compact unit ball in the topology of convergence in λ-measure.

Proof. Define a modified δ-function as follows: for $0 < \varepsilon < 1$ and finite dimensional subspace Z of $L^1(\lambda^{\mathbb{N}})$ let

$$\overline{\delta}(\varepsilon, Z) \equiv \frac{1}{2^{j+1}} \delta(\varepsilon, Z) \quad \text{if} \quad \frac{1}{j+1} < \varepsilon \leqq \frac{1}{j}.$$

Let $X_1 \equiv \text{span}(\{1\}) \subset L^1(\lambda^{\mathbb{N}})$ and $W_1 \equiv \{1\}$. Inductively assume that finite dimensional subspaces X_i and finite norm-one subsets W_i of X_i have been constructed with X_i depending on only finitely many coordinates, $i = 1,\ldots,m$. By Theorem 7.3.5 there is a finite dimensional subspece $X_{m+1} \supset X_m$ with X_{m+1} depending on only finitely many coordinates and for each $\phi \in W_m$ there is a finite set $W_{m+1}(\phi)$ in $S(X_{m+1})$ (where

$$W_{m+1}(\phi) \equiv \{h_{\phi j}\phi: j = 1,\ldots,n\}$$

in the terminology of the proof of 7.3.5) for which

$$(8) \qquad \left|\frac{1}{n}\sum_{j=1}^{n} h_{\phi j} - \phi\right|_1 \leqq \overline{\delta}(\tfrac{1}{m}, X_m)$$

and

$$(9) \qquad d(g, U(X_m)) < \overline{\delta}(\tfrac{1}{m}, X_m) \quad \text{for} \quad g \in U(X_{m+1}).$$

Let

$$W_{m+1} \equiv \{W_{m+1}(\phi): \phi \in W_m\}.$$

(N.B. $W_{m+1}(\phi)$ was called $W_m'(\phi)$ in the statement of 7.3.5.) In this manner X_m and W_m may be defined for each $m \geqq 1$. Observe that by induction, each W_{m+1} consists of positive norm-one elements which are independent from X_m.

Suppose now that $f \in U(X_{n+j})$ for some positive integers n and j. Then an entirely straightforward induction using (9) shows that

$$d(f,U(X_n)) \leqq \sum_{i=n}^{n+j} \overline{\delta}(\tfrac{1}{i},X_i) \leqq \sum_{i=n}^{\infty} \frac{1}{2^{i+1}} \delta(\tfrac{1}{i},X_i) \leqq \sum_{i=n}^{\infty} \frac{1}{2^{i+1}} \delta(\tfrac{1}{n},X_n)$$

$$= \frac{1}{2^n} \delta(\tfrac{1}{n},X_n)$$

since δ is "monotone". More generally, if $f \in U(X)$ find a sequence $\{f_i: i = 1,2,\ldots\}$ in $\bigcup_{i=1}^{\infty} U(X_i)$ such that $\lim_i |f - f_i|_1 = 0$. Then $\lim_i d(f,f_i) = 0$ as well and consequently

$$d(f,U(X_n)) \leqq \liminf_i (d(f,f_i) + d(f_i,U(X_n)) \leqq \frac{1}{2^n} \delta(\tfrac{1}{n},X_n) < \delta(\tfrac{1}{n},X_n).$$

Since $U(X_n)$ is compact in the topology of convergence in measure and since $\delta(\frac{1}{n},X_n) \leqq \frac{1}{n}$ (see (6) and (7)) it follows that $U(X)$ is relatively compact in the topology of convergence in measure induced by d.

It remains to show that X lacks the RNP. The set $\bigcup_{m=1}^{\infty} W_m$ is an "approximate 1-bush which may be perturbed (as described below) into a true $\frac{1}{2}$-bush. (Once such a bounded bush is produced it will follow that X lacks the RNP by, for example, Corollary 2.3.7.) The following terminology will make this and later arguments smoother.

<u>TERMINOLOGY</u> 7.3.10. For $\phi \in W_m$ denote by $D(\phi)$ the collection of all <u>descendants</u> of ϕ. In other words, $\psi \in D(\phi)$ means that for some $k > m$ we have $\psi \in W_k$ and there is a chain of offspring leading from ϕ to ψ: namely, $\psi_1 \in W_{m+1}(\phi)$, $\psi_2 \in W_{m+2}(\psi_1),\ldots, \psi_{k-1-m} \in W_{k-1}(\psi_{k-2-m})$ such that $\psi \in W_k(\psi_{k-1-m})$. Furthermore, for $\phi \in W_m$ and $k > m$ let

$$D_k(\phi) \equiv W_k \cap D(\phi).$$

Observe that in the new notation, $D(\phi) \cap W_{m+1} = W_{m+1}(\phi)$ for each $\phi \in W_m$. Fix $\phi \in W_m$ and for each $k > m$ let

$$\phi_k \equiv \frac{1}{|D_k(\phi)|} \sum_{\psi \in D_k(\phi)} \psi.$$

Then if $k > m+1$ and if n denotes the number $|W_k(\psi)|$ for some (hence all) $\psi \in W_{k-1}$ (cf. the statement of Theorem 7.3.5) it follows from (8) that

$$|\phi_k - \phi_{k-1}|_1 = \left| \frac{1}{n|D_{k-1}(\phi)|} \sum_{\xi \in D_{k-1}(\phi)} \sum_{\psi \in D_k(\xi)} \psi - \frac{1}{|D_{k-1}(\phi)|} \sum_{\xi \in D_{k-1}(\phi)} \xi \right|_1$$

(10) $$\leqq \frac{1}{|D_{k-1}(\phi)|} \sum_{\xi \in D_{k-1}(\phi)} \left| \frac{1}{n} \sum_{\psi \in D_k(\xi)} \psi - \xi \right|_1$$

$$< \overline{\delta}\left(\frac{1}{k-1}, X_{k-1}\right) \leqq \frac{1}{2^k}.$$

Hence for positive integers j, i with $m < j < i$ (and with $\phi \in W_m$) (10) yields

$$|\phi_i - \phi_j|_1 \leqq \sum_{k=j+1}^{i} |\phi_k - \phi_{k-1}|_1 \leqq \sum_{k=j+1}^{i} \frac{1}{2^k} < \frac{1}{2^j}.$$

That is, for each m and each $\phi \in W_m$, $\{\phi_k : k > m\}$ is Cauchy in $L^1(\lambda)$. Let $\tilde{\phi}$ be the $L^1(\lambda)$-limit of $\{\phi_k : k > m\}$ and note that

$$|\tilde{\phi} - \phi|_1 \leqq \frac{1}{2^m}, \quad |\tilde{\phi}|_1 = 1, \quad \text{and} \quad \tilde{\phi} \geqq 0.$$

For each m let

$$\tilde{W}_m \equiv \{\tilde{\phi} : \phi \in W_m\} \quad \text{and} \quad \tilde{W}_{m+1}(\phi) \equiv \{\tilde{\psi} : \psi \in W_{m+1}(\phi)\} \quad \text{for} \quad \phi \in W_m.$$

Evidently $\bigcup_{m=1}^{\infty} \tilde{W}_m$ is a $\frac{1}{2}$- bush in the intersection of the positive cone of $L^1(\lambda)$ and the unit sphere of X. As indicated above, the existence of such a bush guarantees that X lacks the RNP. Hence the proof of Theorem 7.3.9 is complete.

The subspace X of $L^1(\lambda)$ constructed in Theorem 7.3.9 satisfies conditions (a) and (b) of the main Theorem 7.3.2. In Proposition 7.3.12 it will be established that X has the strong Schur property and Proposition 7.3.13 provides a proof that the $L^1(\lambda)$-closed convex hull of $\bigcup_{m=1}^{\infty} \tilde{W}_m$ (in the notation of the proof of 7.3.9 just completed) has no extreme points (whence X lacks the KMP). We tackle next the central issue: part (c) of Theorem 7.3.2.

Proof of 7.3.2(c). The constants $\beta(m,\varepsilon)$, $\gamma(Z,\varepsilon)$ and $\delta(\varepsilon,Z)$ and the increasing sequence of finite dimensional subspaces $\{X_n: n = 1,2,\ldots\}$ whose closed span is X as described just previous to, and in, the proof of Theorem 7.3.9 will be used without notational alteration. Suppose there were an $\varepsilon > 0$ and an X-valued martingale $(F_n,\mathcal{F}_n)$ with difference sequence $\{D_n: n = 1,2,\ldots\}$ such that $\|F_n\|_\infty \leqq 1$ and $\|D_n\|_1 \geqq \varepsilon$ for each n and $\{\mathcal{F}_n: n = 1,2,\ldots\}$ is dominated by $\{q_n: n = 1,2,\ldots\}$. The first step in creating a contradiction is to "shift" $(F_n,\mathcal{F}_n)$ so that the modified martingale takes all its values in $\bigcup_{m=1}^{\infty} X_m$. In this regard pick $\alpha > 0$ so that $\frac{\alpha}{1-\alpha} < \frac{\varepsilon}{4}$. In this paragraph we denote by $F_n|_A$ the value of F_n on an atom A of $\mathcal{F}_n$. For each atom A of $\mathcal{F}_1$ pick $g_A \in \bigcup_{m=1}^{\infty} X_m$ so that $\|g_A - F_1|_A\| < \alpha$ and let $G_1: [0,1] \to \bigcup_{m=1}^{\infty} X_m$ be the $\mathcal{F}_1$-measurable function with value g_A on A for each atom A of $\mathcal{F}_1$. Evidently $\|G_1 - F_1\|_\infty < \alpha$. Suppose now that $G_1,\ldots,G_n$ have been chosen so that each G_i is $\mathcal{F}_i$-measurable and $\bigcup_{m=1}^{\infty} X_m$-valued for $1 \leqq i \leqq n$, $\mathbb{E}[G_{i+1}|\mathcal{F}_i] = G_i$ [a.s. λ] $1 \leqq i \leqq n-1$, and $\|G_i - F_i\|_\infty < \alpha + \alpha^2 + \cdots + \alpha^i$ for $1 \leqq i \leqq n$. Let $A \in \mathcal{F}_n$ be an atom of positive Lebesgue measure, and suppose that $A = \bigcup_{j=1}^{k} A_j$ where $\{A_j\}_{j=1}^{k}$ is the set of atoms of $\mathcal{F}_{n+1}$ contained in A. Assume further that the indexing has been chosen so that $\lambda(A_k) > 0$. Choose $\tau > 0$ such that $(k-1)\tau\frac{\lambda(A)}{\lambda(A_k)} < \alpha^{n+1}$. For each $i = 1,\ldots,k-1$ choose $g_j \in \bigcup_{m=1}^{\infty} X_m$ such that

$$(11) \qquad \|F_{n+1}|_{A_j} + G_n|_A - F_n|_A - g_j\|_1 < \tau$$

and define G_{n+1} on $\bigcup_{j=1}^{k-1} A_j$ by: $G_{n+1}|_{A_j} \equiv g_j$. Furthermore, let

(12) $$g_k \equiv \frac{1}{\lambda(A_k)} [\lambda(A)G_n|_A - \sum_{j=1}^{k-1} \lambda(A_j)g_j]$$

and let $G_{n+1}|_{A_k} \equiv g_k$. Observe that

(13) $$\int_A G_{n+1}d\lambda = \sum_{i=1}^{k} \lambda(A_i)G_{n+1}|_{A_i} = \lambda(A)G_n|_A = \int_A G_n d\lambda.$$

Since $\int_A F_{n+1}d\lambda = \int_A F_n d\lambda$ it follows that

(14) $$F_{n+1}|_{A_k} = \frac{1}{\lambda(A_k)} [\lambda(A)F_n|_A - \sum_{i=1}^{k-1} \lambda(A_i) F_{n+1}|_{A_i}]$$

and hence from (12) and (14) and the definition of τ

(15)
$$\begin{aligned}
&|F_{n+1}|_{A_k} + G_n|_A - F_n|_A - g_k|_1 \\
&= |(\frac{\lambda(A)}{\lambda(A_k)} - 1)(F_n|_A - G_n|_A) + \sum_{i=1}^{k-1} \frac{\lambda(A_i)}{\lambda(A_k)} (g_i - F_{n+1}|_{A_i})|_1 \\
&= |\sum_{i=1}^{k-1} \frac{\lambda(A_i)}{\lambda(A_k)} [F_n|_A - G_n|_A + g_i - F_{n+1}|_{A_i}]|_1 \\
&\leq \sum_{i=1}^{k-1} \frac{\lambda(A_i)}{\lambda(A_k)} \tau \leq (k-1)\tau \frac{\lambda(A)}{\lambda(A_k)} < \alpha^{n+1}.
\end{aligned}$$

Consequently by (11) and (15), for each $j = 1,\ldots,k$

(16) $$|F_{n+1}|_{A_j} - G_{n+1}|_{A_j}|_1 < |F_n|_A - G_n|_A|_1 + \alpha^{n+1} < \alpha + \cdots + \alpha^n + \alpha^{n+1}$$

by inductive hypothesis. In this manner G_{n+1} is defined on the atoms of F_n of positive Lebesgue measure and we let $G_{n+1}|_A \equiv 0$ if $A \in F_n$ is an atom of Lebesgue measure 0. Then G_{n+1} is F_{n+1}-measurable, $\bigcup_{m=1}^{\infty} X_m$-valued, $\mathbb{E}[G_{n+1}|F_n] = G_n$ [a.s. λ] by (13) and $\|G_{n+1} - F_{n+1}\|_\infty < \alpha + \cdots + \alpha^{n+1}$ by (16). Finally, observe that for each n

$$\|G_{n+1} - G_n\|_1 \geq \|F_{n+1} - F_n\|_1 - \|F_{n+1} - G_{n+1}\|_1 - \|F_n - G_n\|_1 \geq \varepsilon - 2\sum_{j=1}^{\infty} \alpha^j \geq \varepsilon/2$$

while

$$\|G_1\|_1 \geqq \|F_1\|_1 - \|F_1 - G_1\|_1 \geqq \varepsilon - \alpha > \varepsilon/2.$$

Thus $(G_n, \mathcal{F}_n)$ is a bounded $\varepsilon/2$ - bush with values in $\bigcup_{m=1}^{\infty} X_m$. (A scalar multiple o this martingale then places its range in $U(\bigcup_{m=1}^{\infty} X_m)$ at the possible cost of replaci $\varepsilon/2$ by some other positive constant.) For notational simplicity we shall hencefort revert to the old notation of $(F_n, \mathcal{F}_n)$ with difference sequence $\{D_n : n = 1,2,\dots\}$ and $\varepsilon > 0$ and assume that for each i

Range $F_i \subset X_n$ for some n, $\|F_i\|_\infty \leqq 1$ and $\|D_i\|_1 \geqq \varepsilon$ for each i.

Since $\|F_i\|_1 \leqq \|F_{i+1}\|_1 \leqq 1$ for each i we may choose an index j_0 such that

$$\|F_{j_0}\|_1 \geqq \sup_i \|F_i\|_1 - \varepsilon/30.$$

Let n be so large that

$$\frac{1}{n} < \frac{\varepsilon}{30}, \quad F_{j_0} \text{ is } X_n\text{-valued and if } m \equiv \dim X_n \text{ then } \beta(m,\varepsilon) \geqq j_0.$$

In what follows make the abbreviations:

$$\Delta \equiv \delta(\tfrac{1}{n}, X_n) \quad \text{and} \quad k \equiv \beta(m,\varepsilon) \quad (m = \dim X_n \text{ as above}).$$

(See (6), (7) and Proposition 7.3.7 for a definition of these constants.) Since F_k is constant on each atom of $\mathcal{F}_k$ there is a $U(X_n)$-valued $\mathcal{F}_k$-measurable function G such that (by using (b) of 7.3.9)

(17) $$d(G(t), F_k(t)) \leqq \Delta \text{ for all } t \in [0,1].$$

Let

$$G_i \equiv \mathbb{E}[G \mid \mathcal{F}_i] \quad \text{for } i = 1,\dots,k.$$

Suppose that $\{h_1,\dots,h_j\} \subset L^1(\lambda)$, $\alpha_i \geqq 0$, $\sum_{i=1}^{j} \alpha_i = 1$ and $h \equiv \sum_{i=1}^{j} \alpha_i h_i$.

Then by an elementary calculation,

$$d(h,0) \leqq j\cdot\max\{d(h_i,0): 1 \leqq i \leqq j\}.$$

Consequently, when (H_i,F_i) is an $L^1(\lambda)$-valued martingale then for each positive integer p

(18) $$\max_t d(H_{p-1}(t),0) \leqq |F_p| \max_t d(H_p(t),0).$$

(In fact, the inequality (18) may be improved by replacing $|F_p|$ on the right side of (18) by the following number:

$$\max\{j: A = \bigcup_{i=1}^{j} A_i \text{ where } A \text{ is an atom of } F_{p-1} \text{ and}$$

$$\{A_i\}_{i=1}^{j} \text{ is the set of atoms of } F_p \text{ contained in } A\}.$$

Thus in the special case where $F_i = \mathcal{D}_i$ - the dyadic σ-algebra - for each i, $|F_p|$ may be replaced by 2. We have no need for this improvement and use (18) in the sequel.) By induction, then,

(19) $$\max_t d(H_i(t),0) \leqq |F_p||F_{p-1}|\cdots|F_{i+1}| \max_t d(H_p(t),0) \quad \text{for } i < p.$$

Since $(G_i - F_i, F_i)_{i=1}^{k}$ is an $L^1(\lambda)$-valued martingale it follows from (17) and (19) that

(20) $$\max_t d((G_{j_o} - F_{j_o})(t),0) \leqq |F_k||F_{k-1}|\cdots|F_{j_o+1}| \max_t d((G_k - F_k)(t),0)$$

$$\leqq |F_k|^{k-j_o} \Lambda \leqq q_k^{k-j_o} \Delta$$

since $|F_i| \leqq |F_{i+1}|$ for each i. This important observation is the only one requiring that the sequence of σ-algebras $\{F_n: n = 1,2,\ldots\}$ be dominated by the fixed sequence $\{q_n: n = 1,2,\ldots\}$ and that the martingale have the F_i's as its sequence of σ-algebras. But

$$q_k^{k-j_o}\Delta \leqq q_k^{k}\Delta = q_k^{k}(q_{\beta(m,1/n)})^{-\beta(m,1/n)}\gamma(X_n,\tfrac{1}{n}). \tag{21}$$

Now $k \equiv \beta(m,\varepsilon)$ and $\frac{1}{n} < \frac{\varepsilon}{30} < \varepsilon$ so that $k \leqq \beta(m,1/n)$. It follows that

$$q_k^{k}(q_{\beta(m,1/n)})^{-\beta(m,1/n)} \leqq 1$$

and hence from (21) that

$$q_k^{k-j_o}\Delta \leqq \gamma(X_n,\tfrac{1}{n}).$$

Let $\Gamma \equiv \gamma(X_n,\frac{1}{n})$. Recapitulating, then, gives

$$d(G_{j_o}(t),F_{j_o}(t)) \leqq \Gamma \quad \text{for all} \quad t \in [0,1]. \tag{22}$$

Using the definition of Γ as a uniform integrability measure on $U(X_n)$, (22) will be turned into an inequality bounding $\|G_{j_o} - F_{j_o}\|_\infty$. Indeed, for fixed t temporarily set $h \equiv G_{j_o}(t) - F_{j_o}(t)$. Then $|h|_1 \leqq 2$. Now (22) states that $\lambda(|h| \geqq \Gamma) \leqq \Gamma$ so that (by definition of Γ as $\gamma(X_n,\frac{1}{n})$)

$$\int_{(|h| \geqq \Gamma)} |h|\, d\lambda \leqq \frac{1}{n}|h|_1 \leqq \frac{2}{n}.$$

On the other hand it is obvious that $\int_{(|h|<\Gamma)} |h|\, d\lambda \leqq \Gamma \leqq \frac{1}{n}$ so that

$$\int |h|\, d\lambda \leqq \frac{2}{n} + \frac{1}{n} = \frac{3}{n}.$$

That is,

$$|G_{j_o}(t) - F_{j_o}(t)|_1 \leqq \frac{3}{n} < \frac{\varepsilon}{10} \quad \text{for each} \quad t \in [0,1].$$

In particular, then, $|G_{j_o}(t)|_1 \geqq |F_{j_o}(t)|_1 - \frac{\varepsilon}{10}$ and hence by choice of j_o,

$$\|G\|_1 \geqq \|G_{j_o}\|_1 \geqq \|F_{j_o}\|_1 - \frac{\varepsilon}{10} \geqq \sup_i \|F_i\|_1 - \frac{\varepsilon}{30} - \frac{\varepsilon}{10} \geqq \sup_i \|F_i\|_1 - \frac{2\varepsilon}{15}. \tag{23}$$

For each $t \in [0,1]$ let

$$A_t \equiv (\|G(t) - F_k(t)\| \leqq \Gamma).$$

Since $d(G(t),F_k(t)) \leqq \Delta \leqq \Gamma$ for each t it follows that $\lambda(A_t{}^c) \leqq \Gamma$, and again from the definition of Γ,

$$\|G(t)\chi_{A_t{}^c}\|_1 \leqq \frac{1}{n} < \frac{\varepsilon}{30}. \tag{24}$$

Combining (23) and (24) yields

$$\begin{aligned}\int \|G(t)\chi_{A_t}\|_1 d\lambda(t) &\geqq \int \|G(t)\|_1 d\lambda(t) - \int \|G(t)\chi_{A_t{}^c}\|_1 d\lambda(t) \\ &\geqq \sup_i \|F_i\|_1 - \frac{\varepsilon}{6}.\end{aligned} \tag{25}$$

The importance of (25) is that it gives rise to the bound

$$\begin{aligned}\int \|F_k(t)\chi_{A_t}\|_1 d\lambda(t) &\geqq \int (\|G(t)\chi_{A_t}\|_1 - \Gamma) d\lambda(t) && \text{(by definition of } A_t) \\ &\geqq \sup_i \|F_i\|_1 - \frac{\varepsilon}{6} - \frac{\varepsilon}{30} && (\text{since } \Gamma \leqq \frac{1}{n} < \frac{\varepsilon}{30}) \qquad (26) \\ &= \sup_i \|F_i\|_1 - \frac{\varepsilon}{5}.\end{aligned}$$

The next calculations are aimed at providing a corresponding lower bound for $\int \|F_k(t)\chi_{A_t{}^c}\|_1 d\lambda(t)$. (See (27).) Indeed, by definition of k as $\beta(m,\varepsilon)$ and by Proposition 7.3.7 we have $\|F_k - G\|_1 \geqq \varepsilon/3$. By the definition of A_t

$$\int \|(F_k(t) - G(t))\chi_{A_t}\|_1 d\lambda(t) \leqq \Gamma < \frac{\varepsilon}{30}$$

so that

$$\int \|F_k(t) - G(t))\chi_{A_t{}^c}\|_1 d\lambda(t) \geqq \frac{\varepsilon}{3} - \frac{\varepsilon}{30} = \frac{3\varepsilon}{10}.$$

On the other hand, for each t

$$|(F_k(t) - G(t))\chi_{A_t^c}|_1 \leqq |F_k(t)\chi_{A_t^c}|_1 + |G(t)\chi_{A_t^c}|_1$$

and we observed that $|G(t)\chi_{A_t^c}|_1 \leqq \frac{\varepsilon}{30}$ in (24). Consequently

$$(27) \qquad \int |F_k(t)\chi_{A_t^c}|_1 d\lambda(t) \geqq \frac{3\varepsilon}{10} - \frac{\varepsilon}{30} = \frac{4\varepsilon}{15}.$$

The two estimates (26) and (27), however, combine to yield the contradiction

$$\|F_k\|_1 = \int |F_k(t)\chi_{A_t}|_1 d\lambda(t) + \int |F_k(t)\chi_{A_t^c}|_1 d\lambda(t)$$

$$\geqq \sup_i \|F_i\|_1 - \frac{\varepsilon}{5} + \frac{4\varepsilon}{15} > \sup_i \|F_i\|_1$$

and the proof of Theorem 7.3.2(c) is thus complete.

The next goal is to demonstrate that 7.3.2(d) holds for the space X construc in 7.3.9. Recall that a Banach space Y satisfies the Schur property if each seque in Y which tends to 0 weakly, tends to 0 in norm; (equivalently: if weakly com pact sets are norm compact).

DEFINITION 7.3.11. (a) Let $\{y_i : i = 1,2,\ldots\}$ be a sequence in $U(Y)$ and let $M > 0$. Then $\{y_i : i = 1,2,\ldots\}$ is M-equivalent to the usual ℓ_1-basis provided

$$M \left\| \sum_{i=1}^{n} t_i y_i \right\| \geqq \sum_{i=1}^{n} |t_i|$$

for each finite collection of scalars $t_1,\ldots,t_n$. ($\{y_i : i = 1,2,\ldots\}$ is equivalent to the usual ℓ_1-basis if it is M-equivalent to the usual ℓ_1-basis for some M.)

(b) Y has the strong Schur property if there is an $M > 0$ such that wheneve $0 < \delta \leqq 2$ and $\{y_i : i = 1,2,\ldots\}$ is a δ-separated sequence in $U(Y)$ (i.e. $\|y_i - y$ $\geqq \delta$ for $i \neq j$) there is a subsequence M/δ-equivalent to the usual ℓ_1-basis.

Johnson and Odell [1974] showed that there are subspaces of $L^1(\lambda)$ which have the Schur property but not the strong Schur property.

PROPOSITION 7.3.12. The subspace X of $L^1(\lambda)$ constructed in Theorem 7.3.9 has th strong Schur property.

Proof. Fix δ, $0 < \delta \leqq 2$ and suppose that $\{f_n : n = 1,2,\ldots\}$ is a δ-separated sequence in $U(X)$. Because $U(X)$ is relatively compact in the topology of convergence in λ-measure there is a subsequence $\{f'_n : n = 1,2,\ldots\}$ which converges in the d-metric. The Subsequence Splitting Lemma of Rosenthal [1983] states that in this case there is a subsequence $\{f''_n : n = 1,2,\ldots\}$ of $\{f'_n : n = 1,2,\ldots\}$ and there are sequences $\{g_n : n = 1,2,\ldots\}$ and $\{h_n : n = 1,2,\ldots\}$ such that for each n

(a) $f''_n = g_n + h_n$;

(b) $(g_n \neq 0) \cap (h_n \neq 0) = \emptyset$;

(c) $(h_n \neq 0) \cap (h_m \neq 0) = \emptyset$ for $n \neq m$;

(d) $\{g_n : n = 1,2,\ldots\}$ is uniformly integrable.

It follows immediately from (c) that $\lim_n d(h_n, 0) = 0$ and consequently, using (a) and the fact that $\{f''_n : n = 1,2,\ldots\}$ converges in measure, $\{g_n : n = 1,2,\ldots\}$ converges in measure as well. It is easy to see that $\{g_n : n = 1,2,\ldots\}$ also converges in $L^1(\lambda)$-norm $|\cdot|_1$, since it is uniformly integrable. Since $\{f''_n : n = 1,2,\ldots\}$ is δ-separated, $\liminf_{\substack{n \neq m \\ n,m\to\infty}} |f''_n - f''_m|_1 \geqq \delta$ and from above, $\liminf_{\substack{n \neq m \\ n,m\to\infty}} |g_n - g_m|_1 = 0$. Thus

(28)
$$\liminf_{\substack{n \neq m \\ n,m\to\infty}} |h_n - h_m|_1 \geqq \delta.$$

One consequence of (28) is that for sufficiently large n, $|h_n|_1 \geqq \delta/4$. For each m let $B_m \equiv (h_m \neq 0)$. Since $B_n \cap B_m = \emptyset$ for $n \neq m$ by (c), by using (d) we can produce an N such that simultaneously $|h_n|_1 \geqq \delta/4$ when $n \geqq N$ and

$$\int_B |g_j|\, d\lambda < \delta/8 \quad \text{for } j \geqq N \text{ where } B \equiv \bigcup_{i=N}^{\infty} B_i.$$

Now let $\{t_i : i = 1,2,\ldots\}$ be a sequence of scalars with only finitely many nonzero terms. Then by (a) and (b), $|g_n|_1 + |h_n|_1 = |f''_n|_1 \leqq 1$ and consequently

$$\left|\sum_{i\geq N} t_i f''_i\right|_1 \geq \int_B \left|\sum_{i\geq N} t_i f''_i\right| d\lambda = \int_B \left|\sum_{i\geq N} t_i g_i + \sum_{i\geq N} t_i h_i\right| d\lambda$$

$$\geq \int_B \left|\sum_{i\geq N} t_i h_i\right| d\lambda - \int_B \left|\sum_{i\geq N} t_i g_i\right| d\lambda$$

$$\geq \frac{\delta}{4} \sum_{i\geq N} |t_i| - \frac{\delta}{8} \sum_{i\geq N} |t_i| = \frac{\delta}{8} \sum_{i\geq N} |t_i| .$$

It follows that $\{f''_n : n \geq N\}$ is $\frac{8}{\delta}$-equivalent to the usual ℓ_1-basis.

The final proposition of this section is due to Elton and independently to Bourgain. (We follow Elton [1982] below.) The notation is that of the proof of 7.3.9. Because we have occasion to take the closure of various sets in X in the topology of convergence in λ-measure as well as in the $L^1(\lambda)$-norm topology it is convenient to denote by d-cl(A) and L^1-cl(A) respectively, the closure of a set $A \subset L^1(\lambda)$ in the d-metric topology and the $|\cdot|_1$-norm topology. (Observe, then, that if $A \subset X$, d-cl(A) need not be a subset of X.) The following easily verifie facts will soon come in handy:

(29) If f_n $n = 1,2,\ldots$ and f belong to $L^1(\lambda)$ and $\lim_n d(f_n,f) = 0$ then $|f|_1 \leq \liminf_n |f_n|_1$. (That is, $|\cdot|_1$ is d-lower semicontinuous.)

(30) The d-metric and $|\cdot|_1$-metric coincide on S(X). That is, if $|f_n|_1 = |f|_1 = 1$ for each n then $\lim_n d(f_n,f) = 0$ if and only if $\lim_n |f_n - f|_1 = 0$.

(31) $U(L^1(\lambda))$ is d-complete. That is, if $\{f_n : n = 1,2,\ldots\}$ is Cauchy in the d-metric and $|f_n|_1 \leq 1$ for each n then there is an $f \in U(L^1(\lambda))$ such tha $\lim_n d(f_n,f) = 0$.

Let $\tilde{W} \equiv \bigcup_{m=1}^{\infty} \tilde{W}_m$ be the bounded 1/2-bush described in the proof of Theorem 7.3.9.

PROPOSITION 7.3.13 (Elton; Bourgain). X does not have the KMP. (In fact, if $C \equiv L^1\text{-cl}(\text{co}(\tilde{W}))$ then $\text{ex}(C) = \emptyset$.)

Proof. (Elton) Let us review briefly the construction of X in Theorem 7.3.9 while we subscript some of the constants along the way. Begin with $X_1 \equiv \mathrm{span}(\{1\})$ and $W_1 \equiv \{1\}$. In general, suppose $X_1,\ldots,X_m$ and $W_1,\ldots,W_m$ have already been chosen. (Recall, $\delta(\varepsilon,Z)$ is defined in (7) and $\overline{\delta}(\varepsilon,Z)$ is defined at the beginning of the proof of 7.3.9.) Let $\varepsilon_m \equiv \overline{\delta}(\frac{1}{m},X_m)$ in Theorem 7.3.5 and let $\eta_m \equiv \eta(X_m,\varepsilon_m)$ in the statement of Lemma 7.3.6. Let

(32)
$$\xi_m \equiv \frac{\eta_m{}^2}{4|W_m|}.$$

In the proof of Theorem 7.3.5, Theorem 7.3.4 was invoked with this ξ_m to construct h and a positive integer n (with the properties listed in the statement of 7.3.4) and we let $\{h_{\alpha j}: \alpha \in W_m,\ 1 \leqq j \leqq n\}$ be an independent collection of functions each with the same distribution as h and we defined

(33)
$$W_{m+1}(\alpha) \equiv \{h_{\alpha j}\alpha: 1 \leqq j \leqq n\} \quad \text{for each } \alpha \in W_m.$$

Then we let

$$W_{m+1} \equiv \cup\{W_{m+1}(\alpha): \alpha \in W_m\} \quad\text{and}\quad X_{m+1} \equiv X_m \oplus \mathrm{span}(W_{m+1}).$$

Recall that in the terminology of 7.3.10, for $\phi \in W_m$, $D(\phi)$ is the set of descendants of ϕ and $D_k(\phi)$ is the set of descendants of ϕ in W_k $(k > m)$. Let $K \equiv d\text{-}\mathrm{cl}(\mathrm{co}(\tilde{W})) \subset L^1(\lambda)$. By Theorem 7.3.2(c), $U(X)$ is relatively d-compact and hence K is d-totally bounded and complete by (31); hence K is d-compact. Most of the proof of this proposition is concerned with showing that the only possible extreme point of K is 0. (K is clearly convex.) Once this is done, the proof that $\mathrm{ex}(C) = \emptyset$ follows easily. Indeed, pick any $f \in C$. Then since $\tilde{W} \subset S(X) \cap$ (positive cone of $L^1(\lambda)$), $C \subset S(X) \cap$ (positive cone of $L^1(\lambda)$) as well. Thus $|f|_1 = 1$. Since $f \neq 0$, $f \notin \mathrm{ex}(K)$, and so there are $f_1,f_2 \in K$ with $f_1 \neq f_2$ and $f = \frac{1}{2} f_1 + \frac{1}{2} f_2$. Hence $|f_1|_1 = |f_2|_1 = 1$ (since $K \subset U(L^1(\lambda))$ by (29)). Thus there are sequences $\{f_1^n: n = 1,2,\ldots\}$ and $\{f_2^n: n = 1,2,\ldots\}$ in $\mathrm{co}(\tilde{W})$ such that $\lim_n d(f_i^n,f_i) = 0$ for $i = 1,2,$. Now $|f_i^n|_1 = 1 = |f_i|_1$ so (30) applies. That is,

$\lim_n \| f_i^n - f_i \|_1 = 0$ whence $f_i \in C$ for $i = 1,2$. But this implies that $f \notin ex(C)$ so that $ex(C) = \emptyset$ as claimed. (As an aside, we remark that the compact convex set $co(K \cup -K)$ in $(L^1(\lambda),d)$ has no extreme points. The first example of such a phenom enon was given by Roberts [1977] and Theorem 7.3.4 is a generalization of some of h work [1983].)

By $co_0(A)$ for a set $A \subset L^1(\lambda)$ we shall mean $co(\{0\} \cup A)$ but in the special case when A is a singleton $\{\phi\}$, let $co_0(\{\phi\})$ be written in the more convention notation $[0,\phi]$. To show that $ex(K) \subset \{0\}$ we will establish the following estimat (Assume $\phi \in W_n$ in (34)-(37).)

(34) If $k > n$ and $\psi \in co_0(D_{k+1}(\phi))$ then $d(\psi, co_0(D_k(\phi))) < \eta_k$.

(35) If $\psi \in co_0(D_{n+1}(\phi))$ then $d(\psi,[0,\phi]) < \eta_n$.

(36) If $\psi \in co_0(D(\phi))$ then $d(\psi,[0,\phi]) \leqq 2^{-n}$.

(37) $d(\phi,0) < 2^{-n}$.

Temporarily assume (34)-(37) have been verified. By analogy with 7.3.10, in the bus $\tilde{W}$ we write $\tilde{D}(\tilde{\phi})$ for the set of descendants of an element $\tilde{\phi} \in \tilde{W}$. Suppose now tha $\phi \in W_n$ and observe that if $\tilde{\psi} \in \tilde{D}(\tilde{\phi})$ then $\psi \in D(\phi)$ and consequently $D(\psi) \subset D(\phi)$ Since $\tilde{\psi} \in L^1\text{-cl}(D(\psi))$ (see the last part of the proof of 7.3.9) it follows that $\tilde{\psi} \in L^1\text{-cl}(D(\phi))$ for each $\tilde{\psi} \in \tilde{D}(\tilde{\phi})$. Thus

$$d\text{-cl}(co_0(\tilde{D}(\tilde{\phi}))) \subset d\text{-cl}(co_0(D(\phi))).$$

Given $\beta \in d\text{-cl}(co_0(\tilde{D}(\tilde{\phi})))$ there is thus a sequence $\{\beta_i: 1 - 1,2,\ldots\}$ in $co_0(D(\Phi))$ such that $\lim_i d(\beta_i,\beta) = 0$. By (36) $d(\beta_i,[0,\phi]) \leqq 2^{-n}$ for each i and hence $d(\beta,[0,\phi]) \leqq 2^{-n}$ as well. Consequently, for each $\tilde{\phi} \in \tilde{W}_n$ and for each $\beta \in d\text{-cl}(co_0(\tilde{D}(\tilde{\phi})))$, by (37)

$$d(\beta,0) \leqq d(\beta,[0,\phi]) + \max_{0 \leqq t \leqq 1} d(t\phi,0) \tag{38}$$

$$\leqq 2^{-n} + d(\phi,0) < 2^{-n} + 2^{-n} = 2^{-(n-1)}.$$

To see that $ex(K) \subset \{0\}$ we can now argue as follows. Suppose $f \in ex(K)$. For $\tilde{\phi} \in \tilde{W}$ let $K(\tilde{\phi}) \equiv d\text{-}cl(co_0(\tilde{D}(\tilde{\phi})))$ and observe that $K(\tilde{\phi})$ is d-compact and convex. Moreover, we claim that for any positive integer n

(39) $$K = co(\bigcup_{\tilde{\phi}\in\tilde{W}_n} K(\tilde{\phi})).$$

Indeed, the set on the right side of (39) is the convex hull of finitely many compact convex sets and is thus compact and convex. It contains each element of $\tilde{W}_m$ for $m > n$ and to see that it also contains each element $\tilde{W}_m$ for $m \leqq n$ just note that the elements of each such set are convex combinations of the elements of $\tilde{W}_{n+1}$. Thus the set on the right of (39) contains the bush $\tilde{W}$ which generates K, and (39) follows directly. Since f is assumed to be an extreme point of K, from (39) f must belong to some one of the sets $K(\tilde{\phi})$ for some $\tilde{\phi} \in \tilde{W}_n$. But then from (38), $d(f,0) \leqq 2^{-(n-1)}$. Since n is arbitrary, $f = 0$. Thus $ex(K) \subset \{0\}$ as was to be shown. Thus in order to complete the proof of 7.3.13 it suffices to demonstrate (34)-(37).

To prove (34) note that $\beta \in D_{k+1}(\phi)$ means that β is the descendant of some element α in $D_k(\phi)$. In particular, $\beta \in W_{k+1}(\alpha)$. Hence if $\psi \in co_0(D_{k+1}(\phi))$ then by combining together the β's which are descendants of a given α we may write

(40) $$\psi \equiv \sum_{\alpha\in D_k(\phi)} t_\alpha \beta_\alpha \quad \text{where } t_\alpha \geqq 0, \quad \sum_{\alpha\in D_k(\phi)} t_\alpha \leqq 1 \quad \text{and}$$

$$\beta_\alpha \in co(W_{k+1}(\alpha)) \quad \text{for some } \alpha \in D_k(\phi).$$

From the construction of $W_{k+1}(\alpha)$ (see (33)) "$\beta_\alpha \in co(W_{k+1}(\alpha))$ for some $\alpha \in D_k(\phi)$" may be written out in the form

$$\beta_\alpha = \sum_{j=1}^{m} s_{\alpha j}(h_{\alpha j}\alpha) \quad \text{where each } s_{\alpha j} \geqq 0, \quad \sum_{j=1}^{m} s_{\alpha j} = 1 \quad \text{and}$$

$\{h_{\alpha j}\}_{j=1}^{m}$ is a collection of independent copies of h

as in Theorem 7.3.4 for the constant ξ_k.

Now let $g_\alpha \equiv \sum_{j=1}^{m} s_{\alpha j} h_{\alpha j}$ so that (40) may be rewritten as

$$(41) \qquad \psi = \sum_{\alpha\in D_k(\phi)} t_\alpha g_\alpha \alpha .$$

By Theorem 7.3.4(a), $g_\alpha \geqq 0$, $|g_\alpha|_1 \leqq 1$ and $g_\alpha \in X_{k+1}$ so that by 7.3.4(b) there is a constant c_α with $c_\alpha \geqq 0$ (since $g_\alpha \geqq 0$) and $c_\alpha \leqq 1$ such that

$$(42) \qquad \lambda(|g_\alpha - c_\alpha| \geqq \xi_k) \leqq \xi_k .$$

Let $\hat{\alpha} \equiv \sum_{\alpha\in D_k(\phi)} t_\alpha c_\alpha \alpha$ and $f \equiv \xi_k \sum_{\alpha\in D_k(\phi)} t_\alpha \alpha$. Then $\hat{\alpha} \in co_0(D_k(\phi))$. Note the similarity between the form of $\hat{\alpha}$ above and ψ in (41). It will be shown that $d(\psi,\hat{\alpha}) < \eta_k$ which will establish (34). First, from (41) and (42),

$$(43) \qquad \lambda(|\psi - \hat{\alpha}| > f) \leqq \lambda\Big(\sum_{\alpha\in D_k(\phi)} |g_\alpha - c_\alpha| t_\alpha \alpha > \xi_k \sum_{\alpha\in D_k(\Phi)} t_\alpha \alpha\Big)$$

$$\geqq \lambda\Big(\bigcup_{\alpha\in D_k(\phi)} (|g_\alpha - c_\alpha| t_\alpha \alpha > \xi_k t_\alpha \alpha)\Big) \leqq |W_k| \xi_k .$$

But

$$(44) \qquad \lambda(|\psi - \hat{\alpha}| > \eta_k) \leqq \lambda(|\psi - \hat{\alpha}| > f) + \lambda(f \geqq \eta_k)$$

and $\lambda(f \geqq \eta_k) \leqq \frac{1}{\eta_k} |f|_1$ (since

$$|f|_1 \geqq \int_{(|f| \geqq \eta_k)} |f| \, d\lambda \geqq \eta_k \lambda(f > \eta_k)).$$

Hence, since $\eta_k \leqq \xi_k \leqq 1$, from (32) and (43), (44) becomes

$$\lambda(|\psi - \hat{\alpha}| > \eta_k) \leqq |W_k| \xi_k + \frac{1}{\eta_k} |f|_1 \leqq \xi_k(|W_k| + \frac{1}{\eta_k}) = \frac{\eta_k{}^2}{4} + \frac{\eta_k}{4|W_k|} < \eta_k .$$

Thus (34) holds as claimed.

The proof of (35) is a much easier version of the proof of (34). Indeed, if $\psi \in co_0(D_{n+1}(\phi))$ for some $\phi \in W_n$ then write ψ in the form

$$\psi = \sum_{j=1}^{m} s_{\phi j} h_{\phi j} \phi \quad \text{where} \quad s_{\phi j} \geq 0, \quad \sum_{j=1}^{m} s_{\phi j} = 1 \quad \text{and} \quad \{h_{\phi j} : 1 \leq j \leq n\}$$

are independent copies of h as chosen in 7.3.4 with ξ_n.

Let $g_\phi \equiv \sum_{j=1}^m s_{\phi j} h_{\phi j}$ so that $\psi = g_\phi \phi$. Note that $|g_\phi|_1 \leq 1$, $g_\phi \in X_{n+1}$ and $g_\phi \geq 0$; hence by 7.3.4(b) there is a constant c_ϕ $0 \leq c_\phi \leq 1$, such that $\lambda(|g_\phi - c_\phi| \geq \xi_n) \leq \xi_n$. Then

$$\begin{aligned}\lambda(|\psi - c_\phi \phi| > \eta_n) &\leq \lambda(|\psi - c_\phi \phi| > \xi_n \phi) + \lambda(\xi_n \phi \geq \eta_n) \\ &\leq \lambda(|g_\phi - c_\phi| \phi > \xi_n \phi) + \frac{1}{\eta_n} |\xi_n \phi|_1 \\ &\leq \xi_n + \frac{\xi_n}{\eta_n} < \eta_n.\end{aligned}$$

Finally, observe that $c_\phi \phi \in [0,\phi]$ so that $d(\psi,[0,\phi]) \leq \eta_n$ as advertised. This is (35).

Consider next inequality (36). Suppose that $\phi \in W_n$ and that $\psi \in co_0(D(\phi))$. Then for some $k \geq n+1$ and scalars $t_i \geq 0$ with $\sum_{i=n+1}^k t_i \leq 1$ we have

$$\psi = \sum_{i=n+1}^{k} t_i \psi_i \quad \text{for some choice of} \quad \psi_i \in co_0(D_i(D_i(\phi)) \quad i = n+1,\dots,k.$$

Now $\psi_k \in co_0(D_k(\phi))$ so by (34) there is a $\beta \in co_0(D_{k-1}(\phi))$ such that $d(\psi_k,\beta) < \eta_{k-1}$. Let $t'_{k-1} \equiv t_{k-1} + t_k$ and let

$$\psi'_{k-1} \equiv \begin{cases} \dfrac{t_{k-1}}{t'_{k-1}} \psi_{k-1} + \dfrac{t_k}{t'_{k-1}} \psi_k & \text{if } t'_{k-1} > 0 \\ \\ 0 & \text{if } t'_{k-1} = 0. \end{cases}$$

Then $\psi'_{k-1} \in co_0(D_{k-1}(\phi))$ and we let

$$\psi' \equiv \sum_{i=n+1}^{k-2} t_i\psi_i + t'_{k-1}\psi'_{k-1}.$$ Observe that

$$d(\psi',\psi) = d(t_k\beta, t_k\psi_k) \leqq d(\beta,\psi_k) < \eta_{k-1}.$$

This procedure for producing ψ' from ψ may be repeated $k-(n+1)$ times to provide $\psi^* \in co_0(D_{n+1}(\phi))$ such that $d(\psi^*,\psi) < \sum_{i=n+1}^{k-1} \eta_i$. Now $d(\psi^*,[0,\phi]) < \eta_n$ by (35) and hence

$$d(\psi,[0,\phi]) < \sum_{i=n}^{k-1} \eta_i \leqq \sum_{i=n}^{k-1} \varepsilon_i \leqq \sum_{i=1}^{k-1} \frac{1}{2^{i+1}} < \frac{1}{2^n}.$$

Finally, to show (37) write ϕ in the form $h\psi$ for $\psi \in W_{n-1}$ and h as in Theorem 7.3.4 for the constant ξ_{n-1}. Then from 7.3.4(a), $\lambda(h \geqq \xi_{n-1}) \leqq \xi_{n-1}$ and hence

$$\lambda(h\phi \geqq \eta_{n-1}) \leqq \lambda(h \geqq \xi_{n-1}) + \lambda(\psi \geqq \frac{\eta_{n-1}}{\xi_{n-1}}) \leqq \xi_{n-1} + \frac{\xi_{n-1}}{\eta_{n-1}} |\psi|_1$$

$$< \eta_{n-1} \leqq 2^{-n}.$$

This completes the proof of (34)-(37) and hence of Proposition 7.3.13 and hence of Theorem 7.3.2.

§4. Pettis integration and the weak Radon-Nikodým Property.

There are many ways to see why the Bochner integral is the integral of choice fo Radon-Nikodym questions of the type examined in these notes. Perhaps the most direc is the fact, central to many of our arguments, that the simple functions are dense i an L^1-sense. This forces the ranges of Bochner integrable functions to be almost separably valued, which in turn reflects the fact that the RNP is a separably determined property. There is a marked deterioration in the geometry when the Bochner integral is replaced by weaker integrals in the definition of the RNP, but there is a compensating gain in other tie-ins with the structure of certain Banach spaces. We briefly outline some of what is known and reference several recent highly readab sources for further details in this rapidly expanding area. (Often the techniques are substantially different from those of the first six chapters, and in order to keep within reasonable space confines no proofs are included.)

In general weak integrals have been studied for functions f from a probability space into a locally convex topological vector space Y by considering the behavior of certain families of real-valued functions associated naturally with f. The Pettis integral defined below, the weak substitute for Bochner's integral in the ensuing discussion, depends on the family $\{F\circ f\colon F \in Y^*\}$.

DEFINITION 7.4.1. Let $(\Omega,\mathcal{F},\mu)$ be a probability space and $f\colon \Omega \to X$.

(a) f is Dunford integrable if $F\circ f \in L^1(\mu)$ for each $F \in X^*$.

(b) f is Pettis integrable if f is Dunford integrable and for each $A \in \mathcal{F}$ there is a point $x_A \in X$ such that $F(x_A) = \int_A F\circ f d\mu$ for each $F \in X^*$. If f is Pettis integrable we denote by $(P) - \int_A f d\mu$ the point x_A just described.

(c) For f Pettis integrable let $\|f\|_p \equiv \sup\{\int |F\circ f| d\mu\colon F \in U(X^*)\}$. $\|\cdot\|_p$ is called the Pettis norm.

Let us examine this definition. Thus suppose that $(\Omega,\mathcal{F},\mu)$ is a probability space and that $f\colon \Omega \to X$ is Dunford integrable. For each $A \in \mathcal{F}$ let $\psi_A\colon X^* \to \mathbb{R}$ be the functional $\psi_A(F) \equiv \int_A F\circ f d\mu$. If Range (f) is bounded, it is obvious that $\psi_A \in X^{**}$. (More generally, even without the boundedness assumption on Range (f), $\psi_A \in X^{**}$. This is a consequence of the closed graph theorem. See Diestel and Uhl [1977], Lemma 1, p. 52.) The question is, when is $\psi_A \in X$? To indicate the level of the problem, suppose that $f\colon [0,1] \to c_0$ is the map $f(t) \equiv (n\chi_{(0,1/n]}(t))$ for $t \in [0,1]$. It may be checked that f is Dunford integrable when μ is Lebesgue measure on $\mathcal{B}([0,1])$ but that $\psi_{(0,1]} = (1,1,\ldots) \in c_0^{**}\backslash c_0$. (See Diestel and Uhl [1977], Example 3, p. 53 for further details). Observe that f is not bounded in this example. Indeed, it is essential that f be unbounded when the range is separable in order to produce an example of this sort since separably valued weakly measurable functions are strongly measurable and if f were bounded as well it would be Bochner integrable (by Theorem 1.1.7) and hence Pettis integrable. Moreoever, even if f is not bounded, examples of the sort indicated above may be eliminated by excluding c_0 from the range of f. (More precisely, if $f\colon \Omega \to X$ is Dunford integrable and strongly measurable and if c_0 does not embed in X then f is Pettis integrable. See Diestel and Uhl [1977] , Theorem 7, p. 54.) (Here and elsewhere, by "embed" we mean "linearly homeomorphic to a subspace of".) The most interesting questions, though, concerning the breakdown of Pettis inegrability, occur when there are no "problems at infinity" due to possibly unbounded functions. Thus, at the nex level of subtlety let us add boundedness to the conditions on f.

Say that a Banach space <u>X has the μ-Pettis Integral Property</u> (μ-PIP) if each $f: \Omega \to X$ which is Dunford integrable and has bounded range is Pettis integrab (That is, each bounded weakly measurable f is Pettis integrable.) <u>X has the PIP</u> if it has the μ-PIP for each probability space (Ω,F,μ). (Thus, clearly, reflexiv spaces have the PIP as do separable spaces - cf. Theorem 1.1.7.) Using the continuu hypothesis Phillips [1940a], Example 10.8, showed that $\ell_\infty([0,1])$ fails the μ-PIP where μ is Lebesgue measure on $\mathcal{B}([0,1])$. Other examples of spaces failing the PI have been given more recently which do not depend on the continuum hypothesis: Edgar [1977], §6, Example 2 for $C([0,\omega_1])$ (ω_1 denotes the first uncountable ordinal); Edgar [1980] for $J(\omega_1)$, the long James space - see Theorem 7.7.1 in these notes; Fremlin and Talagrand [1979], Example 2D for the space ℓ_∞. There is an interesting discussion of these and other examples in Edgar [1979], pages 568-69. The example of Fremlin and Talagrand is a source of considerable pathology. For example it shows that the simple functions need not be $\|\cdot\|_P$-dense in the space of Pettis integrable functions. A theorem of Pettis [1938] states that if f is Pettis integrable and $m_f: F \to X$ is <u>the indefinite Pettis integral of f</u> (i.e. if

$$(1) \qquad m_f(A) \equiv (P) - \int_A f d\mu \quad \text{for } A \in F)$$

then m_f is actually a vector measure. The Fremlin-Talagrand example shows that th range of this measure need not be separable. (This is in striking contrast to the situation for the range of a vector measure which arises as the indefinite Bochner integral of a Bochner integrable function on a probability space. In this case the range is relatively norm compact. See Uhl [1969]. See also Diestel and Uhl [1977] Chapter IX.) Before leaving the pathology behind we remark that questions of Pettis integrability sometimes lead to problems of axiomatic independence. Section 3 of Fremlin and Talagrand [1979] is concerned with various questions relating the μ-PI and certain axiomatic systems, while Talagrand [1980] has shown that, under the assumption of Martin's axiom, the range of an X-valued indefinite Pettis integral is relatively compact provided ℓ_∞ is not a quotient of X.

In contrast to the Bochner integral where the probability space domain plays little role, the Pettis integral depends in an important way on both the domain and the range. To elucidate, let us temporarily restrict attention to perfect probability space domains. Recall that a probability space (Ω,F,μ) is perfect if whenever $g: \Omega \to \mathbb{R}$ is μ-measurable there is a Borel subset B of $\mathbb{R}$ such that $B \subset g(\Omega)$ and $g(\mu)(B) = 1$. (Equivalently, whenever $E \subset \mathbb{R}$ and $g^{-1}(E) \in F$ then for each $\varepsilon > 0$ there is a compact subset D of E with $g(\mu)(E) < g(\mu)(D) + \varepsilon$.) See Sazanov [1965], Koumoullis [1981] and Fremlin [1975] for information about such spaces. (By way of example, suppose that $\Omega \subset [0,1]$ is of outer Lebesgue measure one and inner Lebesgue measure zero. Let λ denote Lebesgue measure on $\mathcal{B}([0,1])$. Let $F \equiv \{B \cap \Omega: B \in \mathcal{B}([0,1])\}$ and define μ on F by: $\mu(A) \equiv \lambda(B)$ for any $B \in \mathcal{B}([0,1])$ for which $B \cap \Omega = A \in F$. Then μ is well defined since the λ-outer measure of Ω is one. Thus (Ω,F,μ) is a probability space and evidently the identity map $id: \Omega \to [0,1]$ is F-measurable. Moreover, $id(\mu) = \lambda$. Nevertheless $\lambda(B) = 0$ for each $B \in \mathcal{B}([0,1])$ which is contained in Ω. Consequently (Ω,F,μ) is not perfect.) Many "reasonable" probability spaces are perfect. Radon probability spaces (see Schwartz [1973] for a definition, and then Sazanov [1965], Theorem 10 and Phillips [1940b], Theorem 10) and in particular Borel subsets of completely metrizable separable spaces, Lusin, and more generally, Souslin spaces (all with $F \equiv$ Borel σ-algebra) are perfect for any probability measure; and (Ω,F,μ) for any tight probability measure μ on the σ-algebra F of Borel subsets of the completely regular space Ω is perfect as well. Some of the advantages of having perfect domains are evidenced by the following results, the first of which is due to Stegall (see Fremlim and Talagrand [1979], Theorem 3J), the second to Geitz [1981] (Theorems 3 and 6) and the third to Musial [1980] (Theorem 1 and Corollary 2).

THEOREM 7.4.2. Let (Ω,F,μ) be a perfect probability space, X a Banach space and $f: \Omega \to X$ Pettis integrable. Let m_f be the indefinite Pettis integral of f (see (1)). Then Range (m_f) is a relatively norm compact subset of X.

THEOREM 7.4.3. Let $(\Omega,\mathcal{F},\mu)$ be a perfect probability space and $f\colon \Omega \to X$. Then f is Pettis integrable if and only if there is a sequence $\{f_n\colon n = 1,2,\ldots\}$ of simpl[e] functions on Ω to X such that both

(a) $\{F\circ f_n\colon F \in U(X^*),\ n = 1,2,\ldots\}$ is uniformly integrable; and

(b) for each $F \in X^*$, $\lim_n F\circ f_n = F\circ f$ [a.s.μ].

In this case, $(P)-\int_A f d\mu = \lim_n \int_A f_n d\mu$ in the weak topology on X, for each $A \in \mathcal{F}$.

THEOREM 7.4.4. Let X be a Banach space.

(a) Given any probability space $(\Omega,\mathcal{F},\mu)$ the simple functions are $\|\cdot\|_p$-dense in the space of all Pettis integrable functions on Ω to X if and only if each X-valued measure on $\mathcal{F}$ which has a Pettis integrable derivative with respect to μ, has relatively compact range.

(b) If $(\Omega,\mathcal{F},\mu)$ is a perfect probability space then the simple functions are $\|\cdot\|_p$-dense in the space of Pettis integrable functions.

In yet another attack on the problem of characterizing Pettis integrability, Riddle, Saab and Uhl [1983], Theorem 6, have provided sufficient conditions for Pettis integrability against every tight probability measure.

THEOREM 7.4.5. Let Ω be compact Hausdorff, X separable, and $f\colon \Omega \to X^*$. Assume that f is weakly universally Lusin measurable. (That is, assume that for each $\varepsilon > 0$, $\psi \in X^{**}$ and $\mu \in \mathcal{P}_t(\Omega)$ there is a compact set $D \subset \Omega$ with $\mu(D) > 1-\varepsilon$ and $\psi\circ f|_D$ continuous.) Then if f is bounded it is Pettis integrable with respec[t] to each $\mu \in \mathcal{P}_t(\Omega)$.)

(Note that, by standard arguments, the "weak universally Lusin measurable" hypothesi[s] above is another way of saying that $F\circ f$ is strongly μ-measurable for each $\mu \in \mathcal{P}_t$ and $F \in X^{**}$.)

Recently Geitz [1982] has characterized Pettis integrability in terms of cert[ain] geometrical constructs connected with the range. For any probability space $(\Omega,\mathcal{F},\mu)$, function $f\colon \Omega \to X$, $A \in \mathcal{F}$ and partition $\Pi \equiv \{B_1,\ldots,B_n\}$ of A into finitely ma[ny] subsets each in $\mathcal{F}$ let

$$S(\Pi,A) \equiv \overline{\sum_{i=1}^{n} \mu(B_i)\overline{co}(f(B_i))}.$$

Note that if f is Pettis integrable and $A \in F$, and if Π is as above then by the Hahn-Banach theorem

$$(P) - \int_A f d\mu = \sum_{i=1}^{n} (P) - \int_{B_i} f d\mu \in S(\Pi,A).$$

As Geitz [1982]. Theorem 3.2 points out, the converse is true as well.

THEOREM 7.4.6. Let (Ω,F,μ) be an arbitrary probability space and suppose that $f\colon \Omega \to U(X)$ is weakly measurable. Then f is Pettis integrable if and only if for each $A \in F$ we have $\bigcap_{\Pi} S(\Pi,A) \neq \emptyset$, where the intersection is indexed by the finite partitions Π of A chosen from F.

An elementary observation is in order concerning Theorem 7.4.3. Indeed, suppose that (Ω,F,μ) is a perfect probability space and that $f\colon \Omega \to U(X)$ is weakly measurable. By considering only $2U(X)$-valued collections of simple functions (in which 7.4.3(a) is automatically satisfied) Theorem 7.4.3 may be restated in this case as follows: A bounded function $f\colon \Omega \to X$ is Pettis integrable if and only if there is a uniformly bounded sequence $\{f_n : n = 1,2,\ldots\}$ of simple functions such that $\lim_n F\circ f_n = F\circ f$ [a.s.μ] for each $F \in X^*$. Of course the exceptional set of measure zero on which $\{F\circ f_n : n = 1,2,\ldots\}$ fails to converge to $F\circ f$ may vary with $F \in X^*$. Should this exceptional set of measure zero remain fixed independently of $F \in X^*$ then by Theorem 1.1.7, f is Bochner integrable. Thus one way of viewing the distinction between Bochner integrability and Pettis integrability among weakly measurable bounded functions is to focus on the difference between fixed and roving sets of measure zero. Other examples of this type of distinction will be made later in this section when contrasting the RNP and the WRNP (defined next).

DEFINITION 7.4.7. Let C be a closed convex subset of a Banach space X. Let (Ω,F,μ) be a probability space and say that C has the weak Radon-Nikodým Property for (Ω,F,μ) if whenever $D \subset C$ is closed bounded and convex and $m\colon F \to X$ is a vector measure with $AR(m) \subset D$, there is a μ-Pettis integrable function $f: \Omega \to C$ such that $m(A) = (P) - \int_A f d\mu$ for each $A \in F$. (N.B. We only assume f is C-valued, not that it is D-valued.) C has the WRNP if C has the weak Radon-Nikodým Property for each complete probability space (Ω, F, μ).

Musial [1979] was the first to systematically study the WRNP and versions of several of the basic results on such spaces may be found there. We begin by summarizing a few general properties before turning to the more fully developed subject o dual WRNP spaces. Unreferenced portions of the next theorem come from Musial [1979] (including his "added in proof" remarks). In the rest of this section λ denotes Lebesgue measure on M_λ, the σ-algebra of Lebesgue measurable subsets of $[0,1]$.

THEOREM 7.4.8. (a) Let Y be a separably complementable space, and let X be a subspace of Y. Then X has the WRNP if and only if X has the RNP. (Y is separably complementable if each separable subspace of Y is contained in a separable complemented subspace of Y.) It follows that if X is either weakly compactly generated or $L^1(Q)$ for some finite measure space (Γ, G, Q) then X has the WRNP if and only if it has the RNP. (If $X = L^1(Q)$ has the RNP then it is of the form $\ell_1(T)$ for some set T, as was noted in 3.3.7 and 4.1.9.)

(b) No Banach space containing an isomorphic copy of c_o has the WRNP. (In historical order Musial [1979] showed that the above result holds if c_o is replaced by ℓ_∞, and later Janicka [1980] proved the stronger result for c_o. Anoth proof of the c_o result appears in Ghoussoub and Saab [1981]. Musial [1980] prov the still stronger result that c_o does not embed in any Banach space with the Compact Range Property. See part (g) of this theorem for definitions and the relevance of the CRP).

(c) $L^1(\lambda)$ cannot be embedded in a Banach space with the WRNP. (See Ghoussoub and Saab [1981].)

(d) If X has the WRNP then X has the RNP if and only if whenever (Ω, F, μ) is a complete probability space and $f: \Omega \to X$ is Pettis integrable then there is a $g: \Omega \to X$ which is strongly measurable such that $F \circ f = F \circ g$ [a.s.μ] for each $F \in X$

(e) If X has the WRNP and $Y \subset X$ is a complemented subspace of X then Y has the WRNP.

(f) If Y is a weak* closed linear subspace of a dual space X^* and X^* has the WRNP then Y has the WRNP. (See also Janicka [1980].)

(g) If X has the WRNP then X has the Compact Range Property. (X has the <u>Compact Range Property</u> for (Ω,F,μ) if each μ-continuous, X-valued measure on F of finite variation has relatively compact range. X has the CRP if it has the CRP for each probability space.) See Musial [1980].

(h) X has the WRNP if and only if, given any probability space (Ω,F,μ), directed set $(A,<)$ and bounded X-valued martingale $(f_\alpha,F_\alpha)_{\alpha\in A}$ of Pettis integrable functions then (f_α,F_α) is convergent in $\|\cdot\|_p$ - norm. See Musial [1980].

(i) X has the WRNP if and only if each bounded martingale of X-valued simple functions $(f_n,F_n)_{n=1}^\infty$ on a probability space (Ω,F,μ) is convergent in $\|\cdot\|_p$ - norm. See Musial [1980].

(j) X has the WRNP if and only if it has the WRNP for $([0,1],M_\lambda,\lambda)$. See Musial [1980], Remark 5.

In connection with 7.4.8 parts (e) and (f) we mention that the WRNP is NOT an hereditary property for subspaces in general. Indeed, Lindenstrauss and Stegall [1975] have constructed a separable Banach space X which lacks the RNP and yet, among other properties, X^{**} has the WRNP. (See the discussion of examples following Theorem 7.4.11.) Since X is separable and lacks the RNP, X lacks the WRNP, and hence $\hat{X} \subset X^{**}$ is a subspace of a WRNP space which lacks the WRNP. This rather unsettling state of affairs is due, of course, to the fact that it is hard to pin down the range of a Pettis integrable function. Janicka [1980] discusses several possible range sets for weak Radon-Nikodým derivatives.

The following theorem of Musial [1979] (see "Notes added in proof") and of Janicka [1980] (both of whom based their proofs in large part on results of Rosenthal [1974], Odell and Rosenthal [1975] and Haydon [1976b] concerning Banach spaces containing ℓ_1) shows that - among other things - dual WRNP spaces, like dual RNP spaces, can be cleanly characterized by their preduals.

<u>THEOREM</u> 7.4.9. Let X be a Banach space. Then the following are equivalent:

(a) X contains no isomorphic copy of ℓ_1.

(b) X^* has the WRNP.

(c) Each separable subspace of X has dual possessing the WRNP.

(d) (Odell, Rosenthal; Haydon). Each weak* compact convex subset of X^* is the norm closed convex hull of its extreme points.

(e) (Haydon). If $D \subset X^*$ is weak* compact then

$$w^*cl(co(D)) = \overline{co}(D).$$

Of course the equivalence 7.4.9 (a) $\Leftrightarrow$ (b) paves the way to several other structural characterizations of the WRNP in duals. Here is a sample. (A nice proof of (b) and (c) below appears in Riddle and Uhl [1983].)

THEOREM 7.4.10. The following are equivalent to the statement that X^* has the WRNP.

(a) (Rosenthal [1974]). Each bounded sequence in X has a weakly Cauchy subsequence.

(b) (Pelczynski [1968]). If $T: L^1[0,1] \to X^*$ is a bounded linear operator then T takes weakly compact sets into norm compact sets. (Such an operator is called a Dunford-Pettis operator.)

(c) (Pelczynski [1968]). X^* contains no copy of $L^1[0,1]$. (See also Theorem 7.4.8(c).)

(d) (Haydon [1976b]). Each $\psi \in X^{**}$ is $\mathcal{U}_t(\mathcal{B}(U(X^*),\text{weak}^*) - \mathcal{B}(\mathbb{R})$ measurable.

(e) (Haydon [1976b]). Each $\psi \in X^{**}$ is $\mathcal{U}_t(\mathcal{B}(U(X^*),\text{weak}^*) - \mathcal{B}(\mathbb{R})$ measurable and satisfies the barycentric calculus. ($\psi \in X^{**}$ satisfies the barycentric calculus if given $\mu \in \mathcal{P}_t(U(X^*),\text{weak}^*)$ and if $F_0 \in U(X^*)$ denotes its barycenter, then $\int \psi d\mu = \psi(F_0)$. Of course each $\mu \in \mathcal{P}_t(U(X^*),\text{weak}^*)$ has a barycenter since $(U(X^*),$ weak*) is a compact convex set.)

(f) (Riddle, Uhl [1983]). Each bounded X^*-valued martingale $(F_n,\mathcal{F}_n)$ is Cauchy in the Pettis norm.

(g) (Riddle, Uhl [1983]). For each L^1-bounded uniformly integrable martingale $(f_n,\mathcal{F}_n)$ there is an $f: [0,1] \to X^*$ such that $\lim_n \psi \circ f_n = \psi \circ f$ [a.s.λ] for each $\psi \in X^{**}$. (The exceptional set of measure zero may vary with ψ.)

(h) (Riddle, Uhl [1983]). X^* contains no bounded δ-Rademacher trees for any $\delta > 0$. (A <u>δ-Rademacher tree</u> $\{F_n : n = 1,2,\ldots\}$ in X^* is a tree - so that $\frac{1}{2} F_{2n} + \frac{1}{2} F_{2n+1} = F_n$ for each n - such that

$$\left\| \sum_{i=2^n}^{2^{n+1}-1} (-1)^i F_i \right\| \geqq 2^n \delta \qquad \text{for } n = 1,2,\ldots .$$

If X^* contains a bounded δ-Rademacher tree the martingale giving rise to the tree fails to be Cauchy in Pettis norm.)

Observe that (f), (g) and (h) of the above theorem correspond to the following well established equivalences for a dual X^* to have the RNP:

(f') Each bounded X^*-valued martingale is convergent in $L^1_{X^*}(\lambda)$ (and hence Cauchy in $L^1_{X^*}(\lambda)$). (See Corollary 2.3.7.)

(g') For each uniformly integrable $L^1_{X^*}$-bounded martingale $(f_n, \mathcal{F}_n)$ there is an $f: [0,1] \to X^*$ and a set $E \subset [0,1]$ with $\lambda(E) = 0$ such that $\lim_n \psi \circ f_n(t) = \psi \circ f(t)$ for each $t \in [0,1] \setminus E$. (To see that (g') is equivalent to the RNP in X^* observe that if X^* has the RNP then each $L^1_{X^*}$-bounded martingale converges almost surely by Theorem 2.2.7. For the converse, pick any $U(X^*)$-valued martingale $(f_n, \mathcal{F}_n)$. This martingale is uniformly integrable and the f guaranteed by (g') is weakly measurable and almost separably valued since each f_n is. The Hahn-Banach theorem guarantees that f is almost surely $U(X^*)$-valued and hence Theorem 1.1.7 shows that f is Bochner integrable. Then $E[f \mid \mathcal{F}_n] = f_n$ [a.s.λ] and the proof is completed by appeal to Theorem 1.3.1 and Corollary 2.3.7.)

(h') X^* contains no bounded δ-trees for any $\delta > 0$. (See Theorem 4.4.1.)

Now suppose that X is separable and that X^* has the WRNP. As might well be anticipated, rather strong measurability conclusions are available as is indicated below. (See Musial [1979] for details for (b)-(e) and Bourgain [1980] for (a).)

<u>THEOREM</u> 7.4.11. Let X be a separable Banach space. Then

(a) If X^* has the WRNP, each closed convex subset of X^* with the RNP is separable.

(b) X^* has the WRNP.

(c) Let $f: [0,1] \to X^*$ be weak* measurable. Then f is weakly measurable. (f is <u>weak* measurable</u> if for each $x \in X$ the function $t \to f(t)(x)$ for $t \in [0,1]$ is measurable.)

(d) Let $f: [0,1] \to X^*$ be weak* integrable. Then f is Pettis integrable. (f is <u>weak* integrable</u> if each of the functions $t \to f(t)(x)$ is integrable for $x \in X$.)

(e) X is weak* sequentially dense in X^{**}.

It should be remarked that 7.4.11 (e) $\Rightarrow$(b) is valid without the separability assumption on X, if "weak* sequentially dense" is replaced by "weak* ω_1-sequentially dense". (Let $X_0 \equiv \hat{X} \subset X^{**}$ and for each $\alpha < \omega_1$, the first uncountable ordinal, define $X_\alpha \subset X^{**}$ by induction: if $\alpha = \beta + 1$ let X_α be the set of weak* sequential limits of sequences in X_β. If α is a limit ordinal let $X_\alpha \equiv \bigcup_{\beta<\alpha} X_\beta$. Then <u>$X$ is weak* ω_1-sequentially dense in X^{**}</u> if $X^{**} = \bigcup_{\alpha<\omega_1} X_\alpha$.) In keeping with the spirit of 7.4.11(d) we also remark that for any Banach space X, X^* has the WRNP if and only if whenever (Γ, G, Q) is a complete probability space and $f: \Gamma \to X^*$ is weak* integrable then there is a Pettis integrable function $g: \Gamma \to X^*$ weak* equivalent to f (i.e. $f(\cdot)(x) = g(\cdot)(x)$ [a.s.Q] for each $x \in X$).

Nontrivial examples based on the ℓ_1-characterization of WRNP preduals (7.4.9(a)) may be given. For example, the James tree space JT (James [1974]) contains no copy of ℓ_1. Hence JT* has the WRNP. On the other hand, JT is separable while JT* is not. Hence JT* lacks the RNP by Theorem 4.4.1. (It should be noted that JT is in fact a separable dual; hence JT has the RNP. Even more is true: all the even duals $(JT)^{2n}$ have the RNP. Of course all the odd duals $(JT)^{2n+1}$ lack the RNP since JT* lacks the RNP. Finally, since ℓ_1 does not embed in any of the JT duals, $(JT)^n$ has the WRNP for each $n \geqq 0$. See Lindenstrauss and Stegall [1975] as well as James [1974] for details.) Here is a second startling example. Starting with the James function space Z constructed by Lindenstrauss and Stegall [1975] the space $X \equiv Z \oplus Z^*$ may be shown to have the following properties: all duals of X (excluding X itself) have the WRNP. None of these duals have the RNP. See

Musial [1979] for a discussion of this and other examples.

Recently Riddle, Saab and Uhl [1983] have taken a major step towards the localization of some of the characterizations listed above for the WRNP in dual spaces. There is a very interesting overriding principle in this paper: namely, the existence of a faithful parallel between the already established duality between the GSP and the RNP - cf. Chapter 5 - and the duality between weakly precompact sets and sets with the WRNP. (Recall: a set $A \subset X$ is _weakly precompact_ if each bounded sequence in A has a weakly Cauchy subsequence.) In order to describe their key result, we need a "local definition" of a Dunford-Pettis operator (cf. 7.4.10(b).) Say that a subset $\underline{K}$ of X _is a set of complete continuity_ if whenever (Ω,F,μ) is a probability space and T is a bounded linear operator from $L^1(\mu)$ into X which maps the positive part of the unit sphere, $S(L^1(\mu))^+$, into K, then T is a Dunford-Pettis operator. Compare the following theorem of Riddle, Saab and Uhl [1983] to our Theorems 5.3.11 and 5.3.5.

THEOREM 7.4.12. The following are equivalent for a bounded linear operator $T: X \to Y$.

(a) $T(U(X))$ is weakly precompact.

(b) $T^*(U(Y^*))$ is a set with the WRNP.

(c) $T^*(U(Y^*))$ is a set of complete continuity.

(d) T factors through a Banach space which contains no isomorph of ℓ_1.

(e) T^* factors through a Banach space with the WRNP.

The analogy between Theorem 7.4.12 above and the results of Chapter 5 may be carried even further. Recall that $K \subset X$ has the GSP if and only if (by Stegall's characterization, Theorem 5.5.4) given any $\varepsilon > 0$, probability space (Ω,F,μ) and bounded linear operator $T: X \to L^\infty(\mu)$ there is a set A in F with $\mu(A) < \varepsilon$ such that whenever $\{x_n : n = 1,2,...\}$ is a sequence in K then $\{T(x_n)\chi_{\Omega\setminus A} : n = 1,2,...\}$ in $L^\infty(\mu)$ has a norm convergent subsequence. The analogous result about weakly precompact sets is as follows: $K \subset X$ is weakly precompact if and only if given any $\varepsilon > 0$, probability space (Ω,F,μ) and bounded linear operator $T: X \to L^\infty(\mu)$, if $\{x_n : n = 1,2,...\}$ is a sequence in K then there is an $A \in F$ with $\mu(A) < \varepsilon$ (the set A possibly depends on $\{x_n : n = 1,2,...\}$) such that $\{T(x_n)\chi_{\Omega\setminus A} : n = 1,2,...\}$

in $L^{\infty}(\mu)$ has a norm convergent subsequence. (Note the distinction, once again, between fixed and roving sets, this time of small measure.) This result appears in Riddle, Saab and Uhl [1983] as Theorem 1.

Haydon's results about the universal measurability of elements of the second dual of spaces not containing ℓ_1 (Theorem 7.4.10(d) and (e)) may be localized as well, at some cost in terms of additional assumptions. (See §7.8 for some of the RNP versions of the following results.)

THEOREM 7.4.13. Let X be a Banach space and suppose that $K \subset X^*$ is weak* compact and convex. If either of the following hypotheses is satisfied then conditions (a)-(e) below are equivalent. Hypothesis 1: X is separable. Hypothesis 2: K is absolutely convex (as well as weak* compact and convex).

(a) K has the WRNP.

(b) The identity map $id: (K,\text{weak*}) \to (K,\|\cdot\|)$ is universally weakly measurable. (That is, $\psi \circ id$ is μ-measurable for each $\mu \in P_t(K,\text{weak*})$ and $\psi \in X^{**}$.)

(c) The identity map $id: (K,\text{weak*}) \to (K,\|\cdot\|)$ is Pettis integrable with respect to each $\mu \in P_t(K,\text{weak*})$.

(d) Each $\psi \in X^{**}$ is $U_t(\mathcal{B}(K,\text{weak*})) - \mathcal{B}(\mathbb{R})$ measurable and satisfies the barycentric calculus. (See 7.4.10(e) for definitions.)

(e) For each weak* compact subset D of K and for each $\psi \in X^{**}$, $\psi|_K$: $(K,\text{weak*}) \to \mathbb{R}$ has at least one point of continuity.

This theorem appears scattered throughout §4 of Riddle, Saab and Uhl [1983]. (It is shown there - see Theorem 14 and Lemma 2 - that in Theorem 7.4.13 above, (e) ⇒ (a) and (d) ⇒ (a) even without either additional hypothesis in effect.)

Theorem 7.4.8(d) represents the tip of a fairly large (and growing) body of results concerned with the following type of problem: Let (Ω,F,μ) be a probability space and $f,g: \Omega \to X$. Say that <u>f and g are weakly equivalent</u> (written $f \simeq g$) if $F \circ f = F \circ g$ [a.s.μ] for each $F \in X^*$. (The exceptional set of measure zero may vary with F.) Suppose that f is weakly measurable. The question is: under what conditions is there a strongly measurable function g with $f \simeq g$? To indicate a simple application we mention that if a weakly measurable bounded functic

f is weakly equivalent to a strongly measurable function then f is Pettis inte-agrable. (Indeed, g must be bounded off a set of measure zero and is hence Bochner integrable.)

Perhaps the most important (and the easiest) situation in which the above question may be answered is when f is Pettis integrable and $\overline{co}$(Range (f)) has the RNP. In this case the indefinite Pettis integral m_f of f has σ-finite variation and its Radon-Nikodým derivative g is a Bochner integrable (hence strongly measurable) function with $f \simeq g$. On the other end of the scale, if $f: [0,1] \to L^\infty(\lambda)$ is the function $f(t) \equiv \chi_{[0,t)}$ then f is weakly measurable and bounded. Nevertheless, it is not weakly equivalent to any strongly measurable function since its range is "essentially nonseparable". (See Edgar [1977] for details.) That the question of weak equivalence is somehow tied to the Baire sets of (X,weak) is perhaps not surprising. The first result about problems of this type represents a clean way of making this link explicit. (See Edgar [1977], Theorem 5.2.)

<u>THEOREM</u> 7.4.14. Let (Ω,F,μ) be a probability space and suppose that $f: \Omega \to X$ is weakly measurable. Then f is weakly equivalent to a strongly measurable function $g: \Omega \to X$ if and only if $f(\mu)$ is tight on Baire(X,weak).

A more geometric flavored theorem, due to Uhl [1978], is somewhat stronger.

<u>THEOREM</u> 7.4.15. Let (Ω,F,μ) be a probability space and suppose that $f: \Omega \to X$ is weakly measurable. Then f is weakly equivalent to a strongly measurable function $g: \Omega \to X$ if and only if:

For each $E \in F$ with $\mu(E) > 0$ there is a $B \subset E$, $B \in F$, $\mu(B) > 0$ such that for some weakly compact convex set $K_B \subset X$,

$$F \circ f \leq \sup F(K_B) \quad [a.s.\mu] \quad \text{for each} \quad F \in X^*.$$

(The exceptional set may vary with F.)

A more directly geometric approach has been taken by Geitz [1982]. As usual, suppose that (Ω,F,μ) is a probability space and that $f: \Omega \to X$. For each $E \in F$ define <u>the core of f relative to E</u> by

$$\mathrm{cor}_f(E) \equiv \bigcap_{\substack{A\in\mathcal{F}\\ \mu(A)=0}} \overline{\mathrm{co}}(f(E\setminus A)).$$

Theorem 2.2 of Geitz [1982] asserts that when f is Pettis integrable then $\mathrm{cor}_f(E) = \overline{\mathrm{co}}(AR(m_f|_E))$ where by m_f we mean the indefinite Pettis integral of f. The remainder of the notation is explained in Theorem 2.2.6. The next theorem appears in various places in Geitz [1982]. (For part (d), due to Talagrand, see p. 548 of Geitz [1982].)

THEOREM 7.4.16. Let $(\Omega,\mathcal{F},\mu)$ be a probability space and $f,g\colon \Omega \to X$ two weakly measurable functions.

(a) If $f \simeq g$ then $\mathrm{cor}_f(E) = \mathrm{cor}_g(E)$ for each E.

(b) Suppose that for each $E \in \mathcal{F}$ with $\mu(E) > 0$ we have $\mathrm{cor}_f(E) \neq \emptyset \neq \mathrm{cor}_g(E)$. If $\mathrm{cor}_f(E) = \mathrm{cor}_g(E)$ for each $E \in \mathcal{F}$ then $f \simeq g$.

(c) If g is strongly measurable then $f \simeq g$ if and only if given $\varepsilon > 0$ inside each set $E \in \mathcal{F}$ of positive μ-measure there is a subset $B \in \mathcal{F}$, $\mu(B) > 0$ such that both $\mathrm{cor}_f(B) \neq \emptyset$ and diameter $(\mathrm{cor}_f(B)) < \varepsilon$.

(d) Suppose that f is bounded and that $\mathrm{cor}_f(E) \neq \emptyset$ for each $E \in \mathcal{F}$, $\mu(E) > 0$. Then f is Pettis integrable.

For a bounded X-valued function f on a probability space $(\Omega,\mathcal{F},\mu)$, the family $\mathcal{A} \equiv \{F\circ f\colon F \in U(X^*)\}$ may be studied with an eye towards relating the measurability properties of f with the topological properties of $\mathcal{A}$ as a subset of an appropriate function space. This is the subject of Geitz and Uhl [1981], where a number of very nice results appear. A sample is reproduced here. (The notation and implicit assumptions throughout Theorem 7.4.17 are as just indicated above. In particular, keep in mind that f is bounded.)

THEOREM 7.4.17. (a) If f is Pettis integrable and $\mathcal{A}$ is relatively weakly compact as a subset of $L^\infty(\mu)$ then $f \simeq g$ for some strongly measurable $g\colon \Omega \to X$. (The converse, however, is false.)

(b) Let $\mathbb{B}(\mathcal{F})$ denote the Banach space of bounded real valued measurable functions on $(\Omega,\mathcal{F},\mu)$ with the supremum norm. If f is weakly measurable and $\mathcal{A}$ is relatively weakly compact as a subset of $\mathbb{B}(F)$ then $f \simeq g$ for some strongly measurable $g: \Omega \to X$. (The converse, however, is false in this case as well.)

(c) Suppose that Ω is compact Hausdorff, $F \equiv \mathcal{B}(\Omega)$, the Borel σ-algebra, and that $\mu \in P_t(\Omega)$. If f is weakly measurable then f is strongly measurable if and only if for each $E \in F$ with $\mu(E) > 0$ there is a $B \in F$, $\mu(B) > 0$ such that $B \subset E$ and

$$A_B \equiv \{F \circ (f\chi_B): F \in U(X^*)\}$$

is a separable subset of $\mathbb{B}(F)$.

Finally, we turn to two other connections between the RNP and WRNP. The first is a result of Bourgain [1980] (which, by the way, yields Theorem 7.4.11(a) directly).

<u>THEOREM</u> 7.4.18. Let C be a closed bounded convex subset of a dual Banach space X^*. If C is not weak* dentable (see 3.5.3) then either ℓ_1 embeds in X or C contains a dyadic δ-tree for some $\delta > 0$. Consequently, if X^* has the WRNP then each closed bounded convex subset of X^* with the RNP is weak* dentable.

Here is one consequence of 7.4.18. Observe that if X^* has the RNP then ℓ_1 cannot embed in X (since $\ell_1{}^*$ is not separable) by Theorem 4.4.1. If K is a weak* compact convex subset of X^* then of course K has no δ-trees and hence, according to 7.4.18, K is weak* dentable. Arguments dating back to Namioka and Phelps [1975] (and which may be found in bits and pieces throughout Chapters 4 and 5) show that X must be an Asplund space. Thus, Theorem 7.4.18 above provides another proof of the long elusive verification that if X^* has the RNP then X is Asplund. (The converse implication, X Asplund implies X^* has the RNP, was historically not nearly so troublesome to verify, and may be found in Namioka and Phelps [1975] as well.)

The assumption that a Banach space be a Banach lattice has been used by several authors to great advantage. (See, for example, Lindenstrauss and Tzafriri [1979] and for an indication of some directions recently taken in connection with the subject matter of these notes, see §7.13). We mention here only the following result of Ghoussoub and Saab [1981] which represents an improvement in some earlier result of Lotz.

THEOREM 7.4.19. Let X be a complemented subspace of a Banach lattice. Then X has the WRNP if and only if X has the RNP.

This result follows from the fact that an order continuous Banach lattice is separably complementable, and in such spaces, according to Theorem 7.4.8(a), the WRNP and the RNP are equivalent properties.

§5. Quasi-martingales and the theorem of Bourgain and Phelps revisited.

The proof of Theorem 3.5.4 presented in Chapter 3 is a geometric one whose central idea involves separating a large flat section of a hyperplane from a given point of a convex set in the search for strongly exposing functionals and strongly exposed points. That idea will also appear here, but rather than being surrounded by purely geometric results as it was in Chapter 3, the underlying theme here will be that of quasi-martingales. Their advantages lie in the ease with which it is possible to go from "approximate" bushes to true bushes (via a quasi-martingale decomposition theorem established as Theorem 7.5.3 below). (For example, the "filter" approach used in the proof of Theorems 2.3.6 and 3.7.2 may be supplanted by the use of quasi-martingales.) This section is drawn in large part from Kunen and Rosenthal [1982], though the decomposition Theorem 7.5.3 goes back to others, notably Bellow [1978] and Edgar and Sucheston [1976].

DEFINITION 7.5.1. Let (Ω,F,P) be a probability space, X a Banach space and $\{\varepsilon_n: n = 0,1,\ldots\}$ a sequence of nonnegative numbers with $\sum_{n=0}^{\infty} \varepsilon_n < \infty$. A sequence $(f_n,F_n)_{n=0}^{\infty}$ for which

(a) $F_0 \subset F_1 \subset \cdots \subset F$ is an increasing sequence of σ-algebras;

(b) $f_n: \Omega \to X$ is $F_n - B(X)$ measurable for each n; and

(c) $\|\mathbb{E}[f_{n+1}|F_n] - f_n\|_1 \leq \varepsilon_n$ for each n;

is said to be a quasi-martingale corresponding to $\{\varepsilon_n: n = 0,1,\ldots\}$. If $(f_n,F_n)_{n=0}^{\infty}$ satisfies (a), (b) and (c') $\sum_{n=0}^{\infty} \| \mathbb{E}[f_{n+1} | F_n] - f_n \|_1 < \infty$

(but the sequence $\{\varepsilon_n : n = 0,1,\ldots\}$ is unspecified) then (f_n, F_n) is simply said to be a quasi-martingale.

Suppose that $1 < p \leqq \infty$, that $f: \Omega \to X$ is strongly measurable and that $\|f\|_p < \infty$ where

$$\|f\|_p \equiv \left(\int_\Omega \|f(\omega)\|^p dP(\omega)\right)^{1/p} \quad \text{if } p < \infty \text{ and}$$

$$\|f\|_\infty \equiv \inf_{\substack{A \in F \\ P(A)=0}} \sup_{\Omega \setminus A} \|f(\omega)\| \quad \text{when } p = \infty.$$

Naturally enough, the set of all equivalence classes of such functions (two functions being in the same equivalence class if they agree almost surely) is denoted by $L_X^p(\Omega, F, P)$ and it is a Banach space when equipped with $\|\cdot\|_p$. Then a sequence $(f_n, F_n)_{n=0}^{\infty}$ is called a p-quasi-martingale corresponding to $\{\varepsilon_n : n = 0,1,\ldots\}$ if it satisfies (a), (b) and

(c") $\| \mathbb{E}[f_{n+1} | F_n] - f_n \|_p \leqq \varepsilon_n$ for each n

and, more casually, a p-quasi-martingale if the sequence $\{\varepsilon_n : n = 0,1,\ldots\}$ is unspecified but

(c"') $\sum_{n=0}^{\infty} \| \mathbb{E}[f_{n+1} | F_n] - f_n \|_p < \infty.$

Of course, a martingale is just a quasi-martingale corresponding to the sequence all of whose terms are 0.

The process of going back and forth between tree-like structures and certain types of martingales is described below in sufficient generality to cover all cases needed in these notes. It is based on the natural association described in the proof of Theorem 2.2.1 for bushes. Suppose then that $A \subset \mathbb{N} \times \mathbb{N}$ is a subset such that

(1) The unique element of A with first coordinate 0 is $(0,1)$;

(2) there is a function $F: A \to (\bigcup_{k=1}^{\infty} A^k, \bigcup_{k=1}^{\infty} (0,1)^k)$, sometimes written as the

pair of functions (F_1,F_2) with respective ranges $\cup A^k$ and $\cup(0,1)^k$, which associate with each $(n,i) \in A$ a j-tuple $\{(n+1,i_1),\ldots,(n+1,i_j)\}$ in A^j for some positive integer j along with positive scalars $t_{n+1,i_1},\ldots,t_{n+1,i_j}$ such that $\sum_{k=1}^j t_{n+1,i_k} = 1$;

(3) for each $n \geqq 0$,

$$\bigcup_{\{i:(n,i)\in A\}} F_1(n,i) = \{(n+1,j): (n+1,j) \in A\}$$

and $F_1(n,i_1) \cap F_1(n,i_2) = \emptyset$ if (n,i_1) and (n,i_2) are distinct elements of A.

Let $\{a_{n,i}: (n,i) \in A\}$ be a sequence of points of X, for A as above, and associate a sequence of pairs $(f_n,F_n)_{n=0}^{\infty}$ with $\{a_{n;i}: (n,i) \in A\}$ as follows:

Let f_0 be the function constantly $a_{0,1}$ defined on $[0,1)$. Let $\pi_0 \equiv \{[0,1)\}$ and $F_0 \equiv \{\emptyset,[0,1)\}$. Assume now that π_n is a finite partition of $[0,1)$ into pairwise disjoint half open subintervals, one for each element of A with first coordinate n. (Denote the element of π_n corresponding to (n,i) by $A_{n,i}$.) Suppose that

$$F(n,i) \equiv ((n+1,i_1),(n+1,i_2),\ldots,(n+1,i_j);t_{n+1,i_1},\ldots,t_{n+1,i_j}).$$

Divide $A_{n,i}$ into j half open pairwise disjoint subintervals, the $k^{\underline{th}}$ one of which, A_{n+1,i_k}, has length $(t_{n+1,i_k})\cdot\text{length}(A_{n,i})$. Let f_{n+1} have constant value a_{n+1,i_k} on A_{n+1,i_k} and let π_{n+1} be the collection of all A_{n+1,i_k}'s as (n,i) varies over all pairs in A with first coordinate n. Finally, let F_{n+1} be the σ-algebra generated by π_{n+1}.

Observe that each F_n is a finite σ-algebra and that f_n is F_n-measurable. Without any further restrictions on the points $a_{n,i}$ and the function F used in describing conditions on the indexing set A, little can be said. Consider, however some of the special cases of interest to us:

(a) When $T \equiv \{x_{n,i}: n = 0,1,\ldots;\ 1 \leqq i \leqq 2^n\}$ is a tree in X (see (4) and (5) of §7.1 for this way of describing a tree) then $A = \{(n,i): n = 0,1,\ldots;\ 1 \leqq i \leqq 2^n\}$ and

$F(n,i) \equiv ((n+1,2i-1),(n+1,2i);1/2,1/2)$ for each $(n,i) \in A$.

(b) When B is a bush, A contains (n,i) if and only if, at the $n^{\underline{th}}$ stage there are at least i elements of B. In that case,

$$F(n,i) \equiv ((n+1,i_1),\ldots,(n+1,i_j);t_{n+1,i_1},\ldots,t_{n+1,i_j})$$

means that the elements of the $(n+1)^{\underline{st}}$ stage of the bush into which the point $a_{n,i}$ has been "split up" are, say, $a_{n+1,i_1},\ldots,a_{n+1,i_j}$ and that

$$\sum_{k=1}^{j} t_{n+1,i_k} a_{n+1,i_k} = a_{n,i}.$$

(c) As the final example for now, suppose that a bounded set $D \subset X$ fails to be dentable. Then for some $\varepsilon > 0$, $x \in \overline{co}(D\setminus U_\varepsilon[x])$ for each $x \in D$ and hence, for each such x and for any $\alpha > 0$ there is a convex combination of points of $D\setminus U_\varepsilon[x]$ which is within α of x. Thus, if $\{\varepsilon_n: n = 0,1,\ldots\}$ is a summable sequence of positive numbers, an "approximate" bush inside D can be built very much as in the Maynard construction in Theorem 2.2.1. Indeed, begin with any $x \in D$ at the $0^{\underline{th}}$ stage. If for some $n \geqq 1$ the points of D at stage n have been specified and are here denoted by $x_1,\ldots,x_j$, find j finite collections $E_1,\ldots,E_j$ of points of D such that for each $i = 1,\ldots,j$, $\|x_i - \sum_{y\in E_i} \lambda_y y\| < \varepsilon_n$ for some collection of positive scalars λ_y, $y \in E_i$, with $\sum_{y\in E_i} \lambda_y = 1$, and $\|y - x_i\| \geqq \varepsilon$ for each $y \in E_i$. Let the points of the $(n+1)^{\underline{st}}$ stage be those of $\bigcup_{i=1}^{j} E_i$. This inductive procedure leads to an "approximate" bush, and the sequence of pairs $(f_n,F_n)_{n=0}^{\infty}$ associated with this structure is evidently an ∞-quasi-martingale corresponding to $\{\varepsilon_n: n = 0,1,\ldots\}$.

Before moving on, we formalize the above association for easy reference.

<u>DEFINITION</u> 7.5.2. Let A be a subset of $\mathbb{N} \times \mathbb{N}$ as described in conditions (1), (2) and (3) above, let $\{a_{n,i}: (n,i) \in A\}$ be a subset of the Banach space X, indexed by A, and suppose that $(f_n,F_n)_{n=0}^{\infty}$ is the sequence of pairs constructed

from $\{a_{n,i}: (n,i) \in A\}$ as indicated just previous to examples (a), (b) and (c) above. Then $(f_n, F_n)_{n=0}^{\infty}$ is said to be <u>the function version of</u> $\{a_{n,i}: (n,i) \in A\}$. When $\{a_{n,i}: (n,i) \in A\}$ is a bush as in example (b) above, A is said to be the set of <u>bush indices</u> for $\{a_{n,i}: (n,i) \in A\}$.

With the aid of the following decomposition theorem it will be possible to replace quasi-martingales (as described in example (c) preceding 7.5.2) with a martingale at least when the containing set D is closed bounded and convex. Hence by considering the range of the martingale as described below, we may replace the approximate ε-bush by a true ε'-bush for any $\varepsilon' < \varepsilon$. See Corollary 7.5.6. It is this replacement potential which, when coupled with the ease of producing quasi-martingales, lies at the heart of the advantage of using quasi-martingales over martingales in this geometric context.

<u>THEOREM</u> 7.5.3. Fix (Ω, F, P) and X. Suppose that $1 \leqq p \leqq \infty$, $\{\varepsilon_n: n = 0,1,\ldots\}$ is a sequence of nonnegative numbers with $\sum_{n=0}^{\infty} \varepsilon_n < \infty$ and $(f_n, F_n)_{n=0}^{\infty}$ is a p-quasi-martingale corresponding to $\{\varepsilon_n: n = 0,1,\ldots\}$. Let C be a closed convex subset of X and assume that $f_n(\omega) \in C$ [a.s.P] for each n. Then there is a martingale $(g_n, F_n)_{n=0}^{\infty}$ such that

(a) $g_n(\omega) \in C$ [a.s.P] for each n;

(b) $\|g_n - f_n\|_p \leqq \sum_{j=n}^{\infty} \varepsilon_j$ for all n; and

(c) $\lim_n (f_n - g_n) = 0$ [a.s.P].

Proof. Observe that when $f: \Omega \to C$ is integrable and $G \subset F$ is a sub σ-algebra then $\mathbb{E}[f|G]$ is also C-valued almost surely (by the Hahn Banach theorem). Fix k temporarily and suppose that $k+1 \leqq j$. Then

$$\mathbb{E}[f_j - f_{j-1}|F_k] = \mathbb{E}[\mathbb{E}[f_j - f_{j-1}|F_{j-1}]|F_k] = \mathbb{E}[\mathbb{E}[f_j|F_{j-1}] - f_{j-1}|F_k]$$

and hence

$$\|\mathbb{E}[f_j - f_{j-1}|F_k]\|_p \leqq \|\mathbb{E}[f_j|F_{j-1}] - f_{j-1}\|_p \leqq \varepsilon_{j-1} \quad \text{for each } j \geqq k+1.$$

It follows that if $k \leqq m < n$ then

$$\text{(I)} \qquad \|\mathbb{E}[f_m|F_k] - \mathbb{E}[f_n|F_k]\|_p \leq \sum_{j=m+1}^{n} \|\mathbb{E}[f_j - f_{j-1}|F_k]\|_p \leq \sum_{j=m+1}^{n} \varepsilon_{j-1} \leq \sum_{j=k}^{\infty} \varepsilon_j .$$

Thus $\{\mathbb{E}[f_n|F_k]: n = 0,1,\ldots\}$ is Cauchy in $L_X^p(P)$ and hence converges to some $F_k - B(C)$ measurable function g_k in $\|\cdot\|_p$. Since some subsequence of $\{\mathbb{E}[f_n|F_k]: n = 0,1,\ldots\}$ converges to g_k almost surely (Theorem 1.1.8(f)), g_k is almost surely C-valued. Moreover, since $\mathbb{E}[f_k|F_k] = f_k$, by (I)

$$\|g_k - f_k\|_p = \lim_n \|\mathbb{E}[f_n|F_k] - \mathbb{E}[f_k|F_k]\|_p \leq \sum_{j=k}^{\infty} \varepsilon_j .$$

Of course, $\|g_k - \mathbb{E}[f_n|F_k]\|_1 \leq \|g_k - \mathbb{E}[f_n|F_k]\|_p$ by the Cauchy-Schwarz inequality. Now suppose that $A \in F_k$. Then

$$\int_A g_{k+1}dP = \lim_n \int_A \mathbb{E}[f_n|F_{k+1}]dP = \lim_n \int_A \mathbb{E}[f_n|F_k]dP = \int_A g_k dP$$

and hence $(g_k,F_k)_{k=0}^{\infty}$ is a martingale.

Thus it remains to show that $\lim_n (f_n - g_n) = 0$ [a.s.P] in order to complete the proof of Theorem 7.5.3. A lemma is required. (The notation of 7.5.3 will be used in 7.5.4.)

<u>LEMMA</u> 7.5.4. Let $(h_n,F_n)_{n=0}^{\infty}$ be an X-valued quasi-martingale corresponding to $\{\varepsilon_n: n = 0,1,\ldots\}$ and suppose that $h_\infty \in L_X^1(P)$ satisfies the condition $\lim_n \|h_n - h_\infty\|_1 = 0$. Let

$$h^*(\omega) \equiv \sup\{\|h_j(\omega)\|: j = 0,1,\ldots\} \quad \text{for} \quad \omega \in \Omega.$$

Then

(a) For each $t > 0$, $t\cdot P(\{\omega: h^*(\omega) > t\}) \leq \|h_\infty\|_1 + \sum_{j=0}^{\infty} \varepsilon_j$; and

(b) $\lim_n h_n = h_\infty$ [a.s.P].

Once the lemma is proved, the proof of Theorem 7.5.3 is straightforward. Indeed, let $h_n \equiv f_n - g_n$ for each n and $h_\infty \equiv 0$. From what has already been established, $(f_j - g_n, F_n)_{n=0}^{\infty}$ is a p-quasi-martingale (and hence a 1-quasi-martingale by Cauchy-

Schwarz) corresponding to $\{\varepsilon_n : n = 0,1,\ldots\}$ which converges in $L^1_X(P)$ to h_∞. Then using the lemma, $\lim_n h_n = \lim_n(f_n - g_n) = 0$ [a.s.P] by 7.5.3(b), which establishes 7.5.3(c). Thus the proof of 7.5.3 rests on the proof of 7.5.4(b), which in turn rests on establishing 7.5.4(a).

Proof of 7.5.4. Let $d_0 \equiv h_0$ and for each $n \geqq 1$ let $d_n \equiv h_n - h_{n-1}$. Then $\|\mathbb{E}[d_{j+1}|F_j]\|_1 \leqq \varepsilon_j$ and $d_0 + \cdots + d_j = h_j$ for $j = 0,1,\ldots$. Fix $t > 0$ and define $\tau : \Omega \to \{0,1,\ldots\} \cup \{\infty\}$ by

$$\tau(\omega) \equiv \begin{cases} \infty & \text{if } \|h_j(\omega)\| \leqq t \text{ for each } j = 0,1,\ldots \\ \\ \inf\{j : \|h_j(\omega)\| > t\} & \text{otherwise.} \end{cases}$$

Observe that τ is a stopping time with respect to $\{F_j : j = 0,1,\ldots\}$ (i.e. $\tau^{-1}(j) \in F_j$ for all $j = 0,1,\ldots$) and we define $h_\tau : \Omega \to X$ by

$$h_\tau(\omega) \equiv h_{\tau(\omega)}(\omega) \quad \text{for} \quad \omega \in \Omega.$$

Intuitively, h_τ is the quasi-martingale stopped whenever it leaves $tU(X)$. Let $E_j \equiv \tau^{-1}(j)$ for each $j = 0,1,\ldots$. Then

$$\int_{E_j} \|h_\tau(\omega)\| dP(\omega) = \int_{E_j} \|(d_0 + \cdots + d_j)(\omega)\| dP(\omega)$$

(II)

$$= \lim_n \int_{E_j} \|\mathbb{E}[d_0 + \cdots + d_n | F_j](\omega) - \mathbb{E}[d_{j+1} + \cdots + d_n | F_j](\omega)\| dP(\omega)$$

$$\leqq \lim_n \{\int_{E_j} \|h_n(\omega)\| dP(\omega) + \sum_{k=j+1}^{n} \int_{E_j} \|\mathbb{E}[d_k | F_j](\omega)\| dP(\omega)\}$$

$$= \int_{E_j} \|h_\infty(\omega)\| dP(\omega) + \sum_{k=j+1}^{\infty} \int_{E_j} \|\mathbb{E}[d_k | F_j](\omega)\| dP(\omega).$$

Now for $j < k$, since χ_{E_j} is F_j measurable and $F_j \subset F_{k-1}$,

$$\text{(III)}\quad \begin{aligned}\int_{E_j} \|\mathbb{E}[d_k | F_j](\omega)\| \, dP(\omega) &= \int_\Omega \|\mathbb{E}[\chi_{E_j} d_k | F_j](\omega)\| \, dP(\omega) \\ &\leq \int_\Omega \|\mathbb{E}[\chi_{E_j} d_k | F_{k-1}](\omega)\| \, dP(\omega) = \int_{E_j} \|\mathbb{E}[d_k | F_{k-1}](\omega)\| \, dP(\omega).\end{aligned}$$

Thus from (II) and (II),

$$\text{(IV)}\quad \begin{aligned}&\int_{\{\omega:\tau(\omega)<\infty\}} \|h_\tau(\omega)\| \, dP(\omega) \\ &\leq \int_{\{\omega:\tau(\omega)<\infty\}} \|h_\infty(\omega)\| \, dP(\omega) + \sum_{k=1}^{\infty} \sum_{j<k}^{\infty} \int_{E_j} \|\mathbb{E}[d_k | F_{k-1}](\omega)\| \, dP(\omega) \\ &\leq \int_{\{\omega:\tau(\omega)<\infty\}} \|h_\infty(\omega)\| \, dP(\omega) + \sum_{k=1}^{\infty} \int_\Omega \|\mathbb{E}[d_k | F_{k-1}](\omega)\| \, dP(\omega) \\ &\leq \int_{\{\omega:\tau(\omega)<\infty\}} \|h_\infty(\omega)\| \, dP(\omega) + \sum_{j=0}^{\infty} \varepsilon_j.\end{aligned}$$

It follows, then, from (IV) that

$$\text{(V)}\quad \begin{aligned}\int_\Omega \|h_\tau(\omega)\| \, dP(\omega) &= \int_{\{\omega:\tau(\omega)<\infty\}} \|h_\tau(\omega)\| \, dP(\omega) + \int_{\{\omega:\tau(\omega)=\infty\}} \|h_\tau(\Omega)\| \, dP(\omega) \\ &\leq \int_{\{\omega:\tau(\omega)<\infty\}} \|h_\infty(\omega)\| \, dP(\omega) + \sum_{j=0}^{\infty} \varepsilon_j + \int_{\{\omega:\tau(\omega)=\infty\}} \|h_\infty(\omega)\| \, dP(\omega) \\ &= \|h_\infty(\omega)\|_1 + \sum_{j=0}^{\infty} \varepsilon_j.\end{aligned}$$

If, now, $h^*(\omega) > t$ then both $\tau(\omega) < \infty$ and $\|h_\tau(\omega)\| > t$ and hence from (V)

$$t \cdot P(\{\omega\colon h^*(\omega) > t\}) \leq \int_\Omega \|h_\tau(\omega)\| \, dP(\omega) \leq \|h_\infty\|_1 + \sum_{j=0}^{\infty} \varepsilon_j.$$

This is part (a) of Lemma 7.5.4.

Now let

$$D_t \equiv \{\omega \in \Omega\colon \; \|h_\infty(\omega) - h_m(\omega)\| > t \;\; \text{for infinitely many } m\}.$$

Fix a positive integer n, and suppose that $m > n$. Then since $\lim_j (h_j - h_m) = h_\infty - h_m$ in $L^1_X(P)$, some subsequence of $\{h_j - h_m : j = m, m+1, \ldots\}$ converges almost surely to $h_\infty - h_m$ by Theorem 1.1.8(f). Hence, up to a set of measure zero,

(VI) $$\{\omega : \|h_\infty(\omega) - h_m(\omega)\| > t\} \subset \{\omega : \sup_{j>m} \|(h_j - h_m)(\omega)\| > t\}.$$

On the other hand, suppose that for some $j > m$, $\|(h_j - h_m)(\omega)\| > t$. Then either $\|(h_m - h_n)(\omega)\| > t/2$ or, failing that and for this same j, $\|(h_j - h_n)(\omega)\|$ $(\geqq \|(h_j - h_m)(\omega)\| - \|(h_m - h_n)(\omega)\|) > t/2$. It follows that

(VII) $$\{\omega : \sup_{j>m} \|(h_j - h_m)(\omega)\| > t\} \subset \{\omega : \sup_{j>n} \|(h_j - h_n)(\omega)\| > t/2\}$$

and thus, by combining (VI) and (VII) with the definition of D_t, up to a set of measure zero

(VIII) $$D_t \subset \{\omega : \sup_{j>m} \|(h_j - h_n)(\omega)\| > t/2\}.$$

Define a sequence of functions as follows:

$$H_j \equiv \begin{cases} 0 & \text{if } 0 \leqq j \leqq n \\ h_j - h_n & \text{if } j \geqq n+1 \end{cases}$$

and let $\xi_j \equiv \varepsilon_j$ for $j \geqq n$ and $\xi_j \equiv 0$ for $j < n$. Then $(H_j, F_j)_{j=0}^\infty$ is a quasi-martingale corresponding to $\{\xi_j : j = 0,1,\ldots\}$ since, for $j \geqq n$

$$\begin{aligned} \|\mathbb{E}[H_{j+1} | F_j] - H_j\|_1 &= \|\mathbb{E}[h_{j+1} - h_n | F_j] - (h_j - h_n)\|_1 \\ &= \|\mathbb{E}[h_{j+1} | F_j] - h_j\|_1 \leqq \varepsilon_j. \end{aligned}$$

Let $H_\infty \equiv h_\infty - h_n \in L^1_X(P)$ and observe that $\lim_j \|H_j - H_\infty\|_1 = 0$. Thus Lemma 7.5.4(a) may be applied to the quasi-martingale $(H_j, F_j)_{j=0}^\infty$, the $L^1_X(P)$ function H_∞ and the summable sequence $\{\xi_j : j = 0,1,\ldots\}$ with result

(IX) $$\frac{t}{2}\cdot P(\{\omega: \sup_{j>n}\|(h_j - h_n)(\omega)\| > \frac{t}{2}\}) \leqq \|h_\infty - h_n\|_1 + \sum_{j=n}^{\infty} \varepsilon_j .$$

When (VIII) and (IX) are combined it follows that

(X) $$\frac{t}{2}\cdot P(D_t) \leqq \|h_\infty - h_n\|_1 + \sum_{j=n}^{\infty} \varepsilon_j$$

and since the left side is independent of n while the right side approaches zero as $n \to \infty$, $P(D_t) = 0$. Since $t > 0$ was arbitrary, $\lim_n h_n = h_\infty$ [a.s.P] and Lemma 7.5.4 is proved.

There are some consequences of Theorem 7.5.3 and the construction preceding it which will be useful to record.

<u>COROLLARY</u> 7.5.5. Let $K \subset X$ be a closed bounded convex set. Then the following are equivalent.

(a) K has the RNP.

(b) Each K-valued quasi-martingale converges almost surely.

(c) Each K-valued quasi-martingale converges in L^1_X-norm.

Proof. (b) ⇒ (a) and (c) ⇒ (a) even without the "quasi" so it suffices to prove that (a) ⇒ (b) and (a) ⇒ (c). But the martingale associated with the given K-valued quasi-martingale as in Theorem 7.5.3 is K-valued (by 7.5.3(a)) and hence it converges both almost surely and in L^1_X-norm since K has the RNP. It follows from Theorem 7.5.3 (b) and (c) that the quasi-martingale converges in $L^1_X(P)$ and pointwise almost surely as well.

The next consequence of Theorem 7.5.3 is hardly new (see, for example, Theorem 2.3.6) but it bares repetition if only to emphasize the orderly fashion in which it may now be derived.

<u>COROLLARY</u> 7.5.6. Let K be a closed bounded convex set and suppose that for some $\delta > 0$, $K = \overline{co}(K \backslash U_\delta[x])$ for each $x \in K$. Then K contains a δ'-bush for each $\delta' < \delta$.

Proof. Fix $0 < \delta' < \delta$ and let $\{\varepsilon_j : j = 0,1,\dots\}$ be a sequence of positive numbers such that $\sum_{j=0}^{\infty} \varepsilon_j \leq \frac{1}{2}(\delta-\delta')$. Let $(f_n, F_n)_{n=0}^{\infty}$ be the ∞-quasi-martingale with values in K corresponding to $\{\varepsilon_j : j = 0,1,\dots\}$ which is the function version of the "approximate bush" constructed just prior to Definition 7.5.2. Let $(g_n, F_n)_{n=0}^{\infty}$ be the martingale associated with this quasi-martingale by Theorem 7.5.3. Then from (a) of that theorem, g_n is K-valued for each n, and for each $t \in [0,1)$

$$\|g_{n+1}(t) - g_n(t)\| \geq \|f_{n+1}(t) - f_n(t)\| - \|f_{n+1}(t) - g_{n+1}(t)\| - \|f_n(t) - g_n(t)\|$$

$$\geq \delta - \sum_{j=n+1}^{\infty} \varepsilon_j - \sum_{j=n}^{\infty} \varepsilon_j \geq \delta'.$$

Finally, since each g_n is F_n-measurable and, by construction, F_n is a finite σ-algebra, each g_n has finite range and may be thought of as the $n^{\underline{th}}$ stage of the δ'-bush $\bigcup_{n=0}^{\infty}$ Range g_n inside K.

The last consequence of Theorem 7.5.3 listed here will play an important role i the proof of the main theorem of this section (see Theorem 7.5.9).

COROLLARY 7.5.7. Fix (Ω,F,P), let K be a closed bounded convex subset of X a assume that K has the RNP. Let $(f_n, F_n)_{n=0}^{\infty}$ be an X-valued quasi-martingale suc that

(1) $\sup\{\|f_n(\omega)\| : \omega \in \Omega;\ n = 0,1,\dots\} \equiv M < \infty$;

(2) $f_0(\omega) \in K$ for all $\omega \in \Omega$; and

(3) if for some $\omega \in \Omega$ and some n, $f_n(\omega) \notin K$ then $f_{n+1}(\omega) = f_n(\omega)$.

Then $\lim_n f_n(\omega)$ exists [a.s.P].

(N.B. An interpretation of 7.5.7 is that if (g_n, G_n) is a uniformly bounded X-valued quasi-martingale which starts in a closed bounded convex set with the RNP an stops whenever it leaves, then it converges almost surely.)

Proof. Without loss of generality assume $0 \in K$. Suppose that $(f_n, F_n)_{n=0}^{\infty}$ i as in the statement of 7.5.7 and that it corresponds to the summable sequence $\{\varepsilon_n : n = 0,1,\dots\}$. For each n let

$$D_n \equiv \{\omega \in \Omega: f_n(\omega) \in K\}.$$

Then $D_n \supset D_{n+1}$ and $D_n \in F_n$ for each n. Moreover, $\sum_{n=0}^{\infty} P(D_n \setminus D_{n+1}) \leq 1$. Now, for any $n \geq 0$

$$\|\mathbb{E}[f_{n+1}\chi_{D_{n+1}} \mid F_n] - f_n\chi_{D_n}\|_1$$

$$\leq \|\mathbb{E}[f_{n+1}\chi_{D_n} \mid F_n] - f_n\chi_{D_n}\|_1 + \|\mathbb{E}[f_{n+1}\chi_{D_{n+1}} \mid F_n] - \mathbb{E}[f_{n+1}\chi_{D_n} \mid F_n]\|_1$$

$$= \|(\mathbb{E}[f_{n+1} \mid F_n] - f_n)\chi_{D_n}\|_1 + \|\mathbb{E}[f_{n+1}(\chi_{D_{n+1}} - \chi_{D_n}) \mid F_n]\|_1$$

$$\leq \|\mathbb{E}[f_{n+1} \mid F_n] - f_n\|_1 + \|f_{n+1}(\chi_{D_{n+1}} - \chi_{D_n})\|_1$$

$$\leq \varepsilon_n + M \cdot P(D_n \setminus D_{n+1})$$

and thus

$$\sum_{n=0}^{\infty} \|\mathbb{E}[f_{n+1}\chi_{D_{n+1}} \mid F_n] - f_n\chi_{D_n}\|_1 = \sum_{n=0}^{\infty} \varepsilon_n + M < \infty.$$

That is, $(f_n\chi_{D_n}, F_n)_{n=0}^{\infty}$ is a quasi-martingale. Since $0 \in K$, it is K-valued whence, by Corollary 7.5.5, it converges almost surely. On the other hand, the sequence $(f_n\chi_{\Omega\setminus D_n}, F_n)_{n=0}^{\infty}$ is pointwise eventually constant by definition and hence it certainly converges almost surely. The Corollary follows.

In order to tackle the Bourgain-Phelps Theorem 3.5.4 we establish an intermediate result first which lends itself naturally to (quasi-) martingale techniques. See the diagram.

<u>LEMMA</u> 7.5.8. Let K be a closed bounded convex subset of $U(X)$ and assume that K has the RNP. Let $f \in S(X^*)$ and suppose that $\alpha > 0$. Then there is a point $x_o \in S(K,f,\frac{\alpha}{2})$ such that $x_o \notin \overline{co}(V \cup (K \setminus U_{\alpha/2}[x_o]))$ where V denotes the "slab"

$$V \equiv \{x \in U_1[x_o]: f(x) = M(K,f) - \alpha\}.$$

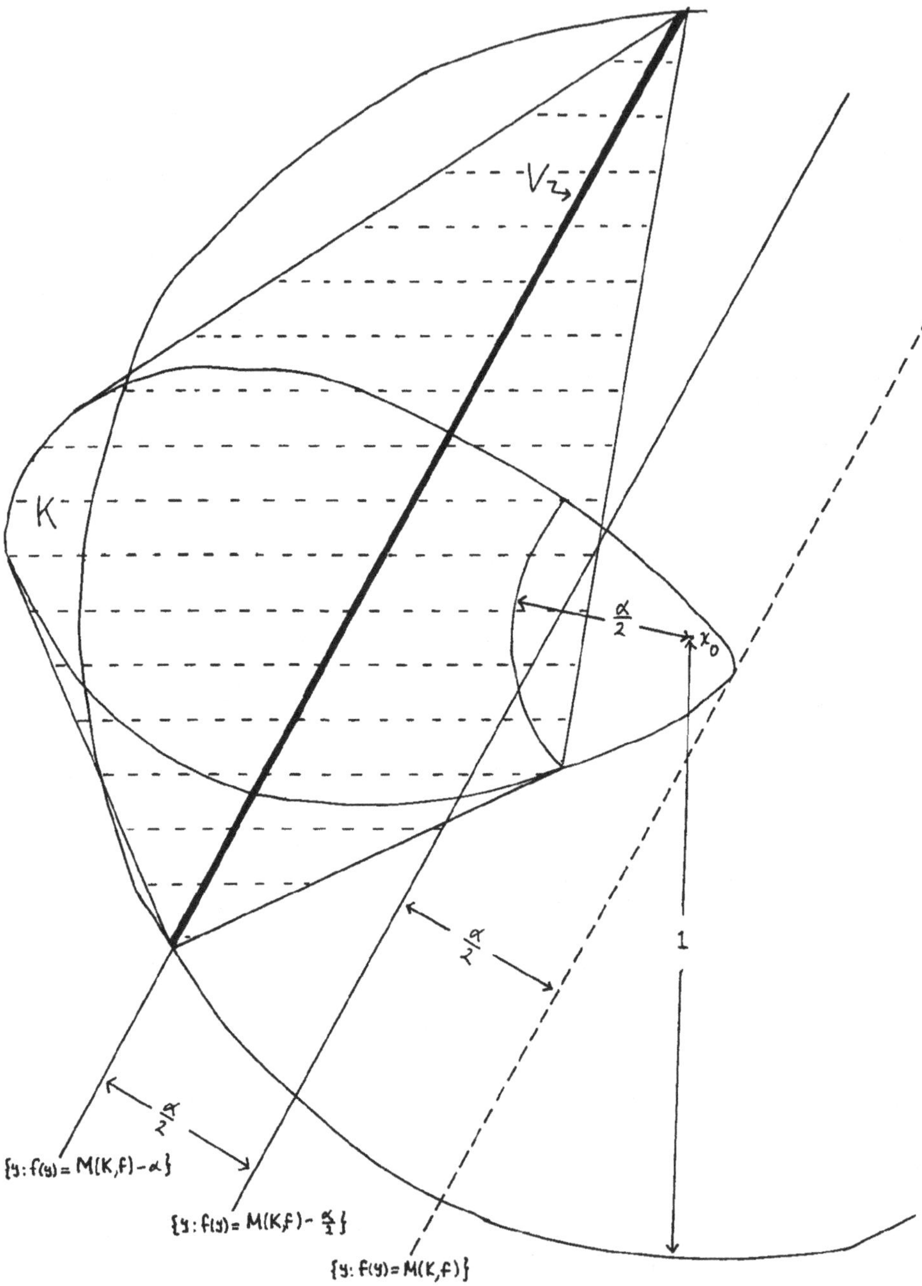

SHADED AREA = $\overline{co}(V \cup (K \setminus U_{\alpha/2}[x_0]))$

DIAGRAM FOR LEMMA 7.5.8.

Proof. Suppose that $K \subset U(X)$, $f \in S(X^*)$ and $\alpha > 0$ have been chosen so that the conclusion of the lemma were false. Pick any $x_o \in K$ for which $f(x_o) > M(K,f) - \frac{\alpha}{2}$. Then x_o is almost a convex combination of elements $x_1,\ldots,x_n$ of X such that for each $i = 1,\ldots,n$ either $x_i \in K$ and $\|x_i - x_o\| \geqq \frac{\alpha}{2}$, or, $f(x_i) = M(K,f) - \alpha$ and $\|x_i - x_o\| \leqq 1$. If $f(x_i) \leqq M(K,f) - \frac{\alpha}{2}$ stop this process, but if $f(x_i) > M(K,f) - \frac{\alpha}{2}$ repeat it with x_i in place of x_o. The procedure just outlined is highly reminiscent of that described in the last example preceding Definition 7.5.2. The precise argument follows.

At the 0th stage pick any $x_o \in K$ such that $f(x_o) > M(K,f) - \frac{\alpha}{2}$ and choose a sequence $\{\varepsilon_n : n = 0,1,\ldots\}$ of positive numbers such that

$$\text{(XI)} \qquad M(K,f) - \frac{\alpha}{2} < f(x_o) - \sum_{j=0}^{\infty} \varepsilon_j .$$

Let $\Omega \equiv [0,1)$, $P \equiv$ Lebesgue measure on Ω, denote by π_o the trivial partition $\{[0,1)\}$ and by $\mathcal{F}_o$ the trivial σ-algebra it generates. Let $F_o : \Omega \to X$ be the constant function: $F_o(t) = x_o$ for $t \in [0,1)$. Now suppose that $n \geqq 0$, that finite partitions $\pi_o,\ldots,\pi_n$ of $[0,1)$ into pairwise disjoint half-open intervals, σ-algebras $\mathcal{F}_o,\ldots,\mathcal{F}_n$ and functions $F_o,\ldots,F_n$ from Ω to X have all been constructed so that for each i, $\mathcal{F}_i$ is the σ-algebra generated by π_i, f_i is $\mathcal{F}_i$-measurable, and for any $j \leqq n-1$ and any $t \in [0,1)$

(A) if $f(F_j(t)) > M(K,f) - \frac{\alpha}{2}$ then $F_j(t) \in K$ and either

(a) $F_{j+1}(t) \in K$ and $\|F_{j+1}(t) - F_j(t)\| \geqq \frac{\alpha}{2}$

or

(b) $f(F_{j+1}(t)) = M(K,f) - \alpha$ and $\|F_{j+1}(t) - F_j(t)\| \leqq 1$;

(B) if $f(F_j(t)) \leqq M(K,f) - \frac{\alpha}{2}$ then $F_{j+1}(t) = F_j(t)$;

(C) $\|\mathbb{E}[F_{j+1} | \mathcal{F}_j] - F_j\|_\infty < \varepsilon_j$.

The construction of F_{n+1}, π_{n+1} and $\mathcal{F}_{n+1}$ is by now routine: F_n is constant on the atoms of $\mathcal{F}_n$ – i.e., on the elements of π_n. Let $B \in \pi_n$ and let $t \in B$. Let $F_{n+t}(t) \equiv F_n(t)$ if $f(F_n(t)) \leqq M(K,f) - \frac{\alpha}{2}$. If, however, $f(F_n(t)) > M(K,f) - \frac{\alpha}{2}$

then it is easy to check from (A) and (B) that $F_n(t) \in K$. Thus by hypothesis, $F_n(t)$ is within ε_n of a convex combination $\sum_{i=1}^k \beta_i y_i$ of points each of which is either a point of K of distance at least $\frac{\alpha}{2}$ from $F_n(t)$ or is a point of distance at most one from $F_n(t)$ on the hyperplane $\{y: f(y) = M(K,f) - \alpha\}$. We assume for convenience that each β_i in the above summation is strictly positive. Let $B_1,\ldots,B_k$ be a partition of B into pairwise disjoint half-open subintervals, the $i^{\underline{th}}$ one of which has length $\beta_i P(B)$ and define F_{n+1} on B by $F_{n+1}(s) \equiv y_i$ if $s \in B_i$. Then let π_{n+1} be the collection of all such new half-open subintervals chosen as B ranges over π_n, let $\mathcal{F}_{n+1}$ denote the σ-algebra generated by π_{n+1} and observe that (A), (B) and (C) hold for $j \leqq n$ and all $t \in [0,1)$. Since the quasi-martingale $(F_n,\mathcal{F}_n)_{n=0}^\infty$ originates in K, always stays within $\overline{co}(K + U(X))$, a bounded set, and stops whenever it leaves K, Corollary 7.5.7 applies and hence $F(t) \equiv \lim_n F_n(t)$ exists [a.s.P]. Note that if $f(F_j(t)) > M(K,f) - \frac{\alpha}{2}$ then either $F_{j+1}(t) \in K$ and $\|F_{n+1}(t) - F_j(t)\| \geqq \frac{\alpha}{2}$ or $f(F_{j+1}(t)) = M(K,f) - \alpha$ in which case

$$\frac{\alpha}{2} < f(F_j(t)) - f(F_{j+1}(t)) \leqq \|F_{j+1}(t) - F_j(t)\|$$

so that

(XII) $$\text{if } f(F_j(t)) > M(K,f) - \frac{\alpha}{2} \text{ then } \|F_{j+1}(t) - F_j(t)\| \geqq \frac{\alpha}{2}.$$

It follows that the only way $\lim_n F_n(t)$ can exist is for $f(F_n(t)) \leqq M(K,f) - \frac{\alpha}{2}$ for sufficiently large n. Thus $f(F(t)) \leqq M(K,f) - \frac{\alpha}{2}$ [a.s.P] and consequently

(XIII) $$f\left(\int_{[0,1)} F\,dP\right) = \int_{[0,1)} f\circ F\,dP \leqq M(K,f) - \frac{\alpha}{2}.$$

On the other hand, from the decomposition Theorem 7.5.3 there is a martingale $(G_n,\mathcal{F}_n)_{n=0}^\infty$ such that $\|G_n - F_n\|_1 \leqq \sum_{j=n}^\infty \varepsilon_j$ for each $n = 0,1,\ldots$ and hence (since $\int F_0\,dP = x_0$)

$$\left\|\int_{[0,1)} F_n dP - x_o\right\|$$

$$\text{(XIV)} \quad \leqq \left\|\int_{[0,1)} (F_n - G_n)dP\right\| + \left\|\int_{[0,1)} (G_n - G_o)dP\right\| + \left\|\int_{[0,1)} (G_o - F_o)dP\right\|$$

$$\leqq \sum_{j=n}^{\infty} \varepsilon_j + 0 + \sum_{j=0}^{\infty} \varepsilon_j .$$

Since $\int_{[0,1)} FdP = \lim_n \int_{[0,1)} F_n dP$ it follows from (XIV) that

$$\text{(XV)} \qquad \left\|\int_{[0,1)} FdP - x_o\right\| \leqq \sum_{j=0}^{\infty} \varepsilon_j .$$

But this is impossible since by (XIII) and (XIV)

$$f(x_o) \leqq f\left(\int_{[0,1)} FdP\right) + \sum_{j=0}^{\infty} \varepsilon_j \leqq M(K,f) - \frac{\alpha}{2} + \sum_{j=0}^{\infty} \varepsilon_j$$

which contradicts the choice of $\{\varepsilon_n : n = 0,1,\ldots\}$ in (XI). This completes the proof of Lemma 7.6.8.

<u>THEOREM</u> 7.5.9. (Theorem 3.5.4 of Bourgain, Phelps). Let K be a closed bounded convex subset of a Banach space X and suppose that K has the RNP. Then $K = \overline{co}(\text{str exp}(K))$ and $SE(K)$ is a dense G_δ subset of X^*.

Proof. It evidently suffices to consider $K \subset U(X)$. If K is a singleton the theorem is trivial, and we suppose henceforth that it is not. Let $f \in S(X^*)$ and pick any $\delta > 0$. We will produce $g \in S(X^*)$ such that both $\|f - g\| < \delta$ and g determines a slice of K of diameter less than δ. Once that is accomplished, appeal is made to Bishop's Lemma 3.2.7 which states that any such K satisfies the conclusions of the Bourgain Phelps theorem. If f is constant on K replace it by one which is not (which is easily done without endangering the desired conclusion since the functionals which are not constant on K are dense in $S(X^*)$ by the Hahn Banach theorem). Choose any α such that

(XVI) $$0 < \alpha < \min\{\frac{\alpha}{2}, \frac{1}{10}(\sup f(K) - \inf f(K))\}.$$

According to Lemma 7.5.8 there is an $x_0 \in S(K,f,\alpha/2)$ such that

$$x_0 \notin C \equiv \overline{co}(V \cup (K\backslash U_{\alpha/2}[x_0])) \quad \text{where} \quad V \equiv \{x \in U_1[x_0]: f(x) = M(K,f) - \alpha\}.$$

Choose any $g \in S(X^*)$ which strictly separates x_0 from C. That is, choose g so that $\sup g(C) < g(x_0)$. Then for some $\beta > 0$, $x_0 \in S(K,g,\beta)$ and $C \cap S(K,g,\beta) = \emptyset$. Hence

(XVII) $$\text{If} \quad x \in S(K,g,\beta) \quad \text{then} \quad \|x - x_0\| < \frac{\alpha}{2} < \frac{\delta}{4}$$

and consequently diameter $S(K,g,\beta) \leqq \frac{\delta}{2} < \delta$. That is, g does determine slices of K of diameter less than δ, and it remains to establish that $\|f - g\| < \delta$.

Since $\sup g(V) < g(x_0)$, g separates 0 from

$$V - x_0 \equiv \{x \in U(X): f(x) = M(K,f) - \alpha - f(x_0)\}.$$

Suppose that $y \in U(X)$. If $f(y) < -\alpha$ then certainly $f(y) < M(K,f) - f(x_0) - \alpha$ so there is a t, $0 < t < 1$ such that $f(ty) = M(K,f) - f(x_0) - \alpha$ and hence $ty \in V - x_0$. Consequently $g(ty) < g(0) = 0$. It follows that $g(y) < 0$ whenever $f(y) < -\alpha$ for $y \in U(X)$ or, otherwise stated, if $y \in U(X)$ and $g(y) \geqq 0$ then $f(y) \geqq -\alpha$. In particular, if $x \in U(X)$ and $g(x) = 0$ then both $f(x) \geqq -\alpha$ and $f(-x) \geqq -\alpha$ so that

(XVIII) $$\text{If} \quad x \in U(X) \quad \text{and} \quad g(x) = 0 \quad \text{then} \quad |f(x)| \leqq \alpha.$$

According to the Bishop Phelps Lemma 3.3.2, either $\|f - g\| \leqq 2\alpha$ or $\|f + g\| \leqq 2\alpha$ follows from (XVIII) since $\|f\| = \|g\| = 1$. As the following argument demonstrates, $\|f + g\| \leqq 2\alpha$ is impossible whence $\|f - g\| \leqq 2\alpha < \delta$, which will complete the proof of the theorem.

Assume for now that $\|f + g\| \leqq 2\alpha$. Since $K \subset U(X)$, for any $x \in K$

(XIX) $$-2\alpha \leqq f(x) - -g(x) \leqq 2\alpha$$

and hence $-2\alpha \leqq M(K,f) - -g(x)$ for each $x \in K$. It follows that

(XX) $$-2\alpha \leqq M(K,f) - M(K,-g).$$

Thus from (XIX) (with x_o for x) and (XX) we obtain

(XXI) $$-2\alpha + (g(x_o) - 2\alpha) - \frac{\alpha}{2} \leqq -2\alpha - f(x_o) - \frac{\alpha}{2} \leqq -2\alpha - M(K,f)$$

$$\leqq -M(K,-g) = \inf g(K).$$

That is,

(XXII) $$-\frac{9}{2}\alpha \leqq \inf g(K) - g(x_o) \leqq 0.$$

Since $\|g\| = 1$ and $M(K,g) = M(S(K,g,\beta),g)$, (XVII) shows that

(XXIII) $$g(x_o) \geqq M(K,g) - \frac{\alpha}{2}$$

and thus, when combined with (XXII), (XXIII) yields

$$M(K,g) - \frac{\alpha}{2} \leqq g(x_o) \leqq \inf g(K) + \frac{9}{2}\alpha$$

or, in other words,

$$\sup g(K) - \inf g(K) \leqq 5\alpha.$$

Consequently, using the fact that $\|f+g\| \leqq 2\alpha$ we obtain

$$\sup f(K) - \inf f(K) \leqq \sup g(K) - \inf g(K) + 4\alpha \leqq 9\alpha$$

which contradicts the choice of α in (XVI) and proves that $\|f-g\| \leqq 2\alpha$ after all. This completes the proof of the Bourgain Phelps theorem.

§6. Approximations of trees, bushes and extreme points.

A closed bounded convex set $K \subset X$ lacks the RNP if and only if it contains an ε-bush for some $\varepsilon > 0$ (cf. Theorem 2.3.6). As the Bourgain-Rosenthal example (see §7.3, and in particular, Theorem 7.3.2(c)) emphasizes, an important distinction must be made between ε-bushes of arbitrary "growth rate" and those of growth rate limited by a given sequence of positive integers $\{q_n: n = 1,2,\ldots\}$. In particular, in the simplest case dealt with by Bourgain and Rosenthal (in which $q_n = 2^n$ for each n) o must be aware of the distinction between ε-bushes and ε-trees, since there are no bounded ε-trees inside their space for any $\varepsilon > 0$ in spite of the fact that it lac the RNP. This difference may be more systematically explored with the aid of the fo lowing terminology (see Kunen and Rosenthal [1982] for (b) and (d) below).

DEFINITION 7.6.1. Let $(\Omega,\mathcal{F},P)$ be a probability space and $\{\mathcal{F}_n: n = 0,1...\}$ an in creasing sequence of finite sub σ-algebras of $\mathcal{F}$. Assume that $\mathcal{F}_0 = \{\emptyset,\Omega\}$, and tha if $P(A) = 0$ for some $A \in \bigcup_{n=0}^{\infty} \mathcal{F}_n$ then $A = \emptyset$.

(a) Then $\{\mathcal{F}_n: n = 0,1,...\}$ is said to be a bush decomposition.

(b) $\{\mathcal{F}_n: n = 0,1,...\}$ is a generalized dyadic decomposition if in addition, each $\mathcal{F}_n$ contains precisely 2^n atoms.

(c) The standard dyadic decomposition has $\Omega \equiv [0,1)$, $P \equiv$ Lebesque measure or $\mathcal{F} \equiv \mathcal{B}([0,1))$ and the nth σ-algebra (denoted by $\mathcal{D}_n$ instead of $\mathcal{F}_n$) is generated by $\{[\frac{j-1}{2^n},\frac{j}{2^n}): j = 1,...,2^n\}$.

(d) A quasi-martingale $(f_n,\mathcal{F}_n)_{n=0}^{\infty}$ is dyadic (respectively bush, generalize dyadic, standard dyadic) if it is a quasi-martingale with respect to a dyadic (respectively bush, generalized dyadic, standard dyadic) decomposition.

Before turning to the main subject matter of this section, we mention a result which points up just how much room there is between ε-trees and ε-bushes.

PROPOSITION 7.6.2. Let $K \subset X$ be a closed bounded convex set and let λ denote Le besgue measure on $\mathcal{B}([0,1))$. Then the following are equivalent.

(a) K lacks the RNP.

(b) There are a K-valued bush martingale $(f_n,\mathcal{F}_n)_{n=0}^{\infty}$ on $([0,1),\mathcal{B}([0,1),\lambda)$ and a $\delta > 0$ such that $\|f_{n+1}(t)-f_n(t)\| \geq \delta$ for each $t \in [0,1)$ and each $n = 0,1...$.

(c) There are a K-valued standard dyadic martingale $(G_n,\mathcal{D}_n)_{n=0}^{\infty}$ and a $\delta >$ such that $\lim_n \sup \|G_{n+1}-G_n\|_1 \geqq \delta$.

(N.B. It follows immediately from this proposition that if K is a closed bounded convex set such that $\lim_n \|G_{n+1}-G_n\|_1 = 0$ for each K-valued standard dyadic martin

$(G_n, \mathcal{D}_n)$ then K has the RNP. Nevertheless, it is possible that K lacks the RNP and yet $\lim_n \inf \|G_{n+1}-G_n\|_1 = 0$ for each K-valued standard dyadic martingale as is the case for $K = U(X)$ where X is one of the spaces constructed by Bourgain and Rosenthal—see §7.3. Hence by part (c) of the above proposition, "$\lim_n \inf \|G_{n+1}-G_n\|_1 = 0$" cannot be strengthened to "$\lim_n \|G_{n+1}-G_n\|_1 = 0$" in Theorem 7.3.2(c).)

Proof of 7.6.2. (a) and (b) are equivalent by Theorem 2.3.6 and certainly if K contains the range of a martingale as in (c) of the statement of the Proposition then K does not have the MCP (and hence K fails to have the RNP). Thus (c) ⇒ (a).

(a) ⇒ (c) Without loss of generality assume $K \subset U(X)$. By virtue of Theorem 2.2.2 there is a bush martingale $(f_n, F_n)_{n=0}^{\infty}$ on $([0,1), \mathcal{B}([0,1)), \lambda)$ such that for some δ, $0 < \delta < \frac{1}{2}$ and each $n = 0,1,\ldots$ both

(1) $\|f_{n+1}(t) - f_n(t)\| \geq 2\delta$ for each $t \in [0,1)$; and

(2) each atom of F_n is a half-open subinterval of $[0,1)$.

If π_n denotes the set of atoms of F_n and if for each $A \in \pi_n$, x_A denotes the value of f_n on A, then $f_n = \sum_{A\in\pi_n} x_A \chi_A$. Then by a straightforward inductive argument, choose a partition of $[0,1)$ into pairwise disjoint half-open intervals, say $\{I_A : A \in \pi_n\}$, such that for each n (and with G_n the algebra generated by $\{I_A : A \in \pi_n\}$) we have

(3) $\lambda(I_A)$ is a dyadic rational for each $A \in \pi_n$ so that each $\lambda(A)$ is of the form $\frac{k}{2^j}$ for positive integers k and j, $k \leq 2^j$;

(4) $\sum_{A\in\pi_n} \lambda(I_A \,\Delta\, A) < \frac{\delta}{16} 2^{-n}$; and

(5) $\|\mathbb{E}[f_{n+1}|G_n] - f_n\|_1 < \frac{\delta}{16} 2^{-n}$.

Let $h_n \equiv \sum_{A\in I_n} x_A \chi_{I_A}$. Then since $K \subset U(X)$, by (4),

(6) $$\|h_n - f_n\|_1 \leq 2 \sum_{A\in\pi_n} \lambda(I_A \,\Delta\, A) < \frac{\delta}{8} 2^{-n}$$

and hence, from (5) and (6),

$$\|\mathbb{E}[h_{n+1}|G_n] - h_n\|_1$$
$$\leq \|\mathbb{E}[h_{n+1}|G_n] - \mathbb{E}[f_{n+1}|G_n]\|_1 + \|\mathbb{E}[f_{n+1}|G_n] - f_n\|_1 + \|f_n - h_n\|_1$$
$$\leq \|h_{n+1} - f_{n+1}\|_1 + \frac{\delta}{16}2^{-n} + \frac{\delta}{8}2^{-n} \leq \frac{\delta}{8}2^{-(n+1)} + \frac{3}{16}2^{-n}$$
$$= \frac{\delta}{4}2^{-n}.$$

Thus $(h_n, G_n)_{n=0}^{\infty}$ is a quasi-martingale corresponding to $\{\frac{\delta}{4}2^{-n}: n = 0,1,\ldots\}$. (See Definition 7.5.1.) Consequently, either by the "filtering down" method of Theorems 2.3.6 and 3.7.2 or by the decomposition Theorem 7.5.3, there is a K-valued martingale $(g_n, G_n)_{n=0}^{\infty}$ (since each h_n is K-valued) such that

(7) $$\|g_n - h_n\|_1 \leq \sum_{j=n}^{\infty} \frac{\delta}{4}2^{-n} \quad \text{for each } n.$$

It follows from (1) and (7) that for each n

$$\|g_{n+1} - g_n\|_1 \geq \|h_{n+1} - h_n\|_1 - \|g_{n+1} - h_{n+1}\|_1 - \|g_n - h_n\|_1$$
$$\geq 2\delta - \sum_{j=n+1}^{\infty} \frac{\delta}{4}2^{-n} - \sum_{j=n}^{\infty} \frac{\delta}{4}2^{-n} > \delta.$$

Finally, choose a strictly increasing sequence of positive integers $\{k_n: n = 1,2,\ldots\}$ such that for each n, $\lambda(I_A)$ is a positive integer multiple of 2^{-k_n} for each $A \in \pi_n$. Let $G_0 \equiv g_0$. For each positive integer i let n be chosen such that $k_n \leq i < k_{n+1}$ and let $G_i \equiv g_n$. Then G_n is measurable with respect to $\mathcal{D}_n$ and for each n

$$\|G_{k_{n+1}} - G_{k_n}\|_1 = \|g_{n+1} - g_n\|_1 \geq \delta.$$

This completes the proof.

There are at least two obvious ways of putting a "δ-separation" type condition on a tree. Indeed, suppose that a point x of the nth stage is divided into two points, say y and z, at stage $n+1$. The two methods are, then

(A) $\|y - x\| \geq \delta$ and $\|z - x\| \geq \delta$; or

(B) $\|y - z\| \geq 2\delta$.

Clearly (A) ⇒ (B) but not visa versa. We know that "not having the RNP" and "having a bounded generalized dyadic martingale with a δ-separations condition of type (A)" are not equivalent (by virtue of the Bourgain-Rosenthal example of Theorem 7.3.2) but with proper care, type (B) separated trees may be found precisely when the RNP is missing. The exact specifications for the type (B) separation conditions are set fort in the next definitions due to Ho [1979], and her characterization of the RNP follows in Theorem 7.6.5. As in §7.1 it will often be convenient in what follows to think of a tree as a doubly indexed sequence $\{x_{n,i}: (n,i) \in I\}$ where I denotes the set

$$I \equiv \{(n,i) \in \mathbb{N}\times\mathbb{N}: n = 0,1,\ldots;\ \ 1 \leq i \leq 2^n\} .$$

DEFINITION 7.6.3. Let $\{x_{n,i}: (n,i) \in I\}$ be a tree. A branch of this tree is a sequence $\{x_{n,i(n)}: n = 0,1,\ldots\}$ where $1 \leq i(n) \leq 2^n$ for each n satisfies: $i(n+1) = 2i(n) - 1$ or $i(n+1) = 2i(n)$ for each n. The branch is eventually left turning (or leftward Ho) if there is a positive integer k such that $i(n+1) = 2i(n) - 1$ for each $n \geq k$.

DEFINITION 7.6.4. A weighted tree in X is a doubly indexed sequence of pairs $\{(x_{n,i},t_{n,i}): (n,i) \in I\}$ in $X\times\mathbb{R}$ such that

(a) $0 \leq t_{n,i} \leq \frac{1}{2}$ for each $(n,i) \in I$;

(b) $\sum_{n=0}^{\infty} t_{n,i(n)} = \infty$ whenever $\{x_{n,i(n)}: n = 0,1,\ldots\}$ is an eventually left turning branch; and

(c) $x_{n,i} = (1 - t_{n,i})x_{n+1,2i-1} + t_{n,i}x_{n+1,2i}$ for $(n,i) \in I$.

The function version (see Definition 7.5.2) $(f_n,F_n)_{n=0}^{\infty}$ of a weighted tree is a special type of generalized dyadic martingale. Such martingales and their quasi-companions will appear in the proof the next theorem.

THEOREM 7.6.5 (Ho). Let K be a closed bounded convex subset of a Banach space X. Then K lacks the RNP if and only if there are a K-valued weighted tree $\{(x_{n,i},t_{n,i}): (n,i) \in I\}$ and an $\varepsilon > 0$ such that

(8) $$\|x_{n+1,2i-1} - x_{n+1,2i}\| \geqq \varepsilon \quad \text{for each} \quad (n,i) \in I .$$

Proof. Suppose that $T \equiv \{(x_{n,i},t_{n,i}): (n,i) \in I\}$ in $K\times\mathbb{R}$ is a weighted tre which is ε-separated in the sense of (8). Let $P_1: X\times\mathbb{R} \to X$ be the first coordina projection. Choose a point $y_1 \in P_1(T)$ such that

$$y_1 \notin C \equiv \overline{co}(P_1(T) \setminus U_{\delta/8}[y_1]).$$

(If no such point of $P_1(T)$ exists, K is not subset dentable, hence lacks the RNP and this half of the proof is complete. Henceforth assume there is such a y_1.) Th pick $f \in X^*$ and α,β with $\beta > 0$ such that

$$\sup f(C) < \alpha < \alpha + \beta < f(y_1) .$$

Evidently $S \equiv \{x \in P_1(T): f(x) \geqq \alpha\}$ is a slice of $P_1(T)$ of diameter at most $\frac{\delta}{4}$. Now $y_1 \equiv x_{n,i}$ for some $(n,i) \in I$, and since

$$\|y_1 - x_{n+1,2i}\| = (1 - t_{n,i})\|x_{n+1,2i-1} - x_{n+1,2i}\| \geqq \frac{\delta}{2}$$

(recall: $t_{n,i} \leqq \frac{1}{2}$) it follows that $x_{n+1,2i} \notin S$. Thus $f(x_{n+1,2i}) < \alpha$. Now let $y_2 \equiv x_{n+1,2i-1}$. Then

$$\begin{aligned} f(y_2) &= f((1 - t_{n,i})^{-1}(x_{n,i} - t_{n,i}x_{n+1,2i})) \\ &> (1 - t_{n,i})^{-1}(\alpha + \beta) - t_{n,i}(1 - t_{n,i})^{-1}\alpha = \alpha + (1 - t_{n,i})^{-1}\beta . \end{aligned}$$

More generally the leftward turning branch in $P_1(T)$ obtained by letting $y_j = x_{n+j,2^j i-(2^j-1)}$ for $j = 1,2,\ldots$ satisfies

(9) $$f(y_j) > \alpha + \beta\cdot\prod_{k=1}^{j}(1 - t_{n+k,2^k i-(2^k-1)})^{-1} \quad \text{for } j \geqq 1 .$$

But $\sum_{k=1}^{\infty} t_{n+k,2^k i-(2^k-1)} = \infty$ since these are the numbers forming the second coordina of a left turning branch. Consequently the products in (9) are unbounded. This is impossible since each y_j is a point of $P_1(T)$, a subset of the bounded set K. It follows that $P_1(T)$ is not dentable whence K lacks the RNP.

In order to establish the converse, the following lemma will be used.

LEMMA 7.6.6 Let K be a closed bounded convex subset of $U(X)$ and suppose that $0 < \varepsilon < \frac{1}{2}$. Let $x_1,\ldots,x_n$ be points of K and $t_1 \geq t_2 \geq \cdots \geq t_n > 0$ positive numbers with $\sum_{i=1}^{n} t_i = 1$. Let $x \equiv \sum_{i=1}^{n} t_i x_i$ and assume that $\|x - x_i\| \geq \varepsilon$ for $i = 1,\ldots,n$. Then there is a positive integer k and there are sequences $\{y_i : 1 \leq i \leq k\}$ and $\{z_i : 1 \leq i \leq k\}$ in $co(\{x_i : 1 \leq i \leq k\})$ together with a finite sequence of positive numbers $\{s_i : 1 \leq i \leq k\}$ such that for each j, $1 \leq j \leq k$,

(a) $\|y_j - z_j\| > \frac{1}{2}\varepsilon$;

(b) $y_{j-1} = (1 - s_j)y_j + s_j z_j$ (where $y_0 \equiv x$) ; and

(c) $\frac{\varepsilon}{9} < \sum_{i=1}^{k} s_i \leq \frac{1}{2}$.

Let us assume the lemma for now and complete the proof of Theorem 7.6.5. (The proof of the lemma appears below.) Assume, then, that K lacks the RNP, and without loss of generality we further assume $K \subset U(X)$. Let $0 < \varepsilon < 1$ and choose $K_1 \subset K$ ($K_1 \neq \emptyset$) so that K_1 is closed bounded and convex and $x \in \overline{co}(K_1 \setminus U_{2\varepsilon}[x])$ for each $x \in K_1$ (see Theorem 2.3.6). For each n choose $\varepsilon_n > 0$ such that $\sum_{n=0}^{\infty} \varepsilon_n < \frac{\varepsilon}{4}$. A weighted tree in K_1 will be built as follows. Fix $p \in K_1$. By choice of K_1 and ε, there are finitely many elements $x_1,\ldots,x_n$ of $K_1 \setminus U_{2\varepsilon}[p]$ such that

$$\|p - \sum_{i=1}^{n} t_i x_i\| \leq \varepsilon_0 , \quad \sum_{i=1}^{n} t_i = 1 \text{ and } t_1 \geq t_2 \geq \cdots \geq t_n > 0.$$

Let $\bar{p} \equiv \sum_{i=1}^{n} t_i x_i$. Then since $\varepsilon_0 < \varepsilon$ we have

$$\|\bar{p} - x_i\| \geq \|p - x_i\| - \|\bar{p} - p\| > \varepsilon \quad \text{for } 1 \leq i \leq n$$

and hence Lemma 7.6.6 applies. Thus for some positive integer $k < n$ there are points $y_i, z_i \in K_1$ and scalars $s_i > 0$ $(1 \leq i \leq k)$ such that $\|y_i - z_i\| > \frac{1}{2}$ for each i, $y_0 = \bar{p}$ and for each i, $1 \leq i \leq k$, $y_{i-1} = (1 - s_i)y_i + s_i z_i$ while $\frac{\varepsilon}{9} < \sum_{i=1}^{k} s_i \leq \frac{1}{2}$.

Start the construction of a "quasi-weighted tree" by letting $x_{0,1} \equiv p$, $x_{1,1} \equiv y_1$, $x_{1,2} \equiv z_1$, $x_{2,1} \equiv y_2$, $x_{2,2} \equiv z_2$, ..., $x_{k,1} \equiv y_k$, $x_{k,2} \equiv z_k$. Since the zeroth and the first stages of construction are complete (i.e. $x_{n,i}$ has been defined for all $(n,i) \in I$ with $n = 0$ or 1) the first incomplete stage is the second one (where

$x_{2,3}$ and $x_{2,4}$ remain to be defined). If the building step described in the previous paragraph is repeated, this time with $p = x_{1,2}$ and with perturbation ε_1 in place of ε_0, points $x_{2,3}$, $x_{2,4}$, $x_{3,5}$, $x_{3,6}$, ... will be obtained. The induct process continues by looking for the smallest incomplete stage, say stage n (for which less than 2^n choices $x_{n,i}$ have been made). Then some point p of stage $(n-1)$ has not been supplied with "branching points" at the nth stage and the procedure of the previous paragraph is applied with this p and perturabtion ε_{n-1}. Evidently the "quasi-weighted tree" so produced may be turned into a function versio $(f_n,F_n)_{n=0}^{\infty}$ (see 7.5.2) and $(f_n,F_n)_{n=0}^{\infty}$ is a generalized dyadic ∞-quasi-martingale corresponding to $\{\varepsilon_n: n = 0,1,\ldots\}$. By either the backwards averaging technique o Theorem 2.3.6 or by application of the decomposition Theorem 7.5.3 there is a genera ized dyadic martingale $(g_n,F_n)_{n=0}^{\infty}$ which takes its values in $K_1 \subset K$ and satisfies $\|g_n - f_n\|_\infty < \frac{\varepsilon}{4}$ for each $n = 0,1,\ldots$. Note that (g_n,F_n) is the function version of a weighted tree $\{y_{n,i}: (n,i) \in I\}$ in K and since $\|x_{n+1,2i-1} - x_{n+1,2i}\| \geq \varepsilon$ for each $(n,i) \in I$ by construction, it follows that

$$\begin{aligned}&\|y_{n+1,2i-1} - y_{n+1,2i}\| \\ &\geq \|x_{n+1,2i-1} - x_{n+1,2i}\| - \|x_{n+1,2i-1} - y_{n+1,2i-1}\| - \|x_{n+1,2i} - y_{n+1,2i}\| \\ &\geq \varepsilon - \frac{\varepsilon}{4} - \frac{\varepsilon}{4} = \frac{\varepsilon}{2}.\end{aligned}$$

Hence the proof of the theorem is complete modulo the proof of Lemma 7.6.6.

Proof of 7.6.6. There are two cases to consider.

Case 1. $t_1 > \frac{\varepsilon}{9}$. Take $k_1 \equiv 1$ and let $w \equiv (1 - t_1)^{-1} \cdot \sum_{i=2}^{n} t_i x_i$. Thus $x = t_1x_1 + (1 - t_1)w$ and

$$\|x_1 - w\| = (1 - t_1)^{-1}\|x - x_1\| > \|x - x_1\| \geq \varepsilon$$

whence (a) of the lemma is satisfied if (y_1,z_1) is either (x_1,w) or (w,x_1). Let $s_1 \equiv \min\{t_1, 1 - t_1\}$ so that automatically, $s_1 \leq \frac{1}{2}$. Observe that $t_1 > \frac{\varepsilon}{9}$ by hypothesis while (since $K \subset U(X)$)

$$1 - t_1 \geq \frac{1}{2}(1 - t_1) + \left\|\sum_{i=2}^{n} t_i x_i\right\| \geq \frac{1}{2}\|x_1 - t_1x_1\| + \|x - t_1x_1\|$$

$$\geq \frac{1}{2}\|x - x_1\| \geq \frac{1}{2}\varepsilon .$$

Thus $\frac{\varepsilon}{9} < s_1 \leq \frac{1}{2}$ whence (c) of the lemma is satisfied as well in this case. If $s_1 = t_1$ let $y_1 \equiv w$ and $z_1 \equiv x_1$. Otherwise (if $s_1 = 1 - t_1$) let $y_1 \equiv x_1$ and $z_1 \equiv w$. In either case, (b) of the lemma is clearly satisfied.

Case 2. $t_1 \leq \frac{\varepsilon}{9}$. Since the t_i's are monotonically decreasing and have sum one there is a positive integer $k < n$ such that

$$\frac{\varepsilon}{9} < t_1 + \dots + t_k \leq \frac{2\varepsilon}{9} .$$

For each j, $1 \leq j \leq k$ let

$$s_j \equiv \frac{t_i}{\sum_{i=j}^{n} t_i} .$$

Then

(10) $$\sum_{i=1}^{k} s_i > \sum_{i=1}^{k} t_i > \frac{\varepsilon}{9}$$

and since

(11) $$s_j < \frac{t_j}{\sum_{i=j+1}^{n} t_i} = \frac{t_j}{1 - \sum_{i=1}^{j} t_i} \leq \frac{t_j}{1 - \frac{2\varepsilon}{9}} \qquad (1 \leq j \leq k)$$

it follows that

(12) $$\sum_{i=1}^{k} s_i < \sum_{i=1}^{k} \frac{t_i}{1 - \frac{2\varepsilon}{9}} \frac{2\varepsilon/9}{1 - \frac{2\varepsilon}{9}} < \frac{1}{2}.$$

Thus condition (c) of the lemma is satisfied by virtue of (10) and (12). Now let $y_0 \equiv x$ and for $j = 1,\dots,k$ let

$$y_j \equiv \frac{\sum_{i=j+1}^{n} t_i x_i}{\sum_{i=j+1}^{n} t_i} \quad \text{and} \quad z_j \equiv x_j .$$

Then

(13) $$y_{j-1} = (1 - s_j)y_j + s_j z_j \quad \text{for } j = 1,\dots,k$$

which is condition (b) of the lemma. It remains to show that $\|y_j - z_j\| > \frac{1}{2}\varepsilon$ for $j = 1,\dots,k$. First observe that $\|y_{i-1} - y_i\| = s_i\|y_i - z_i\|$ by (13). Hence from th fact that $K \subset U(X)$, combined with (11) and the condition $\varepsilon < \frac{1}{2}$ we have

$$\|x - y_j\| \leq \sum_{i=1}^{j} \|y_{i-1} - y_i\| \leq \sum_{i=1}^{j} s_i\|z_i - y_i\| \leq 2\cdot\sum_{i=1}^{j} s_i$$

$$\leq 2\cdot\sum_{i=1}^{j} \frac{t_i}{1 - \frac{2\varepsilon}{9}} \leq \frac{4\varepsilon/9}{1 - \frac{2\varepsilon}{9}} < \frac{1}{2}\varepsilon .$$

Consequently

$$\|y_j - x_j\| \geq \|x - x_j\| - \|x - y_j\| > \varepsilon - \tfrac{1}{2}\varepsilon = \tfrac{1}{2}\varepsilon$$

which shows that (a) of the lemma holds. The proof is complete.

The function version of a weighted tree is a special type of generalized dyadic martingale as was remarked earlier. The next results describe geometric conditions on a set in order for it to contain the range of a generalized dyadic martingale wh: is δ-separated in the usual sense $\|g_{n+1}(\omega) - g_n(\omega)\| \geq \delta$ for all n and all ω . We follow Kunen and Rosenthal [1982] .

<u>DEFINITION</u> 7.6.7. Suppose that $D \subset X$ is bounded and that $\varepsilon > 0$. For each point y in X and for each line segment $[d_1,d_2]$ in X which contains y such that bo endpoints (i.e. d_1 and d_2) belong to D, denote by $m(d_1,d_2,y)$ the shorter of the lengths $\|y - d_1\|$ and $\|y - d_2\|$. (Here $[d_1,d_2]$ denotes the <u>closed</u> line segment from d_1 to d_2 of course.)

(a) $x_0 \in D$ <u>is</u> said to be <u>an</u> <u>ε-strong extreme point of</u> <u>D</u> if for some $\delta >$ • $m(d_1,d_2,y) < \varepsilon$ whenever $y \in U_\delta(x_0)$ and $y \in [d_1,d_2]$ for points d_1,d_2 in D. Let $E_\varepsilon(D) \equiv \{x_0 \in D: x_0$ is an ε-strong extreme point of $D\}$.

(b) $x_0 \in D$ <u>is a strong extreme point of</u> <u>D</u> if x_0 is an ε-strong extreme point for each $\varepsilon > 0$.

(c) For $\varepsilon > 0$ let $G_\varepsilon(D) \equiv \{y \in X:$ for some $d_1, d_2 \in D$ both $y \in [d_1, d_2]$ and $m(d_1, d_2, y) \geqq \varepsilon\}$.

Some elementary observations are in order.

OBSERVATIONS 7.6.8.

(a) Let K be a closed bounded convex set in X. Then

$$E_\varepsilon(D) = \{x \in K: \text{ there is a } \delta > 0 \text{ such that whenever } y,z \in K \text{ satisfy } \|\tfrac{1}{2}(y+z) - x\| < \delta \text{ then } \|y - z\| < 2\varepsilon\} .$$

(b) Let $K \subset X$ be a closed bounded convex in X. Then each denting point of K is a strong extreme point of K and each strong extreme point of K is an extreme point of K.

(c) Let $K \equiv \overline{co}(\{\delta_n: n = 1,2,\ldots\}) \subset \ell_2$ (where δ_n has 1 in the nth coordinate and 0 elsewhere). Then, as mentioned in the first example following 3.2.5, 0 is an exposed point of K which is not strongly exposed. Also, 0 is a strong extreme point of K which is not a denting point.

(d) Let $D_1 \equiv \{\frac{1}{n}\delta \pm \delta_n: n \geqq 2\} \subset \ell_1$ and let $K \equiv w^*cl(co(D_1))$. Then, as was shown in Example 3.2.5, 0 is an exposed point of K which is not strongly exposed. Suppose that $\delta > 0$ and that n is chosen so large that $\frac{1}{n} < \delta$. Then $m(\frac{1}{n}\delta_1 - \delta_n, \frac{1}{n}\delta_1 + \delta_n, \frac{1}{n}\delta_1) = 1$ and hence $0 \notin E_\varepsilon(K)$ for any $\varepsilon < 1$.

(e) Let D be a bounded set. Then $E_\varepsilon(D)$ is a relatively open subset of D for each $\varepsilon > 0$ and hence the set of strong extreme points of D, $\bigcap_{\varepsilon>0} E_\varepsilon(D) = \bigcap_{n=1}^{\infty} E_{1/n}(D)$ is a G_δ set. It also follows from the fact that $E_\varepsilon(D)$ is relatively open that $E_\varepsilon(D) \neq \emptyset$ if $E_\varepsilon(\bar{D}) \neq \emptyset$.

(f) For a bounded set $D \subset X$ and for each $\varepsilon > 0$, $E_\varepsilon(D) = D \setminus \overline{G_\varepsilon(D)}$. Moreover, if K is closed bounded and convex in X then

$$G_\varepsilon(D) = \{\tfrac{1}{2}(y + z): \|y - z\| \geqq 2\varepsilon \ ; \ \ y,z \in K\} .$$

PROPOSITION 7.6.9. Let K be a closed bounded convex subset of X. Then the following are equivalent.

(a) There are a nonempty subset D of K and an $\varepsilon > 0$ such that $E_\varepsilon(D) = \emptyset$.

(b) There are $\varepsilon' > 0$ and a generalized K-valued dyadic martingale $(g_n, F_n)_{n}^{\infty}$ such that $\|g_{n+1}(\omega) - g_n(\omega)\| \geq \varepsilon'$ for each n and each $\omega \in \Omega$.

Proof.

<u>(b) ⇒ (a)</u> Let $(g_n, F_n)_{n=0}^{\infty}$ be a generalized K-valued dyadic martingale with $\|g_{n+1}(\omega) - g_n(\omega)\| \geq \varepsilon'$ for each n and each ω. Let $D \equiv \{g_n(\omega): \omega \in \Omega, n = 0,1,\ldots\}$. Then $G_{\varepsilon'}(D) = D$ whence $E_{\varepsilon'}(D) = \emptyset$ by 7.6.8(f).

<u>(a) ⇒ (b)</u> Suppose that $\emptyset \neq D \subset K$ satisfies $E_\varepsilon(D) = \emptyset$ for some $\varepsilon > 0$. Choose ε' and ε'' such that $\varepsilon' < \varepsilon'' < \varepsilon$. For each $n \geq 0$ choose $\varepsilon_n > 0$ such that $\sum_{n=0}^{\infty} \varepsilon_n \leq \frac{\varepsilon'' - \varepsilon'}{2}$. Given $x \in D$ find $x' \in X$ within ε_0 of x such that, for some $y,z \in D$, $\|y - x'\| \geq \varepsilon$, $\|z - x'\| \geq \varepsilon$ and $x' \in [y,z]$. (This can be done since $D \subset \overline{G_\varepsilon(D)}$ by 7.6.8 (f).) Inductively repeat the above construction. (For example, the second time use y in place of x and the third time use z in place of x, and in both these cases use the perturbation ε_1 in place of ε_0.) The funct version $(f_n, F_n)_{n=0}^{\infty}$ of the resulting approximate tree is easily seen to be a D-val generalized dyadic ∞-quasi-martingale corresponding to $\{\varepsilon_n: n = 0,1,\ldots\}$ such tha $\|f_{n+1}(\omega) - f_n(\omega)\| \geq \varepsilon''$ for each n and each ω. Now, by backwards averaging or by application of the decomposition Theorem 7.5.3, there is a generalized dyadic mar tingale $(g_n, F_n)_{n=0}^{\infty}$ which takes its values in $\overline{co}(D) \subset K$ such that $\|g_n - f_n\|_\infty \leq \sum_{j=}^{\infty}$ for each n. Then for any n and any ω,

$$\|g_{n+1}(\omega) - g_n(\omega)\| \geq \|f_{n+1}(\omega) - f_n(\omega)\| - \|f_n - g_n\|_\infty - \|f_{n+1} - g_{n+1}\|_\infty$$
$$\geq \varepsilon'' - 2\cdot \sum_{n=0}^{\infty} \varepsilon_n \geq \varepsilon' .$$

<u>DEFINITION</u> 7.6.10. Let K be a closed bounded convex subset of X. Then K has the <u>Approximate Krein Milman Property</u> (AKMP) if each nonempty subset D of K ha an ε-strong extreme point for each $\varepsilon > 0$. (X has the AKMP means that U(X) ha the AKMP.)

Although the above definition seems natural enough care must be shown. For example, if K is closed bounded and convex and K has the AKMP it is not known a the time of this writing whether $K = \overline{co}(E_\varepsilon(K))$ for each $\varepsilon > 0$. Some information about this problem may be gleaned from the following.

PROPOSITION 7.6.11. Suppose that $\varepsilon > 0$ and that K is a closed bounded convex set in X. Then

(a) If $E_\varepsilon(K) = \emptyset$ then K contains an ε'-tree for each $\varepsilon' < \varepsilon$.

(b) If $\overline{co}(E_\varepsilon(K)) \neq K$ then for each $\varepsilon' < \varepsilon$ $(\varepsilon' > 0)$ and each $\delta > 0$ there are a standard dyadic K-valued martingale $(g_n, \mathcal{D}_n)_{n=0}^{\infty}$ and a set $A \in \mathcal{B}([0,1))$ with $\lambda(A) > 1 - \delta$ ($\lambda \equiv$ Lebesgue measure) such that $\|g_{n+1}(\omega) - g_n(\omega)\| \geqq \varepsilon'$ for each n and each $\omega \in A$.

Proof. (a) will be established as a special case of (b). Thus suppose that $0 < \varepsilon' < \varepsilon$ and that $0 < \delta < 1$. By hypothesis there are $\alpha > 0$ and $f \in S(X^*)$ such that

$$\overline{co}(E_\varepsilon(K)) \cap S(K,f,\alpha) = \emptyset.$$

Choose $x_0 \in K$ such that $f(x_0) > M(K,f) - \alpha$ (so that $x_0 \in S(K,f,\alpha)$ automatically). Fix ε'' such that $\varepsilon' < \varepsilon'' < \varepsilon$ and set

$$\beta \equiv \alpha - (M(K,f) - f(x_0)). \tag{14}$$

Let $\{\varepsilon_n : n = 0,1,\ldots\}$ be a sequence of positive numbers such that both

$$\sum_{n=0}^{\infty} \varepsilon_n \leqq \frac{\varepsilon'' - \varepsilon'}{2} \quad \text{and} \quad f(x_0) > M(K,f) - \alpha\delta + \sum_{n=0}^{\infty} \varepsilon_n. \tag{15}$$

Create an approximate tree as follows. Since $\overline{S(K,f,\alpha)} \subset \overline{G_\varepsilon(K)}$ by 7.6.8(f) there are points y,z in K such that

$$\|\tfrac{1}{2}(y + z) - x_0\| < \varepsilon_0 \quad \text{and} \quad \min\{\|x_0 - y\|, \|x_0 - z\|\} \geqq \varepsilon''. \tag{16}$$

If $f(y) < M(K,f) - \alpha$ or $f(z) < M(K,f) - \alpha$ then leave it alone, but if, say, $f(y) \geqq M(K,f) - \alpha$ (so that $y \in \overline{S(K,f,\alpha)}$), repeat the procedure of the previous sentence with ε_1 in place of ε_0 and y in place of x_0. In fact, an obvious inductive argument produces an approximate tree whose function version $(h_n, \mathcal{D}_n)_{n=0}^{\infty}$ is a standard dyadic ∞-quasi-martingale corresponding to $\{\varepsilon_n : n = 0,1,\ldots\}$ which satisfies the conditions

(17) $$h_0(\omega) = x_0 \quad \text{for each} \quad \omega \in [0,1) .$$

(18) $$\text{If} \quad f(h_n(\omega)) \geq M(K,f) - \alpha \quad \text{then} \quad \|h_{n+1}(\omega) - h_n(\omega)\| \geq \varepsilon'' .$$

(19) $$\text{If} \quad f(h_n(\omega)) < M(K,f) - \alpha \quad \text{then} \quad h_{n+1}(\omega) = h_n(\omega) .$$

Let $(g_n, \mathcal{D}_n)_{n=0}^{\infty}$ be the standard dyadic K-valued martingale whose existence is guar-teed by Theorem 7.5.3 such that $\|h_n - g_n\|_\infty \leq \sum_{j=n}^{\infty} \varepsilon_j$ for each n.

For each n let

$$A_n \equiv \{\omega \in [0,1) : h_n(\omega) \in \overline{S(K,f,\alpha)}\}$$ (If $E_\varepsilon(K) = \emptyset$ let α be so large that $S(K,f,\alpha) = K$ and hence in this case $A_n = [0,1)$ for each n.)

Then $A_n \supset A_{n+1}$ for each n by (19) and we let $A \equiv \bigcap_{n=1}^{\infty} A_n$. (Note that if $E_\varepsilon($ $= \emptyset$ then $A = [0,1)$.) We will show that $\lambda(A) > 1 - \delta$ (see (23)) and that for eac $\omega \in A$, $\|g_{n+1}(\omega) - g_n(\omega)\| \geq \varepsilon'$ (see (24)). Observe that

$$\int_{[0,1)} f\circ h_n d\lambda = \int_{A_n} f\circ h_n d\lambda + \int_{[0,1)\setminus A_n} f\circ h_n d\lambda$$

(20) $$\leq M(K,f)\lambda(A_n) + (M(K,f) - \alpha)(1 - \lambda(A_n))$$

$$= M(K,f) - \alpha + \lambda(A_n) .$$

Now

$$\left|\int_{[0,1)} (f\circ h_n - f\circ h_0) d\lambda\right| \leq \|h_n - h_0\|_1 \leq \|h_n - g_n\|_1 + \|g_n - g_0\|_1 + \|g_0 - f_0\|_1$$
$$\leq \sum_{j=n}^{\infty} \varepsilon_n + \sum_{j=0}^{\infty} \varepsilon_j$$

whence, from the definition of β in (14) and the fact that h_0 is the function with constant value x_0,

(21) $$\int_{[0,1)} f\circ h_n d\lambda \geq \int_{[0,1)} f\circ h_0 d\lambda - \sum_{j=n}^{\infty} \varepsilon_j - \sum_{j=0}^{\infty} \varepsilon_j$$
$$= M(K,f) + \beta - \alpha - \sum_{j=n}^{\infty} \varepsilon_j - \sum_{j=0}^{\infty} \varepsilon_j .$$

Hence, combining (20) and (21) yields

$$(22) \qquad \lambda(A_n) \geq \frac{\beta - \sum_{j=n}^{\infty} \varepsilon_j - \sum_{j=0}^{\infty} \varepsilon_j}{\alpha} .$$

Then, from (14), (15) and (22)

$$(23) \qquad \lambda(A) \geq \frac{\beta - \sum_{j=0}^{\infty} \varepsilon_j}{\alpha} > 1 - \delta .$$

Now suppose that $\omega \in A$. Then $\|h_{n+1}(\omega) - h_n(\omega)\| \geq \varepsilon''$ by (16), and hence by (15)

$$(24) \quad \|g_{n+1}(\omega) - g_n(\omega)\| \geq \|h_{n+1}(\omega) - h_n(\omega)\| - \|g_{n+1} - h_{n+1}\|_\infty - \|g_n - h_n\|_\infty$$

$$\geq \varepsilon' - \sum_{j=n+1}^{\infty} \varepsilon_j - \sum_{j=n}^{\infty} \varepsilon_j \geq \varepsilon'$$

which completes the proof.

It follows from Proposition 7.6.11 that if $E_\varepsilon(K)$ does not generate K then there is a K-valued standard dyadic martingale $(g_n,\mathcal{D}_n)_{n=0}^{\infty}$ such that $\inf_n \|g_{n+1} - g_n\|_1 > 0$. According to the construction of Bourgain and Rosenthal in section 7.3 there is a Banach space Y which lacks the RNP such that $\inf_n \|g_{n+1} - g_n\|_1 = 0$ for each Y-valued bounded generalized dyadic martingale $(g_n,\mathcal{D}_n)_{n=0}^{\infty}$. Hence such a space satisfies $K = \overline{co}(E_\varepsilon(K))$ for each $\varepsilon > 0$ and each closed bounded convex subset K of Y. Moreover, by Proposition 7.6.9, Y has the AKMP. But according to Theorem 7.3.2(e), Y lacks the KMP. Hence the AKMP does not imply the KMP in general.

The final result of this section concerns the existence of strong extreme points and provides another link with the RNP. Recall that $\hat{x}$ denotes the canonical image in X^{**} of a point x in X.

<u>PROPOSITION</u> 7.6.12. Suppose that $K \subset X$ is closed bounded and convex. If each closed bounded convex subset C of K has a strong extreme point then K has the RNP. In fact, if x_0 is a strong extreme point of C then $\hat{x}_0$ is an extreme point of $w^*cl(\hat{C}) \subset X^{**}$.

Proof. The first part of the proposition follows from the last sentence of its statement using Corollary 3.7.6, parts (1) and (3). (Care should be taken to distinguish between the strong extreme points introduced in this section and the strong extreme points discussed in section 3.7.) Thus it suffices to show that whenever C is a closed bounded convex set and x_0 is a strong extreme point of C then $\hat{x}_0 \in ex(w^*cl(co(\hat{C})))$. If this were not the case for some x_0 then $\hat{x}_0 = \frac{1}{2}(F+G)$ for distinct points $F,G \in w^*cl(\hat{C})$. Choose $f \in X^*$ with $(F-G)(f) = 1$ and nets $(y_\alpha : \alpha \in A)$ and $(z_\alpha : \alpha \in A)$ in C with $\hat{y}_\alpha \to F$ and $\hat{z}_\alpha \to G$ in $(X^{**}, weak^*)$. Then $f(y_\alpha - z_\alpha) \to 1$. Without loss of generality assume

$$f(y_\alpha - z_\alpha) \geqq \frac{1}{2} \quad \text{for each } \alpha \in A .$$

Let $\varepsilon \equiv \frac{1}{4\|f\|}$. Since $x_0 \in E_\varepsilon(C)$ there is a $\delta > 0$ such that for all $y,z \in C$, if $\|\frac{1}{2}(y + z) - x_0\| < \delta$ then $\|y - z\| < 2\varepsilon$. Since $\frac{1}{2}\hat{y}_\alpha + \frac{1}{2}\hat{z}_\alpha \to \hat{x}_0$ in $(X^{**},weak^*)$ then $\frac{1}{2}y_\alpha + \frac{1}{2}z_\alpha \to x_0$ in $(X,weak)$. Hence there are indices $\alpha_1,\ldots,\alpha_n$ in A and positive scalars $t_1,\ldots,t_n$ with $\sum_{i=1}^n t_i = 1$ and $\|\sum_{i=1}^n t_i(\frac{1}{2}y_{\alpha_i} + \frac{1}{2}z_{\alpha_i}) - x_0\| < \delta$. Let $y \equiv \sum_{i=1}^n t_i y_{\alpha_i}$ and $z \equiv \sum_{i=1}^n t_i z_{\alpha_i}$. Then $y,z \in C$ and $\|\frac{1}{2}(y + z) - x_0\| < \delta$. Hence $\|y - z\| < 2\varepsilon = \frac{1}{2\|f\|}$. But then from this inequality together with (25) we obtain the contradiction

$$\begin{aligned} f(y - z) \leqq \|f\| \cdot \|y - z\| < \|f\| \frac{1}{2\|f\|} = \frac{1}{2} \leqq \sum_{i=1}^n t_i f(y_{\alpha_i} - z_{\alpha_i}) \\ = f(\sum_{i=1}^n t_i y_{\alpha_i} - \sum_{i=1}^n t_i z_{\alpha_i}) = f(y - z). \end{aligned}$$

Thus the proof is complete.

§7. A long James space.

The original example of James [1951] is of a nonreflexive space whose natural embedding in its second dual is of codimension one but which is nevertheless isometrically isomorphic to its second dual. That space and variants have served as important examples on which to test conjectures and the variant presented here, called a long James space, has already proved itself as a conjecture cruncher for several RNP related questions. Most of the material of this section is from Edgar [1980], with bits also from Edgar [1977] and [1979]. Other properties of these spaces were considered by Hagler and Odell [1978] . (Appropriate definitions will appear below.)

THEOREM 7.7.1. The long James space $J(\omega_1)$ has the following properties:

(a) $J(\omega_1)$ is isomorphic to a second conjugate Banach space with the RNP;

(b) There is a bounded weakly measurable function on some probability space with values in $J(\omega_1)$ which fails to be Pettis integrable;

(c) $\mathcal{B}(J(\omega_1),\text{weak}) \neq \mathcal{B}(J(\omega_1),\text{weak*})$. (See 4.3.1 for definitions.)

(d) $J(\omega_1)$ is not real compact, not measure-compact, not Lindelöf and not isomorphic to a subspace of a weakly compactly generated space.

(e) $J(\omega_1)$ has no equivalent weakly locally uniformly convex dual norm. (Recall that for any Banach space X, an equivalent norm $\|\cdot\|$ on X is weakly locally uniformly convex if whenever $(x_\alpha:\alpha \in A)$ is a net in $\{x: \|x\| = 1\}$, $\|\bar{x}\| = 1$ and $\|x_\alpha + \bar{x}\| \to 2$ then $x_\alpha \to \bar{x}$ in (X,weak).)

We begin by providing two alternative definitions of the space $J(\eta)$ for η an infinite ordinal, after which the various duals and a predual are identified. At that point enough of the structure of such spaces will be at our fingertips to be able to prove the above theorem with relative ease. As might be expected, the arguments are largely calculational and transfinite induction is pressed into service at almost every turn.

For any set A of two or more ordinal numbers and $f: A \to \mathbb{R}$ define the square variation of f to be

(1) $$\|f\| \equiv \sup\{(\sum_{i=1}^{n-1} |f(\alpha_{i+1}) - f(\alpha_i)|^2)^{\frac{1}{2}} : \alpha_1 < \dots < \alpha_n ;$$
$$\alpha_i \in A \text{ for each } i ; n = 2,3,\dots\} .$$

If η is an ordinal let

$$J(\eta) \equiv \{f: [0,\eta] \to \mathbb{R}: f \text{ is continuous}, \|f\| < \infty, f(0) = 0\} .$$

As may reaily be verified, $(J(\eta),\|\cdot\|)$ is a Banach space.

An alternative description of $J(\eta)$ will also be useful in the sequel. If η is an infinite ordinal and by $NLO(\eta)$ we mean the set

$$NLO(\eta) \equiv \{\alpha\colon \alpha \in [0,\eta],\ \alpha \text{ is not a limit ordinal}\}$$

then there is a unique order preserving surjective map $\psi_\eta\colon [0,\eta) \to NLO(\eta)$. (Construct ψ_η by transfinite induction.) Let $\tilde{J}(\eta)$ be the Banach space

$$\tilde{J}(\eta) \equiv \{f\colon [0,\eta) \to \mathbb{R}\colon \ \|f\| < \infty\ ,\ f(0) = 0\}$$

equipped with the square variation norm. Here is the connection between $\tilde{J}(\eta)$ and $J(\eta)$.

<u>LEMMA</u> 7.7.2. Let η be an infinite ordinal. Then $J(\eta)$ and $\tilde{J}(\eta)$ are isometrically isomorphic.

Proof. Observe that each $f\colon NLO(\eta) \to \mathbb{R}$ with $\|f\| < \infty$ has a unique norm preserving extension to a continuous function on $[0,\eta]$. Hence $\{f\colon NLO(\eta) \to \mathbb{R}\colon \|f\| < \infty\ ,\ f(0) = 0\}$ with square variation norm is a space isometrically isomorphic to $(J(\eta),\|\cdot\|)$. Thus ψ_η induces an isometric isomorphism between $(J(\eta),\|\cdot\|)$ and $\tilde{J}(\eta),\|\cdot\|)$.

Transfinite series in a Banach space X are to be understood as follows. For a given ordinal η and point $x_\alpha \in X$ for each $\alpha < \eta$ the symbol $\sum_{\alpha<\eta} x_\alpha$ is defined recursively:

If $\gamma = 0$ then $\sum_{\alpha<\gamma} x_\alpha \equiv 0$;

If $\gamma = \beta + 1$ is a successor then $\sum_{\alpha<\gamma} x_\alpha \equiv \sum_{\alpha<\beta} x_\alpha + x_\beta$ provided the series on the right converges;

If γ is a limit ordinal then $\sum_{\alpha<\gamma} x_\alpha \equiv \lim_{\beta<\gamma} \sum_{\alpha<\beta} x_\alpha$ where the limit is taken in the norm topology.

<u>PROPOSITION</u> 7.7.3. Given $\eta > 0$, for each $\alpha < \eta$ let

$$h_\alpha \equiv \chi_{(\alpha,\eta]}\ .$$

Then $h_\alpha \in S(J(\eta))$ and $\{h_\alpha\colon \alpha < \eta\}$ is a basis for $J(\eta)$. (That is, for each $f \in J(\eta)$ there is a unique transfinite sequence $\{c_\alpha\colon \alpha < \eta\}$ of real numbers such that $f = \sum_{\alpha<\eta} c_\alpha h_\alpha$.)

Proof. For each $\alpha \leq \eta$ let $P_\alpha : J(\eta) \to J(\eta)$ be the projection

$$P_\alpha(f) \equiv f \cdot \chi_{[0,\alpha]} + f(\alpha)\chi_{(\alpha,\eta]} \qquad \text{for } f \in J(\eta).$$

Suppose that $f \in J(\eta)$, that $\varepsilon > 0$ and that $\gamma \leq \eta$ is a limit ordinal. Then there is a finite sequence $0 \leq \alpha_0 < \alpha_1 < \ldots < \alpha_n \leq \eta$ with

$$\|P_\gamma(f)\|^2 < \sum_{i=1}^{n} |P_\gamma(f)(\alpha_i) - P_\gamma(f)(\alpha_{i-1})|^2 + \varepsilon .$$

Since $P_\gamma(f)$ is constant on $[\gamma,\eta]$ there is no loss of generality in assuming that $\alpha_n \leq \gamma$, and since f is continuous at γ we may (and do) assume even that $\alpha_n < \gamma$. Let $\beta \in (\alpha_n,\gamma)$. Then for the same sequence $\alpha_0,\alpha_1,\ldots,\alpha_n$ as above

$$\|P_\beta(f)\|^2 \geq \sum_{i=1}^{n} |P_\beta(f)(\alpha_i) - P_\beta(f)(\alpha_{i-1})|^2 = \sum_{i=1}^{n} |P_\gamma(f)(\alpha_i) - P_\gamma(f)(\alpha_{i-1})|^2$$
$$> \|P_\gamma(f)\|^2 - \varepsilon .$$

But $P_\beta(f)$ is constant on $[\beta,\gamma]$ and $(P_\gamma - P_\beta)(f)$ is constant on $[0,\beta]$ so

$$\|P_\gamma(f)\|^2 \geq \|P_\beta(f)\|^2 + \|(P_\gamma - P_\beta)(f)\|^2 .$$

That is,

$$\|(P_\gamma - P_\beta)(f)\|^2 \leq \|P_\gamma(f)\|^2 - \|P_\beta(f)\|^2 < \varepsilon.$$

It follows that, for $\gamma \leq \eta$ a limit ordinal and $f \in J(\eta)$

(2) $$\lim_{\beta<\eta} \|P_\beta(f) - P_\gamma(f)\| = 0 .$$

For each $\alpha < \eta$ let $c_\alpha \equiv f(\alpha+1) - f(\alpha)$. Note that $P_0(f) = 0 = \sum_{\alpha<0} c_\alpha h_\alpha$. Now suppose that $\gamma \leq \eta$ and that for each $\beta < \gamma$ we have

$$\sum_{\alpha<\beta} c_\alpha h_\alpha = P_\beta(f).$$

If γ is a limit ordinal then $\sum_{\alpha<\gamma} c_\alpha h_\alpha = P_\gamma(f)$ follows from (2) while, if $\gamma = \beta+1$ then by inductive hypothesis

$$\sum_{\alpha<\gamma} c_\alpha h_\alpha = \sum_{\alpha<\beta} c_\alpha h_\alpha + c_\beta h_\beta = P_\beta(f) + (f(\gamma) - f(\beta))\chi_{[\gamma,\eta]}$$

$$= f\cdot\chi_{[0,\beta]} + f(\beta)\chi_{[\gamma,\eta]} + (f(\gamma) - f(\beta))\chi_{[\gamma,\eta]}$$

$$= P_\gamma(f).$$

Since $P_\eta(f) = f$ it follows that $f = \sum_{\alpha<\eta} c_\alpha h_\alpha$, and since for any $\alpha < \eta$

$$c_\alpha h_\alpha = P_{\alpha+1}(f) - P_\alpha(f) = (f(\alpha+1) - f(\alpha))h_\alpha$$

the uniqueness of $\{c_\alpha : \alpha < \eta\}$ is apparent in the representation of f. This completes the proof.

Arguments similar to that just presented will allow us to construct bases for $J(\eta)^*$ and for a predual of $J(\eta)$. This is the content of the next two results.

PROPOSITION 7.7.4. For each $\alpha \in (0,\eta]$ let $e_\alpha \in J(\eta)^*$ be the evaluation function

$$e_\alpha(f) \equiv f(\alpha) \qquad \text{for } f \in J(\eta).$$

Then $e_\alpha \in S(J(\eta)^*)$ for each $\alpha \in (0,\eta]$ and $\{e_\alpha : \alpha \in (0,\eta]\}$ is a basis for $J(\eta)$

Proof. If $F \in J(\eta)^*$ and $\gamma \leq \eta$ is a limit ordinal then, as is shown directl below,

(3) $$\lim_{\beta<\gamma} F(h_\beta) \text{ exists.}$$

Indeed, if $\lim_{\beta<\gamma} F(h_\beta)$ fails to exist then there are real numbers $a < b$ and ordin $\beta_0 < \beta_1 < \ldots < \gamma$ with $F(h_{\beta_{2i}}) < a$ and $F(h_{\beta_{2i+1}}) > b$ for each i. Note that

$$\left\| \sum_{i=1}^{n} (h_{\beta_{2i+1}} - h_{\beta_{2i}}) \right\| = (2n)^{\frac{1}{2}}.$$

Thus for each n

$$n(b-a) < F\left(\sum_{i=1}^{n} (h_{\beta_{2i+1}} - h_{\beta_{2i}})\right) \leq \|F\| \, (2n)^{\frac{1}{2}}$$

whence $\|F\| = \infty$. It follows that the limit in (3) exists. Thus it is possible to define numbers d_α for $0 < \alpha \leq \eta+1$ by

$$d_\alpha \equiv \begin{cases} 0 & \text{if } \alpha = \eta+1 \\ \lim_{\beta<\alpha} F(h_\beta) & \text{if } \alpha \text{ is a limit ordinal } \leq \eta \\ F(h_\beta) & \text{if } \beta+1 = \alpha \leq \eta . \end{cases}$$

Assume for the time being that $\sum_{\alpha\in(0,\eta]} (d_\alpha - d_{\alpha+1})e_\alpha$ converges. It is a straightforward transfinite induction argument to ascertain that under these circumstances and for each $f \in J(\eta)$

$$\Big(\sum_{\alpha\in(0,\eta]} (d_\alpha - d_{\alpha+1})e_\alpha\Big)(f) = \sum_{\alpha\in(0,\eta]} (d_\alpha - d_{\alpha+1})f(\alpha).$$

In particular, for any $\beta \in [0,\eta)$

(4)
$$\begin{aligned}\Big(\sum_{\alpha\in(0,\eta]} (d_\alpha - d_{\alpha+1})e_\alpha\Big)(h_\beta) &= \sum_{\alpha\in(0,\eta]} (d_\alpha - d_{\alpha+1})h_\beta(\alpha). \\ &= \sum_{\beta+1\leq\alpha\leq\eta} (d_\alpha - d_{\alpha+1}) = d_{\beta+1} = F(h_\beta).\end{aligned}$$

Because $\{h_\alpha: \alpha < \eta\}$ is a basis for $J(\eta)$, if in fact $\sum_{\alpha\in(0,\eta]} (d_\alpha - d_{\alpha+1})e_\alpha$ converges to an element of $J(\eta)^*$ then that element must be F by (4). (This again requires a transfinite induction argument.) Moreover, if $\sum_{\alpha\in(0,\eta]} c_\alpha e_\alpha = F$ for some choice of scalars $\{c_\alpha: \alpha \in (0,\eta]\}$ and if $\beta < \eta$ then $F(h_\beta) = \sum_{\beta+1\leq\alpha\leq\eta} c_\alpha$ whence $F(h_\beta) - F(h_{\beta+1}) = c_{\beta+1}$ for each β. It follows that $c_\beta = d_\beta - d_{\beta+1}$ for each $\beta \in NLO(\eta)$ and if $\beta \leq \eta$ is a limit ordinal then

$$\begin{aligned}c_\beta &= \lim_{\alpha<\beta} \Big(\sum_{\alpha+1\leq\xi\leq\eta} c_\xi - \sum_{\beta+1\leq\xi\leq\eta} c_\xi\Big) = \lim_{\alpha<\beta} (F(h_\alpha) - F(h_\beta)) \\ &= d_\beta - d_{\beta+1}\end{aligned}$$

as well. Hence the uniqueness of representation of $F \in J(\eta)^*$ is established.

It remains to check that for each limit ordinal $\gamma \leq \eta$

$$\Big\{\sum_{\alpha<\beta} (d_\alpha - d_{\alpha+1})e_\alpha : \beta < \gamma\Big\} \quad \text{is norm convergent}$$

under the assumption that $\sum_{\alpha<\beta} (d_\alpha - d_{\alpha+1})e_\alpha$ converges for each $\beta < \gamma$. The firs step in this verification is to ascertain that

$$(5) \qquad F(g) = \sum_{\alpha\in(0,\eta]} (d_\alpha - d_{\alpha+1})g(\alpha) \qquad \text{for each } g \in J(\eta) .$$

Observe that (5) obtains for $g = ch_0$ by the latter part of the calculation in (4 If now (5) holds whenever $g \in J(\eta)$ is of the form $\sum_{\alpha<\beta} c_\alpha h_\alpha$ for some $\beta < \xi \leq \eta$ and choice $\{c_\alpha: \alpha < \beta\}$ of scalars, inductively extend the collection of $g \in J(\eta)$ for which (5) holds as follows.

Case 1. If $\xi = \beta_0 + 1$ is a successor and if $g = \sum_{\alpha<\xi} c_\alpha h_\alpha \in J(\eta)$ let $g_1 \equiv \sum_{\alpha<\beta_0} c_\alpha h_\alpha$. Observe that $g_1 \in J(\eta)$ and that $g = g_1 + c_{\beta_0} h_{\beta_0}$. Now it is easy t see that whenever $\sum_{\alpha<\eta} a_\alpha$ and $\sum_{\alpha<\eta} b_\alpha$ are convergent series then so is $\sum_{\alpha<\eta} (a_\alpha +$ and its value is $\sum_{\alpha<\eta} a_\alpha + \sum_{\alpha<\eta} b_\alpha$. Since $\sum_{\alpha\in(0,\eta]} (d_\alpha - d_{\alpha+1})g_1(\alpha)$ converges tc $F(g_1)$ by inductive hypothesis while

$$\sum_{\alpha\in(0,\eta]} (d_\alpha - d_{\alpha+1})c_{\beta_0} h_{\beta_0}(\alpha) = F(c_{\beta_0} h_{\beta_0})$$

by the latter part of the calculation in (4), it follows that

$$\sum_{\alpha\in(0,\eta]} (d_\alpha - d_{\alpha+1})g(\alpha) = F(g_1) + F(c_{\beta_0} h_{\beta_0}) = F(g) .$$

Case 2. If ξ is a limit ordinal and $g = \sum_{\alpha<\xi} c_\alpha h_\alpha \in J(\eta)$ then by definitio of such a (transfinite) sum, there is a sequence $\{\beta_n: n = 1,2,\ldots\}$ of ordinals β $< \beta_2 < \ldots < \xi$ such that $\lim_n \| \sum_{\alpha<\beta_n} c_\alpha h_\alpha - \sum_{\alpha<\xi} c_\alpha h_\alpha \| = 0$. For each n let $g_n \equiv \sum_{\alpha<\beta_n} c_\alpha h_\alpha$. Then since $\lim_n \| \sum_{\alpha<\beta_n} c_\alpha h_\alpha - \sum_{\alpha<\cup\beta_n} c_\alpha h_\alpha \| = 0$ it follows that

$$\sum_{\alpha<\cup\beta_n} c_\alpha h_\alpha = \sum_{\alpha<\xi} c_\alpha h_\alpha$$

whence $c_\alpha = 0$ for $\alpha \in [\cup\beta_n, \xi)$. In particular, then if $\cup\beta_n < \xi$ the inductive hypothesis already applies to g. Thus assume that $\cup\beta_n = \xi$.

For any n, g_n has constant value $g(\beta_n)$ on $[\beta_n,\eta]$ and $g(\alpha) = g_n(\alpha)$ for $\alpha < \beta_n$ whence, since $d_{\eta+1} = 0$,

$$F(g_n) = \sum_{\alpha\in(0,\eta]} (d_\alpha - d_{\alpha+1})g_n(\alpha) = \sum_{\alpha<\beta_n} (d_\alpha - d_{\alpha+1})g(\alpha) + d_{\beta_n} g(\beta_n).$$

But since g is continuous at ξ, and by virtue of (3), we have $\lim_n d_{\beta_n} g(\beta_n) = d_\xi g(\xi)$. Consequently, $\lim_n (F(g_n) - d_{\beta_n} g(\beta_n)) = F(g) - d_\xi g(\xi)$ exists. That is, since

$$F(g_n) - d_{\beta_n} g(\beta_n) = \sum_{\alpha<\beta_n} (d_\alpha - d_{\alpha+1})g(\alpha)$$

it follows that

$$\sum_{\alpha<\xi} (d_\alpha - d_{\alpha+1})g(\alpha) = F(g) - d_\xi g(\xi)$$

and finally that (5) holds for this g since it has constant value $g(\xi)$ on $[\xi,\eta]$. By transfinite induction, then, (5) holds for all $g \in J(\eta)$.

Now, if $\sum_{\alpha<\beta} (d_\alpha - d_{\alpha+1})e_\alpha$ converges for each $\beta < \gamma$ but $\sum_{\alpha<\gamma} (d_\alpha - d_{\alpha+1})e_\alpha$ fails to converge for some limit ordinal $\gamma \leq \eta$, pick $\varepsilon > 0$ and a sequence $\{\beta_n: n = 1,2,\ldots\}$ in $NLO(\eta)$ such that $0 < \beta_1 < \beta_2 < \ldots < \gamma$ and for each n

$$\Big\| \sum_{\beta_n\leq\alpha<\beta_{n+1}} (d_\alpha - d_{\alpha+1})e_\alpha \Big\| \geq 2\varepsilon .$$

Thus there are functions $f_n \in U(J(\eta))$ such that

$$\Big[\sum_{\beta_n\leq\alpha<\beta_{n+1}} (d_\alpha - d_{\alpha+1})e_\alpha\Big](f_n) \geq \varepsilon \qquad \text{for each } n$$

or, in other words, using the fact that the term in square brackets above converges (by hypothesis),

$$\sum_{\beta_n\leq\alpha<\beta_{n+1}} (d_\alpha - d_{\alpha+1})f_n(\alpha) \geq \varepsilon .$$

Define g_n as follows.

$$g_n(\alpha) \equiv \begin{cases} 0 & \text{if either } \alpha < \beta_n \text{ or } \beta_{n+1} \leq \alpha \leq \eta \\ f_n(\alpha) & \text{if } \beta_n \leq \alpha < \beta_{n+1} . \end{cases}$$

Then g_n is continuous since $\beta_n, \beta_{n+1} \in NLO(\eta)$ and it is easy to check that $\|g_n\| \leq 2\|f_n\| \leq 2$. Moreover,

$$\varepsilon \leq \sum_{\beta_n \leq \alpha < \beta_{n+1}} (d_\alpha - d_{\alpha+1}) f_n(\alpha) = \sum_{\beta_n \leq \alpha < \beta_{n+1}} (d_\alpha - d_{\alpha+1}) g_n(\alpha)$$

for each n. Now define f as follows.

$$f(\alpha) \equiv \begin{cases} 0 & \text{if } \alpha < \beta_1 \text{ or } \cup\beta_n \leq \alpha \leq \eta \\ \frac{1}{n} g_n(\alpha) & \text{if } \beta_n \leq \alpha < \beta_{n+1} \end{cases}$$

and note that $|f(\alpha)| \leq \frac{2}{n}$ whenever $\beta_n \leq \alpha \leq \beta_{n+1}$ for some n. Since each β_n belongs to $NLO(\eta)$, f is easily seen to be continuous. Clearly $f(0) = 0$. We now check that $\|f\| < \infty$, and in that regard, let $0 \leq \alpha_1 < \alpha_2 < \ldots < \alpha_k \leq \eta$ be any finite increasing sequence of ordinals. Observe that if for some n, $\beta_n \leq \alpha_j < \alpha_{j+1} < \ldots < \alpha_p < \beta_{n+1}$ then

$$\sum_{i=j}^{p-1} |f(\alpha_{i+1}) - f(\alpha_i)|^2 = \sum_{i=j}^{p-1} \frac{1}{n^2} \cdot |f_n(\alpha_{i+1}) - f(\alpha_i)|^2$$

$$\leq \frac{1}{n^2} \|f_n\|^2 \leq \frac{4}{n^2} .$$

Thus if the sum $\sum_{i=1}^{k-1} |f(\alpha_{i+1}) - f(\alpha_i)|^2$ is divided into three parts with

<u>part 1</u>: the sum of all terms $|f(\alpha_{i+1}) - f(\alpha_i)|^2$ where both α_i and α_{i+1} belong to some one of the intervals $[\beta_1, \beta_2)$, $[\beta_2, \beta_3)$, $\ldots, [\beta_n, \beta_{n+1})$, $\ldots$;

<u>part 2</u>: the sum of all terms $|f(\alpha_{i+1}) - f(\alpha_i)|^2$ where either $\alpha_{i+1} < \beta_1$ or $\alpha_i \geq \cup\beta_n$;

<u>part 3</u>: the sum of all terms $|f(\alpha_{i+1}) - f(\alpha_i)|^2$ where for some $n < m$, $\beta_n \leq \alpha_i < \beta_{n+1} \leq \beta_m \leq \alpha_{i+1} < \beta_{m+1}$;

then the sum from part 1 is dominated by $\sum_{n=1}^{\infty} \frac{4}{n^2}$ and that from part 2 is 0. The su from the third part is evidently dominated by $\sum_{n=1}^{\infty} (\frac{2}{n} + \frac{2}{n+1})^2$ and hence

$$\|f\| \leq \left[\sum_{n=1}^{\infty} \frac{4}{n^2} + \sum_{n=1}^{\infty} (\frac{2}{n} + \frac{2}{n+1})^2\right]^{\frac{1}{2}} < \infty .$$

But this leads to a contradiction. Indeed, by (5),

$$F(f) = \sum_{\alpha\in(0,\eta]} (d_\alpha - d_{\alpha+1})f(\alpha)$$

whence, in particular, $\sum_{0<\alpha<\gamma} (d_\alpha - d_{\alpha+1})f(\alpha) < \infty$. On the other hand,

$$\sum_{0<\alpha<\gamma} (d_\alpha - d_{\alpha+1})f(\alpha) = \sum_{n=1}^{\infty} \sum_{\beta_n\leq\alpha<\beta_{n+1}} (d_\alpha - d_{\alpha+1})f(\alpha)$$

$$\geq \sum_{n=1}^{\infty} \frac{1}{n} \sum_{\beta_n\leq\alpha<\beta_{n+1}} (d_\alpha - d_{\alpha+1})g_n(\alpha) \geq \sum_{n=1}^{\infty} \frac{1}{n}\varepsilon = \infty .$$

Thus the proof of Proposition 7.7.4 is complete.

The job of identifying a predual of $J(\eta)$ is not nearly so irksome.

PROPOSITION 7.7.5. Let $Y \equiv \overline{\text{span}(\{e_\alpha : \alpha \in (0,\eta] , \alpha \in NLO(\eta)\})} \subset J(\eta)^*$. ($e_\alpha$ is defined as in the statement of 7.7.4.) Then Y^* is isometrically isomorphic to $J(\eta)$.

Proof. Fix $f \in J(\eta)$, $f \neq 0$. Since f is continuous,

(6) $$\|f\| = \sup \{(\sum_{i=1}^{n-1} |f(\alpha_{i+1}) - f(\alpha_i)|^2)^{\frac{1}{2}} : 0 \leq \alpha_1 < \ldots < \alpha_n \leq \eta ;$$
$$\alpha_i \in NLO(\eta) \text{ for each } i ; n = 1,2,\ldots\}.$$

Thus, given $\varepsilon > 0$ there are $\alpha_1 < \alpha_2 < \ldots < \alpha_n \leq \eta$ with each $\alpha_i \in NLO(\eta)$ such that $\sum_{i=1}^{n-1} |f(\alpha_{i+1}) - f(\alpha_i)|^2 \geq \|f\|\cdot(\|f\| - \varepsilon)$. For each $i = 1,\ldots,n-1$ let

$$t_i \equiv \frac{f(\alpha_{i+1}) - f(\alpha_i)}{\|f\|} .$$

Then $\sum_{i=1}^{n-1} t_i^2 \leq 1$ and, using the Schwarz inequality,

$$(7) \qquad \left\| \sum_{i=1}^{n-1} t_i(e_{\alpha_{i+1}} - e_{\alpha_i}) \right\| = \sup \left\{ \sum_{i=1}^{n-1} t_i(g(\alpha_{i+1}) - g(\alpha_i)) : g \in U(J(\eta)) \right\}$$

$$\leq \sup \left\{ \left(\sum_{i=1}^{n-1} t_i^2 \right)^{\frac{1}{2}} \|g\|^2 : g \in U(J(\eta)) \right\} \leq 1 .$$

Consequently,

$$\|f\| \geq \left[\sum_{i=1}^{n-1} t_i(e_{\alpha_{i+1}} - e_{\alpha_i}) \right](f) = \sum_{i=1}^{n-1} \frac{f(\alpha_{i+1}) - f(\alpha_i)}{\|f\|} (f(\alpha_{i+1}) - f(\alpha_i))$$

$$= \frac{1}{\|f\|} \sum_{i=1}^{n-1} (f(\alpha_{i+1}) - f(\alpha_i))^2 \geq \|f\| - \varepsilon$$

and it follows that $\text{span}(\{e_\alpha : \alpha \in (0,\eta]$, $\alpha \in NLO(\eta)\}$ is a norming space of functionals for $J(\eta)$.

Let $\mathcal{T}$ denote the topology on $U(J(\eta))$ of pointwise convergence on the set of functionals $\{e_\alpha : \alpha \in (0,\eta], \alpha \in NLO(\eta)\}$. Evidently $\mathcal{T}$ is the same as the topology on $U(J(\eta))$ of pointwise convergence on $\text{span}(\{e_\alpha : \alpha \in (0,\eta], \alpha \in NLO(\eta)\})$ and, since $U(J(\eta))$ is a bounded set, it even coincides with the topology of pointwise convergence on Y. (We are identifying $J(\eta)$ with its canonical image $\hat{J}(\eta)$ in $J(\eta)^{**}$ of course.) Suppose now that (f_ξ) is a net in $U(J(\eta))$. By going to a subnet if necessary, we may assume that for each $\alpha \in NLO(\eta)$,

$$\lim_\xi f_\xi(\alpha) \equiv f(\alpha) \quad \text{exists.}$$

Note that if $0 \leq \alpha_1 < \ldots < \alpha_n \leq \eta$ and each $\alpha_i \in NLO(\eta)$ then $\sum_{i=1}^{n-1} (f(\alpha_{i+1}) - f(\alpha_i))^2 \leq 1$ so that f, defined on $NLO(\eta)$, has square variation norm at most 1. If $\tilde{f}$ denotes the unique extension of f to a continuous function on $[0,\eta]$ then $\|\tilde{f}\| = \|f\| \leq 1$, and since $\tilde{f}(0) = f(0) = \lim_\xi f_\xi(0) = 0$, $\tilde{f} \in U(J(\eta))$. Since (f_ξ) converges in the $\mathcal{T}$ topology to $\tilde{f}$ it follows that $(U(J(\eta)),\mathcal{T})$ is compact.

Suppose that $F \in Y^*$. Let $\tilde{F} \in J(\eta)^{**}$ be a norm preserving extension of F (possible since $Y \subset J(\eta)^*$) and let (f_ξ) be a net in $\|F\|U(J(\eta))$ such that $(\hat{f}_\xi)$ converges in the weak* topology on $J(\eta)^{**}$ to $\tilde{F}$. By going to a subnet if necessary assume further that (f_ξ) is $\mathcal{T}$-convergent to some $f \in \|F\|U(J(\eta))$. Then clearly $F = \tilde{F}|_Y = \hat{f}|_Y$. That is, the map $f \to \hat{f}|_Y$ maps $J(\eta)$ onto Y^* and since it is linear and Y norms $J(\eta)$, it is an isometric isomorphism.

The final result in this spate of identifications describes $J(\eta)^{**}$.

PROPOSITION 7.7.6. Let η be an infinite ordinal. Then $J(\eta)$ and $J(\eta)^{**}$ are isomorphic. Indeed, $J(\eta)^{**}$ is isometrically isomorphic to $\tilde{J}(\eta+1)$ and the natural set theoretic embedding of $J(\eta)$ into $\tilde{J}(\eta+1)$ corresponds to the canoncial embedding of $J(\eta)$ into $J(\eta)^{**}$.

Proof. Let $e_0 \equiv 0 \in J(\eta)^*$. Then each element of $J(\eta)^{**}$ is completely determined by its behavior on $\{e_\alpha : \alpha \in [0,\eta]\}$ since $\{e_\alpha : \alpha \in (0,\eta]\}$ is a basis for $J(\eta)^*$. Thus the elements of $J(\eta)^{**}$ may be identified with functions on $[0,\eta]$. With this correspondence in mind, if $g\colon [0,\eta] \to \mathbb{R}$ is in $J(\eta)^{**}$ then $g(0) = 0$ (since $e_0 = 0$) and if $0 \leq \alpha_1 < \ldots < \alpha_n \leq \eta$ then (letting

$$t_i \equiv \frac{g(\alpha_{i+1}) - g(\alpha_i)}{\left[\sum_{j=1}^{n-1} (g(\alpha_{j+1}) - g(\alpha_j))^2\right]^{\frac{1}{2}}} \qquad \text{for } i = 1,\ldots,n-1)$$

it follows as in (7) that

$$\left\| \sum_{i=1}^{n-1} t_i(e_{\alpha_{i+1}} - e_{\alpha_i}) \right\| \leq 1$$

and that

$$g\left(\sum_{i=1}^{n-1} t_i(e_{\alpha_{i+1}} - e_{\alpha_i})\right) = \left(\sum_{i=1}^{n-1} (g(\alpha_{i+1}) - g(\alpha_i))^2\right)^{\frac{1}{2}} .$$

Hence necessary conditions for $g\colon [0,\eta] \to \mathbb{R}$ to belong to $J(\eta)^{**}$ are that

(a) $g(0) = 0$; and

(b) $\|g\| < \infty$.

That is, necessary conditions are that $g \in \tilde{J}(\eta+1)$. If, on the other hand, $g \in \tilde{J}(\eta+1)$ and if $\gamma \leq \eta$ is a limit ordinal then it is easy to check that $\lim_{\alpha<\gamma} g(\alpha)$ exists (though it may differ from $g(\gamma)$). Let $g'(\alpha) \equiv g(\alpha)$ for $\alpha \in NLO(\eta)$ and $g'(\gamma) \equiv \lim_{\alpha<\gamma} g(\alpha)$ for each limit ordinal $\gamma \leq \eta$. Then $g' \in J(\eta)$ and $\|g'\| \leq \|g\|$. If $\gamma \leq \eta$ is a limit ordinal for which $\lim_{\alpha<\gamma} g(\alpha) \neq g(\gamma)$ then for each $\xi < \gamma$, $\xi \in NLO(\eta)$,

let

$$g_\xi^\gamma(\alpha) \equiv \begin{cases} g'(\alpha) & \text{if } \alpha \leq \xi \text{ or } \alpha \geq \gamma+1 \\ g(\gamma) & \text{if } \xi < \alpha \leq \gamma . \end{cases}$$

Then $g_\xi^\gamma \in J(\eta)$ and $\|g_\xi^\gamma\| \leq 2\|g\|$. Since $\|g\| < \infty$,

$$A \equiv \{\gamma \in [0,\eta] \setminus NLO(\eta): g(\gamma) \neq g'(\gamma)\}$$

is at most countable whence a net $\{g_\xi^\gamma: \xi < \gamma , \xi \in NLO(\eta) , \gamma \in A)$ may be construc in $J(\eta)$ such that $\|g_\xi^\gamma\| \leq 2\|g\|$ and $\lim_{\xi,\gamma} g_\xi^\gamma(\alpha) = g(\alpha)$ for $\alpha \in [0,\eta]$. Since th net (g_ξ^γ) (considered now as a net in $J(\eta)^{**}$) has a weak* convergent subnet whose limit has value $g(\alpha)$ at each $\alpha \in [0,\eta]$ it follows that $g \in J(\eta)^{**}$. That is, $(\tilde{J}(\eta+1),\|\cdot\|)$ and $(J(\eta)^{**},\|\cdot\|)$ are isometrically isomorphic. Moreover, the natura (identity) embedding of $J(\eta)$ into $\tilde{J}(\eta+1)$ corresponds to the embedding $f \to \hat{f}$ of $J(\eta)$ into $J(\eta)^{**}$. Finally, $\tilde{J}(\eta+1)$ and $J(\eta+1)$ are isometrically isomorphic (Lemma 7.7.2) and since it is easy to construct an isomorphism of $J(\eta+1)$ with $J(\eta$ (use 7.7.3 for example) it follows that $J(\eta)$ and $J(\eta)^{**}$ are isomorphic.

Let us now begin to reap the rewards of these past labors.

<u>THEOREM</u> 7.7.7. Let η be an infinite ordinal.

(1) The density character of $J(\eta)$ agrees with the density character of each of its duals.

(2) $J(\eta)$ is separable if and only if η is countable.

(3) $J(\eta)$ and each of its duals has the RNP.

(4) Y is the unique (up to isometry) predual of $J(\eta)$.

Proof. (1). This follows from the fact that $J(\eta)$ and $J(\eta)^{**}$ are isomorphic (Proposition 7.7.6) together with the inequality $\text{dens}(X) \leq \text{dens}(X^*)$ for every Banac space X.

(2). Since $\{h_\alpha: \alpha < \eta\}$ is a discrete subset of $S(J(\eta))$, if $J(\eta)$ is separable η must be countable. Conversely, since $\{h_\alpha: \alpha < \eta\}$ is a basis for $J(\eta)$ by Prop osition 7.7.3, if η is countable $J(\eta)$ is separable.

(3). It suffices, by Proposition 7.7.6, to establish that both $J(\eta)$ and $J(\eta)^*$ have the RNP, and since the proofs are similar, details are given only for $J(\eta)$. Let Z be a separable subspace of Y (the predual of $J(\eta)$ described in Proposition 7.7.5). The proof is completed by showing that Z^* is separable (see Corollary 4.1.7). Observe that if $F \in J(\eta)^*$ (and in particular, if $F \in Z \subset Y \subset J(\eta)^*$) has the representation $\sum_{\alpha\in(0,\eta]} c_\alpha e_\alpha$ for some choice of scalars c_α, then for all but countably many α's, $c_\alpha = 0$. (If for some $r > 0$ there were infinitely many α_i's with $|c_{\alpha_i}| \geq r$ then let $g_r \in \tilde{J}(\eta+1) \approx J(\eta)^{**}$ be the function

$$g_r(\alpha) \equiv \begin{cases} \frac{1}{n}\operatorname{sgn}(c_{\alpha_n}) & \text{if } \alpha = \alpha_n \\ 0 \text{ otherwise} & . \end{cases}$$

Then $\|g_r\| < \infty$ and $g_r(0) = 0$—so $g_r \in \tilde{J}(\eta+1)$ as stated—but

$$g_r\Big(\sum_{\alpha\in(0,\eta]} c_\alpha e_\alpha\Big) = \sum_{i=1}^{\infty} c_{\alpha_i} g(\alpha_i) = \infty \text{ .)}$$

Thus for each $F \in Z$ there are at most countably many α's used in the basis representation $\sum_{\alpha\in(0,\eta]} c_\alpha e_\alpha$ of F. Since Z is separable, there is a countable set $C \subset [0,\eta] \cap NLO(\eta)$ such that $Y_1 \equiv \overline{\operatorname{span}}(\{e_\alpha : \alpha \in C\}) \supset Z$. Observe that the closure $\overline{C}$ of C in $[0,\eta]$ (with the order topology) is order isomorphic to some countable ordinal η_1. But Proposition 7.7.5 shows that Y_1^* is isometrically isomorphic to $J(\eta_1)$, a separable Banach space by part (2) of this theorem, and since Z^* is the continuous image of Y_1^*, Z^* is separable as well.

(4). Y is a subspace of $J(\eta)^*$ and by part (3) of this theorem, $J(\eta)^*$ has the RNP. Hence Y has the RNP as well. The conclusion now follows from Theorem 5.7.6(a).

We henceforth concentrate on the peculiarities of the space $J(\omega_1)$ where ω_1 denotes the first uncountable ordinal. The notation used below is defined in 4.3.1.

PROPOSITION 7.7.8. Each $F \in J(\omega_1)^*$ is $\mathcal{B}(J(\omega_1),\text{weak}^*)$ measurable. On the other hand, the functional $e_{\omega_1} \in J(\omega_1)^*$ is not $\text{Baire}(J(\omega_1),\text{weak}^*)$ measurable.

Proof. As indicated in the proof of Proposition 7.7.7, each $F \in J(\omega_1)^*$ is of the form $\sum_{\alpha\in(0,\omega_1]} c_\alpha e_\alpha$ where, for all but countably many α, $c_\alpha = 0$. Thus in ord to show that each F is $\mathcal{B}(J(\omega_1),\text{weak*})$ measurable, it suffices to show that this is the case for each e_γ $(\gamma \in (0,\omega_1])$. First observe that if $\gamma \in NLO(\omega_1)$ then e is weak* continuous. Next, if γ is a limit ordinal then for each $r \in \mathbb{R}$, posit integer k and rational number $q > 0$ let

$$A(r,k,q) \equiv \cup\{g \in J(\gamma): \sum_{i=1}^{n} (g(\alpha_{i+1}) - g(\alpha_i))^2 > q^2 - \frac{1}{k^2}; \; g(\alpha_n) > r - \frac{1}{k};$$

$$0 \leq \alpha_1 < \dots < \alpha_n \leq \gamma \; ; \; \alpha_1 \in NLO(\gamma) \text{ for } i=1,\dots,n; \; n = 2,3,\dots\}.$$

Then $A(r,k,q)$ is weak* open for each such triple (r,k,q). Hence the set

$$B(r,k,q) \equiv q\cdot U(J(\gamma)) \cap A(r,k,q)$$

is in $\mathcal{B}(J(\gamma),\text{weak*})$ since it is the intersection of a weak* closed and a weak* ope set. But

$$\{g \in J(\gamma): e_\gamma(g) < r\} = \bigcap_{k=1}^{\infty} \bigcup_{q>0} B(r,k,q)$$

whence e_γ is $\mathcal{B}(J(\gamma),\text{weak*})$ measurable. Finally, if $\gamma < \omega_1$ is a limit ordinal then the restriction map $\psi_\gamma: J(\omega_1) \to J(\gamma)$ is both weak*-weak* continuous and surjective. Thus

$$\{f \in J(\omega_1): e_\gamma(f) < r\} = \{f \in J(\omega_1): e_\gamma(\psi_\gamma(f)) < r\}$$
$$= \psi_\gamma^{-1}(\{g \in J(\gamma): e_\gamma(g) < r\}) \in \mathcal{B}(J(\omega_1),\text{weak*}) \quad .$$

In order to check that e_{ω_1} is not Baire$(J(\omega_1),\text{weak*})$ measurable, observe firs that $e_{\omega_1}(h_{\omega_1}) = 0$ and that $e_{\omega_1}(h_\alpha) = 1$ for $\alpha < \omega_1$. But $(\{h_\alpha: \alpha \in [0,\omega_1]\},\text{wea}$ is homeomorphic to $[0,\omega_1]$ and since each continuous function on $[0,\omega_1]$ is eventually constant, each Baire function on $[0,\omega_1]$ is eventually constant. Since e_{ω_1} does not have this property, e_{ω_1} is not Baire$(J(\omega_1),\text{weak*})$ measurable.

Since $J(\omega_1)$ has the RNP by Theorem 7.7.7 part 3, and since $J(\omega_1)$ is a dual space by Proposition 7.7.5, it follows from Theorems 4.4.4 and 4.4.1(h) that $\mathcal{U}_t(\mathcal{B}(J(\omega_1),\text{norm})) = \mathcal{U}_t(\mathcal{B}(J(\omega_1),\text{weak})) = \mathcal{U}_t(\mathcal{B}(J(\omega_1),\text{weak*}))$. Nevertheless, "erasing" the $\mathcal{U}_t$'s can lead to trouble as is indicated next.

PROPOSITION 7.7.9. $\mathcal{B}(J(\omega_1),\text{weak}) \neq \mathcal{B}(J(\omega_1),\text{weak*})$.

Proof. As indicated in the previous proposition, $(\{h_\alpha: \alpha \in [0,\omega_1]\},\text{weak*})$ is homeomorphic to $[0,\omega_1]$. On the other hand, $(\{h_\alpha: \alpha \in [0,\omega_1]\},\text{weak})$ is discrete since, for $\alpha < \omega_1$

$$\{\beta \in [0,\omega_1]: (e_{\alpha+1} - e_\alpha)(h_\beta) > \tfrac{1}{2}\} = \{h_\alpha\}$$

and (for $\alpha = \omega_1$)

$$\{\beta \in [0,\omega_1]: e_{\omega_1}(h_\beta) < \tfrac{1}{2}\} = \{h_{\omega_1}\} .$$

Thus each subset of $\{h_\alpha: \alpha \in [0,\omega_1]\}$ belongs to $\mathcal{B}(J(\omega_1),\text{weak})$ while there are subsets of $\{h_\alpha: \alpha \in [0,\omega_1]\}$ which fail to belong to $\mathcal{B}(J(\omega_1),\text{weak*})$. (Indeed, if $A \subset [0,\omega_1))$ is such that neither it nor its complement in $[0,\omega_1)$ contains a closed unbounded set then $\{h_\alpha: \alpha \in A\}$ is not in $\mathcal{B}(J(\omega_1),\text{weak*})$.)

The last two results are typical of themes dealt with in detail in Edgar [1977], [1979], and Talagrand [1977a], for example, in which numerous measure theoretic notions in Banach spaces are compared and contrasted. (Each of the terms appearing in Theorem 7.7.1(d) is discussed in Edgar [1977] and [1979].) Briefly, recall that X is <u>real compact</u> if (X,weak) is homeomorphic to a closed subset of a product of lines (with the product topology). Hewitt [1950] characterized real compact spaces Y as those for which every 2-valued measure μ on Baire(Y) is τ-smooth - i.e. whenever (f_α) is a net of continuous real-valued bounded functions on Y decreasing pointwise to 0 then $\lim_\alpha \int f_\alpha d\mu = 0$. X is <u>measure compact</u> if and only if <u>every</u> probability measure on Baire(X,weak) is τ-smooth. It is known (Edgar [1977], Proposition 5.4) that X is measure compact if and only if whenever $(\Omega,\mathcal{F},\mu)$ is a probability space and $\phi: \Omega \to X$ is weakly measurable then there is an $\mathcal{F} - \mathcal{B}(X)$ measurable function $\psi: \Omega \to X$ such that

$$\mu(\{\omega \in \Omega: F\circ\psi(\omega) \neq F\circ\phi(\omega)\}) = 0$$

for each $F \in X^*$. (Edgar [1979] calls this latter property of a Banach space

μ-measure compact.) If (Ω, F, μ) is a probability space and X is μ-measure compa then X has the μ-PIP (Edgar [1979] , Proposition 3.1) whence, if X is measure compact, then it has the PIP. (See the discussion after 7.4.1 for information about the Pettis Integral Property.) Edgar [1979] , Proposition 4.3 showed that if X ha the PIP then X is real compact. Moreover, if X is isomorphic to a subspace of a weakly compactly generated Banach space then (X,weak) is Lindelöf (since this is true of weakly compactly generated spaces in their weak topologies. See Talagrand [1975] . Alternatively, 7.8.7(d) contains a short proof that a weakly compactly gen ated space is weakly K-analytic -cf. Definition 7.8.2(b). Then use Lemma 7.8.8 to s that such a space is automatically Lindelöf. Finally, it is clear that a closed sub set of a Lindelöf space is Lindelöf.) Since Lindelöf spaces are measure compact (Varadarajan [1965] , p. 175) spaces X which are Lindelöf in their weak topologie (and in particular, spaces isomorphic to subspaces of weakly compactly generated spaces) are real compact. Thus, to summarize, if X has the PIP, or if X is measure compact, or if (X,weak) is Lindelöf, or if X is isomorphic to a subspace of a weakly compactly generated space, then X is real compact. This will be helpful in the proof of 7.7.1(d).

One last diversion will be needed in order to prepare for 7.7.1(e). We note th following useful result of W. Schackermeyer (see Edgar [1979] , Theorem 2.1).

PROPOSITION 7.7.10. Let $\mathcal{T}$ be a locally convex topology on X for which U(X) i $\mathcal{T}$-closed. Then the map $(t,x) \to tx$ is a Borel isomorphism of

$$([0,\infty) \times S(X), \mathcal{B}([0,\infty)) \times \mathcal{B}(S(X),\mathcal{T})) \quad \text{onto} \quad (X \setminus \{0\}, \mathcal{B}(X \setminus \{0\},\mathcal{T})) \quad .$$

Proof. First of all the map $(t,x) \to tx$ from $[0,\infty) \times S(X)$ onto $X \setminus \{0\}$ is continuous, hence Borel measurable. It is clearly one-one. Moreover, the inverse m takes $x \in X \setminus \{0\}$ to $(\|x\|, \frac{x}{\|x\|})$. The first coordinate of this map is $x \to \|x\|$ wh is lower semicontinuous since U(X) is $\mathcal{T}$-closed, while the second is the composit of the map $x \to (\|x\|, x)$ with the continuous map $(\|x\|, x) \to \frac{x}{\|x\|}$. Hence each coord nate map for the inverse of $(t,x) \to tx$ is Borel measurable and the proof is compl

There are several conclusions of interest from this seemingly modest result.

(See, for example, Edgar [1979], p. 562 for some examples.) We shall be satisfied with establishing the following corollary here (which extends Proposition 4.4.2.)

COROLLARY 7.7.11. Let X be a Banach space which satisfies the following condition: on $S(X^*)$ the $\mathcal{T}_1$ and $\mathcal{T}_2$ topologies coincide (where the $\mathcal{T}_1$ and $\mathcal{T}_2$ topologies refer to any two locally convex topologies which make $U(X^*)$ closed. Thus $\mathcal{T}_1$ and $\mathcal{T}_2$ may each be any of the three topologies: norm, weak, or weak* on X^*.) Then $\mathcal{B}(X^*,\mathcal{T}_1) = \mathcal{B}(X^*,\mathcal{T}_2)$.

Proof. By hypothesis the identity map $id: (S(X^*),\mathcal{T}_1) \to (S(X^*),\mathcal{T}_2)$ is a homeomorphism. By Proposition 7.7.11 the identity map $id: (X^* \setminus \{0\},\mathcal{T}_1) \to (X^* \setminus \{0\},\mathcal{T}_2)$ is a Borel isomorphism. It is easy to check that this forces $\mathcal{B}(X^*,\mathcal{T}_1) = \mathcal{B}(X^*,\mathcal{T}_2)$.

Here, finally, is the proof of 7.7.1.

Proof of 7.7.1. Part (a) was established in Proposition 7.7.6 and Theorem 7.7.7 part 3, while part (c) of 7.7.1 is Proposition 7.7.9. From the discussion above Theorem 7.7.10, parts (b) and (d) will follow once it is shown that (X,weak) is not homeomorphic to a closed subset of a product of lines (i.e. (X,weak) is not real compact). In fact, by Hewitt's characterization mentioned above 7.7.10, it suffices to produce a 2-valued measure P on Baire($J(\omega_1)$,weak) which is not τ-smooth.

On Baire($[0,\omega_1)$,order topology) define the 2-valued measure μ by

$$\mu(B) \equiv \begin{cases} 0 & \text{if } B \text{ is countable} \\ 1 & \text{if } [0,\omega_1) \setminus B \text{ is countable} \end{cases} .$$

Let $\psi:[0,\omega_1) \to \{h_\alpha: \alpha \in [0,\omega_1]\} \subset J(\omega_1)$ be the map $\psi(\alpha) \equiv h_\alpha$. Now for any $\beta < \omega_1$, $e_\beta \circ \psi(\alpha_1) \le e_\beta \circ \psi(\alpha_2)$ whenever $\alpha_2 \le \alpha_1$ so that $e_\beta \circ \psi$ is a decreasing function, hence Baire($[0,\omega_1)$) measurable for each $\beta < \omega_1$. But if $F \in J(\omega_1)^*$ then $F = \sum_{\alpha\in(0,\omega_1]} c_\alpha e_\alpha$ for some choice of real numbers c_α all but countably many of which are 0 (see the proof of Theorem 7.7.7) and it follows that $F \circ \psi$ is Baire($[0,\omega_1)$) measurable for each $F \in J(\omega_1)^*$. But from Edgar [1977] Theorem 2.3, Baire($J(\omega_1)$, weak) is the smallest σ-algebra for which all the elements of $J(\omega_1)^*$ are measurable, and hence the measure $P \equiv \psi(\mu)$ is a 2-valued measure on $\text{Baire}(J(\omega_1),\text{weak})$. For each $\alpha < \omega_1$ let

$$Z_\alpha \equiv \{f \in J(\omega_1)\colon f(\beta) = 0 \quad \text{for } \beta \leqq \alpha \text{ and } f(\omega_1) = 1\} .$$

Then $Z_\alpha \in \text{Baire}(J(\omega_1),\text{weak})$ and $P(Z_\alpha) = 1$ for each $\alpha < \omega_1$. Moreover, if $\alpha < \beta < \omega_1$ then $Z_\beta \subset Z_\alpha$; and finally, $\bigcap_{\alpha<\omega_1} Z_\alpha = \emptyset$. But then $\{\chi_{Z_\alpha} : \alpha < \omega_1\}$ is a decreasing net of continuous functions converging pointwise to 0 and since $\int_{J(\omega_1)} \chi_{Z_\alpha} dP = P(Z_\alpha) = 1$ it follows that P is not τ-smooth. Thus from the above remarks, the proof of 7.7.1 parts (b) and (d) is complete.

It thus remains to establish 7.7.1(e). If $J(\omega_1)$ had an equivalent dual norm $|||\cdot|||$ which were weakly locally uniformly convex then in the $|||\cdot|||$ unit sphere the weak and weak* topologies would coincide (as is clear from the definition). Hence, from Corollary 7.7.11, $\mathcal{B}(J(\omega_1),\text{weak*}) = \mathcal{B}(J(\omega_1),\text{weak})$. But from 7.7.1 part (c), this is not the case. Hence no such norm exists on $J(\omega_1)$.

§8. More on measurability and the RNP.

In this section the analysis (begun in sections 4.3 and 4.4) of the interplay between various measurability criteria and the RNP is carried forward. Besides a few brief remarks about the use of lifting theorems, some work of Saab [1980a], [1980b] and [1981a] is summarized; further details and results may be found in these papers and their references. Schwartz [1973] is recommended for measure theoretic background.

For a topological space $(D,\mathcal{T})$ the notation $P_t(D,\mathcal{T})$ $(P_t(D))$ and $\mathcal{B}(D,\mathcal{T})$ ($\mathcal{B}$(D)) is as in section 4.3, and for $\mu \in P_t(D,\mathcal{T})$, M_μ denotes the μ-completion of $\mathcal{B}(D,\mathcal{T})$ (cf. 4.3.8). The central notions appear in the next two definitions.

DEFINITION 7.8.1. Let $(D_1,\mathcal{T}_1)$ and $(D_2,\mathcal{T}_2)$ be Hausdorff topological spaces and $f\colon D_1 \to D_2$.

(a) If $\mu \in P_t(D_1,\mathcal{T}_1)$ then f is *μ-Lusin-measurable* if for each $\varepsilon > 0$ and compact subset K of D_1 there is a compact set $K_\varepsilon \subset K$ such that $\mu(K \setminus K_\varepsilon) < \varepsilon$ and $f|_K : K_\varepsilon \to D_2$ is continuous. (Here and elsewhere, $g|_M$ denotes the restriction of a function g to the subset M of its domain.)

(b) f is universally Lusin-measurable (abbreviated universally measurable sometimes) if f is μ-Lusin-measureable for each $\mu \in P_t(D_1, T_1)$.

The standard notation $\sigma(X,X^*)$ and $\sigma(X^*,X)$ will be used in this section when convenient to denote, respectively, the weak topology on X and the weak* topology on X* . Part (a) of the next definition is due to Choquet [1959] .

DEFINITION 7.8.2 (a). Let (D,T) be a topological space. Then D is said to be K-analytic if there is a $K_{\sigma\delta}$ subset C_1 of some compact Hausdorff space C and a continuous surjective map $f: C_1 \to D$. (Recall, C_1 is a $K_{\sigma\delta}$ set if it is a countable intersection of sets each of which is a countable union of compact sets.)
(b). If D is a subset of a Banach space X, D is weakly K-analtyic if $(D,\sigma(X,X^*))$ is K-analytic.

The following elementary consequence of Theorem 12, p. 125 of Schwartz [1973] will be used in several places in this section.

THEOREM 7.8.3. Suppose that on a set D there are two comparable topologies T_1 and T_2 with T_2 finer than T_1 and T_1 completely regular. If (D,T_2) is K-analytic then the identity map $id: (D,T_1) \to (D,T_2)$ is universally measurable.

We shall also have occasion to use a "weak* Radon-Nikodým" theorem. The one appearing in 7.8.4(a) below is a special case of ones in, for example, Dinculeanu [1967] , Theorem 4, p. 263 and Weizsäcker [1978] , Corollary 1.7, p. 210. In 7.8.4(b) we have taken the special case of Weizsäcker's theorem which will be needed in the proof of Theorem 7.8.11. It is significantly stronger than 7.8.4(a) and its proof is technically more intricate than that of 7.8.4(a). (Since the use of lifting theorems is well illustrated in the proof of 7.8.4(a), the proof of 7.8.4(b) is not included. However, we remark for those interested in 7.8.4(b) that Weizsäcker's hypothesis in Corollary 1.7 of [1978] that the cylindrical measure associated with the vector measure be concentrated on C is satisfied for m as in 7.8.4(b) since $(Y^*,\sigma(Y^*,Y))$ and $(Y,\sigma(Y,Y^*))$ are in duality so the fact that $AR(m) \subset C$ allows us to apply Proposition 5, p. 190 of Schwartz [1973] .) Much use of liftings has been made in the search for "weak densities" of vector measures. In addition to the references above, we mention Goldman [1977] and Edgar and Talagrand [1980] (who discuss some limitations of liftings in such a context).

Recall that for $y \in Y$, $\hat{y}$ is the canonical image of y in the Banach space Y^{**}.

THEOREM 7.8.4. Let (Ω, F, P) be a complete probability space, Y a Banach space, $m: F \to Y^*$ a vector measure, $C \subset Y^*$ a weak* compact convex set and assume that $m \ll P$ and $AR(m) \subset C$.

(a) Then there is an $f: \Omega \to C$ such that for each $y \in Y$ the map $\hat{y} \circ f: \Omega \to \mathbb{R}$ is F-measurable and

$$\int_A \hat{y} \circ f dP = m(A)(y) \qquad \text{for each } y \in Y \text{ and } A \in F .$$

(b) Under the general hypotheses of 7.8.4 there is a $g: \Omega \to C$ such that

(i) $g^{-1}(B) \in F$ for each $B \in \mathcal{B}(C, \text{weak*})$;

(ii) $g(P)$, the distribution of P by g (see 4.3.7) belongs to $P_t(C, \text{weak*})$;

(iii) $m(A)(y) = \int_A \hat{y} \circ g dP$ for each $y \in Y$ and $A \in F$.

Proof of (a). For each $y \in Y$ let $m_y: \Omega \to \mathbb{R}$ be the scalar measure on F

$$m_y(A) \equiv M(A)(y) \quad \text{for } A \in F .$$

Then $m_y \ll P$ and in fact

$$AR(m_y) \subset C(y) \equiv \{h(y): h \in C\} \subset \mathbb{R} . \tag{1}$$

It follows that there is a Radon-Nikodým derivative g_y of m_y with respect to P which takes <u>all</u> its values in $C(y)$. That is, there is an F-measurable function $g_y: \Omega \to C(y) \subset \mathbb{R}$ such that

$$\int_A g_y dP = m(A)(y) \qquad \text{for each } y \in Y \text{ and } A \in F . \tag{2}$$

The idea is to somehow define $f: \Omega \to Y^*$ using the g_y's. Observe that for each $r, s \in \mathbb{R}$ and $x, y \in Y$

$$r g_x + s g_y = g_{rx+sy} \quad [a.s.P] \tag{3}$$

follows easily from (2). The obvious choice for f: $f(\omega)(x) \equiv g_x(\omega)$, cannot work as the g_y's are now defined since for some ω's , there are $x,y \in Y$ and $r,s \in \mathbb{R}$ with $rg_x(\omega) + sg_y(\omega) \neq g_{rx+sy}(\omega)$ and hence $f(\omega)$ would not be linear. It is easy enough to fix this up by an elementary application of Zorn's lemma to produce a new set of g_y's which differ from their old counterparts on sets of measure zero and for which (3) holds at each $\omega \in \Omega$. The cost of producing these modified g_y's is that their ranges no longer necessarily lie in the respective sets $C(y)$. Consequently although $f(\omega)(x) \equiv g_x(\omega)$ now makes $f(\omega)$ linear on X for each ω , there is no guarantee that $f(\omega) \in C$ or even that $f(\omega)$ is continuous. In order, then, to simultaneously fix (3) up so that the "[a.s.P]" may be eliminated without increasing the size of the range of the g_y's , we may invoke a countability argument which goes back to Dunford and Pettis (see Diestel and Uhl [1977] , p. 80) in the special case when C is norm separable, or, in case C is not necessarily separable, appeal to a lifting theorem. We adopt this latter approach. According to Schwartz [1973] , Theorem 15, p. 131 (or to Dinculeanu [1967] , Theorem 1, p. 206, or to Ionescu Tulcea [1969] , Theorem 3, p. 46) there is a lifting on $M^\infty(P)$. (For $f: \Omega \to \mathbb{R}$ let

$$\text{ess sup}(f) \equiv \inf \{\sup \{|f(\omega)| : \omega \in \Omega \setminus N \ ; N \in F;\ P(N) = 0\}.$$

Then define

$$M^\infty(P) \equiv \{f: \Omega \to \mathbb{R}: f \text{ is } F\text{-measurable and } \text{ess sup}(f) < \infty\} .$$

Let $\pi: M^\infty(P) \to L^\infty(P)$ be the natural quotient map. A <u>lifting</u> on $M^\infty(P)$ is a mapping $\rho: M^\infty(P) \to M^\infty(P)$ with the following—somewhat redundant set of—properties:

(A) $\pi(\rho(f)) = \pi(f)$ for $f \in M^\infty(P)$;

(B) If $\pi(f) = 0$ then $\rho(f) = 0$ (i.e. $\rho(f)(\omega) = 0$ for each $\omega \in \Omega$) ;

(C) ρ is linear and multiplicative on $M^\infty(P)$;

(D) $\|\rho(f)\|_\infty \equiv \sup \{|\rho(f)(\omega)| : \omega \in \Omega\} < \infty$ for $f \in M^\infty(P)$;

(E) $\text{ess sup}(f) = \|\rho(f)\|_\infty$ for each $f \in M^\infty(P)$ (so that ρ is an isometry from the ess sup norm to the sup norm) ;

(F) $\rho(1) = 1$ and $\rho(f) \geq 0$ if $f \geq 0$ for $f \in M^\infty(P)$.)

The proof of the theorem now proceeds in a straightforward manner. Indeed, for each $y \in Y$ replace g_y by $\rho(g_y)$. Since $\rho(g_y)$ and g_y differ on a set of P-measure zero by (A) it follows that $\rho(g_y)$ is a Radon-Nikodým derivative of m_y wi respect to P. Let $\beta_y \equiv \sup \{h(y) : h \in C\}$ for each $y \in Y$ and observe from (1) that

(4) $$g_y \leqq \beta_y \cdot 1 \ .$$

We conclude from (F) above that $\rho(g_y) \leqq \beta_y \cdot 1$ for each $y \in Y$. Now let $f(\omega)(x) \equiv \rho(g_x)(\omega)$ for $x \in Y$ and $\omega \in \Omega$. To check that $f(\omega)$ is linear on Y for each ω pick $x, y \in Y$ and $r, s \in \mathbb{R}$. Then (3) may be rephrased as $\pi(rg_x + sg_y - g_{rx+sy}) =$ and the linearity of ρ together with (B) above yields $r\rho(g_x) + s\rho(g_y) = \rho(g_{rx+s}$ In other words,

$$f(\omega)(rx + sy) = rf(\omega)(x) + sf(\omega)(y) \ .$$

Finally, if for some $\omega \in \Omega$, $f(\omega) \notin C$ then there is a $y \in Y$ such that $f(\omega)(y) >$ Since this contradicts (4), and since $\rho(g_y) = \frac{dm_y}{dP}$ from above, the proof is complete

The following proposition contains some measure theoretic background which will soon be used. The proof of part (a) may be found in Schwartz [1973] , p. 26, Theore 5 while parts (b) and (c) follow in order by elementary manipulation of the relevant definitions. Part (d) is a disguised form of our Theorem 4.4.4, and, like that theorem, is due to Schwartz [1973] , p. 162.

<u>PROPOSITION</u> 7.8.5. Let $(D_i, \mathcal{T}_i)$ be a completely regular topological space for $i =$ 1,2,3.

(a) If $f: D_1 \to D_2$ is μ-Lusin-measurable (for some $\mu \in \mathcal{P}_t(D_1)$) then $f^{-1}(B) \in \mathcal{M}_\mu$ for each $B \in \mathcal{B}(D_2, \mathcal{T}_2)$. Moreover, if $\nu \in \mathcal{P}_t(D_1)$, $g: D_1 \to D_2$ and $(D_2, \mathcal{T}_2)$ i separable metric then g is ν-Lusin-measurable if and only if $g^{-1}(B) \in \mathcal{M}_\nu$ for ea $B \in \mathcal{B}(D_2, \mathcal{T}_2)$.

(b) If $\mu \in \mathcal{P}_t(D_1)$ and $f: D_1 \to D_2$ is μ-Lusin-measurable then $f(\mu) \in \mathcal{P}_t(D_2$ ($f(\mu)$ is the distribution of μ by f as in 4.3.7.)

(c) If $\mu \in \mathcal{P}_t(D_1)$, if $f: D_1 \to D_2$ is μ-Lusin-measurable and $g: D_2 \to D_3$ i

$f(\mu)$-Lusin-measurable then $g \circ f$ is μ-Lusin-measurable. Consequently, if both f and g are universally measurable then so is $g \circ f$.

(d) Let D be a weakly closed convex subset of a Banach space. Then the identity map $\mathrm{id}: (D,\text{weak}) \to (D,\|\cdot\|)$ is universally measurable.

Proof. See the remarks preceding the statement of the proposition for parts (a), (b) and (c). As for the proof of (d), let K be a weakly compact subset of D and $\mu \in P_t(D,\text{weak})$. According to Theorem 4.4.4 there is a norm separable and closed subset A of K with $\mu(K \setminus A) = 0$. Evidently $\mathcal{B}(A,\|\cdot\|) = \mathcal{B}(A,\text{weak})$ and by Lemma 4.3.10 the restriction of μ to A, $\mu|_A$, is inner regular with respect to norm compact sets. Hence given $\varepsilon > 0$ there is a norm compact set $K_\varepsilon \subset A$ with $\mu(A \setminus K_\varepsilon) < \varepsilon$. Then $\mathrm{id}: (K_\varepsilon,\text{weak}) \to (K_\varepsilon,\|\cdot\|)$ is a homeomorphism and is, in particular, continuous. Since $\mu(K \setminus K_\varepsilon) = \mu(A \setminus K_\varepsilon) < \varepsilon$, the proof is complete.

The main theorem of Saab [1980a] appears next. It is very close in spirit and in proof to parts of Theorem 4.3.11.

THEOREM 7.8.6. Let C be a weak* compact convex subset of X^*. Then the following are equivalent:

(a) C has the RNP;

(b) The identity map $\mathrm{id}_1: (C,\text{weak*}) \to (C,\|\cdot\|)$ is universally measurable;

(c) For each separable subspace Y of X, the identity map $\mathrm{id}_2: (C|_Y,\sigma(Y^*,Y)) \to (C|_Y,\|\cdot\|)$ is universally measurable. (Recall, $C|_Y \equiv \{h|_Y : h \in C\} \subset Y^*$.)

(d) For every subspace Y of X, the identity map $\mathrm{id}_3: (C|_Y,\sigma(Y^*,Y)) \to (C|_Y,\|\cdot\|)$ is universally measurable.

Proof. (a) ⇒ (b) Let $\mu \in P_t(C,\sigma(X^*,X))$. According to Theorem 4.3.11(b) there is a norm separable closed bounded convex subset C_1 of C with $\mu(C_1) = 1$ and by taking $Y \equiv \overline{\mathrm{span}}(C_1)$ and using Lemma 4.3.6 we see that $Y \in \mathcal{B}(X^*,\text{weak*})$ is a separable subspace of X^* such that

$$\mu(C \cap Y) = 1 . \tag{5}$$

Since $(Y,\|\cdot\|)$ is a complete separable metric space it is homeomorphic to a G_δ

subset of a compact metric space and is thus K-analytic. According to Theorem 7.8. then,

$$id_4: (Y,\sigma(X^*,X)) \to (Y,\|\cdot\|)$$

is universally measurable and consequently so is the map

$$\text{(6)} \qquad I: (Y,\sigma(X^*,X)) \to (X,\|\cdot\|) ,$$

the composition of id_4, a universally measurable map, with the injection of $(Y,\|\cdot\|)$ into $(X,\|\cdot\|)$. It follows from (5) and (6) and the definition of μ-Lusin-measurability that $id_1: (C,\sigma(X^*,X)) \to (C,\|\cdot\|)$ is also μ-Lusin-measurable. Since $\mu \in P_t(C,\sigma(X^*,X))$ was arbitrary, id_1 is universally measurable.

<u>(b) ⇒ (d)</u> Let Y be a subspace of X, let $J: (Y,\|\cdot\|) \to (X,\|\cdot\|)$ denote the natura injection and let J^* be its adjoint. By hypothesis

$$id_1: (C,\sigma(X^*,X)) \to (C,\|\cdot\|)$$

is universally measurable. Consequently $J^*\circ id_1: (C,\sigma(X^*,X)) \to (J^*(C),\|\cdot\|)$ is universally measurable as well. Suppose now that $\mu \in P_t(J^*(C),\sigma(Y^*,Y))$. Since $J^*(C$ is compact and the map $J_1^*: (C,\sigma(X^*,X)) \to (J^*(C),\sigma(Y^*,Y))$ defined by $J_1^*(h) \equiv J^*($ for all $h \in C$ is both continuous and surjective, there is a $\nu \in P_t(C,\sigma(X^*,X))$ wit $J_1^*(\nu) = \mu$ by Lemma 4.3.9(b). Observe that $J^*\circ id_1 = id_3\circ J_1^*$ (where id_3 is the map defined in the statement of 7.8.6(d)). Thus $id_3\circ J_1^*$ is universally measurable and in particular, $id_3\circ J_1^*$ must be ν-Lusin-measurable. Since C is $\sigma(Y^*,Y)$ compact it is easy to check that id_3 is in fact $J_1^*(\nu)$-Lusin-measurable and since $J_1^*(\nu)$ $= \mu \in P_t(J^*(C),\sigma(Y^*,Y))$ was arbitrary, id_3 is universally measurable as was to be proved.

<u>(d) ⇒ (c)</u> is obvious.

<u>(c) ⇒ (a)</u> Let K be a separable closed bounded convex subset of C. Choose a separable subspace Y of X such that the restirction map $J^*: K \to K|_Y \subset Y^*$ is an isometry. Let λ denote Lebesgue measure on $[0,1]$ and choose any vector measure $m: M_\lambda \to Y^*$ with $m \ll \lambda$ and $AR(m) \subset K|_Y$. (Recall: M_λ denotes the Lebesgue measurable subsets of $[0,1]$.) Since $K|_Y$ is $\sigma(Y^*,Y)$ compact, according to Theorem

7.8.4(a) there is an $f: [0,1] \to K|_Y$ such that $\hat{y}\circ f$ is $\mathcal{M}_\lambda$ measurable for each $y \in Y$ and

(7) $$\int_A \hat{y}\circ f d\lambda = m(A)(y) \quad \text{for } A \in \mathcal{M}_\lambda \text{ and } y \in Y .$$

Note that $\hat{y}\circ f$ is λ-Lusin-measurable by 7.8.5(a). In fact, as we show next, $f: [0,1] \to (K|_Y, \text{weak*})$ is also λ-Lusin-measurable. To see this, let $S \equiv \{y_n: n = 1,2,...\}$ be a countable dense subset of $U(Y)$ and let $\mathcal{T}$ denote the topology on $K|_Y$ of pointwise convergence on S. Clearly $f: [0,1] \to (K|_Y, \mathcal{T})$ is λ-Lusin-measurable. But $\mathcal{T}$ is Hausdorff on $K|_Y$, and formally weaker than the weak* topology which makes $K|_Y$ compact. Hence the $\mathcal{T}$ and $\sigma(Y^*,Y)$ topologies agree on $K|_Y$. Thus $f: [0,1] \to (K|_Y, \sigma(Y^*,Y))$ is indeed λ-Lusin-measurable.

Since $id_2: (K|_Y, \sigma(Y^*,Y)) \to (K|_Y, \|\cdot\|)$ is universally measurable by hypothesis, it follows that $f: [0,1] \to (K|_Y, \|\cdot\|)$ is λ-Lusin-measurable by 7.8.5(c). Note that f is then $\mathcal{M}_\lambda - \mathcal{B}(K|_Y, \|\cdot\|)$ measurable by 7.8.5(a) and since f has bounded and separable range, f is Bochner integrable. Consequently, from (7), $\int_A f d\lambda = m(A)$ for each $A \in \mathcal{M}_\lambda$. But this means that $K|_Y$ has the RNP for $([0,1], \mathcal{M}_\lambda, \lambda)$. It follows that $K|_Y$ has the RNP for $([0,1], \mathcal{B}([0,1]), \lambda)$ and hence, by Theorem 2.3.6 part (2), that $K|_Y$ has the RNP. Since K and $K|_Y$ are affinely isometric, K has the RNP as well. Finally, since K was an arbitrary separable closed bounded convex subset of C, C has the RNP (see 2.3.6 parts (8) and (11)). This completes the proof of Theorem 7.8.6.

A sample of the easy consequences of 7.8.6 indicates its usefulness. (Each of the following has been previously established by other means. See Saab [1980a] for further consequences and examples.)

COROLLARIES 7.8.7. (a) X* has the RNP if and only if the identity mapping $id: (X^*, \text{weak*}) \to (X^*, \|\cdot\|)$ is universally measurable. (Originally proved by Schwartz [1973], p. 162.)

(b) A weak* compact convex subset C of X* which is weakly K-analytic has the RNP. (This is a special case of Corollary 4.2.17. See Lemma 7.8.8 below.)

(c) If X* is weakly K-analytic then X* has the RNP.

(d) If X^* is isomorphic to a subspace of a weakly compactly generated Banach space Y then X^* has the RNP.

(e) Let $C \subset X^*$ be weak* compact and convex. Then C has the RNP if and onl if $T^*(C) \subset Y^*$ has the RNP whenever Y is a Banach space and $T: Y \to X$ is a continuous linear map. (This is Corollary 4.2.14.)

(f) If X^* has the RNP and Y is a subspace of X then Y^* has the RNP. (This follows directly from Theorem 4.4.1 (a) and (g).)

Proofs. (a) Take $C = U(X^*)$ in 7.8.6.

(b) The identity $I: (C,\text{weak*}) \to (C,\|\cdot\|)$ is universally measurable since both identities on C from the weak* to weak topologies (see 7.8.3) and from the weak to nor topologies (see 7.8.5(d)) are universally measurable (see 7.8.5(c)). Hence C has the RNP by 7.8.6 (a) and (b).

(c) $U(X^*)$ is weakly K-analytic since $U(X^*)$ is weakly closed in X^*. Now apply 7.8.7(b).

(d) We first observe that Y is weakly K-analytic since Y is WCG. Indeed, pick a weakly compact absolutely convex subset D of Y such that $\bigcup_{n=1}^{\infty} nD$ is dense in Y. Then it is easy to check that

$$\hat{Y} = \bigcap_{j=1}^{\infty} \bigcup_{n=1}^{\infty} [n\hat{D} + \tfrac{1}{j}U(Y^{**})] .$$

Since $n\hat{D} + \frac{1}{j}U(Y^{**})$ is weak* compact for each n and j, $(\hat{Y},\text{weak*})$ is a $K_{\sigma\delta}$ su set of $(Y^{**},\text{weak*})$. Now embed $(Y^{**},\text{weak*})$ in its Stone-Cech compactification and note that $(\hat{Y},\text{weak*})$ is affinely homeomorphic to (Y,weak). Thus Y is weakly K-analytic. (This argument is due to Talagrand [1975] .) The proof is completed by application of 7.8.7(c).

(e) One direction is trivial. (Take $Y \equiv X$ and $T \equiv$ identity on X.) Suppose tha Y is a Banach space and $T: Y \to X$ is continuous and linear. By 7.8.6(b), $id: (K,\text{weak*}) \to (K,\|\cdot\|)$ is universally measurable, and a look at the proof of 7.8.6 $\Rightarrow$ (d) shows that by mechanically replacing J there by T, $id_3: (T^*(K),\sigma(Y^*,Y)) \to (T^*(K),\|\cdot\|)$ is universally measurable. Apply 7.8.6(b) again. Hence $T^*(K) \subset Y^*$ h the RNP.

(f) Let $J: (Y,\|\cdot\|) \to (X,\|\cdot\|)$ be the inclusion map. Then $U(Y^*) = J^*(U(X^*))$. Since this latter set has the RNP by 7.8.7(e), the proof of 7.8.7 is complete.

<u>LEMMA</u> 7.8.8. Let $(D,\mathcal{T})$ be a completely regular space. If $(D,\mathcal{T})$ is K-analytic then $(D,\mathcal{T})$ is Lindelöf.

Proof. First assume D is a $K_{\sigma\delta}$ subset of a compact Hausdorff space B. Then $D = \bigcap_{n=1}^{\infty} \bigcup_{j=1}^{\infty} K_{n,j}$ where each $K_{n,j}$ is compact in B. Use induction on n starting with $n = 2$ to replace each $K_{n,j}$ of "stage n" by the countable collection of sets obtained by intersecting $K_{n,j}$ with each set in the previously refined stage n-1. (Leave the stage 1 sets alone.) If these newly formed sets are again denoted by $K_{n,j}$'s then not only is $D = \bigcap_{n=1}^{\infty} \bigcup_{j=1}^{\infty} K_{n,j}$ as before, but whenever $x \in D$ there is a sequence of indices $\{(n,i_n): n = 1,2,\ldots\}$ such that $x \in \bigcap_{n=1}^{\infty} K_{n,i_n}$. Let $\mathcal{I}$ denote the set of all sequences (i_n) of positive integers such that for each n, $K_{n+1,i_{n+1}} \subset K_{n,i_n}$. Then

$$D = \bigcap_{n=1}^{\infty} \bigcup_{j=1}^{\infty} K_{n,j} = \bigcup_{(i_n)\in\mathcal{I}} \bigcap_{n=1}^{\infty} K_{n,i_n} .$$

Now suppose that $\{O_\alpha: \alpha \in A\}$ is an open cover of D. Let $(i_n) \in \mathcal{I}$ be temporarily fixed and observe that $\bigcap_{n=1}^{\infty} K_{n,i_n}$ is a compact subset of D. Consequently there is a finite subset of A, say F, such that $\bigcup_{\alpha\in F} O_\alpha \supset \bigcap_{n=1}^{\infty} K_{n,i_n}$ and since $K_{n+1,i_{n+1}} \subset K_{n,i_n}$ for each n, there is an N such that $K_{N,i_N} \subset \bigcup_{\alpha\in F} O_\alpha$. That is, for each $(i_n) \in \mathcal{I}$ there are an N and a finite set $F \subset A$ such that

$$\bigcap_{n=1}^{\infty} K_{n,i_n} \subset K_{N,i_N} \subset \bigcup_{\alpha\in F} O_\alpha .$$

Now let

$M \equiv \{(n,j)$: n,j are positive integers and $K_{n,j}$ is contained in some finite union of elements of $\{O_\alpha: \alpha \in A\}\}$.

Then $\bigcup_{(n,j)\in M} K_{n,j} \supset D$ from the above argument. On the other hand, M is countable, and it follows that there is a countable subset C of A such that $\bigcup_{\alpha\in C} O_\alpha \supset \bigcup_{(n,j)\in M} K_{n,j}$

$\supset D$. Thus D is Lindelöf. This establishes the lemma for such sets D .

To complete the proof, observe that if $(D,\mathcal{T})$ is the continuous image of a Lindelöf space then it is Lindelöf. Since D is the continuous image of a $K_{\sigma\delta}$ se[t] if $(D,\mathcal{T})$ is K-analytic, and since $K_{\sigma\delta}$ sets are Lindelöf from above, the proof is complete.

Theorem 7.8.6 provides a means of describing the controlled relationship betwee[n] the weak* and norm topologies on a dual space in order for the RNP to hold. In a slightly different direction, but with the same philosophy, the positioning of a Banach space in its second dual leads to further results on the RNP. (We follow par[t] of Saab [1980b] and [1981a] below.)

As in 5.7.5, the natural injection of X into X^{**} which associates $\hat{x}$ to x is denoted i_X . Thus $i_X^*: X^{***} \to X^*$ is the natural "restriction to X" projection. To set the stage we begin with the easiest case, that of a dual.

THEOREM 7.8.9. X^* has the RNP if and only if $i_X^*: (X^{***},\text{weak*}) \to (X^*,\text{weak})$ is universally measurable.

Proof. If X^* has the RNP then the identity $id_1: (X^*,\text{weak*}) \to (X^*,\|\cdot\|)$ is universally measurable by 7.8.7(a) and consequently so is $id: (X^*,\text{weak*}) \to (X^*,\text{weak})$. Now i_X^* in the statement of this theorem is the composition of the map $I_X^*: (X^{***},\text{weak*}) \to (X^*,\text{weak*})$ (where $I_X^* = i_X^*$ pointwise) with id. Clearly I_X^* is continuous. Since id is universally measurable, so is i_X^* .

We use the same notation as in the previous paragraph to establish the converse. Suppose we can show that id is universally measurable. Since the identity from (X^*,weak) to $(X^*,\|\cdot\|)$ is always universally measurable by 7.8.5(d), a combinatio[n] of 7.8.5(c) with 7.8.7(a) will yield the conclusion that X^* has the RNP. In orde[r] to show that $id: (X^*,\text{weak*}) \to (X^*,\text{weak})$ is universally measurable, pick any $\mu \in P_t(X^*,\text{weak*})$ and write $\mu = \sum \mu_n$ (a finite or countable sum) where each μ_n is a positive multiple of some element of $P_t(X^*,\text{weak*})$ and has weak* compact suppo[rt] E_n . For each n find a weak* compact subset D_n of X^{***} such that $I_X^*(D_n) = E_n$ and use Lemma 4.3.9 part (1) to find a measure ν_n on $\mathcal{B}(D_n,\text{weak*})$ inner regular with respect to weak* compact sets, such that $I_X*(\nu_n) = \mu_n$. With obvious identifications, let $\nu \in P_t(X^{***},\text{weak*})$ be the measure $\nu \equiv \sum \nu_n$. Then $I_X*(\nu) = \mu$,

$$[\nu_n(D_n)]^{-1}\nu_n \in P_t(X^{***},\text{weak*}) \quad \text{and} \quad I_X*([\nu_n(D_n)]^{-1}\nu_n) = [\mu_n(E_n)]^{-1}\mu_n$$

for each n. Because $\text{id}\circ I_X* = i_X*$ is $[\nu_n(D_n)]^{-1}\nu_n$ - Lusin-measurable it follows from the compactness of D_n that id is $I_X*([\nu_n(D_n)]^{-1}\nu_n)$ - Lusin-measurable. That is, id is $[\mu_n(E_n)]^{-1}\mu_n$ - Lusin-measurable for each n, and hence id is μ-Lusin-measurable. Since $\mu \in P_t(X^*,\text{weak*})$ was arbitrary, id is universally measurable as was to be shown.

The situation is not quite so nice when X is not a dual space. First of all, X need not be complemented in X** . (For example, c_0 is not complemented in $c_0^{**} = \ell_\infty$. See Lindenstrauss [1967] for the much stronger result that whenever Y is a complemented infinite dimensional subspace of ℓ_∞ then Y is isomorphic to ℓ_∞ .) Secondly, even if there is a (norm-norm continuous linear surjective) projection π: X** → X , it may be weak*-weak universally measurable without X having the RNP. (Edgar, von Weizsäcker and Talagrand have established results which, combined, show that if λ denotes Lebesgue measure on [0,1] , S is the set of extreme points of the unit ball of $L^1[0,1]^{**}$ which are nonnegative, in the weak* topology—so that $L^\infty[0,1]$ is isometrically isomorphic to C(S)—and if $\hat{\lambda}$ is considered as an element of $P_t(S) \subset C(S)^*$, then the projection $\pi: L^1[0,1]^{**} \to L^1[0,1]$ which assigns $\mu \in C(S)^*$ its absolutely continuous part with respect to the measure $\hat{\lambda}$ in its Lebesgue decomposition is $\mathcal{B}(L^1[0,1]^{**},\text{weak*}) - \mathcal{B}(L^1[0,1],\text{weak})$ measurable and hence universally measurable. Nevertheless, of course, $L^1[0,1]$ does not have the RNP.) Thus for those X complemented in their second duals, something more than universal measurability of a projection map is needed in general in order to ensure the RNP. One such extra condition is described in the next theorem.

Suppose that $(D,\mathcal{T})$ is a topological space, that $f: D \to U(X^{**})$ for some Banach space X and that $\mu \in P_t(D,\mathcal{T})$. If for each $F \in X^*$, $\hat{F}\circ f: D \to \mathbb{R}$ is $\mathcal{B}(D,\mathcal{T})$-measurable and for each $A \in \mathcal{B}(D,\mathcal{T})$ the linear functional on X* whose value at $F \in X^*$ is $\int_A \hat{F}\circ f d\mu$ actually belongs to U(X**) then we denote that element of

$U(X^{**})$, reasonable enough, by $w^*\text{-}\int_A f d\mu$ and we say that <u>f is weak* integrable wi</u> <u>respect to μ</u>. Observe that by the Hahn-Banach theorem, the identity map $I: (U(X^{**}),\text{weak*}) \to (U(X^{**}),\text{weak*})$ is weak* integrable with respect to μ for each $\mu \in P_t(U(X^{**}),\text{weak*})$.

<u>DEFINITION</u> 7.8.10. Fix $\mu \in P_t(U(X^{**}),\text{weak*})$ and let $\Psi: (U(X^{**}),\text{weak*}) \to \mathbb{R}$ be μ Lusin-measurable. If Ψ is affine on $U(X^{**})$ then <u>Ψ satisfies the barycentric formula for μ</u> if for each $A \in B(U(X^{**}),\text{weak*})$

$$\Psi(w^*\text{-}\int_A I d\mu) = \int_A \Psi d\mu .$$

The first part of the next theorem continues in the vein of Theorems 4.3.11 and 7.8.6, while the second resembles Theorem 7.8.9. The two results are listed under c heading since their proofs overlap considerably. See Saab [1981a] for examples.

<u>CONVENTIONS FOR THEOREM 7.8.11.</u>

1. $I: (U(X^{**}),\text{weak*}) \to (U(X^{**}),\text{weak*})$ denotes the identity map.
2. $P_X \equiv \{\nu \in P_t(U(X^{**}),\text{weak*}): w^*\text{-}\int_A I d\nu \in X$ for each $A \in B(U(X^{**}),\text{weak*})\}$.

   (N.B. X is identified here and below with $\hat{X} \subset X^{**}$.)

<u>THEOREM</u> 7.8.11. (a) The following are equivalent for any Banach space X.

(i) X has the RNP ;

(ii) The identity map $id_1: (U(X^{**}),\text{weak*}) \to (U(X^{**}),\|\cdot\|)$ is ν-Lusin measur for each $\nu \in P_X$.

(b) Let X be a Banach space which is complemented in its second dual, and let $\pi_1: (X^{**},\|\cdot\|) \to (X,\|\cdot\|)$ denote a continuous linear surjective projection. Then th following are equivalent:

(i) X has the RNP ;

(ii) Let $\nu \in P_X$. Then $\pi: (U(X^{**}),\text{weak*}) \to (U(X),\|\cdot\|)$ is ν-Lusin-measurab (where $\pi_1 = \pi$ pointwise) and for each $F \in X^*$, the affine map $\hat{F}\circ\pi: U$ $\to \mathbb{R}$ satisfies the barycentric formula for ν.

The following lemma will be useful in the proof of the above theorem.

<u>LEMMA</u> 7.8.12. Let $\mu \in P_t(U(X^{**}),\text{weak*})$. If g,h are μ-Lusin-measurable functions from $(U(X^{**}),\text{weak*})$ to $(U(X^{**}),\text{weak*})$ such that for each $A \in \mathcal{B}(U(X^{**}),\text{weak*})$ we have $w^*\text{-}\int_A g d\mu = w^*\text{-}\int_A h d\mu$ then $g = h$ [a.s.μ] .

Proof. If the lemma were false, by some elementary reductions we could produce a compact subset K of $(U(X^{**}),\text{weak*})$ on which both g and h are continuous, and a $\phi_0 \in \text{support}(\mu|_K)$ at which $g(\phi_0) \neq h(\phi_0)$. Find $F \in X^*$ such that $\hat{F}(g(\phi_0) - h(\phi_0)) = 1$ and a relatively weak* open subset V of K such that each ϕ in V satisfies $\hat{F}(g(\phi) - h(\phi)) \geqq \frac{1}{2}$. Since $\phi_0 \in \text{support}(\mu|_K)$, $\mu(V) > 0$ whence

$$0 = \int_V \hat{F}\circ g d\mu - \int_V \hat{F}\circ h d\mu = \int_V \hat{F}(g(\phi) - h(\phi))d\mu(\phi) \geqq \frac{1}{2}\mu(V) > 0 .$$

This contradiction establishes the lemma.

<u>Proof of 7.8.11</u>. Assume first that X has the RNP. Fix $\nu \in P_X$ and define $m: \mathcal{B}(U(X^{**}),\text{weak*}) \to U(X)$ by

$$m(A) \equiv w^*\text{-}\int_A I d\nu \qquad \text{for } A \in \mathcal{B}(U(X^{**}),\text{weak*}) .$$

Evidently $\|m(A)\| \leqq \nu(A)$ and it follows that m must be countably additive (since it is clearly finitely additive). Thus there is a separably valued function $f: U(X^{**}) \to U(X)$ which is Bochner integrable with respect to ν such that

$$(8) \qquad m(A) = \int_A f d\nu = w^*\text{-}\int_A I d\nu \in X \quad \text{for each } A \in \mathcal{B}(U(X^{**}),\text{weak*}) .$$

Evidently f is $(U(X^{**}),\text{weak*})$ - $(U(X^{**}),\text{weak*})$ ν-Lusin-measurable as well (by 7.8.5(a)) and hence by 7.8.12 and formula (8), $f = I$ [a.s.ν] . Because f (hence I) is almost separably valued and $\text{Range}(f) \subset U(X)$ there is a separable subspace Y of X such that $\nu(U(Y)) = 1$. Since the restriction of $\mathcal{B}(U(X^{**}),\text{weak*})$ to Y is $\mathcal{B}(U(Y),\|\cdot\|)$ (cf. the argument in the third paragraph of the proof of 4.1.1, for example) it follows from Lemma 4.3.10 that there is a sequence $\{K_n: n = 1,2,\ldots\}$ of norm compact subsets of U(Y) (and hence of U(X)) with $\nu(\cup K_n) = 1$. Thus

$$(9) \qquad id_1: (U(X^{**}),\text{weak*}) \to (U(X^{**}),\|\cdot\|)$$

is ν-Lusin-measurable. This establishes (i) ⇒ (ii) of 7.8.11(a). In order to prove

(i) ⇒ (ii) for 7.8.11(b) note that π is the composition of the ν-Lusin-measurable mapping id_2: $(X^{**},weak^*) \to (X^{**},\|\cdot\|)$ (use (9)) with the original continuous projection π_1: $(X^{**},\|\cdot\|) \to (X,\|\cdot\|)$. Hence π, too, is ν-Lusin-measurable.

To show that $\hat{F}\circ\pi$ satisfies the barycentric formula (for $F \in X^*$) we calculate as follows:

$$\hat{F}\circ\pi(w^*\text{-}\int_A Id\nu) = \hat{F}(w^*\text{-}\int_A Id\nu) \quad \text{(from (8))}$$

$$= \hat{F}(m(A)) = \sum \hat{F}(m(A\cap K_n)) = \sum \int_{A\cap K_n} \hat{F}d\nu$$

$$= \sum \int_{A\cap K_n} \hat{F}\circ\pi d\nu = {}_A\ \hat{F}\circ\pi d\nu \quad .$$

The proof of 7.8.11 thus reduces to showing that (ii) ⇒ (i) for parts (a) and (b). Let λ be Lebesgue measure on $[0,1]$ (and M_λ the Lebesgue measurable subsets of $[0,1]$). Pick any vector measure $m: M_\lambda \to X$ such that $m << \lambda$ and $AR(m) \subset U(X)$. By Theorem 7.8.4(b) there is a $g: [0,1] \to (U(X^{**}),weak^*)$ such that

(a) $g^{-1}(B) \in M_\lambda$ for each $B \in \mathcal{B}(U(X^{**}),weak^*)$;

(b) $g(\lambda) \in P_t(U(X^{**}),weak^*)$;

(c) $m(A) = w^*\text{-}\int_A gd\lambda$ for each $A \in M_\lambda$.

Since $AR(m) \subset U(X)$ it follows from (b) and (c) that

$$w^*\text{-}\int_B Idg(\lambda) \in U(X) \qquad \text{for each } B \in \mathcal{B}(U(X^{**}),weak^*)$$

Thus $g(\lambda) \in P_X$. According to the hypothesis (ii) of 7.8.11(a), id_1 as defined in (9) is $g(\lambda)$-Lusin-measurable. (We show in the next paragraph that the hypothesis (ii) of 7.8.11(b) also establishes that id_1 is ν-Lusin-measurable for each $\nu \in P_X$ and in particular, id_1 is $g(\lambda)$-Lusin-measurable. Let us assume this for the time being and complete the implications (ii) ⇒ (i) for both parts (a) and (b).) By definition of id_1 being $g(\lambda)$-Lusin-measurable, there is a sequence $\{K_n: n = 1,2,\ldots\}$ of norm compact sets in $U(X^{**})$ such that $g(\lambda)(\cup K_n) = 1$. Hence $id_1\circ g: [0,1] \to (U(X^{**}),\|\cdot\|)$ is λ-almost separably valued. Using this fact, (a) above yields the conclusion

$$(id_1\circ g)^{-1}(O) \in M_\lambda \qquad \text{for each norm open set } O \text{ in } U(X^{**}) .$$

Consequently $id_1 \circ g$ is Bochner integrable by Theorem 1.1.7. Thus

$$m(A) = \int_A id_1 \circ g d\lambda = \int_A g d\lambda \qquad \text{for each } A \in M_\lambda .$$

That is, g is X-valued almost surely, and equals $\frac{dm}{d\lambda}$. Hence X has the RNP for $([0,1], M_\lambda, \lambda)$ and hence it has the RNP (by Theorem 2.3.6, for example).

It remains to show that if X satisfies the hypothesis (ii) of part (b) of 7.8.11 then id_1 as defined in (9) is ν-Lusin-measurable for each $\nu \in P_X$. Pick any $\nu \in P_X$, let $\bar{m}(B) \equiv w^*\text{-}\int_B Id\nu$ for $B \in \mathcal{B}(U(X^{**}), \text{weak}^*)$ and note that $\bar{m}(B) \in X$ for each such B by definition of P_X. For $F \in X^*$ it follows from our hypotheses that

$$\int_B \hat{F} d\nu = \hat{F}(\bar{m}(B)) = \hat{F} \circ \pi_1(\bar{m}(B)) = \int_B \hat{F} \circ \pi_1 d\nu$$

and thus $w^*\text{-}\int_B id_1 d\nu = w^*\text{-}\int_B \pi_1 d\nu$ for each $B \in \mathcal{B}(U(X^{**}), \text{weak}^*)$. By Lemma 7.8.12, then, $id_1 = \pi_1$ [a.s.ν]. Thus there is a sequence $\{C_n : n = 1,2,\ldots\}$ of weak* compact subsets of $U(X^{**})$ such that $\nu(\cup C_n) = 1$ and $id_1 = \pi_1$ on each C_n. It follows that $C_n \subset X$ and hence C_n is compact in the $\sigma(X,X^*)$ topology on X. But then ν may be considered as a measure in $P_t(U(X), \sigma(X,X^*))$ and by Theorem 4.4.4(a) ν has norm separable support. Thus there is a sequence $\{K_n : n = 1,2,\ldots\}$ of norm compact subsets of $U(X)$ with $\nu(\cup K_n) = 1$. It now follows from the definition that $id_1 : (U(X^{**}), \text{weak}^*) \to (U(X^{**}), \|\cdot\|)$ is ν-Lusin-measurable. This completes the proof.

<u>COROLLARY</u> 7.8.13. If there is a continuous linear projection $\pi_1 : (X^{**}, \|\cdot\|) \to (X, \|\cdot\|)$ which is weak*-weak Baire-1 then X has the RNP. (Recall: a function is Baire-0 if it is continuous, and Baire-n if it is the pointwise limit of a sequence of functions in Baire-(n-1) for $n > 0$.)

Proof. If π_1 is Baire-1 it is certainly weak*-weak μ-Lusin-measurable for each $\mu \in P_X$. Hence it is also weak*-norm μ-Lusin-measurable for each $\mu \in P_X$ by 7.8.5(d). Note that for each $F \in X^*$, $\hat{F} \circ \pi_1$ is a weak*-weak Baire-1 affine function on the weak* compact convex set $(U(X^{**}), \text{weak}^*)$. Thus by the well known theorem of Choquet [1969] (see alternatively, Phelps [1966], p. 100, Theorem) $\hat{F} \circ \pi_1$ satisfies the barycentric formula for μ. Thus X satisfies hypotheses (b)(ii) of

Theorem 7.8.11, and it follows from that theorem that X has the RNP. (It should b remarked that the hypotheses of 7.8.13 cannot be weakened without significant alter ation of the proof, since an example of Choquet [1962] —or see Phelps [1966] , p. 10 Example—shows that Baire-2 affine real-valued functions on compact convex sets do not in general satisfy the barycentric formula.)

In closing we remark that space limitations do not allow us to consider the theory of weakly K-analytic Banach spaces as developed by Talagrand [1975] , [1977b] [1977c], and [1979] (see also Saab [1981b]) although there is some consolation in the fact that the interplay between such spaces and those with the RNP is less signi ficant than for the spaces considered in this section.

§9. The partial orderings $<_{SE}$ and $<_D$ for integral representations.

In this section we will provide a detailed analysis of the orderings $<_{SE}$ and $<_D$ (see Definitions 6.1.1 and 6.3.7(b)) which will round out some of the material of section 6.3. The notation shall be that of Chapter 6 with the additions:

α denotes the first infinite ordinal; and

$\tilde{F}$ denotes the $\tilde{P}$-completion of $\mathcal{B}(\tilde{\Omega})$.

Each of the results below is from Bourgin [1979] . The main link between the orderings is as follows.

THEOREM 7.9.1. Let K be a closed bounded convex subset of a Banach space X and assume that K has the RNP. If $f,g \in L^1_K(P)$ and $f <_{SE} g$ then $f(P) <_D g(P)$. Conversely, suppose that $\mu_1,\mu_2 \in P_t(K)$ and that $\mu_1 <_D \mu_2$. Then there are functi f and g in $L^1_K(P)$ such that $f(P) = \mu_1$, $g(P) = \mu_2$ and $f <_{SE} g$.

Proof. One direction is easy. Indeed, if $f,g \in L^1_K(P)$ and $f <_{SE} g$ then fi $\beta < \omega_1$ such that $f = \mathbb{E}[g|F_\beta]$. If $F \in C_b(K)$ is convex, it follows from Jensen' inequality (see Meyer [1966] , p. 29) that

$$\int_K F df(P) = \int_\Omega F\circ f dP = \int_\Omega F\circ \mathbb{E}[g|F_\beta]dP \leq \int_\Omega E[F\circ g|F_\beta]dP$$
$$= \int_\Omega F\circ g dP = \int_K f dg(P)$$

whence $f(P) <_C g(P)$. The fact that $f(P) <_D g(P)$ is a consequence of Theorem 6.3.9.

Conversely, suppose that $\mu_1, \mu_2 \in P_t(K)$ and that $\mu_1 <_D \mu_2$ on K. Let $\tilde{K}$ denote a separable closed convex subset of K such that both $\mu_1(\tilde{K}) = \mu_2(\tilde{K}) = 1$ and $\tilde{K} = \overline{\text{span}}(\tilde{K}) \cap K$. Let $\tilde{f} \in L^1_{\tilde{K}}(\tilde{P})$ satisfy $\tilde{f}(\tilde{P}) = \mu_1$ (see Proposition 6.1.2) and observe that since $\mu_1 <_D \mu_2$ on K , $\mu_1 <_D \mu_2$ on $\tilde{K}$ as well. Hence there is a μ_1-dilation $T: \tilde{K} \to P_t(\tilde{K})$ such that $\int_{\tilde{K}} T d\mu_1 = \mu_2$. If necessary, modify T on a set of μ_1-measure 0 in order to be assured that T(x) has barycenter x for each $x \in \tilde{K}$. Observe that $T(\tilde{f}(\tilde{\omega})) \in P_t(\tilde{K})$ for each $\tilde{\omega} \in \tilde{\Omega}$ and hence, again by Proposition 6.1.2, given $\tilde{\omega} \in \tilde{\Omega}$ there is at least one $\tilde{h} \in L^1_{\tilde{K}}(\tilde{P})$ such that $\tilde{h}(\tilde{P}) = T(\tilde{f}(\tilde{\omega}))$. It will be important in what follows to know that the correspondence $\tilde{\omega} \to \tilde{h}$ can be made in a measurable fashion, and this may be accomplished in the manner indicated below. The map $\tilde{g} \to \tilde{g}(\tilde{P})$ from $L^1_{\tilde{K}}(\tilde{P})$ into $P_t(\tilde{K})$ is surjective by Proposition 6.1.2. It is continuous since if $f_n \to f$ in $L^1_{\tilde{K}}(\tilde{P})$ and $g \in C_b(\tilde{K})$ then

$$\int_{\tilde{K}} g df_n(\tilde{P}) = \int_{\tilde{\Omega}} g \circ f_n d\tilde{P} \to \int_{\tilde{\Omega}} g \circ f d\tilde{P} = \int_{\tilde{K}} g df(\tilde{P})$$

whence $f_n(\tilde{P}) \to f(\tilde{P})$ in $P_t(\tilde{K}) \subset C_b(\tilde{K})^*$. Moreover, both $L^1_{\tilde{K}}(\tilde{P})$ and $P_t(\tilde{K})$ are completely metrizable and separable. (Indeed, $L^1_{\tilde{K}}(\tilde{P})$ clearly has these properties, the separability of $P_t(\tilde{K})$ follows from the fact that $co(\{\varepsilon_x: x \in \tilde{K}\})$ is dense in $P_t(\tilde{K}) \subset C_b(\tilde{K})^*$, while its complete metrizability is a consequence of Theorem 6.3.3.) It follows from Theorem 6.2.7 that there is a function $S: P_t(\tilde{K}) \to L^1_{\tilde{K}}(\tilde{P})$ which is $\mathcal{U}(\mathcal{B}(P_t(\tilde{K}))) - \mathcal{B}(L^1_{\tilde{K}}(\tilde{P}))$ measurable such that $S(\nu)(\tilde{P}) = \nu$ for each $\nu \in P_t(\tilde{K})$. Now consider the map

$$S \circ T \circ f: \tilde{\Omega} \to L^1_{\tilde{K}}(\tilde{P}) .$$

If O is a relatively open subset of $L^1_{\tilde{K}}(\tilde{P})$ then $S^{-1}(O) \in \mathcal{U}(\mathcal{B}(P_t(\tilde{K})))$. It follows from Lemma 6.2.5 parts (1) and (3) that T is $\mathcal{U}(\mathcal{B}(\tilde{K})) - \mathcal{U}(\mathcal{B}(P_t(\tilde{K})))$ measurable since it is $\mathcal{U}(\mathcal{B}(\tilde{K})) - \mathcal{B}(P_t(\tilde{K}))$ measurable, and thus $T^{-1}(S^{-1}(O)) \in \mathcal{U}(\mathcal{B}(\tilde{K}))$. Since $\tilde{f}$ is $\tilde{\mathcal{F}}$-$\mathcal{B}(\tilde{K})$ measurable, it is $\tilde{\mathcal{F}} - \mathcal{U}(\mathcal{B}(\tilde{K}))$ measurable (again, by Lemma 6.2.5 (1) and (2)) . It follows from Theorem 1.1.7 , then, that $S \circ T \circ \tilde{f} \in L^1_{L^1_{\tilde{K}}(\tilde{P})}(\tilde{\Omega}, \tilde{\mathcal{F}}, \tilde{P})$.

Define $\tilde{g}_1: \tilde{\Omega} \times \tilde{\Omega} \to K$ by

$$\tilde{g}_1(\tilde{\omega},\tilde{\omega}') \equiv [S\circ T\circ \tilde{f}(\tilde{\omega})](\tilde{\omega}').$$

It is established in Proposition 7.9.2 below that under these conditions there is a $\tilde{g}\colon \tilde{\Omega}\times\tilde{\Omega} \to \tilde{K}$ such that $\tilde{g}(\tilde{\omega},\cdot) = \tilde{g}_1(\tilde{\omega},\cdot)$ [a.s.$\tilde{P}$] for each $\tilde{\omega} \in \tilde{\Omega}$ and $\tilde{g}$ is $\tilde{F}\times\tilde{F}$ - $B(\tilde{K})$ measurable. Assume the existence of such a $\tilde{g}$ for now. It will be convenien in what follows to "factor out $\tilde{\Omega}$" from Ω and write $\Omega = \tilde{\Omega}\times\Omega_1$ (and even $\Omega = \tilde{\Omega}\times\tilde{\Omega}\times\Omega_2$). The probability measure P will similarly be factored as $P = \tilde{P}\times P_1$ and $P = \tilde{P}\times\tilde{P}\times P_2$. Functions on $\tilde{\Omega}$ will be identified with functions on $\Omega = \tilde{\Omega}\times\Omega_1$ which depend only on the $\tilde{\Omega}$ coordinates so that, for example, $\tilde{f}\colon \tilde{\Omega} \to X$ corresponds to $f\colon \Omega = \tilde{\Omega}\times\Omega_1 \to X$ given by $f(\tilde{\omega},\omega) \equiv \tilde{f}(\tilde{\omega})$. Functions on $\tilde{\Omega}\times\tilde{\Omega}$ may similarly be ident fied with functions on $\Omega = \tilde{\Omega}\times\tilde{\Omega}\times\Omega_2$ and in particular, $\tilde{g}\colon \tilde{\Omega}\times\tilde{\Omega} \to \tilde{K}$ will be identifi with $g\colon \Omega \to \tilde{K}$.

It will be shown first that $\mathbb{E}[g|F_\alpha] = f$. (Recall: α is the first infinite ordinal.) Certainly f is F_α measurable by construction. If now $A \in F_\alpha$, there is a set $B \in B(\tilde{\Omega})$ such that $P(A \,\Delta\, (B\times\Omega_1)) = 0$. Consequently

$$\begin{aligned}
\int_A g\,dP &= \int_{B\times\Omega_1} g\,dP = \int_B\int_{\tilde{\Omega}}\int_{\Omega_2} g(\tilde{\omega},\tilde{\omega}',\omega_2)\,dP_2(\omega_2)\,d\tilde{P}(\tilde{\omega}')\,d\tilde{P}(\tilde{\omega})\\
&= \int_B\int_{\tilde{\Omega}} \tilde{g}(\tilde{\omega},\tilde{\omega}')\,d\tilde{P}(\tilde{\omega}')\,d\tilde{P}(\tilde{\omega}) = \int_B\int_{\tilde{\Omega}} \tilde{g}_1(\tilde{\omega},\tilde{\omega}')\,d\tilde{P}(\tilde{\omega}')\,d\tilde{P}(\tilde{\omega})\\
&= \int_B\int_{\tilde{\Omega}} [S\circ T\circ \tilde{f}(\tilde{\omega})](\tilde{\omega}')\,d\tilde{P}(\tilde{\omega}')\,d\tilde{P}(\tilde{\omega}) = \int_B\int_{\tilde{K}} x\,d[S\circ T\circ \tilde{f}(\tilde{\omega})](\tilde{P})(x)\,d\tilde{P}(\tilde{\omega})\\
&= \int_B\int_{\tilde{K}} x\,d[T(\tilde{f}(\tilde{\omega}))](x)\,d\tilde{P}(\tilde{\omega}) = \int_B \tilde{f}(\tilde{\omega})\,d\tilde{P}(\tilde{\omega})\\
&= \int_B\int_{\Omega_1} f(\tilde{\omega},\omega')\,dP_1(\omega')\,d\tilde{P}(\tilde{\omega}) = \int_{B\times\Omega_1} f\,dP = \int_A f\,dP.
\end{aligned}$$

Hence $\mathbb{E}[g|F_\alpha] = f$.

Note that for any $\mu \in P_t(\tilde{K})$ we have $\int_{\tilde{K}} \varepsilon_x\,d\mu(x) = \mu$. It is shown below that $g(P) = g(\tilde{P}\times\tilde{P}\times P_2) = \mu_2$, which, evidently, is equivalent to the desired conclusion $\tilde{g}(\tilde{P}\times\tilde{P}) = \mu_2$.

$$\mu_2 = \int_{\tilde{K}} T(x)\,d\mu_1(x) = \int_{\tilde{K}} T(x)\,d\tilde{f}(\tilde{P})(x) = \int_{\tilde{\Omega}} T\circ \tilde{f}(\tilde{\omega})\,d\tilde{P}(\tilde{\omega})$$

$$= \int_{\tilde{\Omega}} [S\circ T\circ \tilde{f}(\tilde{\omega})](\tilde{P})d\tilde{P}(\tilde{\omega}) = \int_{\tilde{\Omega}}\int_{\tilde{\Omega}} \varepsilon_{[S\circ T\circ \tilde{f}(\tilde{\omega})](\tilde{\omega}')} d\tilde{P}(\tilde{\omega}')d\tilde{P}(\tilde{\omega})$$

$$= \int_{\tilde{\Omega}}\int_{\tilde{\Omega}} \varepsilon_{\tilde{g}_1(\tilde{\omega},\tilde{\omega}')} d\tilde{P}(\tilde{\omega}')d\tilde{P}(\tilde{\omega}) = \int_{\tilde{\Omega}\times\tilde{\Omega}} \varepsilon_{\tilde{g}(\tilde{\omega},\tilde{\omega}')} d[\tilde{P}\times\tilde{P}](\tilde{\omega},\tilde{\omega}')$$

$$= \int_{\tilde{K}} \varepsilon_x d\tilde{g}(\tilde{P}\times\tilde{P})(x) = \int_{\tilde{K}} \varepsilon_x dg(P)(x) = g(P).$$

That is, $g(P) = \mu_2$, $f(P) = \mu_1$ and $\mathbb{E}[g|F_\alpha] = f$ so that $f <_{SE} g$. Thus the proof of this theorem rests on the existence of g as asserted in the previous paragraphs.

PROPOSITION 7.9.2. Let (Γ,G,Q) be a probability space whose σ-algebra G is contained in the Q-completion of a countably generated σ-algebra. (That is, assume (Γ,G,Q) is separable.) Assume further that $\tilde{K}$ has the RNP. If $g\colon \Gamma \to L^1_{\tilde{K}}(Q)$ is Bochner integrable (i.e. if $g \in L^1_{L^1_{\tilde{K}}(Q)}(Q)$) then there is an $h\colon \Gamma \to L^1_{\tilde{K}}(Q)$ such that both

(a) $g(\omega,\cdot) = h(\omega,\cdot)$ [a.s.Q] for each $\omega \in \Gamma$ where $g(\omega,\omega')$ is shorthand for $[g(\omega)](\omega')$, and similarly for $h(\omega,\omega')$.

(b) $h\colon \Gamma\times\Gamma \to \tilde{K}$ is $G\times G - B(\tilde{K})$ measurable.

Proof. Without loss of generality assume that $0 \in \tilde{K}$. Let $\{B_i\colon i = 1,2,\ldots\}$ be a sequence in G such that for each $A \in G$ there is a B in the σ-algebra generated by $\{B_i\}_{i=1}^{\infty}$ with $Q(A \,\Delta\, B) = 0$. Let G_n denote the σ-algebra generated by $\{B_i\colon i = 1,2,\ldots,n\}$ for $n = 1,2,\ldots$. For each n and each $\omega \in \Gamma$ there is a unique atom of G_n, henceforth denoted by $A_n(\omega)$, which contains ω. For each $\omega \in \Gamma$ and $A \in G$ let

$$P(\omega,A) \equiv \int_A [g(\omega)](\omega')dQ(\omega').$$

Thus for each $\omega \in \Gamma$, $P(\omega,\cdot)$ is an X-valued measure absolutely continuous with respect to Q, whose average range is in $\tilde{K}$. For each n let

$$h_n(\omega,\omega') \equiv \begin{cases} \dfrac{P(\omega,A_n(\omega'))}{Q(A_n(\omega'))} & \text{if } Q(A_n(\omega')) \neq 0 \\ 0 & \text{if } Q(A_n(\omega')) = 0. \end{cases}$$

It is not difficult to see that $h_n: \Gamma\times\Gamma \to X$ is $G\times G$ measurable and that for any fixed $\omega \in \Gamma$, $(h_n(\omega,\cdot), G_n)$ is a $\tilde{K}$-valued martingale. Since $\tilde{K}$ has the RNP it follows that for each $\omega \in \Gamma$,

(1) $$\lim_n h_n(\omega,\omega') \text{ exists for } Q\text{-almost all } \omega'.$$

Let $A \equiv \{(\omega,\omega'): \lim_n h_n(\omega,\omega') \text{ exists}\}$ and for each $(\omega,\omega') \in A$ let $h(\omega,\omega') \equiv \lim_n h_n(\omega,\omega')$. From (1)

$$\int_{\Gamma\times\Gamma} \chi_A dQ\times Q = \int_\Gamma\int_\Gamma \chi_A(\omega,\omega')dQ(\omega')dQ(\omega) = \int_\Omega 1dQ(\omega) = 1$$

so that h is defined [a.s.Q×Q] as well as being jointly measurable.

In order to check that $h(\omega,\cdot) = g(\omega,\cdot)$ [a.s.Q] for each $\omega \in \Gamma$ suppose initially that C is an atom of G_n for some n. Then

$$\int_C h(\omega,\omega')dQ(\omega') = \int_C h_n(\omega,\omega')dQ(\omega') = P(\omega,C) = \int_C g(\omega,\omega')dQ(\omega').$$

It follows that

(2) $$\int_D h(\omega,\omega')dQ(\omega') = \int_D g(\omega,\omega')dQ(\omega') \quad \text{for } D \in \bigcup_{n=1}^{\infty} G_n$$

and from the separability assumption on G, it is straightforward that (2) maintains more generally for each $D \in G$. Thus $g(\omega,\cdot) = h(\omega,\cdot)$ [a.s.Q] as was to be shown.

Essentially the same argument as was used to prove Theorem 7.9.1 gives the somewhat stronger result below.

<u>THEOREM</u> 7.9.3. Assume that K has the RNP. If $\mu_1,\mu_2 \in P_t(K)$ and if $\mu_1 <_D \mu_2$ then for each $f \in L^1_K(P)$ such that $f(P) = \mu_1$ there is a $g \in L^1_K(P)$ with $g(P) =$ and $f <_{SE} g$.

Theorems 7.9.1 and 7.9.3 may be viewed from a different perspective. Introduce the equivalence relation $\simeq$ on $L^1_K(P)$ by saying that $f \simeq g$ if and only if $f(P)$ $g(P)$. Then let $[f] \equiv \{g \in L^1_K(P): f \simeq g\}$ and $L \equiv \{[f]: f \in L^1_K(P)\}$. Then define $I: L \to P_t(K)$ as might be expected: $I([f]) \equiv f(P)$. The following Corollary is an immediate consequence of Theorem 7.9.1.

COROLLARY 7.9.4. Suppose that K has the RNP. Write $[f] < [g]$ if there are $f_1 \in [f]$ and $g_1 \in [g]$ such that $f_1 <_{SE} g_1$. Then I as defined above is an order isomorphism between $(L,<)$ and $(P_t(K),<_D)$.

The last results of this section provide information about the structure of the equivalence classes $[f]$ for $f \in L^1_K(P)$.

PROPOSITION 7.9.5. Let $f,g \in L^1_K(P)$. If $f <_{SE} g$ and $f \neq g$ on a set of positive measure then $f(P) \neq g(P)$. Hence no two elements of $[f]$ are commensurable.

Proof. Suppose that $f = \mathbb{E}[g|F_\beta]$ for some $\beta < \omega_1$, let $\hat{\Omega} \equiv \{0,1\}^\beta$, let $\hat{P}$ denote the corresponding completion of the product probability measure on F_β and, as convenient, factor Ω and P as $\Omega = \hat{\Omega}\times\Omega'$ and $P = \hat{P}\times P'$. By modifying f and g on sets of measure 0 assume that each is separably valued with Range(f) $\cup$ Range(g) $\subset K$ and that g is $B(\Omega)$ measurable. Note that g is not F_β measurable since f and g differ on a set of positive measure by assumption. Thus there must be a set $\hat{B} \in B(\hat{\Omega})$ with $\hat{P}(\hat{B}) > 0$ such that whenever $\hat{\omega} \in \hat{B}$ then $g(\hat{\omega},\cdot)\colon \Omega' \to K$ is not [a.s.P'] a constant function. Indeed, assume the contrary and pick a set $\hat{A}$ of $\hat{P}$-measure one such that $g(\hat{\omega},\cdot)$ is almost surely constant for each choice of $\hat{\omega} \in \hat{A}$. Let $h(\hat{\omega}) \equiv \int_{\Omega'} g(\hat{\omega},\omega')dP'(\omega')$ for $\hat{\omega} \in \hat{\Omega}$. Since g is $B(\Omega)$ measurable and hence $B(\hat{\Omega})\times B(\Omega')$ measurable, the Fubini theorem asserts that h is $B(\hat{\Omega})$ measurable. Clearly $h(\hat{\omega})$ is the value which $g(\hat{\omega},\cdot)$ assumes almost surely with respect to P' for $\hat{\omega} \in \hat{A}$. Let

$$C \equiv \{(\hat{\omega},\omega') \in \Omega\colon h(\hat{\omega}) \neq g(\hat{\omega},\omega')\}.$$

Then $C \in B(\Omega)$ and

$$\int_{\hat{\Omega}} \chi_C dP = \int_{\hat{A}}\int_{\hat{\Omega}} \chi_C(\hat{\omega},\omega')dP'(\omega')d\hat{P}(\hat{\omega}) = 0$$

so that g is F_β measurable, a contradiction which establishes the claim.

Since $\overline{co}$(Range(f) $\cup$ Range(g)) is a closed bounded convex separable subset of K there is a countable set $\{F_i\}_{i=1}^\infty \subset S(X^*)$ which separates its points and hence, by taking $F \equiv \sum_{i=1}^\infty 2^{-i}F_i^2$, there is a convex $F \in C_b(K)$ which is strictly convex on $\overline{co}$(Range(f) $\cup$ Range(g)). Next observe that one version of $\mathbb{E}[g|F_\beta]$ is given by

(3) $$\mathbb{E}[g|F_\beta](\hat{\omega},\omega') \equiv \int_{\Omega'} g(\hat{\omega},\omega')dP'(\omega')$$

since the function $(\hat{\omega},\omega') \to \int_{\Omega'} g(\hat{\omega},\omega')dP'(\omega')$ is $B(\hat{\Omega})\times\{\emptyset,\Omega'\}$ measurable by Fubini's theorem and, if $D \in F_\beta$ (and if $\hat{D} \in B(\hat{\Omega})$ is chosen so that $P(D \,\Delta\, (\hat{D}\times\Omega'))$ $= 0)$ then

$$\int_D\int_{\Omega'} g(\hat{\omega},\omega')dP'(\omega')dP(\omega) = \int_{\hat{D}\times\Omega'} g(\hat{\omega},\omega')d[\hat{P}\times P'](\hat{\omega},\omega') = \int_D gdP.$$

Finally, note that for $\hat{\omega} \in \hat{B}$ ($\hat{B}$ as in the previous paragraph) there is a set $A' \in B(\Omega')$ with $0 < P'(A') < 1$ and

$$\frac{1}{P'(A')}\int_{A'} g(\hat{\omega},\omega')dP'(\omega') \neq \frac{1}{1-P'(A')}\int_{\Omega'\setminus A'} g(\hat{\omega},\omega')dP'(\omega').$$

Hence, by the strict convexity of F on $\overline{co}(\text{Range}(f) \cup \text{Range}(g))$ as well as by Jensen's inequality (Meyer [1966], p. 29) we have

$$\begin{aligned} &F(\int_{\Omega'} g(\hat{\omega},\omega')dP'(\omega')) \\ &\quad < P'(A')F[\int_{A'} gd\frac{P'}{P'(A')}] + (1-P'(A'))F[\int_{\Omega'\setminus A'} gd\frac{P'}{1-P'(A')}] \\ &\quad \leqq P'(A')\cdot\int_{A'} F\circ gd\frac{P'}{P'(A')} + (1-P'(A'))\cdot\int_{\Omega'\setminus A'} F\circ gd\frac{P'}{1-P'(A')} \\ &\quad = \int_{\Omega'} F\circ g(\hat{\omega},\omega')dP'(\omega'). \end{aligned}$$

It follows that, using (3),

$$\begin{aligned} \int_K Fdf(P) &= \int_\Omega F\circ fdP = \int_\Omega F\circ\mathbb{E}[g|F_\beta]dP = \int_{\hat{\Omega}} F[\int_{\Omega'} g(\hat{\omega},\omega')dP'(\omega')]d\hat{P}(\hat{\omega}) \\ &= \int_{\hat{B}} F[\int_{\Omega'} gdP']d\hat{P} + \int_{\hat{\Omega}\setminus\hat{B}} F[\int_{\Omega'} gdP']d\hat{P} \\ &< \int_{\hat{B}}\int_{\Omega'} F\circ gdP'd\hat{P} + \int_{\hat{\Omega}\setminus\hat{B}} F[\int_{\Omega'} gdP']d\hat{P} \\ &\leqq \int_{\hat{\Omega}}\int_{\Omega'} F\circ gdP'd\hat{P} = \int_\Omega F\circ gdP = \int_K Fdg(P). \end{aligned}$$

Thus $f(P) \neq g(P)$ as was to be shown.

In Theorem 6.2.9 it was shown that if $f \in L^1_K(P)$ is $<_{SE}$ - maximal then $f(P)$ $<_D$ - maximal. The converse is now an easy matter.

COROLLARY 7.9.6. Assume that K has the RNP. Then $f \in L^1_K(P)$ is $<_{SE}$ - maximal if and only if $f(P)$ is $<_D$ - maximal.

Proof. If $\mu \in P_t(K)$ is $<_D$ - maximal and $f \in L^1_K(P)$ satisfies $f(P) = \mu$ and if $f <_{SE} g$ for some $g \in L^1_K(P)$ then $f(P) <_D g(P)$ by Theorem 7.9.1. Hence $f(P) = g(P)$. But then, from Proposition 7.9.5, $f = g$ [a.s.P]. That is, f is $<_{SE}$ - maximal.

§7.10. The Bishop Phelps Property for operators.

The Bishop Phelps Theorem 3.3.1 asserts that the set of support functionals of a closed bounded convex subset K of a Banach space X is norm dense in X*. Since the elements of X* may be viewed as linear operators with range in the Banach space $\mathbb{R}$ a natural question deals with the truth of operator versions of the Bishop Phelps theorem when $\mathbb{R}$ is replaced by an arbitrary Banach space Y. Of course some modification in the statement of the question is required: while, in the statement of the Bishop Phelps theorem the statement

(1) $$F(x_o) = \sup \{F(x): x \in K\}$$

makes perfectly good sense for $K \subset X$, $F \in X^*$ and $x_o \in K$ (true or not), the operator version must be amended to read

(2) $$\|T(x_o)\| = \sup \{\|T(x)\|: x \in K\}$$

for $T: X \to Y$ a bounded linear operator. Note that even when $Y = \mathbb{R}$ in (2) we obtain the somewhat weaker and "symmetrized" version of the Bishop Phelps condition (1)

(1') $$|F(x_o)| = \sup \{|F(x)|: x \in K\}.$$

As we shall see, problems of symmetry - or, rather, the lack of it - play an annoying but manageable role in the proof of the main theorem of this section, Theorem 7.10.4. In contrast to the classical Bishop Phelps results where the key is the fact that the range of the operators lies in $\mathbb{R}$, the operator version depends not on the range space but on the set K in the domain, and the pivotal property for K is the RNP. Bourgain's proof [1977] of the localized version of the Phelps Theorem (called Theorem

3.5.4 and the Bourgain Phelps theorem in these notes) appeared as a direct conseque of his work on the operator Bishop Phelps property and we will present a portion of his argument in Theorem 7.10.2. The most illuminating result of this section, thoug is due to Stegall [1978b] who showed that practically any Banach-valued function o a closed bounded convex set K with the RNP has arbitrarily small perturbations by linear rank one operators which are norm attaining on K. (See 7.10.4.) We begin with a definition and theorem due to Bourgain [1977].

DEFINITION 7.10.1. Let D be a closed and bounded subset of X.

(a) For any Banach space Y let $L(X,Y)$ denote the Banach space of bounded linear operators on X into Y with the usual norm: $\|T\| \equiv \sup \{\|T(x)\|: x \in U(X)\}$.

(b) D has the Bishop Phelps Property (BPP) if whenever Y is a Banach space and $T \in L(X,Y)$ then there are a sequence $\{T_n: n = 1,2,\ldots\}$ in $L(X,Y)$ and a sequence $\{x_n: n = 1,2,\ldots\}$ in D such that for each n, both

(i) $\|T_n\| < \frac{1}{n}$; and

(ii) $\|(T + T_n)(x_n)\| = \sup \{\|(T + T_n)(y)\|: y \in D\}$.

THEOREM 7.10.2. Let K be a closed bounded convex subset of a Banach space X. If each separable closed bounded convex subset of K has the BPP then K has the RNP.

The proof of this theorem is an immediate consequence of Theorem 2.3.6 parts (1 (2) and (11) and the following proposition.

PROPOSITION 7.10.3. Let C be a separable nonempty closed bounded convex subset of X. If C has the BPP then C is dentable.

Proof. We will establish the contrapositive. Thus assume that C is not dent and, without loss of generality, assume $C \subset U(X)$. According to Lemma 3.7.4 there i an $\varepsilon > 0$ such that whenever D_0 and D_1 are subsets of C with $C = \overline{co}(D_0 \cup D_1)$ and diameter$(D_0) \leq 2\varepsilon$ then $C = \overline{co}(D_1)$. A straightforward inductive argument shows that

$$(3) \qquad C = \overline{co}(C \setminus \bigcup_{i=1}^{k} U_\varepsilon[y_i]) \text{ for any finite subset } \{y_i\}_{i=1}^{k} \text{ of } C.$$

Let $Z \equiv \operatorname{span}(C)$ and denote by $\{z_n: n = 1,2,\ldots\}$ a dense sequence in Z. For eac n let $\langle z_n \rangle$ be shorthand notation for the one dimensional space $\operatorname{span}(\{z_n\})$ and define a nonlinear operator $\Psi: X \to \ell_2$ by

$$\Psi(x) \equiv (\|x\|, \tfrac{1}{2}\, d(x,\langle z_1\rangle),\ldots,\tfrac{1}{2^n}\, d(x,\langle z_n\rangle),\ldots) \in \ell_2$$

where $d(x,\langle z_n\rangle)$ denotes the $\|\cdot\|$ distance from x to $\langle z_n\rangle$. Let $|||x||| \equiv \|\Psi(x)\|_2$. ($\|\cdot\|_2$ is the usual ℓ_2 norm.) Evidently $|||\cdot|||$ defines an equivalent norm on X. Since $C \subset X$ has the BPP, the identity operator $I: (X,\|\cdot\|) \to (X,|||\cdot|||)$ may be perturbed by an operator $T \in L((X,\|\cdot\|),(X,|||\cdot|||))$ with $\|T\| < \varepsilon/4$ in such a way that for some $x_0 \in C$

(4) $$|||(I + T)(x_0)||| = \sup\{|||(I + T)(y)|||: y \in C\}.$$

Pick j so that $\|x_0 - z_j\| < \varepsilon/4$. Let $\{y_1,\ldots,y_k\}$ be a finite $\varepsilon/8$ net for $\{y \in \langle z_j\rangle: \|y\| \leqq 1 + \varepsilon\}$ and observe that

(5) $$D \equiv \{y \in C: d(y,\langle z_j\rangle) < \tfrac{7\varepsilon}{8}\} \subset \bigcup_{i=1}^{k} U_\varepsilon[y_i].$$

Thus from (3), $C = \overline{\text{co}}(C \setminus D)$. It follows that

$$2|||(I + T)(x_0)||| = \sup\{|||(I + T)(x_0) + (I + T)(y)|||: y \in C \setminus D\}.$$

If $\{v_i: i = 1,2,\ldots\}$ is a sequence in $C \setminus D$ such that

(6) $$2|||(I + T)(x_0)||| = \lim_n |||(I + T)(x_0) + (I + T)(v_n)|||$$

then, as we show next,

(7) $$d((I + T)(x_0),\langle z_j\rangle) = \lim_n d((I + T)(v_n),\langle z_j\rangle).$$

Indeed, for any point p of X and sequence of points $\{p_n: n = 1,2,\ldots\}$ in X with $\|\Psi(p_n)\|_2 \leqq \|\Psi(p)\|_2$ for each n and $\lim_n \|\Psi(\tfrac{1}{2}p + \tfrac{1}{2}p_n)\|_2 = \|\Psi(p)\|_2$ we have

$$\|\Psi(p)\|_2 = \lim_n \|\Psi(\tfrac{1}{2}p + \tfrac{1}{2}p_n)\|_2 \leqq \lim_n \|\tfrac{1}{2}\Psi(p) + \tfrac{1}{2}\Psi(p_n)\|_2 \leqq \|\Psi(p)\|_2$$

and from the uniform convexity of the ℓ_2 norm, it follows that $\|\Psi(p_n) - \Psi(p)\|_2$ converges to 0 as $n \to \infty$. In particular, the $(j+1)^{\text{st}}$ coordinate of $\Psi(p_n)$ converges to the $(j+1)^{\text{st}}$ coordinate of $\Psi(p)$. Now let p be the point $(I + T)(x_0)$

and p_n the point $(I + T)(v_n)$ for each n. Formula (7) follows.

On the other hand,

$$(8) \qquad d((I + T)(x_0), \langle z_j \rangle) \leqq \|(I + T)(x_0) - x_0\| + d(x_0, \langle z_j \rangle) < \frac{\varepsilon}{4} + \frac{\varepsilon}{4} = \frac{\varepsilon}{2}$$

while for each n

$$(9) \qquad d((I + T)(v_n), \langle z_j \rangle) \geqq d(v_n, \langle z_j \rangle) - \|(I + T)(v_n) - v_n\| \geqq \frac{7\varepsilon}{8} - \frac{\varepsilon}{4} = \frac{5\varepsilon}{4} .$$

The combination of (8) and (9) is impossible in light of (7). This contradiction proves the proposition.

In the next paragraphs, and culminating in Theorem 7.10.4 below, the converse o Theorem 7.10.2 is considered. Thus let K be a closed bounded convex subset of X and assume that K has the RNP. If $\psi: K \to Y$ is any function such that both $x \to \|\psi(x)\|$ is upper semicontinuous and $\sup_{x \in K} \|\psi(x)\| < \infty$, perturbations of ψ by ra one linear operators T from X to Y of arbitrarily small norm will be produced such that each perturbation, $\psi + T$ of ψ, attains its supremum in norm at either one or possibly two points of K. The argument for K and ψ as above consists of a somewhat more technically intricate version, to make up for the lack of symmetry, of the following beautiful application of duality and the Bourgain Phelps Theorem 3.5.4 for the special case in which K is assumed symmetric about 0 and $\|\psi(x)\| = \|\psi(-x)\|$ for $x \in K$. (The general argument appears in the proof of Theorem 7.10.4.)

Thus assume for now that $K \subset X$ is absolutely convex, closed and bounded, that K has the RNP, that $g(x) \equiv \|\psi(x)\|$ is symmetric about 0 and upper semicontinuou and that, without loss of generality, $\sup_{x \in K} g(x) = 1$. The perturbations we seek are of the form $\psi + y_0 F$ for some $y_0 \in Y$ and $F \in X^*$. The idea will be to maximize $g + F$ on K for choices of F of small norm and then to choose the direction y_0 to be of norm one so that $\|(\psi + y_0 F)(\cdot)\| = \|\psi(\cdot)\| + F(\cdot)$ at point(s) of K where this latter function is maximized. Thus it is natural to consider the function $\xi: X^* \to \mathbb{R}$ given by

$$(10) \qquad \xi(F) \equiv \sup \{(g + F)(x) : x \in K\} - 2 \quad \text{for} \quad F \in X^*$$

where subtracting 2 from the above supremum conveniently normalizes ξ so that

$\xi(0) = -1$. Observe that ξ is convex, continuous and symmetric on X^*. Let C denote the epigraph of ξ in $X^* \times \mathbb{R}$. Thus C is the set

$$C \equiv \{(F,s) \in X^* \times \mathbb{R}\colon s \geq \xi(F)\}. \tag{11}$$

Note first that C is a closed convex set. In fact, C is a weak* closed convex subset of $X^* \times \mathbb{R}$ as may be checked directly using the definition of ξ. Thus C is the intersection of the weak* closed halfspaces containing it and we are, again naturally, let to evaluate the (asymmetric) lower polar $C_\circ$ of C:

$$C_\circ \equiv \{(x,r) \in X \times \mathbb{R}\colon (F,s)(x,r) = F(x) + sr \leq 1 \text{ for each } (F,s) \in C\}$$

since its (asymmetric) upper polar, $C_\circ{}^\circ$ is just C again. While the calculations below do not provide a complete representation for $C_\circ$ they do provide motivation for the set D defined in (12) below which will form an adequate surrogate for $C_\circ$ in the future since $D^\circ = C$ (and $\overline{\text{co}}(D) = C_\circ$).

Let (x,r) be a point of $C_\circ$. If $(F,s) \in C$ then $(F,s') \in C$ for $s' \geq s$ so that $(F,s')(x,r) = F(x) + s'r \leq 1$ for each $s' \geq s$. It follows that $r \leq 0$. On the other hand, $(0,-1) \in C$ from which we obtain $-r = (0,-1)(x,r) \leq 1$ and hence $-1 \leq r \leq 0$. If $r = 0$ (i.e. if $(x,0) \in C_\circ$) then $F(x) \leq 1$ for each $F \in X^*$ so $x = 0$. Now suppose $(x,r) \in C_\circ$ with $-1 \leq r < 0$. Then for each $F \in X^*$, $F(x) + r\xi(F) \leq 1$ or, in other words,

$$\begin{aligned} F(x) &\leq 1 + (-r)[\sup\{(g + F)(y)\colon y \in K\} - 2] \\ &\leq 1 + (-r)[\sup\{g(y)\colon y \in K\} + \sup\{F(y)\colon y \in K\} - 2] \\ &= 1 + r - r[\sup\{F(y)\colon y \in K\}]. \end{aligned}$$

Since $-r > 0$,

$$F(\tfrac{x}{-r}) \leq (-1 - \tfrac{1}{r}) + \sup\{F(y)\colon y \in K\} \quad \text{for each} \quad F \in X^*$$

and the Hahn Banach theorem guarantees that $\frac{x}{-r} \in K$. Thus if $(x,r) \in C_\circ$, $r \neq 0$, then $(x,r) = (-r)(\frac{x}{-r},-1)$ is of the form $p(y,-1)$ for some $p \in [0,1]$ and $y \in K$. The precise relationship between p and y may be complicated but observe that a sufficient condition that $p(y,-1)$ belong to $C_\circ$ is that p be so small that the following calculation hold true for each $(F,s) \in C$:

$$(F,s)(p(y,-1)) = p(F(y) - s) \leqq p(F(y) - \xi(F)) \quad \text{[always true if } (F,s) \in C \text{ since } p \geqq 0]$$

$$\leqq p(F(y) - \{(g + F)(y) - 2\}) \quad \text{[always true since } p \geqq 0]$$

$$= p(2 - g(y)) \leqq 1.$$

That is, if $p \leqq \frac{1}{2-g(y)}$ then $p(y,-1) \in C_\circ$. This is precisely the condition defini[ng]

D:

(12) $$D \equiv \{\frac{t}{2-g(y)}(y,-1) : 0 \leqq t \leqq 1, \quad y \in K\}.$$

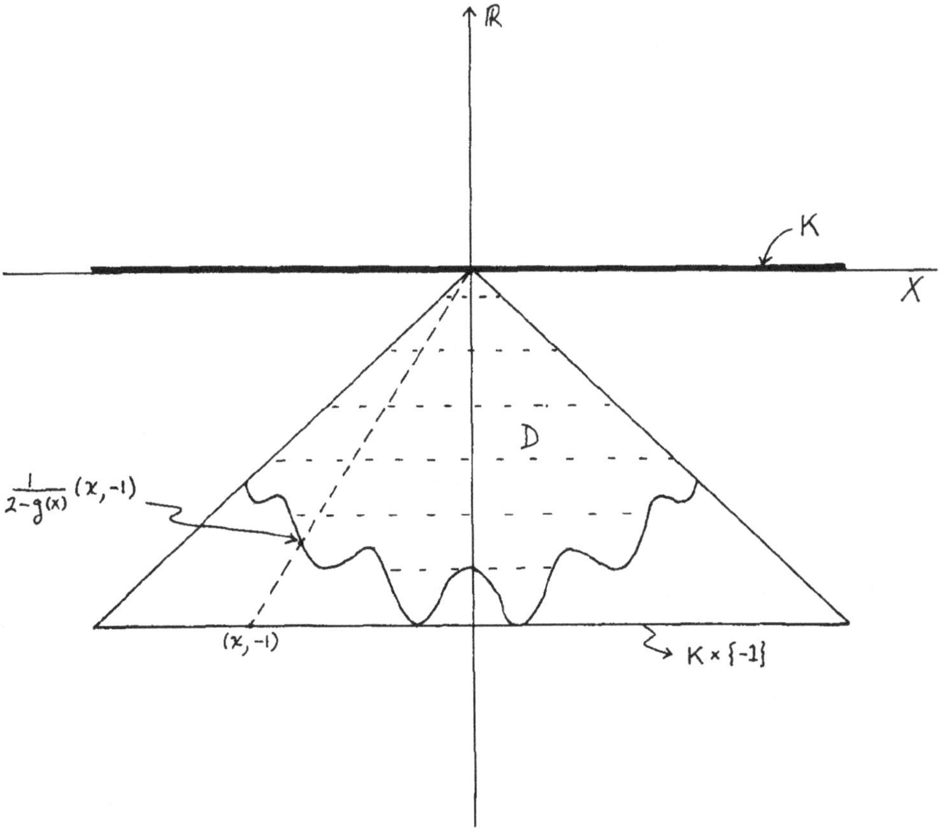

D is closed since g is upper semicontinuous but it is easy to see that D need n[ot]
be convex in general. Nevertheless $D^\circ = C$ as is shown in the next paragraph.

Since $D \subset C_\circ$, evidently $D^\circ \supset C_\circ{}^\circ = C$. On the other hand, suppose that $F \in X^*$ and that $s < \xi(F)$ (i.e. suppose that $(F,s) \notin C$). Choose a sequence $\{y_n : n = 1,2,\ldots\}$ in K such that $\lim_n (g + F)(y_n) - 2 = \xi(F)$. Observe that $\{g(y_n): n = 1,2,\ldots\}$ is bounded since F is bounded on K. Note that

$$(F,s)\left(\frac{1}{2-g(y_n)}(y_n,-1)\right)$$

$$= \frac{F(y_n) - s}{2-g(y_n)} = \frac{F(y_n) - [(g + F)(y_n) - 2]}{2-g(y_n)} + \frac{(g + F)(y_n) - 2 - s}{2-g(y_n)} \tag{13}$$

$$= 1 + \frac{(g + F)(y_n) - 2 - \xi(F)}{2-g(y_n)} + \frac{\xi(F) - s}{2-g(y_n)} .$$

Now the middle fraction of the right side of (13) approaches 0 as $n \to \infty$ while the right most fraction is bounded away from 0 and is positive as n ranges over $\mathbb{N}$. Hence for some n the left hand side of (13) is strictly greater than one. That is, $(F,s) \notin D^\circ$, and we conclude $D^\circ = C$ as was to be shown.

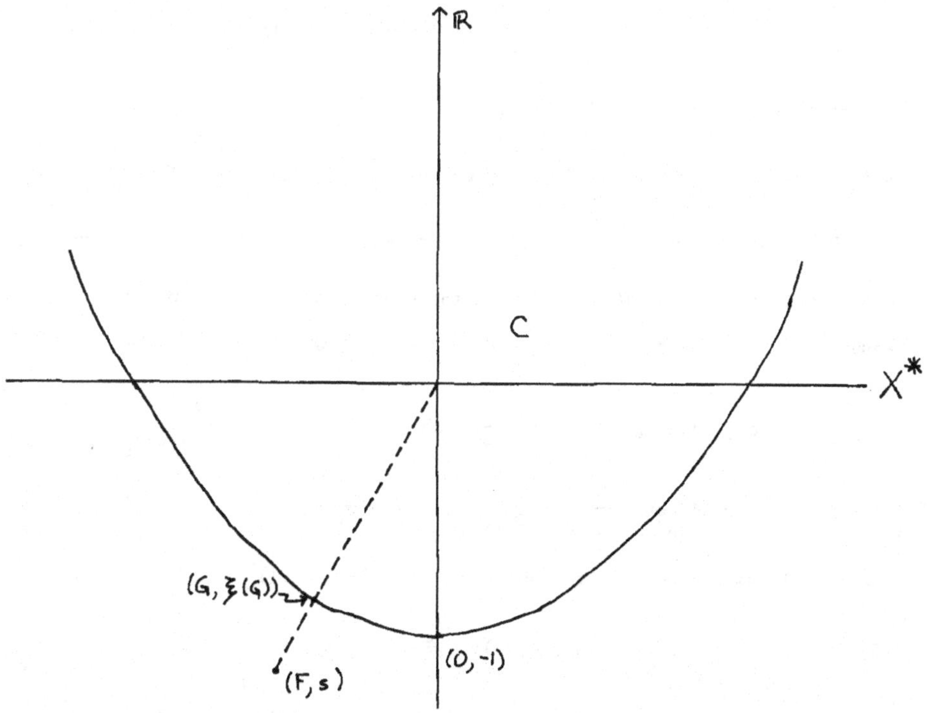

There is a rather direct way to pull needed information about the behavior of $\|\psi + y_0 F\|$ on K from information about $(F,\xi(F))$ on D which is central to Stega argument. Indeed, let $M \equiv \sup\{\|x\|: x \in K\}$ and fix $\delta > 0$ such that $\delta < \frac{1}{2M}$. No $D \subset K \times [-1,0]$ and hence the set of strongly exposing functionals of D, $SE(D)$, which is evidently the same as $SE(\overline{co}(D))$, is a dense G_δ subset of $X^* \times \mathbb{R}$ by th Bourgain Phelps Theorem 3.5.4. (Clearly $K \times [-1,0]$ has the RNP since K does. Alternatively, consult the last paragraph of Chapter 5, section 8.) Of course positive multiples of elements of $SE(D)$ again belong to $SE(D)$ so that by choosing $(F,s) \in SE(D)$ sufficiently close to $(0,-1)$, a point on the graph of the continuou function ξ - see the picture on the previous page - and by multiplying by the posit scalar $\frac{\xi(F)}{s}$, we obtain an element of $SE(D)$ of the form $(G,\xi(G))$ with $\|G\| < \delta$ Suppose that $(G,\xi(G))$ strongly exposes D at $\frac{t_0}{2-g(x_0)}(x_0,-1)$. Since

$$(14)\qquad \begin{aligned} (G,\xi(G))\left(\frac{1}{2-g(x_0)}(x_0,-1)\right) &= \frac{G(x_0) - \xi(G)}{2-g(x_0)} \geqq \frac{G(x_0) - [1 + \|G\|M - 2]}{2-g(x_0)} \\ &\geqq \frac{1 - 2\|G\|M}{2-g(x_0)} > \frac{1 - 2\delta M}{2-g(x_0)} > 0 = (G,\xi(G))(0,0) \end{aligned}$$

it follows that $t_0 = 1$. That is,

$$(15)\qquad (G,\xi(G)) \text{ strongly exposes } D \text{ at } \frac{1}{2-g(x_0)}(x_0,-1).$$

Note that $(G,\xi(G))\left(\frac{1}{2-g(x_0)}(x_0,-1)\right) = 1$. Indeed, let $\alpha \equiv (G,\xi(G))\left(\frac{1}{2-g(x_0)}(x_0,-1)\right)$. Then $\alpha \leqq 1$ since $(G,\xi(G)) \in D^\circ$. On the other hand, if $\{y_n: n = 1,2,\dots\}$ is a sequence in K such that $\lim_n (g + G)(y_n) - 2 = \xi(G)$ then

$$\begin{aligned} \alpha &\geqq \limsup_n\ (G,\xi(G))\left(\frac{1}{2-g(y_n)}(y_n,-1)\right) = \limsup_n \frac{G(y_n) - \xi(G)}{2-g(y_n)} \\ &\geqq \limsup_n \frac{G(y_n) - [(g + G)(y_n) - 2]}{2-g(y_n)} = 1. \end{aligned}$$

Thus $\alpha = 1$ as stated. It follows that $2 - g(x_0) = G(x_0) - \xi(G)$ or, by rearrangi

$$(16)\qquad \sup\{(g + G)(y): y \in K\} = g(x_0) + G(x_0).$$

Since K and g are symmetric about 0 it follows that $G(x_0) = |G|(x_0)$ and that

(17) $$\sup \{(g + G)(y): y \in K\} = \sup \{(g + |G|)(y): y \in K\}$$
$$= g(x_0) + G(x_0) = g(x_0) + |G|(x_0).$$

In order to check that $g + G$ strongly exposes K at x_0 pick any sequence $\{x_n: n = 1,2,\ldots\}$ in K for which $\lim_n (g + G)(x_n) = (g + G)(x_0)$. Then

(18) $$\lim_n (G, \xi(G))\left(\frac{1}{2-g(x_n)}(x_n,-1)\right) = \lim_n \frac{G(x_n) - G(x_0) - g(x_0) + 2}{2-g(x_n)} = 1,$$

and since $(G, \xi(G))$ strongly exposes D at $\frac{1}{2-g(x_0)}(x_0,-1)$ it follows that $\frac{1}{2-g(x_n)}(x_n,-1)$ approaches $\frac{1}{2-g(x_0)}(x_0,-1)$ as $n \to \infty$. By considering the sequence of second coordinates, $\frac{1}{2-g(x_n)} \to \frac{1}{2-g(x_0)}$ and hence convergence in the first coordinate yields $x_n \to x_0$. That is, $g + G$ strongly exposes K at x_0 and it follows easily from this and the symmetry of g and K that $g + |G|$ also "strongly exposes" K at x_0 in the broadened sense that if $(g + |G|)(x_n)$ approaches $(g + |G|)(x_0)$ for a sequence $\{x_n: n = 1,2,\ldots\}$ in K, then $\lim_n \min \{\|x_n - x_0\|, \|x_n + x_0\|\} = 0$.

In order to find rank one linear operators from X to Y of arbitrarily small norm which, when added to ψ yield norm attaining perturbations of ψ, fix $\delta > 0$, choose x_0 and G as above and define $T_\delta: X \to Y$ by

(19) $$T_\delta(x) \equiv G(x)\frac{\psi(x_0)}{\|\psi(x_0)\|} \quad \text{for } x \in X. \quad (\text{If } \psi(x_0) = 0 \text{ let } T_\delta \equiv 0.)$$

Then for any $x \in K$

(20) $$\begin{aligned} \|\psi(x) + T_\delta(x)\| &\leq \|\psi(x)\| + |G|(x) = g(x) + |G|(x) \\ &\leq g(x_0) + |G|(x_0) \qquad [\text{equality only if } x = \pm x_0] \\ &= \|\psi(x_0) + T_\delta(x_0)\|. \end{aligned}$$

Thus $\psi + T_\delta$ attains its supremum in norm at x_0 and possibly at $-x_0$ but nowhere else on K.

The argument just presented contains all the central ideas for the general case where symmetry of g and of K is not assumed. The proof presented with Theorem 7.10.4 below, then, is detailed only insofar as increased technicalities require it to be. It should be noted that Bourgain has an example of a closed bounded convex set K with the RNP such that $\overline{co}(K \cup -K)$ does not have the RNP. Hence eliminating symmetry from the hypothesis cannot be dealt with by "symmetrization" in the most naïve sense at least.

<u>THEOREM</u> 7.10.4. Let $K \subset X$ be a closed bounded convex set and assume that K has the RNP. Let Y be a Banach space and $\psi: K \to Y$ any function such that both $x \to \|\psi(x)\|$ is upper semicontinuous and $\sup_{x\in K} \|\psi(x)\| < \infty$. Then given $\delta > 0$ there is a rank one linear operator $T_\delta \in L(X,Y)$ such that $\|T_\delta\| < \delta$ and $\|\psi + T_\delta\|$ attains i supremum on K at either one or two points.

Proof. Let $f(x) \equiv \|\psi(x)\|$ and assume without loss of generality that $\sup_{x\in K} f(x$ $= 1$. Define $g: K\cup(-K) \to \mathbb{R}$ by

$$(21) \qquad g(x) \equiv \begin{cases} \max\{f(x),f(-x)\} & \text{for } x \in K \cap (-K) \\ f(x) & \text{for } x \in K \setminus (-K) \\ f(-x) & \text{for } x \in (-K) \setminus K. \end{cases}$$

Then g is upper semicontinuous, symmetric and $\sup\{g(x): x \in \pm K\} = 1$. Define $\xi': X^* \to \mathbb{R}$ almost as before (cf. (10))

$$(22) \qquad \xi'(F) \equiv \sup\{(g + F)(x): x \in \pm K\} - 2 \quad \text{for } F \in X^*$$

and observe that ξ' is convex, symmetric, continuous and $\xi'(0) = -1$. Let

$$(23) \qquad C' \equiv \{(F,s) \in X^* \times \mathbb{R}: F \in X^*, \quad s \geqq \xi'(F)\}$$

be the epigraph of ξ' and define D' by

$$(24) \qquad D' \equiv \{\tfrac{t}{2-g(x)}(x,-1): x \in \pm K, \quad 0 \leqq t \leqq 1\}.$$

Then D' is closed since g is upper semicontinuous and

$$(25) \qquad D' \subset \{(tx,-t): 0 \leqq t \leqq 1; \quad x \in K\} \cup \{(tx,-t): 0 \leqq t \leqq 1; \quad x \in -K\}.$$

It follows from Lemma 7.10.5 below that each of the sets on the right side of (25) h the RNP since K does, and from Lemma 7.10.6 that, as D' is a closed subset of th union of two closed bounded convex sets with the RNP, $SE(D')$ contains a dense G_δ subset of $X^* \times \mathbb{R}$. It may be shown (see the argument in the paragraph containing (1 that $(D')^\circ = C'$. Since $(0,0) \in \text{interior}(C')$ and $(0,-1) \in C'$, given $\delta > 0$, exactly as was argued for the symmetric case before (5), there is a $G \in X^*$, $\|G\| < \delta$ such that $(G, \xi'(G)) \in S\mathcal{E}(D')$, and the point strongly exposed by this functional has the form $\frac{1}{2-g(x_0)}(x_0,-1)$ for some $x_0 \in \pm K$. At this point we must begin to take notice of + and - signs. Let $\varepsilon_1 \equiv +1$ if $x_0 \in K$ and $\varepsilon_1 \equiv -1$ if $x_0 \in (-K) \setminus K$. Then $\varepsilon_1 x_0 \in K$ and from the symmetry of g

$$(26) \qquad 1 = (G, \xi'(G))\left(\frac{1}{2-g(x_0)}(x_0,-1)\right) = \frac{G(x_0) - \xi'(G)}{2-g(x_0)} = \frac{\varepsilon_1 G(\varepsilon_1 x_0) - \xi'(G)}{2-g(\varepsilon_1 x_0)} .$$

Thus $2 - g(\varepsilon_1 x_0) = \varepsilon_1 G(\varepsilon_1 x_0) - \xi'(G)$ or, in other words,

$$\sup\{(g + G)(x) : x \in \pm K\} = g(\varepsilon_1 x_0) + \varepsilon_1 G(\varepsilon_1 x_0).$$

Evidently, then, $G(x_0) = \varepsilon_1 G(\varepsilon_1 x_0) = |G|(x_0)$ so that, as in the symmetric case - see (17) - and using the definition of g as the "symmetrization of f on K" we obtain the various equalities

$$(27) \qquad \begin{aligned} &\sup\{(g + G)(y) : y \in \pm K\} = \sup\{(g + |G|)(y) : y \in \pm K\} \\ &= \sup\{(f + |G|)(y) : y \in \pm K\} = g(\varepsilon_1 x_0) + \varepsilon_1 G(\varepsilon_1 x_0) \\ &= g(\varepsilon_1 x_0) + |\varepsilon_1 G|(\varepsilon_1 x_0). \end{aligned}$$

Now either $g(\varepsilon_1 x_0) = f(\varepsilon_1 x_0)$ or possibly, if $-\varepsilon_1 x_0 \in K$, $g(\varepsilon_1 x_0) = f(-\varepsilon_1 x_0)$. Let $\varepsilon_2 \equiv +1$ if $g(\varepsilon_1 x_0) = f(\varepsilon_1 x_0)$ but if $g(\varepsilon_1 x_0) > f(\varepsilon_1 x_0)$ then let $\varepsilon_2 \equiv -1$. In either case, $\varepsilon_1 \varepsilon_2 x_0 \in K$ and $g(\varepsilon_1 \varepsilon_2 x_0) = f(\varepsilon_1 \varepsilon_2 x_0)$. Then

$$(28) \qquad \begin{aligned} &f(\varepsilon_1 \varepsilon_2 x_0) + \varepsilon_1 \varepsilon_2 G(\varepsilon_1 \varepsilon_2 x_0) = g(\varepsilon_1 x_0) + \varepsilon_1 G(\varepsilon_1 x_0) \\ &= g(\varepsilon_1 x_0) + |\varepsilon_1 G|(\varepsilon_1 x_0) = \sup\{(g + G)(y) : y \in \pm K\} \\ &= \sup\{(f + |G|)(y) : y \in \pm K\}. \end{aligned}$$

Hence $\varepsilon_1 \varepsilon_2 x_0$ is a point of K at which the perturbation $f + \varepsilon_1 \varepsilon_2 G$ of f attains its supremum.

To check that $f + \varepsilon_1 \varepsilon_2 G$ actually strongly exposes K at $\varepsilon_1 \varepsilon_2 x_0$ choose any sequence $\{x_n : n = 1,2,\ldots\}$ in K with

$$(29) \qquad \lim_n (f + \varepsilon_1 \varepsilon_2 G)(x_n) = (f + \varepsilon_1 \varepsilon_2 G)(\varepsilon_1 \varepsilon_2 x_0).$$

Note that

$$(30)\qquad f(x_n) + \varepsilon_1\varepsilon_2 G(x_n) \leqq g(x_n) + \varepsilon_1\varepsilon_2 G(x_n)$$
$$\leqq f(\varepsilon_1\varepsilon_2 x_0) + \varepsilon_1\varepsilon_2 G(\varepsilon_1\varepsilon_2 x_0) = g(\varepsilon_1 x_0) + \varepsilon_1 G(\varepsilon_1 x_0).$$

From (29) all parts of the inequality (30) approach $g(\varepsilon_1 x_0) + G(x_0)$ as $n \to \infty$ and hence

$$(31)\qquad \lim_n \frac{-g(\varepsilon_1 x_0) - \varepsilon_1 G(\varepsilon_1 x_0) + \varepsilon_1\varepsilon_2 G(x_n) + 2}{2-g(x_n)} = 1.$$

That is,

$$(32)\qquad \lim_n (G,\xi'(G))\left(\frac{1}{2-g(\varepsilon_1\varepsilon_2 x_n)}(\varepsilon_1\varepsilon_2 x_n,-1)\right) = 1.$$

But $(G,\xi'(G))$ strongly exposes $\frac{1}{2-g(x_0)}(x_0,-1)$ in $K \cup (-K)$ and hence

$$\frac{1}{2-g(\varepsilon_1\varepsilon_2 x_n)}(\varepsilon_1\varepsilon_2 x_n,-1) \to \frac{1}{2-g(x_0)}(x_0,-1) \quad \text{as} \quad n \to \infty.$$

As in the symmetric case, $\varepsilon_1\varepsilon_2 x_n \to x_0$ or, in other words, $x_n \to \varepsilon_1\varepsilon_2 x_0$. Thus $\varepsilon_1\varepsilon_2 x_0$ is strongly exposed in K by $f + \varepsilon_1\varepsilon_2 G$.

Finally, to produce perturbations of the original ψ by rank one linear operat T_δ, $\|T_\delta\| < \delta$ (for any given $\delta > 0$) such that $\|\psi + T_\delta\|$ attains its supremum on K at one or two points, we proceed as in the symmetric case. (See (19) and (20).) For the x_0, ε_1, ε_2, g and G produced above for a given $\delta > 0$ let

$$(33)\qquad T_\delta(x) \equiv \varepsilon_1\varepsilon_2 G(x)\frac{\psi(\varepsilon_1\varepsilon_2 x_0)}{\|\psi(\varepsilon_1\varepsilon_2 x_0)\|} \quad \text{and} \quad T_\delta \equiv 0 \quad \text{if} \quad \psi(\varepsilon_1\varepsilon_2 x_0) = 0.$$

Then, as in (20), $\|f + T_\delta\|$ attains its supremum on K at $\varepsilon_1\varepsilon_2 x_0$ and possibly at $-\varepsilon_1\varepsilon_2 x_0$ but nowhere else. Thus the proof of this theorem rests on establishing the next two lemmas.

LEMMA 7.10.5. Let $K \subset X$ be a closed bounded convex set with the RNP. Then $L \equiv \{t(x,-1): 0 \leqq t \leqq 1;\ x \in K\}$ has the RNP.

Proof. Suppose that $(\Omega,\mathcal{F},P)$ is a probability space and $m: \mathcal{F} \to X \times \mathbb{R}$ is a vector measure with $AR(m) \subset L$. Write m in the form (m_1,m_2) where $m_1: \mathcal{F} \to X$ a $m_2: \mathcal{F} \to \mathbb{R}$ are measures. Since $AR(m_2) \subset [-1,0]$ there is an $f \in L^1_{[-1,0]}(P)$ with

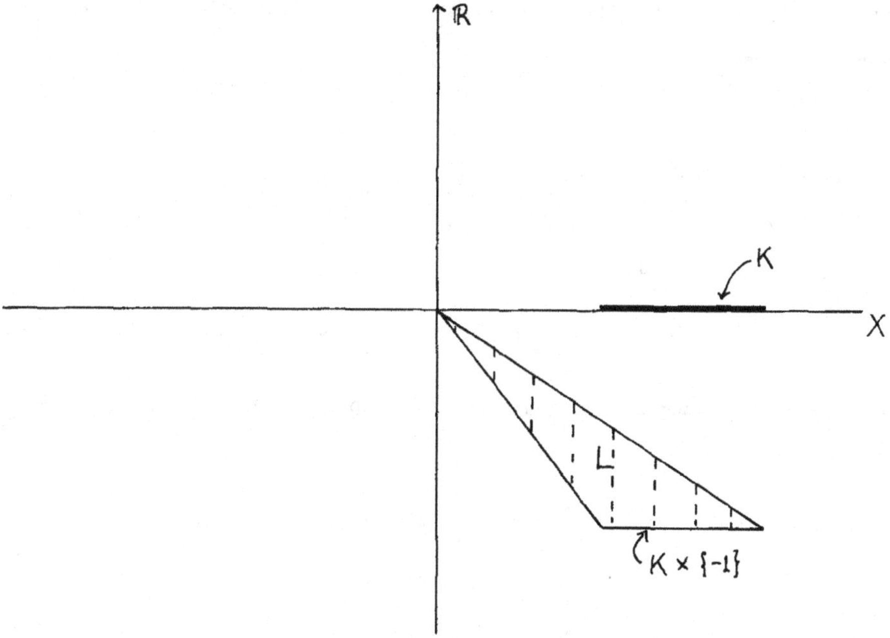

$\int_A fdP = m_2(A)$ for $A \in F$ by the classical Radon-Nikodým theorem. Now suppose $A \in F$ and $m_2(A) \neq 0$. Then

(34) $$\frac{m(A)}{P(A)} = \left(\frac{m_1(A)}{P(A)}, \frac{m_2(A)}{P(A)}\right) = \frac{-m_2(A)}{P(A)}\left(\frac{m_1(A)}{-m_2(A)}, -1\right).$$

Note that $\frac{-m_2(A)}{P(A)} \in [0,1]$ and $\frac{m(A)}{P(A)} \in AR(m) \subset L$. But for each point z of L with $z \neq 0$ there is a unique choice of $p \in [0,1]$ and $x \in K$ such that $z = p(x,-1)$. Hence from (34) we conclude $\frac{m_1(A)}{-m_2(A)} \in K$. Thus

$$\{\frac{m_1(A)}{-m_2(A)}: A \in F, \ -m_2(A) > 0\} \subset K$$

and since K has the RNP there is a $g \in L^1_K(-m_2)$ with $\int_A gd(-m_2) = m_1(A)$ for each $A \in F$. Consequently

$$\int_A -fgdP = m_1(A) \quad \text{for each } A \in F$$

whence $\frac{dm}{dP}$ exists and is $(-fg, f)$. The proof is complete.

LEMMA 7.10.6. Let K_1 and K_2 be two closed bounded convex subsets of X and suppose that each has the RNP. If $D \subset K_1 \cup K_2$ is closed then $SE(D)$ contains a dense G_δ subset of X^*.

Proof. If either $D \subset K_1$ or $D \subset K_2$ the result follows from Theorem 3.5.4. Thus assume $D \cap K_1 \neq \emptyset \neq D \cap K_2$ and let $Q_i \equiv SE(D \cap K_i)$ for $i = 1,2$. If $D \cap K_1 \cap K_2 \neq \emptyset$ let $Q_3 \equiv SE(D \cap K_1 \cap K_2)$, but if $D \cap K_1 \cap K_2 = \emptyset$ let $Q_3 \equiv X^*$. Then Q_i is a dense G_δ set in X^* for $i = 1,2,3$. For $j = 1$ or 2 let

$$V_j = \{F \in X^*: M(D,F) > M(D \cap K_j,F)\} .$$

(See 2.3.1 for the notation $M(\cdot,\cdot)$ above.) If $D \cap K_1 \cap K_2 = \emptyset$ let $V_3 = \emptyset$. Othe wise let

$$V_3 \equiv \{F \in X^*: M(D,F) = M(D \cap K_1 \cap K_2,F\}$$
$$= \bigcap_{n=1}^{\infty} \{F \in X^*: M(D,F) < \frac{1}{n} + M(D \cap K_1 \cap K_2,F)\} .$$

Evidently V_1 and V_2 are open and V_3 is a G_δ set. Let

$$W \equiv (\bigcap_{j=1}^{3} Q_j) \cap (\bigcup_{j=1}^{3} V_j) .$$

Observe that W is a G_δ set. The proof is completed by showing that $W \subset SE(D)$ and that W is dense in X^*. If $F \in W$ then $F \in SE(K_1 \cap D) \cap SE(K_2 \cap D)$. Hence if $M(K_1 \cap D,F) \neq M(K_2 \cap D,F)$ then clearly $F \in SE(D)$. Thus suppose $F \in W$ and $M(K_1 \cap D,F) = M(K_2 \cap D,F)$. Then $F \notin V_1 \cup V_2$ so $F \in V_3$. Since $F \in Q_1 \cap Q_2$ the are points x_1 and x_2, $x_i \in K_i \cap D$ $i = 1,2$ such that F strongly exposes $K_i \cap$ at x_i. But since $F \in V_3$, $M(K_i \cap D,F) = M(D \cap K_1 \cap K_2,F)$ and hence $x_i \in D \cap K_1 \cap K_2$ for $i = 1,2$. It follows that $x_1 = x_2$ whence $F \in SE(D)$. Thus $W \subset SE(D)$. The argument that W is dense is somewhat similar. Indeed, suppose that $F \in X^*$ i arbitrary. Since $\bigcap_{j=1}^{3} Q_j$ is dense, given $\varepsilon > 0$ there is an $F_1 \in \bigcap_{j=1}^{3} Q_j$ with $\|F - F_1\| < \varepsilon/3$. If $K_1 \cap K_2 \cap D = \emptyset$ and $F_1 \notin \bigcup_{j=1}^{3} V_j$ $(= \bigcup_{j=1}^{2} V_j)$ then $M(K_1 \cap D,F_1$ $= M(K_2 \cap D,F_1)$. If F_1 strongly exposes $K_i \cap D$ at x_i $i = 1,2$ then $x_1 \notin K_2$ c course, so there is an $F_2 \in X^*$, $\|F_2\| < \varepsilon/3$, such that $F_2(x_1) > M(K_2 \cap D,F_2)$. Bu then $F_1 + F_2 \in V_2$. Since V_2 is open and $\bigcap_{j=1}^{3} Q_j$ is a dense G_δ there is an $F_3 \in X^*$, $\|F_3\| < \varepsilon/3$ and $F_1 + F_2 + F_3 \in V_2 \cap (\bigcap_{j=1}^{3} Q_j) \subset W$. Since $\|F_1 + F_2 + F_3$ -

$< \varepsilon$ it follows that when $K_1 \cap K_2 \cap D = \emptyset$ then given $\varepsilon > 0$ and $F \in X^*$ there is a $G \in W$ with $\|F - G\| < \varepsilon$. Suppose, finally, that $K_1 \cap K_2 \cap D \neq \emptyset$ and that $F \in X^*$ is arbitrary. As before, given $\varepsilon > 0$ pick $F_1 \in \bigcap_{j=1}^{3} Q_j$ with $\|F - F_1\| < \varepsilon/3$ and assume that $F \notin \bigcup_{j=1}^{3} V_j$ (since otherwise the argument is complete). Then $M(K_1 \cap D, F_1) = M(K_2 \cap D, F_1) > M(K_1 \cap K_2 \cap D, F_1)$. If F_1 strongly exposes $K_i \cap D$ at x_i then $x_1 \in K_1 \cap D$ and hence $x_1 \notin K_2$. Produce F_2 and F_3 exactly as above and conclude, as before, that $F_1 + F_2 + F_3 \in W$ and that $\|F_1 + F_2 + F_3 - F\| < \varepsilon$. Thus the proof is complete.

As an immediate corollary to 7.10.4 note that when $K \subset X$ is closed bounded convex and has the RNP and Y is any Banach space then

$$\mathcal{D}(K;Y) \equiv \{T \in L(X,Y): \sup \{\|T(x)\|: x \in K\} = \|T(x_0)\| \text{ for some } x_0 \in K\}$$

is dense in $L(X,Y)$. In fact, it follows from Bourgain's approach [1977] that the set is a dense G_δ when K is absolutely convex (as well as being closed bounded convex and having the RNP). But even in the simplest cases for Y the situation remains somewhat unresolved. Indeed, suppose that C is a norm closed bounded convex subset of a dual Banach space X^* and that we take $Y \equiv \mathbb{R}$. It would be interesting to know, for example, that $\mathcal{D}(C;\mathbb{R}) \cap \hat{X} \neq \emptyset$ when C has the RNP. Morris [1981] has conjectured that X^* has the RNP if

$$\{F \in \mathcal{D}(C;\mathbb{R}): F \text{ is nonconstant on } C\} \cap \hat{X} \neq \emptyset \tag{35}$$

whenever C is a closed bounded convex subset of X^*. He has proved that if X is weakly compactly generated and (35) holds for each norm closed bounded convex subset of X^* then X^* has the WRNP. (See section 7.4 for a definition of the weak Radon-Nikodým Property. His result is equivalent to establishing that ℓ_1 does not embed in X.) Finally, Huff [1980] has, in one important case at least, taken the separability hypothesis out of Proposition 7.10.3. Namely, if a Banach space X does not have the RNP then there are two equivalent norms $\|\cdot\|$ and $|||\cdot|||$ such that the identity operator $\mathrm{id}: (X,\|\cdot\|) \to (X,|||\cdot|||)$ is not in the norm closure of the norm attaining linear operators between the two spaces.

§11. Bush structures.

Recall that a closed bounded convex set $K \subset X$ lacks the RNP if and only if it contains a ε-bush for some $\varepsilon > 0$. Of course, if such a set were to have the KMP then the closed convex hull of the bushes in K would be generated by their extreme points and one might hope that, in testing whether an arbitrary closed bounded conve set K has the KMP it actually suffices to check that $C = \overline{co}(ex(C))$ as C range over all sets of the form $\overline{co}(B)$ for $B \subset K$ a bush. While we know of no such theo rem, recent results of Ho [1982] and of James [1981] have shed some light on the extremal structure of the closed convex hull of certain bush subsets of non RNP sets By virtue of examples like that due to Bourgain and Rosenthal (see §7.3) there are non RNP sets K in which certain ε-bush structures cannot appear. Nevertheless, a is indicated below, inside any such non RNP set certain other generic types of ε-bushes must be present. (See Theorems 7.11.4 and 7.11.8.)

We shall need the formal definition of an ε-bush as was given in Definition 6.6.3. Also, the generalized Cantor set $\Delta_{0,1}$ associated with a given bush and the natural probability measure μ on the Borel subsets of $\Delta_{0,1}$ (as discussed after 6.6.3) will play central roles in the arguments of this section. The following cor ventions will be adhered to.

CONVENTIONS 7.11.1.

(a) Throughout this section "bush" will always mean "bounded ε-bush for some $\varepsilon < 0$". The number ε will sometimes be called the separation constant for the bus

(b) If B is a bush, the coefficients $r_{n,i}$ (as in 6.6.3) will often be sup-pressed. Thus, when convenient we shall write $B = \{b_{n,i}: (n,i) \in A\}$.

(c) The symbols $\Delta_{0,1}$ and μ will refer to the generalized Cantor set and natural probability measure on $\mathcal{B}(\Delta_{0,1})$ as defined in and before §6.6, formula (XX for the bush under discussion.

(d) If $B \equiv \{b_{n,i}: (n,i) \in A\}$ is a bush, write $(n+1,j) > (n,i)$ if $(n+1,j) \in A(n,i)$. (See 6.6.3(e) for this notation.) Extend ">" by transitivity, and by abuse of language write $b_{m,k} > b_{n,i}$ if $(m,k) > (n,i)$. In this way B may be considered a partially ordered set with least element $b_{0,1}$. As with the index set A, if $b_{n+1,j} > b_{n,i}$ then $b_{n+1,j}$ is an immediate successor of $b_{n,i}$. More generally, $b_{m,k}$ is a successor of $b_{n,i}$ if $b_{m,k} > b_{n,i}$.

Let $B \equiv \{b_{n,i}: (n,i) \in A\}$ be a bush. Then α is a branch of B if α is a sequence of the form

(1) $$\alpha \equiv \{b_{n,i(n)}: (n+1,i(n+1)) > (n,i(n)) \quad \text{for} \quad n = 0,1,\ldots\}.$$

If α is a branch of B (as in (1)) the sequence $\{I_{n,i(n)}: n = 0,1,\ldots\}$ is decreasing, where the $I_{n,i(n)}$'s are closed intervals used in the definition of $\Delta_{0,1}$. See §6.6, formula (XXVI). Since each $I_{n,i(n)}$ is a closed subinterval of $[0,1]$ there is a unique point $t \in \bigcap_{n=1}^{\infty} I_{n,i(n)}$. It is called the branch point corresponding to α. Note that $\Delta_{n,i}$ (see (XXVII) of §6.6) then consists precisely of the branch points for branches containing $b_{n,i}$ (and intuitively, $\mu(\Delta_{n,i})$ is the probability that a branch of B will contain $b_{n,i}$). Hence arguments concerning sets of branches of B and their "density" in the set of all branches of B may be translated into a discussion of the μ-measures of corresponding sets of branch points of $\Delta_{0,1}$. (Compare this with the definition of $\overset{=}{\Delta}_{0,1}$ given after 6.6.3.)

DEFINITION 7.11.2. Let B and $B_0 \equiv \{(b_{n,i}, r_{n,i}): (n,i) \in A\}$ be bushes in X with $B_0 \subset \overline{\text{co}}(B)$. Then B_0 is an asymptotic subbush of B if there is a 1-1 correspondence ψ from B_0 into B which preserves the partial orderings of the two bushes (as discussed in 7.11.1(d)) such that

$$\lim_n \sup_i \{\|b_{n,i} - \psi(b_{n,i})\|: (n,i) \in A\} = 0 .$$

Suppose now that B_0 is an asymptotic subbush of B (with map ψ as above) and that α is a branch of B_0. Then the branch point of α is the branch point of the unique branch β of B which contains $\{\psi(b_{n,i}): b_{n,i} \in \alpha\}$. Thus, in particular, the branch point of α is a point of the Cantor set $\Delta_{0,1}$ determined by B.)

The next results (except where explicitly mentioned otherwise) are due to James [1981]. The first typifies the types of conclusions James has drawn, and the technique of proof is a prototype for many of the arguments in that paper.

THEOREM 7.11.3. Assume that B is a bush in X with separation constant ε (see 7.11.1(e)) and that $0 < \bar{\varepsilon} < \varepsilon < 1$. Let S be a subset of $\Delta_{0,1}$ with $\bar{\mu}(S) < 1$. (Here $\bar{\mu}$ denotes the outer measure of μ.) Then B has an asymptotic subbush B_0 with separation constant $\frac{1}{2}\bar{\varepsilon}$ which satisfies the following property:

If T denotes the set of branch points of B_0 then both $S \cap T = \emptyset$ and $\mu(T) > 0$.

Proof. Let A denote the set of bush indices for B. Choose a subset A_1 of A such that the following three conditions are met:

$$\text{(2)}\quad \begin{aligned} &S \subset \cup\{\Delta_{m,j} : (m,j) \in A_1\} \equiv S_1; \\ &\mu(S_1) < 1; \text{ and} \\ &\Delta_{m,j} \cap \Delta_{r,k} = \emptyset \text{ whenever } (m,j) \text{ and } (r,k) \text{ are distinct elements of } A_1 . \end{aligned}$$

Without loss of generality assume $B \subset U(X)$ and pick a sequence $\{\varepsilon_i : i = 1,2,\ldots\}$ of positive numbers with $\sum_{i=1}^{\infty} \varepsilon_i < \frac{1}{4}(\varepsilon - \bar{\varepsilon})$. An "approximate bush" B_1 will be constructed below with separation constant $\frac{1}{2}\varepsilon$ corresponding to the "error sequence" $\{\varepsilon_i : i = 1,2,\ldots\}$ (see Definition 7.5.1). B_0, the sought for asymptotic subbush of B, is then obtained by taking the function version of B_1 (see 7.5.2)—evidently an ∞-quasi-martingale corresponding to $\{\varepsilon_i : i = 1,2,\ldots\}$ as described in 7.5.1—applying the decomposition Theorem 7.5.3 to produce an associated martingale, and letting B_0 be the bush whose function version is that martingale. (Note that, according to the decomposition Theorem 7.5.3, the bush B_0 will have separation constant at least $\frac{1}{2}\varepsilon - 2\cdot\frac{1}{4}(\varepsilon - \bar{\varepsilon}) = \frac{1}{2}\bar{\varepsilon}$ as desired.)

The construction of the approximate bush B_1 is by induction. Since S_1 is an open subset of $\Delta_{0,1}$ of μ-measure less than one there is a finite subset I of A such that each of the following three conditions obtains:

$$\text{(3)}\quad \begin{aligned} &\text{(a)}\ \cup\{\Delta_{n,i} : (n,i) \in I\} \supset S_1^c \ (\equiv \Delta_{0,1} \setminus S_1); \\ &\text{(b)}\ \Delta_{n,i} \cap \Delta_{r,s} = \emptyset \text{ if } (n,i) \neq (r,s) \text{ both belong to } I; \text{ and} \\ &\text{(c)}\ \sum\{\mu(\Delta_{n,i}) : (n,i) \in I\} < \mu(S_1^c) + \tfrac{1}{4}\varepsilon_1\mu(S_1^c) . \end{aligned}$$

Observe that for some $(n,i) \in I$

(4) $$\mu(\Delta_{n,i} \cap S_1) \leqq \tfrac{1}{4}\varepsilon_1 \mu(\Delta_{n,i}) .$$

(Indeed, if the contrary were true, using (3c) above we would obtain the contradiction

$$\tfrac{1}{4}\varepsilon_1 \mu(S_1{}^c) \leqq \tfrac{1}{4}\varepsilon_1 \sum_I \mu(\Delta_{n,i}) < \sum_I \mu(\Delta_{n,i} \cap S_1) = \sum_I \mu(\Delta_{n,i}) - \mu(S_1{}^c)$$
$$< \tfrac{1}{4}\varepsilon_1 \mu(S_1{}^c) .)$$

(N.B. Inequality (4) may be viewed as a quantitative version of the statement that the vast majority of branches of B through $b_{n,i}$ correspond to branch points in $S_1{}^c$.) Now fix an $(n,i) \in A$ which satisfies (4) and denote the element $b_{n,i}$ of B by $d_{0,1}$. Then $d_{0,1}$ will be the sole member of stage 0 of the new approximate bush B_1.

The elements of the first stage of B_1 will each belong to the original bush B. In fact, an integer M will be constructed below (see (6)) which depends on $d_{0,1}$, and each element of stage 1 of B_1 will be chosen from

$$(\text{stage } M \text{ of } B) \cup (\text{stage } (M+1) \text{ of } B).$$

Evidently, if the elements of stage 1 of B_1 were exactly the stage M elements of B, $d_{0,1}$ would be a (natural) convex combination of the elements of stage 1 of B_1. As is likely in the construction, though, some elements of the $M^{\underline{th}}$ stage of B will not be chosen. But each "missing element" of stage M is itself a convex combination of its immediate successors in stage $(M+1)$ and hence if, whenever an $M^{\underline{th}}$ stage element of B were missing from stage 1 of B_1, each of its immediate successors in stage $(M+1)$ of B belonged to stage 1 of B_1, again $d_{0,1}$ would, in a natural way, be a convex combination of the elements of stage 1 of B_1. However, it is possible that, in the construction of stage 1 of B_1, some of the immediate successors of missing $M^{\underline{th}}$ stage elements of B will also be missing from stage 1 of B_1 so that there will be "gaps" left when taking convex combinations to attempt to regain $d_{0,1}$ from the chosen stage 1 elements of B_1. If the "density" of the gaps is sufficiently low, the norm of the difference between the natural convex combinations of stage 1 elements of B_1 and $d_{0,1}$ will be dominated by ε_1. It is this implicit density notion which is at the heart of the argument below. Before turning to the detailed proof observe that if $d_{0,1}$ happens to be within $\frac{1}{2}\varepsilon$ of some element x of stage M of B then (since B has separation constant ε) $d_{0,1}$ will be at least $\frac{1}{2}\varepsilon$ from each of the immediate successors of x in stage (M+1) of B. Thus, insofar as separation constant considerations are concerned, we may freely eliminate from stage 1 of B_1 tho elements of stage M of B which are within $\frac{1}{2}\varepsilon$ of $d_{0,1}$, replace them by their immediate successors in stage (M+1) of B, and maintain a separation constant of $\frac{1}{2}\varepsilon$ The detailed argument follows.

Recall that $d_{0,1}$ of B_1 is $b_{n,i}$ of B. Let

(5) $$A_2 \equiv \{(p,q) \in A_1 : b_{p,q} > b_{n,i}\}$$

and choose M so large that

(6) $$\sum \{\mu(\Delta_{p,q}) : (p,q) \in A_2,\ p \geqq M\} \leqq \frac{1}{16}\varepsilon_1\varepsilon_2\mu(\Delta_{n,i}).$$

It is an easy matter to pick the elements of stage 1 of B_1 now. (See also the diagram below in which the small circles and dotted lines refer to missing elements.

<u>A</u> Eliminate from consideration each element x of stage M of B such that $\|x - d_{0,1}\| \leqq \frac{1}{2}\varepsilon$;

<u>B</u> If $b_{M,k}$ was not eliminated in <u>A</u>, add it to the collection of stage 1 elements of B_1 provided it satisfies the following, formula (4) - like condition

(7) $$\mu(\Delta_{M,k} \cap S_1) \leqq \frac{1}{4}\varepsilon_2\mu(\Delta_{M,k}) .$$

<u>C</u> If $b_{M,k}$ has been eliminated because of <u>A</u> and if $b_{M+1,q} > b_{M,k}$ the add $b_{M+1,q}$ to the collection of stage 1 elements of B_1 if

(7') $$\mu(\Delta_{M+1,q} \cap S_1) \leqq \frac{1}{4}\varepsilon_2\mu(\Delta_{M+1,q}) .$$

In order to check that there really are elements in the first stage of B_1 observe that stage 1 of B_1 cannot be empty if the sum of the μ-measures of all $\Delta_{m,k}$ such that $b_{m,k}$ is in stage 1 of B_1 is positive. This is accomplished bel by showing that the sum of the μ-measures of the $\Delta_{m,k}$'s corresponding to the "gap (as described and pictured) is small. It is convenient to first single out the

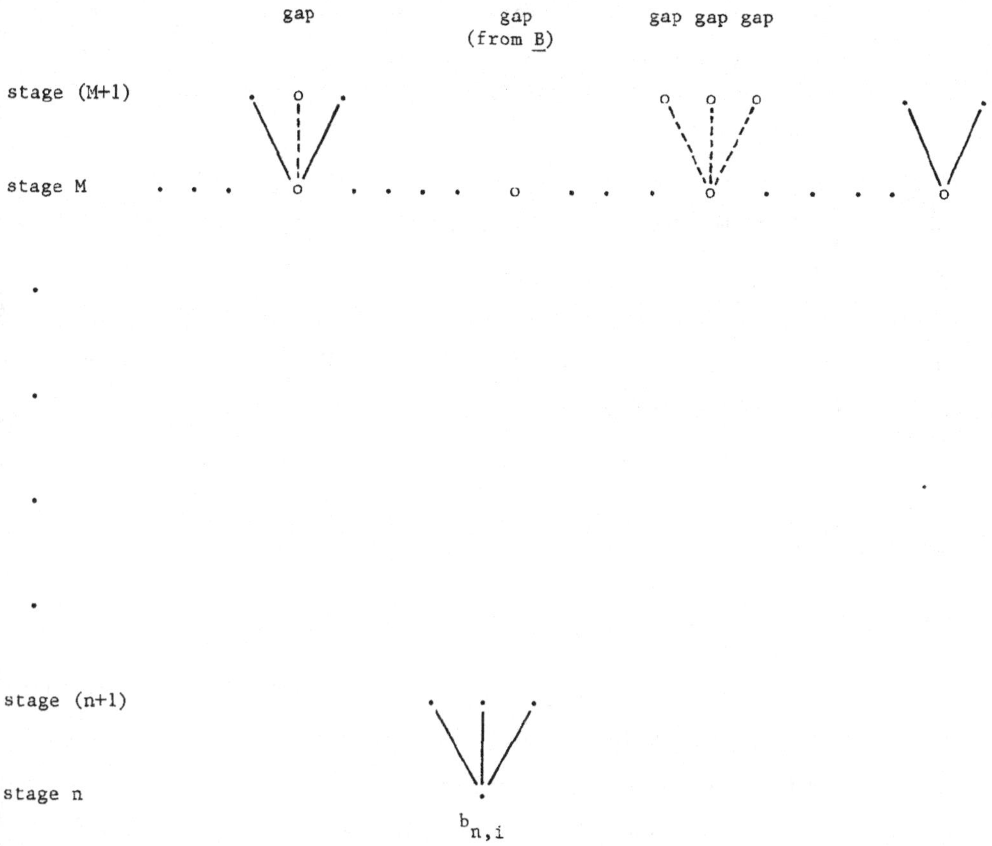

following class of missing $M^{\underline{th}}$ stage elements of B: If $b_{M,k}$ is in stage M of B and if $\Delta_{M,k} \subset \Delta_{r,s}$ for some $(r,s) \in A_1$ then clearly (7) fails and hence $b_{M,k}$ is missing from stage 1 of B_1. An upper bound for the sum, σ, of the μ-measures of all such $\Delta_{M,k}$'s may be found using (4). Indeed, it is easy to check that

$$\sigma \leqq \frac{1}{4}\varepsilon_1 \mu(\Delta_{n,i}). \tag{8}$$

For the remainder of this paragraph assume that $b_{M,k}$ fails to belong to stage 1 of B_1 , and, since we have already taken account in (8) of those elements for which $\Delta_{M,k}$ is a subset of some one of the building blocks of S_1, also assume for the remainder of this paragraph that $\Delta_{M,k} \not\subset \Delta_{r,s}$ for any $(r,s) \in A_1$. Then one of two possibilities occurs:

(i) inequality (7) holds for $\Delta_{M,k}$ and $\|b_{M,k} - d_{0,1}\| \leq \frac{1}{2}\varepsilon$;

(ii) inequality (7) fails for $\Delta_{M,k}$.

Let $b_{M+1,q} > b_{M,k}$. In case (i) either there is no gap at $b_{M+1,q}$ (i.e. $b_{M+1,q}$ belongs to stage 1 of B_1) or there is a gap. In this latter case (7') must fail for $\Delta_{M+1,q}$. From the previous sentences together with case (ii) we see that the co lection of all $b_{M,k}$'s and $b_{M+1,q}$'s for which either (7) or (7') fails for the corresponding Δ sets includes all possible remaining gap locations. Let J denot the collection of indices in A , unaccounted for by (8), for the $b_{M,k}$'s and $b_{M+1,}$ for which (7) or (7') fails and let

$$\tau \equiv \sum_J \mu(\Lambda_{m,j}) .$$

Then directly from the negation of (7) and (7')

(9) $$\tfrac{1}{4}\varepsilon_2\tau < \mu(\bigcup_J \Delta_{M,j} \cap S_1).$$

Now we use the hypothesis about $\Delta_{M,k} \not\subset \Delta_{r,s}$ for any $(r,s) \in A_1$. Indeed, suppose for the moment that $\Delta_{M,k} \cap S_1 = \emptyset$. Then $\Delta_{M,k}$ satisfies (7), $\Delta_{M+1,q} \subset \Delta_{M,k}$ (whe $\Delta_{M+1,q} \cap S_1 = \emptyset$) and hence $\Delta_{M+1,q}$ satisfies (7'). Thus "$\Delta_{M,k} \cap S_1 = \emptyset$" is impos sible and it follows that there is an $(r,s) \in A_1$ such that $\Delta_{M,k} \cap \Delta_{r,s} \neq \emptyset$. Sinc $\Delta_{M,k} \not\subset \Delta_{r,s}$ by hypothesis we conclude that $r \geq M+1$ and hence each index in J ap pears as an index used in the summation in (6) which determined M in the first pla Thus by (6) we have

(10) $$(\bigcup_J \Delta_{M,j} \cap S_1) \leq \sum \{\mu(\Delta_{p,q}): b_{p,q} > b_{n,i} \text{ and } p \geq M\} \leq \tfrac{1}{16}\varepsilon_1\varepsilon_2\mu(\Delta_{n,i}).$$

Combining (9) and (10) yields

(11) $$\tau < \tfrac{1}{4}\varepsilon_1\mu(\Delta_{n,i}) .$$

Consequently the sum of the μ-measures of all $\Delta_{m,j}$ for which $b_{m,j}$ is in stage 1 of B_1 is, from (8) and (11), at least as large as

(12) $$\mu(\Delta_{n,i})(1 - \sigma - \tau) > \mu(\Delta_{n,i})(1 - \tfrac{1}{2}\varepsilon_1) > 0$$

and it follows that stage 1 of B_1 is indeed nonempty.

The requirements on stage 1 of B_1 are twofold: first that a separation of $\frac{1}{2}\varepsilon$ or more be in force between $d_{0,1}$ and each element of stage 1, and secondly, that the difference between $d_{0,1}$ and the "natural" convex combinations of stage 1 elements be dominated in norm by ε_1. The first requirement was taken care of by construction. In the calculations below (to cerify that the second requirement is met) assume all summations are over the full set of indices (M,k) and $(M+1,q)$ corresponding to stage 1 elements of B_1. From (8) and (11)

$$|\mu(\Delta_{n,i}) - \sum \mu(\Delta_{m,j})| = \sigma + \tau < \tfrac{1}{2}\varepsilon_1\mu(\Delta_{n,i}) \tag{13}$$

so that

$$\left|\frac{1}{\sum \mu(\Delta_{m,j})} - \frac{1}{\mu(\Delta_{n,i})}\right| = \left|\frac{\mu(\Delta_{n,i}) - \sum \mu(\Delta_{m,j})}{\mu(\Delta_{n,i})\sum \mu(\Delta_{m,j})}\right| < \frac{\frac{1}{2}\varepsilon_1\mu(\Delta_{n,i})}{\mu(\Delta_{n,i})\sum \mu(\Delta_{m,j})}. \tag{14}$$

Now from (14)

$$\left\|d_{0,1} - \sum b_{m,j}\frac{\mu(\Delta_{m,j})}{\sum \mu(\Delta_{m',j'})}\right\| < \left\|b_{n,i} - \sum b_{m,j}\frac{\mu(\Delta_{m,j})}{\mu(\Delta_{n,i})}\right\| + \left\|\frac{\sum b_{m,j}\mu(\Delta_{m,j})\frac{1}{2}\varepsilon_1}{\sum \mu(\Delta_{m,j})}\right\|. \tag{15}$$

The last term on the right side of (15) is dominated by $\frac{1}{2}\varepsilon_1$ since each $b_{m.j} \in U(X)$ by assumption. The first term on the right side of (15) is easily estimated as well. If there were no gaps this term would be 0, while in general, it is dominated by $\frac{\sigma + \tau}{\mu(\Delta_{n,i})}$. (This again uses the fact that $B \subset U(X)$.) Hence by using (13), (15) becomes

$$\left\|d_{0,1} - \sum b_{m,j}\frac{\mu(\Delta_{m,j})}{\sum \mu(\Delta_{r,s})}\right\| < \tfrac{1}{2}\varepsilon_1 + \tfrac{1}{2}\varepsilon_1 = \varepsilon_1.$$

The construction of B_1 proceeds by induction, the next steps being to repeat the above construction for each element of stage 1 of B_1 (and with ε_2 in place of ε_1) in order to produce the elements of stage 2 of B_1. (Observe that the condition on $d_{0,1}$ as stated in (4) corresponds at the next level of construction of B_1 to conditions (7) and (7') on the elements of stage 1 of B_1.) This completes the argument for the construction of B_1, and the asymptotic subbush B_0 may be found from B_1 as was indicated in the first paragraph of this proof. It therefore remain only to check that B_0 has the properties listed in the statement of the theorem. Note that from the discussion in the first paragraph of this proof, B_0 has the desired separation constant. Next, if T denotes the set of branch points of B_0 th by the inductive version of (12),

$$\mu(T) \geq \prod_{j=1}^{\infty} \mu(\Delta_{n,i})(1 - \tfrac{1}{2}\varepsilon_j) .$$

Since $\sum_{j=1}^{\infty} \varepsilon_j < 1$ this product is strictly positive. Finally, let x be a branch point of B_0 and denote by $\{b_{n,i(n)}\}$ the branch of B_0 corresponding to x. For each n let $\psi(b_{n,i(n)})$ be the point of B corresponding to $b_{n,i(n)}$. Since the branch point determined by the unique branch containing $\{\psi(b_{n,i(n)}): n = 0,1,\ldots\}$ is x and since S_1 is open, if x were an element of S_1 then for sufficiently large n, $x \in \Delta_{n,i(n)} \subset S_1$. But this is impossible since

$$\mu(\Delta_{n,i(n)} \cap S_1) \leq \tfrac{1}{4}\mu(\Delta_{n,i(n)}) \quad \text{for each } n = 0,1,\ldots$$

by construction. Thus $x \notin S_1$. Since $S_1 \supset S$, $S \cap T = \emptyset$ and the proof is comple

Below is an especially satisfying consequence of Theorem 7.11.3.

<u>THEOREM</u> 7.11.4. Let B be a bush in X with separation constant ε and suppose th $0 < \bar{\varepsilon} < \varepsilon$. Then B has an asymptotic subbush B_0 for which

$$\|x - y\| > \frac{1}{4}\bar{\varepsilon}$$

whenever x and y are elements of distinct stages of B_0.

Proof. Without loss of generality assume $B \subset U(X)$. Let V denote a ball of radius $\frac{1}{2}\varepsilon - \delta$ for some $\delta > 0$, $\delta < \frac{\varepsilon}{2}$, and for each $k = 1,2,\ldots$ let

$$J_k \equiv \cup \{\Delta_{m,j}: m \geq k,\ b_{m,j} \in V\} .$$

Then $J_k \supset J_{k+1}$ for each k and we denote by J their intersection $\bigcap_{k=1}^{\infty} J_k$. It first shown that $\mu(J) = 0$. Suppose to the contrary, $\mu(J) > 0$. Let A denote t set of bush indices for B and choose $A_1 \subset A$ such that

$$\bigcup_{A_1} \Delta_{n,i} \supset J, \quad \mu(\bigcup_{A_1} \Delta_{n,i}) < \mu(J)(1 + \tfrac{\delta}{5}) ,$$

and the elements $\Delta_{n,i}$ for $(n,i) \in A_1$ are pairwise disjoint. Then it is easy to see that for some $(n,i) \in A_1$

$$(16) \qquad \mu(\Delta_{n,i} \setminus J) < \tfrac{\delta}{5}\mu(\Delta_{n,i}) .$$

Moreover, it is a direct consequence of (16) that for some immediate successor $b_{n+1,j}$ of $b_{n,i}$,

$$(17) \qquad \mu(\Delta_{n+1,j} \setminus J) < \tfrac{\delta}{5}\mu(\Delta_{n+1,j})$$

as well. Choose $M > n$ so that

$$(18) \qquad \mu(J_M \setminus J) < \tfrac{\delta}{5}\mu(\Delta_{n,i})$$

and then choose a positive integer p so large that, if D denotes the union of the $\Delta_{p,j}$ sets contained in J_M then

$$(19) \qquad \mu(J \setminus D) < \tfrac{\delta}{5}\mu(\Delta_{n,i}) .$$

Of course,

$$\sum_j \{b_{p,j}\frac{\mu(\Delta_{p,j})}{\mu(\Delta_{n,i})}: b_{p,j} > b_{n,i}\} = b_{n,i} .$$

Since (16) and (19) give

$$(20) \qquad \mu(\Delta_{n,i} \setminus D\cap\Delta_{n,i}) \quad (= \mu(\Delta_{n,i} \setminus D)) < \tfrac{2\delta}{5}\mu(\Delta_{n,i}) ,$$

"almost all" of $\Delta_{n,i}$ is contained in D. More quantitatively,

$$\left\|\sum_j \{b_{p,j}\frac{\mu(\Delta_{p,j})}{\mu(D)}: \Delta_{p,j} \subset D\} - b_{n,i}\right\|$$

$$(21) \qquad = \left\|\sum_j \{b_{p,j}\frac{\mu(\Delta_{p,j})}{\mu(D)}: \Delta_{p,j} \subset D\} - \sum_j \{b_{p,j}\frac{\mu(\Delta_{p,j})}{\mu(\Delta_{n,i})}: b_{p,j} > b_{n,i}\}\right\|$$

$$\leq \left\|\sum_j \{b_{p,j}\mu(\Delta_{p,j})[\frac{1}{\mu(D)} - \frac{1}{\mu(\Delta_{n,i})}]: \Delta_{p,j} \subset D\}\right\| + \left\|\sum_j \{b_{p,j}\frac{\mu(\Delta_{p,j})}{\mu(\Delta_{n,i})}: \Delta_{p,j} \subset \Delta_{n,i} \setminus D\}\right\| .$$

Since $\left|\frac{1}{\mu(D)} - \frac{1}{\mu(\Delta_{n,i})}\right| \leq \frac{2\delta}{5\mu(D)}$ by (20) and since $B \subset U(X)$, the first term on th right side of (21) is dominated by $\frac{2\delta}{5}$ while the second term may also be estimated (using (20)) to by no larger than $\frac{2\delta}{5}$. Since each $b_{p,j}$ is an element of V for $\Delta_{p,j} \subset D$, it follows from the previous sentence and from (21) that there is an eleme $v \in V$ with

(22) $$\|v - b_{n,i}\| < \frac{4\delta}{5} .$$

By repeating the argument just presented, but starting with inequality (17) in place of inequality (16), it follows that there is a $w \in V$ with

(23) $$\|b_{n+1,j} - w\| < \frac{4\delta}{5} .$$

Combining (22) and (23) (recall: diameter $V = \varepsilon - 2\delta$) yields

$$\|b_{n+1,j} - b_{n,i}\| < \frac{4\delta}{5} + \|v - w\| + \frac{4\delta}{5} \leq \varepsilon - 2\delta + \frac{8\delta}{5} < \varepsilon$$

which contradicts the fact that B has separation constant ε. Consequently $\mu(J)$

Now fix γ, $\frac{1}{2}\bar{\varepsilon} < \gamma < \frac{1}{2}\varepsilon$, and let $\{V_i: i = 1,2,\ldots\}$ be a sequence of balls of radius γ whose union contains $\overline{\text{co}}(B)$. According to the previous paragraph, for each i it is possible to find an integer k_i and a countable subset A_i of A such that

(I) $\mu(\bigcup_{A_i} \Delta_{r,s}) < 2^{-i-1}$; and

(II) If $b_{n,t} \in V_i$ and $\Delta_{n,t}$ is not a subset of one of the sets $\Delta_{r,s}$ for $(r,s) \in A_i$ then $n < k_i$.

In particular, (II) states that for each i there are only finitely many indices $(n,t) \in A$ with $b_{n,t} \in V_i$ and $\Delta_{n,t}$ not a subset of some $\Delta_{r,s}$ for some $(r,s) \in$

Let

$$S \equiv \bigcup_{i=1}^{\infty} \bigcup_{A_i} \Delta_{r,s} .$$

Then $\mu(S) < \frac{1}{2}$ and according to Theorem 7.11.3 there is an asymptotic subbush B_1 of B with separation constant $\frac{1}{2}\bar{\varepsilon}$ such that, if T denotes the set of branch points of B_1 then both $\mu(T) > 0$ and $S \cap T = \emptyset$. For each i let W_i be a ball of radius smaller than γ with the same center as V_i. If W_i contained infinitely many elements of B_1 then (since B_1 is an asymptotic subbush of B) V_i would contain infinitely many elements of B (corresponding to the elements of B_1 contained in W_i which are "sufficiently far out"). In particular, V_i would contain at least one element $b_{n,t} \in B$ with $n > k_i$ which corresponds to an element $b'_{m,j}$ of B_1 with $b'_{m,j} \in W_i$. But then $\Delta_{n,t} \subset S$ and since the branch points corresponding to branches of B_1 which contain $b'_{m,j}$ are all points of $\Delta_{n,t} \cap T$ we would obtain the contradiction $S \cap T \neq \emptyset$. It follows that W_i contains only finitely many elements of B_1 for each i. In particular, no closed ball of radius $\frac{1}{4}\bar{\varepsilon}$ in X can contain infinitely many members of B_1.

The construction of B_0 from B_1 is now straightforward. Indeed, pick any element $b_{n,i}$ of B_1 and henceforth denote it by $d_{0,1}$, the sole member of stage 0 of B_0. Because only finitely many elements of B_1 belong to the $\frac{1}{4}\bar{\varepsilon}$ ball centered at $d_{0,1}$ there is an integer p_1 such that

$$\|b_{m,j} - d_{0,1}\| > \frac{1}{4}\bar{\varepsilon} \quad \text{if} \quad m \geq p_1 .$$

Let stage 1 of B_0 consist of those successors of $b_{n,i}$ in B_1 which belong to stage p_1 of B_1. By repeating this procedure for each element in the newly formed first stage of B_1 an integer p_2 may be found such that

$$\|b_{p_2,s} - b_{p_1,j}\| > \frac{1}{4}\bar{\varepsilon} \quad \text{whenever } b_{p_1,j} \text{ belongs to stage 1 of } B_0 \text{ and}$$
$$b_{p_2,s} \text{ is a successor of some stage 1 element of } B_0 .$$

In this way B_0 may be inductively constructed. It clearly has the desired properties, so the proof of Theorem 7.11.4 is complete.

The fact that "uniformly separated" bushes exist inside any non RNP set easily implies the next corollary, which is a modest form of our Theorem 3.7.2, originally due to Huff and Morris [1976] . (See the comments following the statement of Theorem

7.11.8 below for the indications of an alternative proof of the full strength Theorem 3.7.2.)

COROLLARY 7.11.5. For a Banach space X the following are equivalent.

(a) X has the RNP.

(b) Each nonempty closed bounded subset D of X has an extreme point.

(c) If D is a bounded closed subset of X and D contains at least two points then there is an $f \in X^*$ which is not constant on D and which attains its supremum on D.

Proof. not (a) ⇒ not (b) If X contains a bounded ε-bush it also contains a closed ε' bush for some $\varepsilon' > 0$ by Theorem 7.11.4. Since bushes have no extreme points this contradicts (b).

(a) ⇒ (b) and (a) ⇒ (c) If D is a closed bounded set in X then $\overline{co}(D)$ has a strongly exposed point by Theorem 3.5.4. If $f \in X^*$ strongly exposes x in $\overline{co}(D)$ it is easy to check both that $x \in D$ and that f cannot be constant on D if D not a singleton. Evidently x is an extreme point of D.

not (a) ⇒ not (c) If X lacks the RNP it contains a closed bush $D \equiv \{b_{n,i}: (n,i) \in A\}$ with separation constant ε for some $\varepsilon > 0$ by Theorem 7.11.4. Assume without loss of generality that $b_{0,1} = 0$ and let $\{s_i: i = 1,2,\dots\}$ be a strictly increasing sequence of positive numbers with $\lim_i s_i = 1$. Suppose that $f \in X^*$ is not constant on D. Since $b_{0,1}$ is a convex combination of the elements of stage n for each n there must be at least one element $b_{n,i} \in D$ such that $f(b_{n,i}) > 0$. Then there is at least one element $b_{n+1,j}$ of D such that $b_{n+1,j} > b_{n,i}$ and $f(b_{n+1,j}) \geq f(b_{n,i})$ (since $b_{n,i}$ is a convex combination of its immediate successors). Consequently $f(s_{n+1}b_{n+1,j}) > f(s_n b_{n,i})$ and it follows that the modified form of D:

$$D' \equiv \{s_n b_{n,i}: (n,i) \in A\} \tag{24}$$

is closed, bounded, contains infinitely many points and fails 7.11.5(c). Thus the proof is complete.

Let $B \equiv \{b_{n,i}: (n,i) \in A\}$ be a bush in X. For each $(n,i) \in A$ denote by

$W(b_{n,i})$ the wedge of $b_{n,i}$, namely, the set

$$W(b_{n,i}) \equiv \overline{co}(\{b_{m,j}: b_{m,j} > b_{n,i}\}) .$$

Thus $W(b_{n,i})$ is a closed bounded convex set which contains $b_{n,i}$. Suppose now that $x \in ex(\overline{co}(B))$. Then x cannot be a point in $co(\{b_{n,i}: (n,i) \in A$ and $n \leq k\})$ for any positive integer k. Hence whenever $x = \lim_j x_j$ (for $x_j \in co(B)$) then some of the points of B used in expressing x_j as a convex combination must come from successively larger stages of B as j increases. Among the most natural candidates for extreme points, then, are the points which are "at the end of the branches" of B, that is, the points in

$$\bigcap_{n=0}^{\infty} W(b_{n,i(n)}) \quad \text{for} \quad \{b_{n,i(n)}: n = 0,1,\ldots\} \text{ a branch of } B.$$

These sets are singled out in the next definition.

DEFINITION 7.11.6. With the notation of the previous paragraph, for any branch $\alpha \equiv \{b_{n,i(n)}: n = 0,1,\ldots\}$ of B, the wedge intersection determined by α is the set $\bigcap_{n=0}^{\infty} W(b_{n,i(n)})$.

COROLLARY 7.11.7. Let $B \subset X$ be a bush and suppose that the μ-outer measure of the set of branch points corresponding to branches with nonempty wedge intersections is strictly less than one. Then B has an asymptotic subbush all of whose wedge intersections are empty.

Proof. Let S denote the set of branch points of B corresponding to branches with nonempty wedge intersection. If $\bar{\mu}(S) < 1$ then by Theorem 7.11.3 there is an asymptotic subbush B_0 of B whose set T of branch points has empty intersection with S (and which satisfies $\mu(T) > 0$). Suppose that β is a branch of B_0 and that α is the branch of B containing the image of β under the order preserving map from B_0 into B as in Definition 7.11.2. Then it is clear both that the wedge intersection determined by β (as a branch in B_0) is a subset of the wedge intersection determined by α (as a branch in B) and that both branches have the same branch point t. Thus, if $t \in T$ then the wedge intersection determined by α is empty (since $t \notin S$) and hence the wedge intersection determined by β is also empty. That is, B_0 is an asymptotic subbush of B all of whose wedge intersections are empty.

It turns out that the μ-outer measure hypothesis of the above corollary can be eliminated (at the cost of considerably more work). The end result of a detailed argument reminiscent of those presented here in the proofs of Theorems 7.11.3 and 7.11.4 is the following powerful theorem of James [1981] .

THEOREM 7.11.8. Let $K \subset X$ be a closed bounded convex set and assume that K lacks the RNP. Then K contains a bush all of whose wedge intersections are empty. (In fact, if B is any bush in a Banach space, B contains an asymptotic subbush all of whose wedge intersections are empty.)

As a measure of the strength of Theorem 7.11.8 we remark first that if B_0 is bush all of whose wedge intersections are empty then B_0 is weakly closed. (Indeed suppose x were a weak cluster point of $B_0 \equiv \{b_{n,i}: (n,i) \in A\}$. Since there are only finitely many elements of stage 1 and all but the elements of stages 0 and 1 are successors of stage 1 elements, x is a weak cluster point of the set of successors of $b_{1,i(1)}$ for some $b_{1,i(1)}$ in stage 1. Similarly, among the immediate successors of $b_{1,i(1)}$ must be at least one $b_{2,i(2)}$ such that x is a weak cluster point of the set of successors of $b_{2,i(2)}$. The inductive procedure for choosing $b_{n,i(n)}$ is clear, and a branch $\alpha \equiv \{b_{n,i(n)}: n = 0,1,\ldots\}$ is determined in this way. Evidently though, x belongs to the wedge intersection determined by α, in contradiction to the assumption on B_0. Thus B_0 is weakly closed.) Consequently, by using Theorem 7.11.8 as above, we may substitute "weakly closed" for "closed" throughout the statement and proof of Corollary 7.11.5 and provide, in this manner, an alternative proof of the full strength Theorem 3.7.2.

There is a class of bushes, studied by Ho [1982] , whose wedge intersections contain the set of extreme points of their closed convex hulls. Indeed, say that the bush $B \equiv \{b_{n,i}: (n,i) \in A\}$ is <u>complemented</u> if there is a positive number Θ such that for each n and for each i such that $(n,i) \in A$,

$$\|x - y\| \geq \Theta\|x\|$$

whenever $x \in \overline{\text{span}}(W(b_{n,i}))$ and $y \in \overline{\text{span}}(\bigcup_{j\neq i} \{z \in W(b_{n,j}): (n,j) \in A\})$.

Thus, for example, if

(25) $$A \equiv \{(n,i): n = 0,1,\dots ;\ 1 \leq i \leq 2^n\}$$

is the index set for a tree and if

$$b_{n,i}(t) \equiv \begin{cases} 2^n & \text{for } t \in [\frac{i-1}{2^n},\frac{i}{2^n}) \\ 0 & \text{elsewhere on } [0,1] \end{cases}$$

then $\{b_{n,i}: (n,i) \in A\}$ is a complemented 1-tree in $L^1[0,1]$ with complementation constant $\Theta = 1$. On the other hand, with A as in (25) again, let $B_1 \subset c_o$ denote the tree of initial segments of $+1$'s and -1's defined inductively by the conditions:

$$b_{0,1} \equiv (1,0,\dots);\quad b_{n+1,2i-1} \equiv b_{n,i} + \delta_{n+1};\quad b_{n+1,2i} \equiv b_{n,i} - \delta_{n+1}$$

where as usual δ_n is the point of c_o which has a 1 in the nth slot and zeros elsewhere. Then B_1 is a tree with separation constant 1 which is not complemented since

$$b_{n+1,2i-1} - b_{n+1,2i} = 2\delta_{n+1} = b_{n+1,2j-1} - b_{n+1,2j} \quad \text{for } 1 \leq i \neq j \leq 2^n$$

whence $\delta_{n+1} \in \overline{\text{span}}(W(b_{n,i})) \cap \overline{\text{span}}(W(b_{n,j}))$.

<u>PROPOSITION</u> 7.11.9. Let B be a complemented bush and suppose that $x \in ex(B)$. Then there is a branch α of B such that x belongs to the wedge intersection determined by α.

Proof. Temporarily fix $(n,i) \in A$ and suppose that $x \in ex(\overline{co}(B))$. Note that

(26) $$\overline{co}(B) = \overline{co}[W(b_{n,i}),\ \overline{co}(\bigcup_{\substack{j\neq i \\ (n,j)\in A}} W(b_{n,j}))].$$

If

$$x \in \overline{co}(B) \setminus co[W(b_{n,i}) , \overline{co}(\bigcup_{\substack{j \neq i \\ (n,j) \in A}} W(b_{n,j}))] \tag{27}$$

then for some r, $0 < r < 1$ and some pair of sequences $\{y_k: k = 1,2,\ldots\}$ in $W(b_{n,i})$ and $\{z_k: k = 1,2,\ldots\}$ in $\overline{co}(\bigcup_{\substack{j \neq i \\ (n,j) \in A}} W(b_{n,j}))$ we have $\lim_k \|ry_k + (1 - r$ $- x\| = 0$. Consequently

$$r(y_k - y_m) + (1 - r)(z_k - z_m) \to 0 \quad \text{as} \quad k,m \to \infty$$

and since B is complemented, both sequences $\{y_k: k = 1,2,\ldots\}$ and $\{z_k: k = 1,2,\ldots\}$ are Cauchy. Let y and z be their respective limits. Then $y \in W(b_{n,i})$ and $z \in \overline{co}(\bigcup_{\substack{j \neq i \\ (n,j) \in A}} W(b_{n,j}))$. Evidently, $x = ry + (1 - r)z$ which is impossible since $x \in ex(\overline{co}(B))$ by hypothesis. Hence by (27),

$$x \in co[W(b_{n,i}) , \overline{co}(\bigcup_{\substack{j \neq i \\ (n,j) \in A}} W(b_{n,j}))$$

and it follows that either $x \in W(b_{n,i})$ or that $x \in \overline{co}(\bigcup_{\substack{j \neq i \\ (n,j) \in A}} W(b_{n,j}))$. Judicious repetition of the above argument shows that, at each stage n, there is an index i such that $x \in W(b_{n,i(n)})$ and $b_{n+1,i(n+1)} > b_{n,i(n)}$. Thus $x \in \bigcap_{n=0}^{\infty} W(b_{n,i(n)})$, the wedge intersection determined by the branch $\{b_{n,i(n)}: n = 0,1,\ldots\}$.

<u>COROLLARY</u> 7.11.10. Let $K \subset X$ be a closed bounded convex set. If K contains a complemented bush than K lacks the KMP.

Proof. According to Theorem 7.11.8, if B is a bush in K there is an asymptotic subbush B_0 of B all of whose wedge intersections are empty. If B is complemented it is straightforward to verify that B_0 is complemented as well. By Proposition 7.11.9, $\overline{co}(B_0)$ has no extreme points.

§7.12. <u>The RNP in locally convex spaces and another look at integral representatio</u>

Several attempts have been made to widen the RNP horizons to incorporate more general classes of spaces than Banach spaces. E. Saab [1975] (Fréchet spaces), D Gilliam [1976] (locally convex spaces with the strict Mackay convergence property)

E. Saab [1976] and [1978] (quasicomplete locally convex spaces of type (BM)), L. Egghe [1977] and [1980] (sequentially complete locally convex spaces) have considered both the breakdown and the stability of some of the connections between the RNP and other notions found throughout these notes, in these less restrictive environments. (Kalton [1979] even managed to mimic some results in a special class of non locally convex spaces.) As pointed out by Saab [1978], there is good reason for the Radon-Nikodým Property and several of its Banach space equivalents to remain equivalent at least in quasicomplete locally convex spaces of type (BM) (i.e. in locally convex Hausdorff topological vector spaces in which each closed and bounded set is complete and metrizable). Indeed, suppose that $(E,\mathcal{T})$ is such a locally convex space. Let $C \subset E$ be a bounded closed convex subset, let $K \equiv \overline{co}(C \cup -C)$ and let $E_K \equiv \sum_{n=1}^{\infty} nK$. Then there is a norm ρ on E_K such that the uniformities on K induced by ρ and by $\mathcal{T}$ coincide. In particular, then, the ρ and $\mathcal{T}$ topologies on K coincide. Since $(K,\mathcal{T})$ is complete, K is a closed bounded convex subset of (E_K',ρ'), the completion of the normed space (E_K,ρ). To construct ρ first find a sequence $\{V_n: n = 1,2,\dots\}$ of closed absolutely convex neighborhoods of 0 in $(E,\mathcal{T})$ such that both $V_{n+1} + V_{n+1} \subset V_n$ for each n and $\{V_n \cap (K + K): n = 1,2,\dots\}$ is a neighbood base for 0 in $(K + K,\mathcal{T})$. Let $\mathcal{T}_1$ denote the topology on E_K determined by these neighborhoods. For each n find $t_n > 0$ such that $K \subset t_n V_n$, let ρ_n be the Minkowski gauge functional for V_n and for $x \in E_K$ let

$$\rho(x) \equiv \sum_{n=1}^{\infty} (2^n t_n)^{-1} \rho_n(x).$$

It may be verified that ρ is a norm on E_K and that the uniform structures on K induced by $\mathcal{T}$, $\mathcal{T}_1$ and ρ all coincide. Thus there is a mechanical vehicle for associating Banach space notions with corresponding notions in C, and the job of "proving" locally convex space versions of Radon-Nikodým Property type results reduces to making the correct locally convex space definitions so that their Banach space counterparts are the ones already well known to us. For example, if $(\Omega,\mathcal{F},P)$ is a probability space then a function $f: \Omega \to E$ is said to be P-integrable if there is a sequence of simple functions $s_n: \Omega \to E$ which converges pointwise almost surely to f such that $\lim_n \int_\Omega q(s_n(\omega)-f(\omega))dP(\omega) = 0$ for each continuous seminorm q on E. Then $\int_A fdP \equiv \lim_n \int_A s_n dP$ $(A \in F)$ makes sense independently of the approximating sequence $\{s_n: n = 1,2,\ldots\}$ aince (E,T) is quasicomplete. C is said to have the Radon-Nikodým Property if whenever (Ω,F,P) is a probability space, $m: F \to E$ is a vector measure with $m \ll P$ and $AR(m) \subset C$ there is a P-integrable $f: \Omega \to E$ suc that $\int_A fdP = m(A)$ for each $A \in F$. The notion of subset dentability in C remai unaltered except that ε-balls are replaced by neighborhoods of 0 in E in the us definition (cf. 2.1.5), and s-dentability, exposed and strongly exposed points also have obvious definitions in this setting. It follows then that the assertion "C ha the RNP" is equivalent to C being subset dentable, to C being subset s-dentabl and to the assertion that each closed bounded convex subset of C is the closed con vex hull of its denting points. In fact, Rieffel's characterization of sets with th RNP (Theorem 2.2.6) carries over to this context. It is even routine to establish t existence of integral representations as well as uniqueness in the case of simplexes cf. Theorems 6.2.9 and 6.4.2—for these sets C. A word of caution is necessary, though. The equivalences indicated here are not unexpected because each relevant de finition is internal to K (or at least to E_K). However, for example, the notion of a (strongly) exposed point requires information about the dual space as well as t space E_K and is thus hardly "internal". Not surprisingly, perhaps, there is a com plete breakdown of Phelp's Theorem (3.5.4) in this setting. Indeed, take $E \equiv \mathbb{R}^{\mathbb{N}}$ a let $C \equiv [-1,1]^{\mathbb{N}}$. Then C has no exposed points.

A deeper use has been made of the locally convex setting by various authors dealing with the noncompact theory of integral representations. In Edgar [1978a] ou Theorem 6.3.9 was placed in a locally convex setting and further generalized so that the convex set in question need not even be assumed closed. Weizsäcker and Winkler [1979] , [1980] have continued along these lines as well as those established by Thomas [1977] and [1978] , and have arrived at some results which may be successfully applied in a range of probabilistic settings. In order to describe their mair results we employ some of the notation of §4.3. Thus, for a topological space (D,T $B(D)$ denotes the Borel σ-algebra and $P_t(D)$ the collection of probability measure on $B(D)$ inner regular with respect to T-compact sets. If M is a subset of P_t

and $A \in \mathcal{B}(D)$, let $f_A: M \to \mathbb{R}$ be the function $f_A(\nu) \equiv \nu(A)$ for $\nu \in M$. Then let $\Sigma(M)$ be the σ-algebra of sets generated by $\{f_A^{-1}(V): V$ is open in $\mathbb{R}$, $A \in \mathcal{B}(D)\}$. Next, suppose that F is a collection of $\mathcal{B}(D)$-measurable real valued functions on D and that M is as above. Let $M|_F \equiv \{\nu \in M: F \subset L^1(\nu)\}$. For any measurable function $g: D \to \mathbb{R}$ for which the following integrals make sense, define the evaluation functional $e_g: M|_F \to \mathbb{R}$ by $e_g(\nu) \equiv \int g d\nu$ for $\nu \in M|_F$. Let T_1 denote the weakest topology on $M|_F$ for which e_g is lower semicontinuous whenever g is lower semicontinuous and bounded on D, and let T be the weakest topology on $M|_F$ containing T_1 such that each e_g is continuous for $g \in F$.

THEOREM 7.12.1. Let F be a countable family of real valued Borel measurable functions on (D,T). Suppose that $M \subset P_t(D)|_F$ is convex and closed in the T topology described above. Then for each $\mu \in M$ there is a probability measure P on $\Sigma(\text{ex}(M))$ such that

$$\mu(A) = \int_{\text{ex}(M)} \nu(A) dP(\nu) \qquad \text{for each } A \in \mathcal{B}(D).$$

The following corollary indicates the usefulness of 7.12.1 above in moment problems.

COROLLARY 7.12.2. Let C be a closed convex subset of $\mathbb{R}^{\mathbb{N}}$. (For example, C may be a countable product of possibly degenerate closed intervals in $\mathbb{R}$.) Let $F \equiv \{f_i: i = 1,2,\ldots\}$ be a sequence of Borel measurable real valued functions on $[0,1]$ and let

$$M \equiv \{\mu \in P_t([0,1])|_F: (\int f_i d\mu)_{i\in\mathbb{N}} \in C\}.$$

Then for each $\mu \in M$ there is a probability measure P on $\Sigma(\text{ex}(M))$ such that

$$\int_{\text{ex}(M)} \nu(A) dP(\nu) = \mu(A) \quad \text{for each } A \in \mathcal{B}([0,1]).$$

Several probabalistic applications of 7.12.1 and 7.12.2 appear in Weizsäcker and Winkler [1980] . It is interesting to note that the proof of the theorem is quite classical in that well-capped cones reappear in a central role. Very briefly, the proof goes as follows. Suppose M satisfies the hypotheses of 7.12.1 and let $\{g_i: i = 0,1,\ldots\}$ be a sequence such that each element of $F \cup \{1\}$ appears infinit often and $g_0 \equiv 1$. Fix $\mu \in M$. Choose an increasing sequence of compact subsets $\{D_n: n = 0,1,\ldots\}$ of D with $D_0 \equiv \emptyset$ such that $g_i|_{D_n}$ is bounded for all i and all n and $\int_{D\setminus D_n} |g_n| d\mu \leq 2^{-2n}$ for each n. Let T' denote the weakest topology on D which contains T such that each function in $F \cup \{\chi_{D_n} : n = 1,2,\ldots\}$ is con tinuous. Now let $L: (\mathbb{R}^+ \cdot P_t(D,T')) \to [0,\infty]$ be the function

$$L(\nu) \equiv \frac{1}{\alpha} \sum_{i=0}^{\infty} 2^i \int_{D\setminus D_i} |g_i| d\nu \quad \text{where} \quad \alpha \equiv \sum_{i=0}^{\infty} 2^i \int_{D\setminus D_i} |g_i| d\mu.$$

It may be shown that $C_\mu \equiv \{\nu \in \mathbb{R}^+ \cdot M: L(\nu) \leq 1\}$ is a cap of $\mathbb{R}^+ \cdot M$ in the topolog T'. That is, C_μ is T'-compact and convex, and $\mathbb{R}^+ \cdot M \setminus C_\mu$ is convex. Moreover, whenever $\rho \in \mathbb{R}^+ \cdot M$ and P_1 is a measure on $\Sigma(\mathbb{R}^+ \cdot M)$ which has ρ as its center of mass (in the sense that $\rho(A) = \int \nu(A) dP_1(\nu)$ for each $A \in \mathcal{B}(D)$) then L satis fies the barycentric calculus:

$$L(\rho) = \int L(\nu) dP_1(\nu). \tag{1}$$

(The proof of these facts is technical but not difficult. See Weizsäcker and Winkle [1979] p. 25, Proposition 1 for details.) According to an earlier result of Winkle [1978] , p. 72, there is a probability measure P' on $\Sigma(\mathrm{ex}(C_\mu))$ which has μ as its center of mass and because L satisfies the formula (1), P' is supported by

$$(\mathbb{R}^+ \cdot \mathrm{ex}(M)) \cap \{\nu \in \mathbb{R}^+ \cdot P_t(D): L(\nu) = 1\}.$$

P' may then be "pushed over" onto $\Sigma(\mathrm{ex}(M))$ where it becomes the sought for P.

Although there is no explicit use of a Radon-Nikodým theorem here, there is a close connection between martingale convergence and results such as 7.12.1 and 7.12. See, for example, the direct proof of Corollary 7.12.2 appearing in Weizsäcker and Winkler [1980] , pages 134-5.

§13. Banach lattices.

A partially ordered Banach space $(X,\|\cdot\|)$ is a Banach lattice if

(1)
(a) $x \leq y$ implies $x + z \leq y + z$ for each x,y,z in X;

(b) $0 \leq tx$ whenever $0 \leq x \in X$ and $0 \leq t \in \mathbb{R}$:

(c) for each x,y in X there is a least upper bound, written $x \vee y$;

(d) $\|x\| \leq \|y\|$ provided $|x| \leq |y|$ where $|x|$ means $x \vee (-x)$.

It is easy to check that each two elements x and y of a Banach lattice also have a greatest lower bound, denoted henceforth by $x \wedge y$. There is extensive information about the structure of such spaces and no attempt will be made to review even all the relevant background for the ensuing discussion. Comprehensive sources include Lindenstrauss and Tzafriri [1979] and Schaefer [1974]. What we will discuss in detail, however, are two recent results which point up the extraordinary power of the lattice assumption. The first is due to Bourgain and Talagrand [1981] while the second is from Talagrand [1983b].

<u>THEOREM</u> 7.13.1. Let X be a Banach lattice. Then X has the RNP if and only if X has the KMP.

<u>THEOREM</u> 7.13.2. Let X be a separable Banach lattice. If X has the RNP then X is isometrically isomorphic to the dual of a Banach lattice.

The proofs of both of these results involve the notion of order dentability, first studied in Ghoussoub and Talagrand [1979] and defined in part c below.

<u>DEFINITION</u> 7.13.3. (a) Let X be a Banach lattice with positive cone X_+. (Thus $X_+ \equiv \{x \in X: 0 \leq x\}$.) An element u of X_+ is a <u>quasi interior point</u> if for each point x of X_+ we have $\lim_n \|x - x \wedge nu\| = 0$.

(b) Let u be a quasi interior point of X_+. Then define $P_n(u) \subset X_+$ for $n = 1,2,\ldots$ as follows:

(2)
$$P_n(u) \equiv \{x \in X_+: \|x \wedge u\| \geq \frac{1}{n}\}.$$

(c) Fix a quasi interior point u in X_+. Let D be a closed convex subset of X_+. Then <u>D is order dentable</u> if there are an $n \in \mathbb{N}$ and a slice S of D such that $S \subset P_n(u)$. A Banach lattice <u>X is order dentable</u> if each nonzero closed bounded convex subset of X_+ is order dentable. (It is shown in Ghoussoub and Talagrand [1979] that order dentability of X does not depend on the choice of quasi interio point u.)

Let $u \equiv (1,1) \in \ell^2_\infty$, two-dimensional ℓ_∞. Then u is a quasi interior point o ℓ^2_∞ with the usual coordinatewise partial ordering and $P_n(u)$ is not convex.

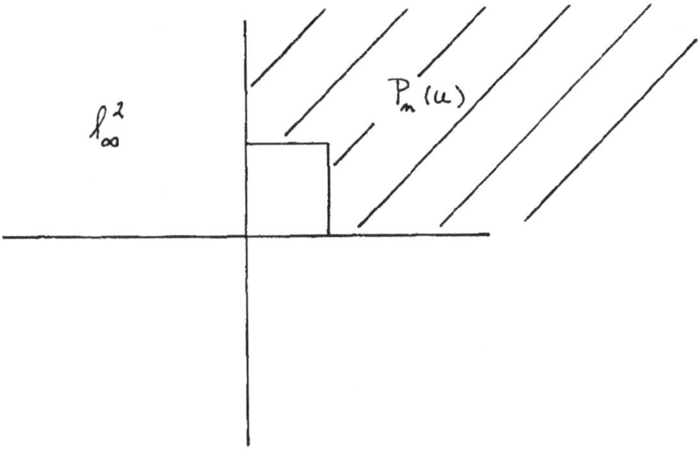

As a second example, consider $L^1[0,1]$ with the usual pointwise ordering. Define a tree in the positive unit sphere as follows:

(3)
$$f_1 \equiv \chi_{[0,1)}. \text{ For each } k \geq 0 \text{ and } n,\ 2^k \leq n < 2^{k+1} \text{ let}$$
$$f_{2n} \equiv 2^k \chi\left[\frac{2n-2^{k+1}}{2^{k+1}}, \frac{2n+1-2^{k+1}}{2^{k+1}}\right) \text{ and } f_{2n+1} \equiv 2^k \chi\left[\frac{2n+1-2^{k+1}}{2^{k+1}}, \frac{2n+2-2^{k+1}}{2^{k+1}}\right).$$

Then $D \equiv \overline{co}(\{f_n\}_{n=1}^\infty)$ is a closed bounded convex set in $S(L^1[0,1]) \cap L^1[0,1]_+$. Le $u \equiv \chi_{[0,1)}$ be the function identically 1. Then u is a quasi interior point. Not that for any $k \geq 0$ and n, $2^k \leq n < 2^{k+1}$ we have

$$\| f_{2n} \wedge u \|_1 = \| f_{2n+1} \wedge u \|_1 = \frac{1}{2^{k+1}} .$$

Since the convex hull of a tree is the same as the convex hull of that part of a tr consisting of the elements of stage k onward for any k, D is not order dentable. This elementary example serves as a prototype for what happens in a non order denta closed bounded convex set. In a rough sense, its elements are "infinitely divisible" into components lying "further and further towards the edge" of the positive cone. Such an example is not possible in spaces like c_0, ℓ_1 and ℓ_∞ and not surprisingly, each of these spaces is order dentable. In Lemma 7.13.9 below some of these comments are made more precise, at least for separable Banach lattices not containing c_0.

It is easy to see that a Banach lattice with the RNP is order dentable. Indeed, if $D \subset X_+$ is a nonzero closed bounded convex set, pick any strongly exposed point $x_0 \neq 0$ of D. Then

$$\|x_0 \wedge u\| = \frac{1}{n}\|nx_0 \wedge nu\| \geqq \frac{1}{n}\|x_0 \wedge nu\| > 0$$

for n sufficiently large and hence for some k, $x_0 \in P_k(u)$. Then there is a slice S of D containing x_0 of sufficiently small diameter that $S \subset P_{2k}(u)$ since the lattice operations are continuous. In the proof of Theorem 7.13.2 below it will be established that a separable order dentable Banach lattice which does not contain a Banach lattice subspace lattice isomorphic to c_0 is a dual space and hence, according to Theorem 4.4.1, has the RNP. A much more direct proof of the following, formally stronger result, appears in Ghoussoub and Talagrand [1979] , Theorem 2.

<u>THEOREM</u> 7.13.4. Let X be a Banach lattice. Then the following are equivalent.

(a) X has the RNP.

(b) X is order dentable and c_0 is not lattice isomorphic to a sublattice of X.

Some standard facts in the theory of Banach lattices are gathered below for easy reference. They will be needed in what follows.

<u>THEOREM</u> 7.13.5. Let X be a Banach lattice and assume that no Banach sublattice of X is lattice isomorphic to c_0. Then each of the following is true.

(a) Each norm bounded increasing sequence in X is norm convergent.

(b) Each directed family in X under $\leqq$ which has an upper bound in X is norm convergent. Its limit coincides with its supremum. (A similar statement holds if "$\leqq$" is replaced by "$\geqq$" and "supremum" by "infimum".)

(c) Order intervals in X are weakly compact. (An order interval is a set of the form $\{x: a \leq x \leq b\}$ for some $a, b \in X$.)

Parts (a) and (b) of 7.13.5 are sometimes referred to as Dini's theorem. These results appear, for example, in Schaefer [1974] as portions of Proposition 5.10, p. and Proposition 5.15, p. 95. It should be noted here as well that when X is a sep rable Banach lattice, its positive cone contains many quasi interior points. Indeed if $\{x_i: i = 1,2,\ldots\}$ is dense in $X_+ \setminus \{0\}$ then $u \equiv \sum_{i=1}^{\infty} 2^{-i}\|x_i\|^{-1}x_i$ is a quasi interior point as is easily checked. If $\{x_i: i = 1,2,\ldots\}$ happens instead to be a maximal collection of mutually disjoint elements in X_+ then u as just defined above is a <u>weak unit</u> of X. That is, $u \wedge x = 0$ implies $x = 0$ for any $x \in X$.

It will be convenient to use the notation $M(D,f)$ and $S(D,f,\lambda)$ introduced in 2.3.1 for the supremum of the values $f(x)$ as x ranges over $D \subset X$ for $f \in X^*$, and the slice of D determined by $f \in X^*$ and $\lambda > 0$, respectively. Before turnin to the detailed proofs of Theorems 7.13.1 and 7.13.2, we remark that the papers of Ghoussoub [1982], Ghoussoub and Talagrand [1979] and that of Davis, Ghoussoub and Lindenstrauss [1981] provide some interesting recent results in this area which we shall not have space to develop here.

<u>DISCUSSION FOR THEOREM 7.13.1.</u>

The fact that the RNP implies the KMP does not require X to be a Banach latti See Theorem 3.3.6. For the converse, observe that if X lacks the RNP so does some separable subspace of X. Thus some separable Banach lattice subspace of X lacks the RNP as well. It therefore suffices to establish that a separable Banach lattice with the KMP has the RNP and consequently, without loss of generality we henceforth assume that X is separable. The following lemmas constitute the bulk of the work in establishing Theorem 7.13.1. The first is a special case of a result due to Ghoussoub [1983], though the proof is from Bourgain and Talagrand [1981]. Note tha Lemma 7.13.10 below is reminiscent of Lemma 3.7.5. Since X has the KMP by hypothe c_0 does not embed in X (since $U(c_0)$ has no extreme points) and hence by part of Theorem 7.13.5, order intervals of X are weakly compact. Thus the standing hy theses in the rest of the discussion are:

(4) (a) X is a separable Banach lattice;

(b) X does not contain c_0 and hence X has weakly compact order intervals.

LEMMA 7.13.6. Let $u \in X_+$ be any element. Then 0 is a denting point of $[0,u]$. That is, $0 \notin \overline{co}([0,u] \setminus U_\varepsilon(0))$ for any $\varepsilon > 0$.

Proof. We first show that, given $\varepsilon > 0$, there is a point v in $[0,u]$ such that

$$\|v\| \geqq \|u\| - \varepsilon \quad \text{and} \quad 0 \notin \overline{co}([0,v] \setminus U_\varepsilon(0)). \tag{5}$$

Suppose no such v exists. A decreasing sequence $\{v_n : n = 0,1\ldots\}$ in $[0,u]$ is constructed below such that for each n both

(6) (a) $\|v_n\| \geqq \|u\| - \varepsilon(1 - 2^{-n})$; and

(b) $\|v_n - v_{n+1}\| \geqq \varepsilon$.

Let $v_0 \equiv u$. If v_n has been constructed subject to the inductive conditions then, since v_n fails to satisfy (5) by hypothesis, there are finitely many points x_i in $[0,v_n]$ and numbers t_i in $[0,1]$ with $\sum t_i = 1$, $\|x_i\| \geqq \varepsilon$ and $\|\sum t_i x_i\| \leqq \varepsilon 2^{-(n+1)}$. Note that $\|v_n - \sum t_i x_i\| \geqq \|v_n\| - \varepsilon 2^{-(n+1)}$. Hence for at least one i, $\|v_n - x_i\| \geqq \|v_n\| - \varepsilon 2^{-(n+1)}$ and we let $v_{n+1} \equiv v_n - x_i$ for this index i. Evidently v_{n+1} satisfies (6) and the inductive step is complete. Thus $\{v_n : n = 0,1,\ldots\}$ is a decreasing sequence of elements with lower bound 0 which satisfies (6b). This is impossible by Theorem 7.13.5(b). Hence a point v satisfying (5) may always be found for a given $\varepsilon > 0$.

In the obvious manner use (5) to construct a decreasing sequence $\{w_n : n = 1,2,\ldots\}$ in $[0,u]$ such that for each n

(7) (a) $\|w_n\| \leqq \|u\| - \frac{1}{2}(1 - 2^{-n})\|u\|$; and

(b) $0 \notin \overline{co}([0,v_n] \setminus U_{2^{-n}}(0))$.

Let $w \equiv \inf_n w_n$ which exists by 7.13.5(b). Evidently

$$\|w\| \geq \frac{1}{2}\|u\| \quad \text{and} \quad 0 \ \text{is a denting point of} \ [0,w]. \tag{8}$$

We now bootstrap (8). Suppose that v_1 and v_2 are two points of $[0,u]$ and that 0 is a denting point of $[0,v_i]$ for $i = 1,2$. It is shown next that 0 is a dent point of $[0,v_1 + v_2]$. Indeed, let $x \in [0,v_1 + v_2]$. Then $x + (v_1 + v_2 - x) = v_1 + v_2$. By the Riesz decomposition Lemma 6.4.3 there are points x_1 and x_2 such that $x_i \in [0,v_i]$ for $i = 1,2$ and $x_1 + x_2 = x$. If $\|x\| \geq 2\varepsilon$ then either $\|x_1\| \geq \varepsilon$ or $\|x_2\| \geq \varepsilon$ and hence

$$\begin{aligned}[0,v_1 + v_2] \setminus U_{2\varepsilon}(0) &\subset (\{[0,v_1] \setminus U_\varepsilon(0)\} + [0,v_2]) \cup ([0,v_1] + \{[0,v_2] \setminus U_\varepsilon(0)) \\ &\subset \mathrm{co}(\{\overline{\mathrm{co}}([0,v_1] \setminus U_\varepsilon(0)) + [0,v_2]\} \cup \{[0,v_1] + \overline{\mathrm{co}}([0,v_2] \setminus U_\varepsilon(0))\}).\end{aligned} \tag{9}$$

Note that $0 \notin \overline{\mathrm{co}}([0,v_1] \setminus U_\varepsilon(0))$ by hypothesis. Since order intervals are weakly compact, $\overline{\mathrm{co}}([0,v_1] \setminus U_\varepsilon(0)) + [0,v_2]$ is a weakly compact convex set which does not contain 0. A similar argument holds for $[0,v_1] + \overline{\mathrm{co}}([0,v_2] \setminus U_\varepsilon(0))$ and hence the set on the right side of (9) is a weakly compact convex subset of $X_+ \setminus \{0\}$. It follows from (9) that $0 \notin \overline{\mathrm{co}}([0,v_1 + v_2] \setminus U_{2\varepsilon}(0))$ whence 0 is a denting point of $[0,v_1 + v_2]$ since $\varepsilon > 0$ was arbitrary.

To complete the proof of the lemma, let

$$V \equiv \{v \in [0,u]\colon 0 \ \text{is a denting point of} \ [0,v]\}.$$

Then V is a directed set under $\leq$ by (8). Hence $w \equiv \sup V$ exists by Theorem 7.13.5(b). First observe that $w = u$. Otherwise, from (8) with $u - w$ in place of u, there would exist $0 \neq v \in [0,u - w]$ such that $v \in V$. This is impossible, however, since it would imply $\sup V = w < w + v \in V$. Thus it remains to establish $w \in V$. Given $\varepsilon > 0$ there is (again by 7.13.5(b)) a point v in V with $\|u - v\| \leq \varepsilon$. Pick any finite collection of points x_i in $[0,u]$ with $\|x_i\| \geq 2\varepsilon$ and for each i choose any $t_i \geq 0$ with $\sum t_i = 1$. Observe that

$$\|x_i \wedge v\| \geq \|x_i\| - \|x_i - x_i \wedge v\| \geq 2\varepsilon - \|x_i \wedge u - x_i \wedge v\| \geq 2\varepsilon - \|u - v\| \geq \varepsilon.$$

Thus since 0 is a denting point of $[0,v]$, there is a $\delta > 0$ depending only on ε and v such that

$$\|\sum t_i x_i\| \geqq \|\sum t_i(x_i \wedge v)\| > \delta > 0.$$

Hence 0 is a denting point of $[0,u]$ and the proof is complete.

LEMMA 7.13.7. Let u be a quasi interior point of X_+. Then for each n there is an integer k such that $\overline{co}(P_n(u)) \subset P_k(u))$.

Proof. Choose any finite collection of points $x_i \in P_n(u)$ and numbers $t_i \geqq 0$ for which $\sum t_i = 1$. Let $B \equiv \overline{co}(\{v \in [0,u]: \|v\| \geqq \frac{1}{n}\})$. Since 0 is a denting point of $[0,u]$ by Lemma 7.13.6, there is an integer k such that $U_{1/k}(0) \cap B = \emptyset$. Now $\sum t_i(x_i \wedge u) \in B$ and hence $\|\sum t_i(x_i \wedge u)\| \geqq \frac{1}{k}$. But

$$0 \leqq \sum t_i(x_i \wedge u) \leqq (\sum t_i x_i) \wedge u.$$

Hence $\|(\sum t_i x_i) \wedge u\| \geqq \frac{1}{k}$. It follows that $co(P_n(u)) \subset P_k(u)$ and since $P_k(u)$ is closed, $\overline{co}(P_n(u)) \subset P_k(u)$ as well.

LEMMA 7.13.8. Let C and D be two closed bounded convex subsets of X_+. Then $co(C \cup D) = \overline{co}(C \cup D)$. That is, $co(C \cup D)$ is closed.

Proof. Let x be a point of $\overline{co}(C \cup D)$ and pick sequences $\{c_n: n = 1,2,...\}$ in C, $\{d_n: n = 1,2,...\}$ in D and $\{t_n: n = 1,2,...\}$ in $[0,1]$ with

$$\lim_n \|x - (t_n c_n + (1 - t_n)d_n\| = 0.$$

Without loss of generality assume that $t \equiv \lim_n t_n$ exists. Then $\lim_n \|x - (tc_n + (1 - t)d_n)\| = 0$. For each n let $c_n' \equiv tc_n$ and $d_n' \equiv (1 - t)d_n$. Of course $\{c_n' \wedge x: n = 1,2,...\}$, being a sequence in the weakly compact order interval $[0,x]$, has a weak cluster point, say $\bar{c}$. Observe that

$$(10) \qquad \begin{aligned} 0 \leqq c_n' - c_n' \wedge x &\leqq c_n' + d_n' - (c_n' \wedge x + d_n' \wedge x) \\ &\leqq c_n' + d_n' - x \wedge (c_n' + d_n'). \end{aligned}$$

But $\lim_n \|c_n' + d_n' - x\| = 0$. Hence from (10),

$$(11) \quad \lim_n \|c_n' - c_n' \wedge x\| \leqq \lim_n \|c_n' + d_n' - x \wedge (c_n' + d_n')\| = \|x - x \wedge x\| = 0$$

and it follows from (11) that, since $\bar{c}$ is a weak cluster point of $\{c_n' \wedge x: n = 1,2,\ldots\}$, it is also a weak cluster point of $\{c_n': n = 1,2,\ldots\}$. Since C is weakly closed and $\{c_n': n = 1,2,\ldots\}$ is a sequence in tC, it follows that $\bar{c} \in t$ Then $\{d_n': n = 1,2,\ldots\}$ must have a weak cluster point $\bar{d}$ in $(1 - t)D$ such that $x = \bar{c} + \bar{d}$. Note that $\bar{c} + \bar{d} \in co(C \cup D)$. The proof is thus complete.

<u>LEMMA</u> 7.13.9. Fix a quasi interior point u in X_+. Let $D \subset X_+$ be a closed conv set and assume that D is not order dentable. If $C \subset D$ is a closed convex set an if $D = \overline{co}((D \cap P_n(u)) \cup C)$ for some (hence all sufficiently large) n, then C =

Intuitively, only an "insignificant" part of a non order dentable convex subset of X_+ lies in any given $P_n(u)$.

<u>Proof of 7.13.9</u>. Suppose the lemma were false and that C is a proper subset of D. Then there are an $f \in X^*$, $\|f\| = 1$ and $\lambda > 0$ with $0 < 4\lambda < M(D,f) - M(C,$ Choose k so that $\overline{co}(P_n(u)) \subset P_k(u)$ (by Lemma 7.13.7). We will show that $S(D,f$ $\subset P_{4k}(u)$ from which it follows that D is order dentable in contradiction to the hypothesis. Fix $x \in S(D,f,\lambda)$. Then by assumption there is a point $y \in co(D \cap P_n($ $\cup C)$ with $\|x - y\| \leq \min\{\lambda, \frac{1}{4k}\}$. Write y in the form $rd + (1 - r)c$ where $0 \leq$ ≤ 1, $d \in co(D \cap P_n(u))$ and $c \in C$. Since $f(x) > M(D,f) - \lambda$ and $f(y) \geq f(x) - \|x - y\| \geq f(x) - \lambda$ we have

$$M(D,f) - 2\lambda \leq f(y) \leq rM(D,f) + (1 - r)M(C,f). \tag{12}$$

Since $M(D,f) - 4\lambda \geq M(C,f)$ it follows from (12) that $r \geq \frac{1}{2}$. Thus $y = rd + (1 -$ $\geq r \geq \frac{1}{2}d$. By the choice of k,

$$\|y \wedge u\| \geq \frac{1}{2}\|d \wedge u\| \geq \frac{1}{2k}.$$

It follows that $\|x \wedge u\| \geq \|y \wedge u\| - \|x - y\| \geq \frac{1}{2k} - \frac{1}{4k} = \frac{1}{4k}$. That is, $x \in P_{4k}(u)$. Since $x \in S(K,f,\lambda)$ was arbitrary, the proof is complete.

<u>LEMMA</u> 7.13.10. Let u be a quasi interior point of X_+. Let D be a closed conve subset of X_+ and assume that D is not order dentable. Let $\bar{x} \in D$ and $f \in X^*$ satisfy $f(\bar{x}) = M(D,f)$. Pick any $\varepsilon > 0$, $\lambda > 0$ and positive integer n. Let

$S \equiv S(D,f,\lambda)$. Then there are finite sequences $\{d_i: 1 \leq i \leq p\}$ in D, $\{f_i: 1 \leq i \leq p\}$ in X^*, $\{\lambda_i: 1 \leq i \leq p\}$ and $\{s_i: 1 \leq i \leq p\}$ in $\mathbb{R}$ such that each of the following is true for $1 \leq i \leq p$:

(a) $f_i(d_i) = M(D,f_i)$;

(b) $S(D,f_i,\lambda_i) \subset S$;

(c) $S(D,f_i,\lambda_i) \cap P_n(u) = \emptyset$;

(d) $\sum s_i = 1$, $s_i \geq 0$ and $\|\bar{x} - \sum_{i=1}^{p} s_i d_i\| < \varepsilon$.

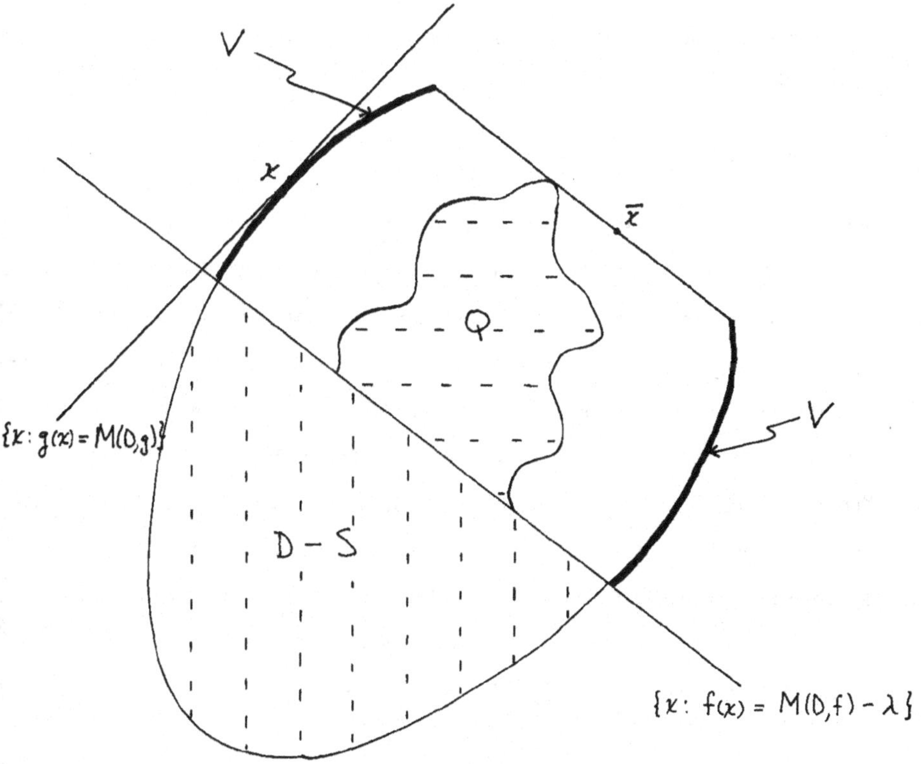

Proof. Let $Q \equiv P_n(u) \cap D$ and

(13) $$V \equiv \{x \in D: \text{for some } g \in X^*,\ g(x) = M(D,g) > M(\{D \setminus S\} \cup Q, g)\}.$$

The Bishop Phelps Theorem 3.3.1 asserts that

$$D = \overline{co}(\{D \setminus S\} \cup Q \cup V)$$

and from Lemma 7.13.9 it follows that

$$D = \overline{co}(\{D \setminus S\} \cup V).$$

Since $\overline{x} \in D$ there are sequences $\{d_n: n = 1,2,\ldots\}$ in $\overline{co}(D \setminus S)$, $\{v_n: n = 1,2,\ldots$ in $\overline{co}(V)$ and $\{q_n: n = 1,2,\ldots\}$ in $[0,1]$ with $\lim_n \|q_n d_n + (1 - q_n)v_n - \overline{x}\| = 0$. By going to a subsequence assume that $q \equiv \lim_n q_n$ exists. Then

$$\lim_n \|qd_n + (1 - q)v_n - \overline{x}\| = 0.$$

Since $f(\overline{x}) = M(D,f)$ and $M(D \setminus S,f) = M(D,f) - \lambda$ by definition of S, it follows that $q = 0$. Hence $\overline{x} \in \overline{co}(V)$.

Here is the proof of Theorem 7.13.1.

Proof of 7.13.1. We must show that if X has the KMP then it has the RNP under the assumptions in (4). If X lacks the RNP then, according to Theorem 7.13.4, X cannot be order dentable. Let u be a quasi interior point of X_+ and let D be closed bounded convex non order dentable subset of X_+. Apply Lemma 7.13.10 inductively as follows. At the $0^{\underline{th}}$ stage pick any point $\overline{x}$ of D and $f \in X^*$ with $f(\overline{x}) = M(D,f)$. Take $\varepsilon \equiv 2^{-0}$, $n \equiv 1$ and $\lambda \equiv 1$. For each $t \in [0,1)$ let $F^0(t) \equiv \overline{x}$, $f^0(t) \equiv f$ and $\lambda^0(t) \equiv \lambda$. Let $\mathcal{F}_0 \equiv \{\emptyset,[0,1)\}$. Apply Lemma 7.13.10 to construct f_i,d_i,λ_i and s_i for $1 \leqq i \leqq p$ as in the statement of that lemma, let $I_1,\ldots,I_p$ be pairwise disjoint half open subintervals of $[0,1)$ with length $(I_i) \equiv s_i$ and let $\mathcal{F}_1$ be the σ-algebra whose atoms are $I_1,\ldots,I_p$. Define $F^1: [0,1) \to D$, $f^1: [0,1) \to X^*$ and $\lambda^1: [0,1) \to \mathbb{R}$ as follows:

$$F^1(t) \equiv d_i, \quad f^1(t) \equiv f_i, \quad \text{and} \quad \lambda^1(t) \equiv \tfrac{1}{2}\lambda_i \quad \text{if } t \in I_i \text{ for } 1 \leqq i \leqq p.$$

This completes stage 1. At the second stage, apply 7.13.10 consecutively to each of the p slices $S(D,f^1(t),\lambda^1(t))$, points d_i, $n \equiv 2$ and $\varepsilon \equiv 2^{-1}$. In this manner we create an increasing sequence of finite sigma algebras $\mathcal{F}_n$ of subsets of $[0,1)$, functions $F^n: [0,1) \to D$, $f^n: [0,1) \to X^*$ and $\lambda^n: [0,1) \to \mathbb{R}$ with the following properties for each $t \in [0,1)$:

$$(14)\quad \begin{array}{ll} \text{(a)} & F^n(t) \in S(D,f^n(t),\lambda^n(t)); \\ \text{(b)} & \|F^n(t) - \mathbb{E}[F^{n+1}|F_n](t)\| \leq 2^{-n}; \\ \text{(c)} & P_n(u) \cap S(D,f^n(t),2\lambda^n(t)) = \emptyset; \text{ and} \\ \text{(d)} & S(D,f^{n+1}(t),2\lambda^{n+1}(t)) \subset S(D,f^n(t),\lambda^n(t)). \end{array}$$

Note that (F^n,F_n) is an ∞-quasi martingale corresponding to the error sequence $\{2^{-n}: n = 0,1,\ldots\}$. (See Definition 7.5.1.) Apply the decomposition Theorem 7.5.3. Thus there is a martingale (G_n,F_n) with values in D such that

$$(15)\qquad G_n(t) \in \overline{S(D,f^n(t),\lambda^n(t))} \subset S(D,f^n(t),2\lambda^n(t))$$

for each $t \in [0,1)$ and each n. (Alternatively, use (14b) to see directly that $G_k(t) \equiv \lim_k \mathbb{E}[F^k|F_n](t)$ exists almost surely with respect to Lebesgue measure. It is easy to check that (G_k,F_k) is a martingale satisfying (15).)

Let $C \equiv \overline{co}(\{G_n(t): t \in [0,1), n = 0,1,\ldots\})$. We will show that $ex(C) \subset \{0\}$, which, in conjunction with Proposition 3.1.1, establishes that K lacks the KMP. In this regard, pick any $x \in C$, $x \neq 0$. Note that $x \wedge u = \frac{1}{n}(nx \wedge nu) \geq \frac{1}{n}(x \wedge nu) > 0$ for n sufficiently large. Hence for some k, $x \in P_k(u)$. Let $I_1,\ldots,I_p$ be the atoms of F_k and for each i in $1,\ldots,p$ let $C_i \equiv \overline{co}(\{G_n(t): n \geq k, t \in I_i\})$. Then from (15) and (14c) we have $C_i \cap P_k(u) = \emptyset$ for each $i = 1,\ldots,p$. In particular, $x \notin C_i$ for $i = 1,\ldots,p$. But (G_n,F_n) is a martingale and hence $C = \overline{co}(\bigcup_{i=1}^p C_i)$. According to Lemma 7.13.8, though, $C = \overline{co}(\bigcup_{i=1}^p C_i) = co(\bigcup_{i=1}^p C_i)$. Thus $x \in co(\bigcup_{i=1}^p C_i) \setminus \bigcup_{i=1}^p C_i$ and hence $x \notin ex(C)$. The proof of Theorem 7.13.1 is complete.

DISCUSSION FOR THEOREM 7.13.2.

The construction of a predual for X depends on first representing X as a function space. Note that since X has the RNP it has the KMP and hence c_o is not isomorphic to a subspace of X. It follows that X is "σ-complete", "σ-order continuous" and has a weak unit. (See, for example, Lindenstrauss and Tzafriri [1979]. These statements differ in terminology only from parts of Theorem 7.13.5 and the brief discussion following it.) According to the representation Theorem 1.b.14 of Lindenstrauss and Tzafriri [1979], p. 25, then, X is lattice isomorphic and isometric to a subspace of $L^1(\Omega,F,\mu)$ for some probability space (Ω,F,μ), with the properties listed in (16) below. Henceforth $(X,\|\cdot\|)$ will be identified with its image in $L^1(\mu)$. Thus $\|\cdot\|$ will refer to the norm on X; $\|\cdot\|_1$ and $\|\cdot\|_\infty$ will refer to the $L^1(\mu)$ and $L^\infty(\mu)$ norms respectively.

(16)
(a) X is a not necessarily closed ideal of $(L^1(\mu),\|\cdot\|_1)$;
(b) $(X,\|\cdot\|)$ is a Banach lattice with the RNP;
(c) $L^\infty(\mu)$ is a $\|\cdot\|_1$ - dense subspace of X;
(d) X is a $\|\cdot\|_1$ - dense subspace of $L^1(\mu)$;
(e) $\|f\|_1 \leqq \|f\| \leqq 2\|f\|_\infty$ for each $f \in L^\infty(\mu)$;
(f) Each $G \in X^*$ may be considered as a μ-measurable real valued function on Ω whose value at $f \in X$ is $\int_\Omega fGd\mu$. Consequently

$$\|G\|^* = \sup \{\int_\Omega fGd\mu\colon \|f\| \leqq 1,\ f \in X\}$$

where $\|\cdot\|^*$ denotes the $\|\cdot\|$ - dual norm on X^*.

Note that for $f \in X$ and for each positive integer n, $\||f| \wedge n\|_1 \leqq \||f| \wedge n\| \leqq \|f\|$ from (16e) and hence, by taking the limit as $n \to \infty$,

(17) $$\|f\|_1 \leqq \|f\| \quad \text{for each} \quad f \in X.$$

Since $(X,\|\cdot\|)$ is separable, so is $(X,\|\cdot\|_1)$ by (17) and thus so is $(L^1(\mu),\|\cdot\|_1)$ by (16d). There is consequently no loss of generality in assuming that F is countably generated. Moreover, it follows directly from (17) that $L^\infty(\mu) \subset X^*$ and that $\|G\|^* \leqq \|G\|_\infty$ for G in $L^\infty(\mu)$. Let A denote a fixed countable subalgebra of F with the property that F is the smallest σ-algebra containing A. (Special extra conditions will be placed on A in (28).) According to the Stone representation theorem for Boolean algebras with unit (see Dunford and Schwartz [1958], Theorem 1, p. 41) there are a totally disconnected compact Hausdorff space K and a Boolean algebra isomorphism ψ from A onto $A(K)$, the algebra of clopen subsets of K. ("Clopen" means simultaneously closed and open.) ψ induces a lattice isometry of $L^1(\mu)$ into $C(K)^*$ which is described in detail next.

Let $Z(K)$ denote the algebra of real valued functions on K which can be wri in the form $\sum_{i=1}^{n} r_i \chi_{B_i}$ for some choices of $n = 1,2,\ldots$, $B_i \in A(K)$ and $r_i \in \mathbb{R}$. Observe that $Z(K) \subset C(K)$. By the Stone Weierstrass theorem, $Z(K)$ is dense in $C(K)$. For $f \in L^1(\mu)$ define $\mu_f\colon Z(K) \to \mathbb{R}$ by

$$\mu_f\left(\sum_{i=1}^{n} r_i \chi_{B_i}\right) \equiv \sum_{i=1}^{n} r_i \int_{\psi^{-1}(B_i)} f d\mu \; . \tag{18}$$

It is easy to check that μ_f is well defined and that it is a continuous linear functional on $Z(K)$. Therefore μ_f has a unique extension to an element of $C(K)^*$ which, by the Riesz representation theorem, may be identified with a regular Borel measure on K. This measure will also be denoted by μ_f. In the special case $f \equiv 1 \in L^1(\mu)$, μ_f will sometimes be denoted by $\overline{\mu}$. Evidently the association $f \to \mu_f$ is positive, linear and injective from $L^1(\mu)$ into $C(K)^*$. Since it is positive, it is also continuous. (See, for example, Schaefer [1974], p.84, Theorem 5.3.) If f has the special form $\sum_{i=1}^{n} r_i \chi_{A_i}$ for some choices of $r_i \in \mathbb{R}$, $A_i \in A$ and $n = 1,2,\ldots$ then $|\mu_f| = \mu_{|f|}$ follows easily. More generally we have $|\mu_f| = \mu_{|f|}$ for each $f \in L^1(\mu)$ by virtue of the continuity of the lattice operations, the continuity of the map $f \to \mu_f$, and the density of functions of the above special form in $L^1(\mu)$. Next observe that $\mu_f << \overline{\mu}$ when $f \in L^1(\mu)$. Indeed, assume the contrary. Then there is a decreasing sequence $\{B_n : n = 1,2,\ldots\}$ in $A(K)$ with $\lim_n \mu_{|f|}(B_n) > 0$ and $\lim_n \overline{\mu}(B_n) = 0$. Since $\mu(\psi^{-1}(B_n)) = \overline{\mu}(B_n) \to 0$ as $n \to \infty$ we have $\int_{\psi^{-1}(B_n)} |f| d\mu \to 0$ as $n \to \infty$. But this is impossible since $\int_{\psi^{-1}(B_n)} |f| d\mu = \mu_{|f|}(B_n)$ is bounded away from 0 by hypothesis. Thus $\mu_f << \overline{\mu}$. Denote the Radon Nikodým derivative of μ_f with respect to $\overline{\mu}$ by $\overline{f}$. Thus $\overline{f} \in L^1(\overline{\mu})$ and we let $\overline{f}d\overline{\mu}$ denote the measure in $C(K)^*$ whose value at $B \in \mathcal{B}(K)$ is $\int_B \overline{f} d\overline{\mu}$. Define $\Psi\colon L^1(\mu) \to C(K)^*$ by

$$\Psi(f) \equiv \overline{f}d\overline{\mu} \quad \text{for} \quad f \in L^1(\mu). \tag{19}$$

A certain degree of caution is necessary. First of all, the Boolean algebra isomorphism ψ between A and $A(K)$ usually can not be extended to a σ-algebra isomorphism between F and $\mathcal{B}(K)$ as easy examples show, and thus a direct identification of the measure $fd\mu$ on F and $\overline{f}d\overline{\mu}$ on $\mathcal{B}(K)$ is out of the question. Secondly, ψ does not come from a point mapping between Ω and K in most cases and hence there is no direct way to identify $\overline{f}$ from f as functions. A useful way of gleaning information about Ψ comes from restating (18) in the form

(20) $$\int_B \overline{f}d\overline{\mu} = \int_{\psi^{-1}(B)} fd\mu \text{ for each } B \in A(K) \text{ and } f \in L^1(\mu).$$

We list some properties of Ψ below for convenience:

(21)
- (a) Ψ is a positive linear mapping of $L^1(\mu)$ onto $\{\overline{f}d\overline{\mu}: \overline{f} \in L^1(\overline{\mu})\} \subset C(K)^*$
- (b) $\Psi(1) = d\overline{\mu}$;
- (c) Ψ is an isometry;
- (d) $\Psi(L^1(\mu))$ is a band in $C(K)^*$.

The first three properties in (21) may be checked directly. In order to check (21d), first observe that $\Psi(L^1(\mu))$ is an ideal in $C(K)^*$. Indeed, suppose that $\nu \in C(K)^*$ and $|\nu| << |\overline{f}|d\overline{\mu}$ for some $\overline{f} \in L^1(\overline{\mu})$. Then $\nu << \overline{\mu}$ whence $\nu = \overline{g}d\overline{\mu}$ f some $\overline{g} \in L^1(\overline{\mu})$ and hence $\overline{g}d\overline{\mu} = \Psi(\Psi^{-1}(\overline{g}d\overline{\mu})) \in \Psi(L^1(\overline{\mu}))$. Now to show that $\Psi(L^1(\mu))$ is a band, suppose (f_α) is a net in $L^1(\overline{\mu})$ and $\nu \equiv \bigvee_\alpha \overline{f}_\alpha d\overline{\mu}$ exists in $C(K)^*$. It evidently suffices to establish that $\nu << \overline{\mu}$. If this were not the case, ν could be written in the form $\nu_1 + \nu_2$ with $\nu_1 << \overline{\mu}$, $\nu_2 \perp \overline{\mu}$ and $\nu_2 \neq 0$. Evidently $\nu_1 \geqq \overline{f}_\alpha d\overline{\mu}$ for each α whence $\nu_1 \geqq \nu$. It follows that $\nu_2 \leqq 0$ and since $\nu_2 \neq 0$ by assumption, there is a $B \in \mathcal{B}(K)$ with $\overline{\mu}(B) = 0$ and $\nu_2(B) < 0$. Hence for any α,

$$0 > \nu_2(B) = (\nu_1 + \nu_2)(B) = \nu(B) \geqq (\overline{f}_\alpha d\overline{\mu})(B) = 0$$

which is a contradiction. Thus (21d) follows.

Let Z denote the space of functions on Ω to $\mathbb{R}$ which assume only finitely many values, each value being taken on some element of A. Equip Z with the supre norm. (There is NO identification of functions in Z which differ on sets of μ-me sure 0.) Thus Z naturally corresponds to its counterpart in $C(K)$, the dense sub space $Z(K)$. Since each element of $C(K)^*$ is determined by its behavior on $Z(K)$, this is true in particular of the elements of $\Psi(X) \subset \Psi(L^1(\mu)) \subset C(K)^*$. It will tur out that the predual of X constructed below may be built in an inductive manner fr the restriction to appropriate sets in F of the functions in Z. Recall that a su set C of X_+ is <u>hereditary</u> if whenever $f \in C$, $g \in X_+$ and $g \leqq f$ then $g \in C$.

It will be our practice to think of elements G of X^* as real valued functions on Ω subject to the restrictions indicated in (16f). In particular, then, to say that $G \in X^*$ assumes only finitely many values means that $G: \Omega \to \mathbb{R}$ has finite range.

<u>LEMMA</u> 7.13.11. Let C be a closed bounded convex hereditary subset of X_+. Then there is a countable set $D \subset X^*$ such that each element of D assumes only finitely many values and

$$C = \{f \in L^1(\mu): G(f) \leqq 1 \text{ for each } G \in D\}.$$

Proof. Let $\{f_n: n = 1,2,\dots\}$ be a dense sequence in $X_+ \setminus C$. For each n let $d_n \equiv \inf \{\|f_n - f\|: f \in C\}$. In order to check that

$$d_n = \inf \{\|f_n - h\|: h \in C - X_+\} \tag{22}$$

observe first that whenever $g \in X_+ \setminus C$ and $f \in C$ then $|g - f \wedge g| = g - f \wedge g \leq |g - f|$ so that $\|g - f \wedge g\| \leq \|g - f\|$. Suppose now that $f \in C$ and $h \in X_+$. Then $f \wedge (f_n + h) \in C$ and, from the previous remark,

$$\|f_n + h - f \wedge (f_n + h)\| \leqq \|f_n + h - f\|. \tag{23}$$

The Riesz decomposition Lemma 6.4.3 applied to $f_n + h = [f_n + h - f \wedge (f_n + h)] + f \wedge (f_n + h)$ asserts the existence of $h_1, h_2 \in X_+$ with $h_1 + h_2 = f \wedge (f_n + h)$, $h_1 \leqq f_n$ and $h_2 \leqq h$. Since $h_1 \leq f$, $h_1 \in C$ since C is hereditary. Using (23) and the fact that $h - h_2 \geqq 0$ we obtain

$$d_n \leqq \|f_n - h_1\| \leqq \|f_n - h_1 + (h - h_2)\| \leqq \|f_n + h - f\| = \|f_n - (f - h)\|.$$

Thus (22) follows since $f \in C$ and $h \in X_+$ were arbitrary.

Let $V(X)$ denote the open unit ball of X. Then the open convex sets $C - X_+ + \frac{1}{2} d_n V(X)$ and $f_n + \frac{1}{2} d_n V(X)$ are disjoint so there are an element H_n of X^* and numbers $a < b$ such that

$$\begin{aligned} &H_n(f) \leqq a \quad \text{for } f \in C - X_+ + \tfrac{1}{3} d_n V(X) \quad \text{and} \\ &H_n(g) > b \quad \text{for } g \in f_n + \tfrac{1}{3} d_n V(X). \end{aligned} \tag{24}$$

Evidently $0 < a$ and $H_n \geqq 0$. By multiplying by the appropriate scale factor we henceforth also assume $a < 1 < b$. Although H_n is not bounded in general (as a function on Ω) given $\varepsilon > 0$ it may be uniformly approximated within ε by an F-measurable positive function G_n on Ω which assumes only finitely many values on any finite interval of $\mathbb{R}_+$. Note that for $f \in X$, $\int G_n f d\mu = H_n(f) + \int (G_n - H_n) f d\mu$. Since C is bounded in X, it is $\|\cdot\|_1$ bounded as well by (17). Hence, by choosi[ng] $\varepsilon > 0$ sufficiently small and using (24) we can guarantee both that

$$\text{(25)}\qquad \begin{aligned} G_n(f) &\leqq a + \frac{1-a}{2} < 1 \quad \text{for} \quad f \in C; \quad \text{and} \\ G_n(g) &\geqq b - \frac{b-1}{2} > 1 \quad \text{for} \quad g \in f_n + \tfrac{1}{3} d_n V(X). \end{aligned}$$

Let $D_1 \equiv \{G_n \wedge k: k,n \in \mathbb{N}\}$ and $D_2 \equiv \{-n\chi_A: n = 1,2,\ldots , A \in A\}$. Finally, let $D \equiv D_1 \cup D_2$.

To see that D has the desired properties suppose first that $f \in C$. Then clearly $(G_n \wedge k)(f) \leqq G_n(f) \leqq 1$ and $-n \int_A fd\mu \leqq 0 < 1$ for all choices of $n,k \in$ [$\mathbb{N}$] and $A \in A$ so that $C \subset \{f \in L^1(\mu): G(f) \leqq 1$ for each G in $D\}$. On the other ha[nd] pick any $f \in L^1(\mu)$ for which $G(f) \leqq 1$ for each $G \in D$. Note that $f \geqq 0$ since $-n \int_A fd\mu \leqq 1$ for $A \in A$ and $n \in \mathbb{N}$ by definition of D_2. Now fix a positive inte[-] ger p. Then $f \wedge p \in L^\infty(\mu) \subset X$. If $f \wedge p \notin C$ then there is an n such that $f \wedge p \in f_n + \frac{1}{3} d_n V(X)$ and hence $G_n(f \wedge p) > 1$. Thus for some k, $(G_n \wedge k)(f \wedge p)$ > 1 since evaluation of an element of X^* at an element of X is by integration. But since $f \wedge p \leqq f$ and $G_n \wedge k \geqq 0$, it follows that $(G_n \wedge k)(f) > 1$ which is impossible. Thus $\{f \wedge p: p = 1,2,\ldots\}$ is an increasing sequence in the norm boun[ded] set C. Since such a sequence necessarily converges by Theorem 7.13.5, and since t[he] limit evidently coincides with the supremum f, it follows that $f \in C$. The proof [of] the lemma is complete.

We return now to the problem of providing a full set of defining conditions fo[r] the algebra A. The idea is to use the above lemma on a countable set of closed bounded convex subsets of X_+ which reflect the order dentability properties of X In this regard, let $U \equiv U(X) \cap X_+$. Then U is a closed bounded convex hereditary subset of X_+. For each ordinal number α and for each integer $n \geqq 0$ define the set $U(\alpha,n)$ by transfinite induction:

(26)
(a) $U(0,0) \equiv U$;

(b) For each α and each n, $U(\alpha,n) \equiv \overline{co}(\{f \in U(\alpha,0): \|f \wedge 1\| \leqq \frac{1}{n}\})$;

(c) For each α, $U(\alpha+1,0) \equiv \bigcap_n U(\alpha,n)$;

(d) If α is a limit ordinal then $U(\alpha,0) \equiv \bigcap_{\beta<\alpha} U(\beta,0)$.

Clearly $U(\alpha,n)$ is a closed bounded hereditary convex subset of X_+ and since X is order dentable, if $U(\alpha,0) \neq \{0\}$ then $U(\alpha+1,0) \neq U(\alpha,0)$. Because $(X,\|\cdot\|)$ is second countable, there is a countable ordinal α_o such that $U(\beta,0) \neq \{0\}$ if $\beta < \alpha_o$ and $U(\alpha_o,0) = \{0\}$. According to Lemma 7.13.11, for each $\beta \leqq \alpha_o$ and each $n \geqq 0$ there is a countable subset $D(\beta,n)$ of X^* consisting of functions each of which is F-measurable and each of which assumes only finitely many values, such that

(27) $$U(\beta,n) = \{f \in L^1(\mu): G(f) \leqq 1 \text{ for each } G \text{ in } D(\beta,n)\}.$$

Let

(28) A be a countable subalgebra of F which generates F such that $G^{-1}(O) \in A$ whenever O is open in $\mathbb{R}$ and $G \in D(\beta,n)$ for some $\beta \leqq \alpha_o$ and $n \geqq 0$.

According to the previous discussion - see, for example, (21) - there is an order isometry Ψ between $L^1(\mu)$ and $\{\bar{f}d\bar{\mu}: \bar{f} \in L^1(\bar{\mu})\} \subset C(K)^*$. We shall henceforth identify $L^1(\mu)$ and this subspace of $C(K)^*$. In particular, μ and $\bar{\mu}$ will be identified in the subsequent discussion and X will be considered a subspace of $C(K)^*$. Also, with the exception of the proof of Lemma 7.13.13 below, all subsequent references to the weak* topology are to the weak* topology on $C(K)^*$ induced by $C(K)$, and, as usual, $w^*cl(W)$ will denote the weak* closure of a subset W of $C(K)^*$. By Lemma 7.13.11, $U(\beta,n) \subset L^1(\mu)$ is closed in the weakest topology on $L^1(\mu)$ for which each function in Z is continuous. Now translate this statement into the corresponding one for subsets of $C(K)^*$. Then Z corresponds to $Z(K)$ and since $Z(K) \subset C(K)$, it follows that

(29) $U(\beta,n)$ is a relatively weak* closed subset of $L^1(\mu) \subset C(K)^*$ for each (β,n).

We make the following notational conventions:

(30)
$$C_\Gamma \equiv \{f \in C: f|_{\Omega\setminus\Gamma} \equiv 0\} \quad \text{for any} \quad \Gamma \subset \Omega \quad \text{and} \quad C \subset X.$$
$$A_\infty \equiv \{\bigcap_{n=1}^{\infty} A_n : A_n \in A \quad \text{for each} \quad n\}.$$

Recall that $X^\perp$, the collection of elements in $C(K)^*$ orthogonal to X, a band in $C(K)^*$. (See, for example, Schaefer [1974], p.61.) As our final bit of additional notation for now, when C in (30) above is one of the sets $U(\alpha,n)$ as defined in (26) and $\Gamma \subset \Omega$, the set $U(\alpha,n)_\Gamma$ will be written $U_\Gamma(\alpha,n)$. The next proposition is pivotal in the proof of Theorem 7.13.2. Its proof will be postponed until after the proof of the main theorem below. (At that time we will establish th formally stronger result stated in Proposition 7.13.14.)

<u>PROPOSITION</u> 7.13.12. Given $\varepsilon > 0$ there is a set $\Delta \in A_\infty$ with $\mu(\Delta) \geq 1 - \varepsilon$ such that $X^\perp \cap w^*cl(U_\Delta) = \{0\}$.

The next lemma is standard fare. It will be used in several places in the identification of a predual for X.

<u>LEMMA</u> 7.13.13. Let E be a Banach lattice and let F be a norming subspace, not necessarily closed, of E^*. (F is <u>norming</u> if $\|x\| = \sup\{f(x): \|f\| \leq 1, \ f \in F\}$ for each $x \in E$.) Assume that the positive part of the unit ball of E, $U(E) \cap E_+$ is $\sigma(E,F)$ - compact. (As usual, the $\sigma(E,F)$ topology is the weakest topology on such that each element of F is continuous.) Then F^* is isometrically isomorphic to E.

Proof. It is easy to check that $(U(E),\sigma(E,F))$ is compact since $(U(E) \cap E_+, \sigma(E,F))$ is compact by hypothesis. Let $i: F \to E^*$ be the identity injection. Since F is norming, $i^*|_{\hat{E}}: \hat{E} \to F^*$ is an isometry into F^*. Of course, $i^*: (E^{**},\text{weak}^*) \to (F^*,\text{weak}^*)$ is continuous and $U(\hat{E})$ is weak* dense in $U(E^{**})$. Since $i^*(U(E^{**})) = U(F^*)$ it follows that $i^*(U(\hat{E}))$ is weak* dense in $U(F^*)$. On the other hand, $i^*(U(\hat{E}))$ is weak* compact from above. Thus $i^*(U(\hat{E})) = U(F^*)$. The sought for isometry is thus $i^*|_{\hat{E}} \circ j: E \to F^*$ where $j(x) \equiv \hat{x}$ denotes the natural injection of into E^{**}.

The proof of Talagrand's Theorem 7.13.2 is now within reach.

Proof of 7.13.2. According to Proposition 7.13.12 there is a set Δ in A_∞ such that $\mu(\Delta) \geqq \frac{1}{2}$ and $X^\perp \cap w^*cl(U_\Delta) = \{0\}$. Evidently U_Δ is hereditary and it follows that $w^*cl(U_\Delta)$ is hereditary as well. Pick any $\nu \in w^*cl(U_\Delta)$ and write ν in the form $\nu = fd\mu + \nu_1$ where ν_1 is mutually singular with respect to μ. Since $\nu \geqq 0$ it follows that $0 \leqq \nu_1 \leqq \nu$ and hence that $\nu_1 \in w^*cl(U_\Delta)$. But $\nu_1 \in X^\perp$. According to the choice of Δ, the only element of $X^\perp$ in $w^*cl(U_\Delta)$ is 0 and hence $\nu_1 = 0$. That is,

(31) $$w^*cl(U_\Delta) \subset L^1(\mu) \subset C(K)^*.$$

It should be remarked here that the careful choice of A in (28), which set the stage for Proposition 7.13.12, was necessary in order to establish the inclusion in (31). (31) will soon be applied in (33) to show that U_Δ is weak* compact, the key to the entire argument.

Because X is an ideal in $L^1(\mu)$ by (16a), given $f \in X$, $f\chi_\Delta \in X$ as well. That is, X_Δ is a subspace of X. In fact, as we show next (see (32)) X_Δ is a relatively weak* closed subspace of X. Write $\Omega \setminus \Delta$ in the form $\bigcup_{n=1}^{\infty} \Gamma_n$ for some increasing sequence $\{\Gamma_n: n = 1,2,...\}$ chosen from A. Then clearly

(32) $$X_\Delta = \bigcap_{n=1}^{\infty} \bigcap_{\Gamma \in A} \{f \in X: \int_{\Gamma \cap \Gamma_n} fd\mu = 0\}.$$

From (29), U is a relatively weak* closed subset of $L^1(\mu)$. Since $U \cap w^*cl(X_\Delta) = U_\Delta$ from above, we conclude with the aid of (31) that

(33) $$w^*cl(U_\Delta) \subset \left[L^1(\mu) \cap w^*cl(U_\Delta)\right] \cap w^*cl(X_\Delta) \subset U \cap w^*cl(X_\Delta) = U_\Delta.$$

That is, U_Δ is weak* closed in $C(K)^*$. Hence U_Δ is weak* compact.

Recall that Z is the space of A-measurable real valued functions on Ω which assume only finitely many values. Let

$$Z^\Delta \equiv \{G\chi_\Delta: G \in Z\}.$$

Bear in mind that, since Δ is not necessarily in A, Z^Δ is not generally a subspace of Z and in particular, $Z^{\Delta} \neq Z_{\Delta}$. Let X^*_{Δ} denote the subspace of X^* defined b

$$X^*_{\Delta} \equiv \{G \in X^*: G|_{\Omega \setminus \Delta} \equiv 0\} = \{G\chi_{\Delta}: G \in X^*\},$$

and we remark that $Z^{\Delta} \subset X^*_{\Delta}$. Provide Z^{Δ} with the relative norm topology as a sub space of X^*. Then U_{Δ}, which is evidently the positive part of the unit ball of X_{Δ}, is weak* compact according to the calculations in (33). Since U_{Δ} is bounded, clearly the weak* topology on U_{Δ} coincides with the $\sigma(X_{\Delta}, Z^{\Delta})$ topology on U_{Δ}. Lemma 7.13.13 asserts that $(Z^{\Delta})^*$ is isometrically isomorphic to X_{Δ}.

Let $\Omega_1 \equiv \Omega \setminus \Delta$, let $F_1 \equiv \{\Gamma \in F: \Gamma \subset \Omega_1\}$ and denote by μ_1 the restriction μ to F_1. Then

$$X_1 \equiv \{f\chi_{\Omega_1}: f \in X\} \subset L^1(\Omega_1, F_1, \mu_1)$$

is a subspace of X since X is an ideal in $L^1(\mu)$ by (16a). In fact, it is easy to check that each of the properties listed for X as a subspace of $L^1(\mu)$ in (16) carries over to the corresponding property of X_1 as a subspace of $L^1(\mu_1)$. Hence the entire discussion from (16) onward may be repeated with X_1 in place of X and (Ω_1, F_1, μ_1) in place of (Ω, F, μ). In particular, there is a set $\Delta_1 \in F$ disjoint from Δ such that $\mu(\Delta_1) \geq \frac{1}{2}\mu(\Omega_1) = \frac{1}{2}\mu(\Omega \setminus \Delta)$ and X_{Δ_1} is isometrically isomorphic $(Z^{\Delta_1})^*$. Denote Δ henceforth by Δ_0. Inductively repeat the argument just given. A sequence $\{\Delta_n: n = 0,1,\ldots\}$ of pairwise disjoint sets in F is thus produced whi satisfies the following conditions:

(a) $\mu(\bigcup_{n=1}^{\infty} \Delta_n) = 1$; and

(b) If for each n we let

$$Z_n \equiv \{G\chi_{\Delta_n} : G \in Z\}$$

and if Z_n is considered as a not necessarily closed subspace of $(X^*, \|\cdot\|^*)$ then X_{Δ_n} is isometrically isomorphic to Z_n^*.

Let $Y \equiv \overline{\mathrm{span}}(\bigcup_{i=0}^{\infty} Z_n) \subset (X^*, \|\cdot\|^*)$. Evidently Y is a Banach lattice subspace of $(X^*, \|\cdot\|^*)$ and the proof will be completed by showing that Y^* is isometrically iso morphic to X. According to Lemma 7.13.13, it suffices to demonstrate that U, the positive part of the unit ball of X, is $\sigma(X,Y)$ compact since Y clearly norms X. Because Y is separable, bounded subsets of X are $\sigma(X,Y)$ metrizable and hence it suffices to establish that each sequence $\{f_n: n = 1,2,\ldots\}$ in U has a $\sigma(X,Y)$ convergent subsequence. Let $\{f'_n: n = 1,2,\ldots\}$ be a subsequence of $\{f_n: n = 1,2,\ldots\}$ chosen so that for each nonnegative integer p, the sequence $\{f'_n\chi_{\Delta_p} : n = 1,2,\ldots\}$, which lies in the positive part of the unit ball of X_{Δ_p}, is $\sigma(X_{\Delta_p}, Z_p)$ convergent to some element g_p. Evidently g_p also lies in the positive part of the unit ball of X_{Δ_p}. Assume for now that $h_k \equiv \sum_{p=0}^{k} g_p$ belongs to U for each $k = 0,1,\ldots$. (The proof appears below.) Then $h_0 \leqq h_1 \leqq \cdots$ and hence $\{h_k: k = 0,1,\ldots\}$ is norm convergent by Theorem 7.13.5(a). Let $h \equiv \lim_k h_k$. Since U is norm closed, $h \in U$. Evidently h is the $\sigma(X,Y)$ limit of the sequence $\{f'_n: n = 1,2,\ldots\}$ as was to be shown.

Thus it remains to show that $h_k \in U$ for each k. Suppose not. Since $h_k \notin U$, precisely as in the proof of Lemma 7.13.11 (with U in place of C) there is an F - measurable nonnegative function $G \in X^*$ which assumes only finitely many values on any finite interval of $\mathbb{R}$ such that $G(f) < 1$ for each $f \in U$ while $G(h_k) > 1$. Note that from $G(f) < 1$ for $f \in U$ it follows that $\|G\|^* \leqq 1$. Hence for some positive number m, $(G \wedge m)(h_k) > 1$ while $\|G \wedge m\|^* \leqq \|G\|^* \leqq 1$. Then G may be approximated by an element H of Z which, with a little care, will satisfy: $\|H\|^* \leqq 1$ and $H(h_k) > 1$. Let $\Delta' \equiv \bigcup_{p=0}^{k} \Delta_p$ and $H' \equiv H\chi_{\Delta'}$. Then $H' \in Y$ and $H'(h_k) > 1$. On the other hand, $H'(f'_n) \leqq \|H'\|^*\|f'_n\| \leqq \|H\|^*\|f'_n\| \leqq 1$ for each n. That is,

$$1 \geqq \int (H\chi_{\Delta'})f'_n d\mu = \sum_{p=0}^{k} \int (H\chi_{\Delta_p})(f'_n\chi_{\Delta_p})d\mu$$

and since $\lim_n (H\chi_{\Delta_p})(f'_n\chi_{\Delta_p}) = (H\chi_{\Delta_p})(g_p)$ for each $p = 0,\ldots,k$ it follows that

$$1 \geqq \sum_{p=0}^{k} (H\chi_{\Delta_p})(g_p) = H(\sum_{p=0}^{k} g_p) = H(h_k),$$

a contradiction. Thus $h_k \in U$ and the proof of Theorem 7.13.2 rests on establishing Proposition 7.13.12.

It is clear that Proposition 7.13.12 is an immediate corollary of the following somewhat more technical result.

PROPOSITION 7.13.14. For any ordinal $\alpha \leq \alpha_0$, set $\Gamma \in A_\infty$ and number $\varepsilon > 0$ the is a set Δ in A_∞ with $\mu(\Delta) \geq 1 - \varepsilon$ such that

$$X^\perp \cap w^*cl(U_{\Gamma\cap\Delta}) \subset w^*cl(U_\Gamma(\alpha,0)).$$

The next two lemmas will be needed in the proof of the above proposition. Reca first that for any Banach space E, subset A of E and $f \in E^*$, $M(K,f)$ denotes the number $\sup\{f(x): x \in A\}$.

LEMMA 7.13.15. Let $\{C^n: n = 1,2,\ldots\}$ be a decreasing sequence of bounded convex subsets of X. Assume that $0 \in C^n$ and that C^n is relatively weak* closed in X for each n. Suppose that $G \in Z$. Then for each $\varepsilon > 0$ there is a set $\Delta \in A$ wit $\mu(\Delta) \geq 1 - \varepsilon$ such that

$$M(\bigcap_n w^*cl(C^n{}_\Delta),\hat{G}) \leq M(\bigcap_n C^n,G) + \varepsilon.$$

Proof. We will show that for any $r \geq M(\bigcap_n C^n,G)$ and $\varepsilon > 0$ there is a set $\Delta \in A$ with $\mu(\Delta) \geq 1 - \varepsilon$ and $M(\bigcap_n w^*cl(C^n{}_\Delta),\hat{G}) \leq r + \varepsilon$. Since $0 \in \bigcap_n C^n$ it follow that $r \geq 0$ and by shuffling r and ε slightly we may assume $r > 0$. By taking the appropriate scale factor of G we further assume that $r + \frac{\varepsilon}{2} = 1$. Let W deno the closed unit ball of $(L^\infty(\mu),\|\cdot\|_\infty)$. We have $L^\infty(\mu) \subset X \subset C(K)^*$ and the weak* topology on W coincides with the $\sigma(X,Z(K))$ topology on W since W is bounded in $C(K)^*$ and $Z(K)$ is dense in $C(K)$. By our accepted abuse of notation, the $\sigma(X,Z(K))$ topology is the $\sigma(X,Z)$ topology, and since $Z \subset L^1(\mu)$, the $\sigma(L^\infty(\mu), L^1(\mu))$ topology on W is at least as strong as the weak* topology on W. Because W is $\sigma(L^\infty(\mu),L^1(\mu))$ compact and the weak* topology is Hausdorff, these two topolc agree on W. That is, W is weak* compact.

Pick $t \in \mathbb{R}$ such that $\varepsilon^2 t \geq 2$ and let S denote the closed slice of tW determined by G:

$$S \equiv \{h \in tW: G(h) \geq 1\}.$$

Since $G \in Z$, S is weak* compact and since $M(\bigcap_n C^n,G) \leq r < 1$, it follows that

$S \cap \bigcap_n C^n = \emptyset$. The sequence $\{C^n: n = 1,2,\ldots\}$ is decreasing and each C^n is relatively weak* closed in $X \supset L^\infty(\mu)$ by hypothesis. Hence for some positive integer N, $S \cap C^N = \emptyset$. According to the Hahn-Banach theorem on $(X,\text{weak*})$ there are $H \in C(K)$ and $s \in \mathbb{R}$ with

$$M(C^N,H) < s < \inf \{H(h): h \in S\}. \tag{34}$$

Evidently the collection of H's in $C(K)$ satisfying (34) is open and since $Z(K)$ is dense in $C(K)$ there is a $G' \in Z$ satisfying (34). (As usual, Z and $Z(K)$ are identified here.) Since $0 \in C^N$, $s > 0$ and by scaling G' as necessary we may assume that $s = 1$. Thus

$$M(C^N,G') < 1 < \inf \{G'(h): h \in S\}. \tag{35}$$

Suppose for a moment that $G = G'$. Let $\Delta \equiv \Omega$ in this case and observe that for $f \in C^N = C^N_\Omega$, $G(f) = G'(f) \leq 1 < 1 + \frac{\varepsilon}{2} = r + \varepsilon$. Hence if $\nu \in w^*cl(C^N)$ then $\nu(G) \leq r + \varepsilon$ and consequently $M(\bigcap_n w^*cl(C^n),\hat{G}) \leq r + \varepsilon$ under these special assumptions. The proof of the lemma reduces to showing that, in the manner important to us here, $G - \lambda G'$ is "small" for an appropriate choice of $\lambda \in [0,1]$. Observe that

(36) if $h \in tW$ and $G'(h) \leq 1$ then $h \in tW \setminus S$ and hence $G(h) \leq 1$.

For any set $T \subset X$ let $T_\circ \subset Z$ denote the polar of T. Thus

$$T_\circ \equiv \{H \in Z: H(f) \leq 1 \text{ for each } f \in T\}.$$

Now let $V \equiv \{f \in X: G'(f) \leq 1\}$. From (36) we have $G \in (tW \cap V)_\circ$. But $(tW \cap V)_\circ = w^*cl(co((tW)_\circ \cup V_\circ))$ by a calculation like that in Lemma 5.4.2(d). Since $V_\circ = \{\lambda G': \lambda \in [0,1]\}$ is $\sigma(Z,X)$ compact it follows that $G \in co((tW)_\circ \cup V_\circ)$ and hence there are $\lambda \in [0,1]$ and $G'' \in (tW)_\circ$ such that $G = G'' + \lambda G'$. The sought for set Δ is then

$$\Delta \equiv \{\omega \in \Omega: |G'(\omega)| \leq \tfrac{\varepsilon}{2}\}. \tag{37}$$

Indeed, since $G'' \in (tW)_\circ \subset Z$, $\Delta \in A$ and $\|G''\|_1 \leq \frac{1}{t} \leq \frac{\varepsilon^2}{2}$ whence $\mu(\Delta) \geq 1 - \varepsilon$. Moreover, suppose $f \in C^N_\Delta$. Recall that $C^N_\Delta \subset U(X)$. From (35) and (17) we obtain

(38) $G(f) \leq \lambda G'(f) + \int_{\Delta} G'' f d\mu \leq \lambda G'(f) + \frac{\varepsilon}{2}\|f\|_1 \leq \lambda + \frac{\varepsilon}{2}\|f\| \leq \lambda + \frac{\varepsilon}{2} \leq 1 + \frac{\varepsilon}{2} = r + \varepsilon.$

Hence $\nu(G) \leq r + \varepsilon$ for $\nu \in w^*cl(C^N_\Delta)$. It follows that $M(\bigcap_n w^*cl(C^n_\Delta), \hat{G}) \leq r + \varepsilon$ and the lemma follows.

<u>LEMMA</u> 7.13.16. Let $\{C^n: n = 1,2,\ldots\}$ be a decreasing sequence of bounded convex subsets of X. Assume that $0 \in C^n$ and that C^n is relatively weak* closed in X for each n. Given $\varepsilon > 0$ there is a set Δ in A_∞ such that $\mu(\Delta) \geq 1 - \varepsilon$ and

$$\bigcap_n w^*cl(C^n_\Delta) \subset w^*cl(\bigcap_n C^n).$$

Proof. Let $\{G_p: p = 1,2,\ldots\}$ be an enumeration of the elements of Z which have rational range. By Lemma 7.13.15, for each positive integer k there is a set $\Delta(p,k) \in A$ with $\mu(\Delta(p,k)) \geq 1 - \varepsilon 2^{-p-k}$ and

$$M(\bigcap_n w^*cl(C^n_{\Delta(p,k)}), \hat{G}_p) \leq M(\bigcap_n C^n, G_p) + 2^{-k}.$$

Let $\Delta \equiv \bigcap_{p,k} \Delta(p,k)$. Then $\Delta \in A_\infty$ and $\mu(\Delta) \geq 1 - \varepsilon$. Moreover, for each p,

$$M(\bigcap_n w^*cl(C^n_\Delta), \hat{G}_p) \leq M(\bigcap_n C^n, G_p) = M(w^*cl(\bigcap_n C^n), G_p).$$

Since $\{G_p: p = 1,2,\ldots\}$ is dense in Z it follows from the separation theorem tha

$$\bigcap_n w^*cl(C^n_\Delta) \subset w^*cl(\bigcap_n C^n)$$

as was to be shown.

Here, finally, is the proof of Proposition 7.13.14, the last step in the proof Theorem 7.13.2.

<u>Proof of 7.13.14</u>. The proof is by induction on α. When $\alpha = 0$ we may take $\Delta \equiv \Omega$. Now assume the statement of the Proposition holds for each ordinal $\beta < \alpha$.

<u>Case 1</u>. α is a limit ordinal. Let $\alpha_1 < \alpha_2 < \ldots$ be chosen with $\lim_n \alpha_n =$ Fix $\Gamma \in A_\infty$ and $\varepsilon > 0$. It follows from Lemma 7.13.16 that there is a set $\Delta_0 \in A_\infty$ with $\mu(\Delta_0) \geq 1 - \frac{\varepsilon}{2}$ and

(39) $$\bigcap_n w^*\mathrm{cl}(U_{\Delta_0\cap\Gamma}(\alpha_n,0)) \subset w^*\mathrm{cl}(\bigcap_n U_\Gamma(\alpha_n,0)) = w^*\mathrm{cl}(U_\Gamma(\alpha,0)).$$

From the inductive hypothesis, for each n there is a set $\Delta_n \in A_\infty$ with

$$\mu(\Delta_n) \geq 1 - 2^{-(n+1)}\varepsilon \quad\text{and}\quad X^\perp \cap w^*\mathrm{cl}(U_{\Delta_n\cap(\Delta_0\cap\Gamma)}) \subset w^*\mathrm{cl}(U_{\Delta_0\cap\Gamma}(\alpha_n,0)).$$

Now let $\Delta \equiv \bigcap_{n=0}^{\infty} \Delta_n$. Then $\Delta \in A_\infty$, $\mu(\Delta) \geq 1 - \varepsilon$ and, from (39),

$$X^\perp \cap w^*\mathrm{cl}(U_{\Delta\cap\Gamma}) \subset X^\perp \cap \bigcap_n w^*\mathrm{cl}(U_{\Delta_n\cap(\Delta_0\cap\Gamma)}) \subset X^\perp \cap \bigcap_n w^*\mathrm{cl}(U_{\Delta_0\cap\Gamma}(\alpha_n,0))$$
$$\subset X^\perp \cap w^*\mathrm{cl}(U_\Gamma(\alpha,0)) \subset w^*\mathrm{cl}(U_\Gamma(\alpha,0)).$$

Thus the inductive step is established when α is a limit ordinal.

<u>Case 2</u>. $\alpha = \beta + 1$. We will need the following fact about the embedding of X in $L^1(\mu)$:

(40) $$\limsup_{\delta\to 0} \{\|g\|: g \in X,\ 0 \leq g \leq 1,\ \|g\|_1 \leq \delta\} = 0.$$

Indeed, if (40) were not true then for some $r > 0$ we could create a sequence $\{g_n: n = 1,2,\dots\}$ in X such that for each n, $0 \leq g_n \leq 1$, $\|g_n\|_1 \leq 2^{-n}$, and $\|g_n\| \geq r$. Because X is σ-complete (see, for example, Lindenstrauss and Tzafriri [1979], Theorem 1.0.5, p.6) we conclude that $h_k \equiv \sup\{g_n: n \geq k\}$ exists in X for each k. Then $\{h_k: k = 1,2,\dots\}$ is a decreasing sequence in X_+ with pointwise limit 0 and yet $\|h_k\|_1 \geq \|g_k\|_1 \geq r$ for each k. This is impossible by Dini's Theorem 7.13.5b. Thus (40) maintains.

Since $\alpha = \beta + 1$, $U(\alpha,0) = \bigcap_n U(\beta,n)$. Hence, given $\Gamma \in A_\infty$ we have $U_\Gamma(\alpha,0) = \bigcap_n U_\Gamma(\beta,n)$. We remark here that for any set $\Pi \in A_\infty$, $U_\Pi(\alpha,0)$ is relatively weak* closed in X since $U_\Pi(\alpha,0) = U(\alpha,0) \cap X_\Pi$, $U(\alpha,0)$ was shown relatively weak* closed (even in $L^1(\mu)$) in (29) and X_Π was shown relatively weak* closed in X in the discussion ending with (32). Hence Lemma 7.13.16 applies to the sequence $\{U_\Gamma(\beta,n): n = 0,1,\dots\}$. Thus there is a set Δ_1 in A_∞ with $\mu(\Delta_1) \geq 1 - \frac{\varepsilon}{2}$ such that $\bigcap_n w^*\mathrm{cl}(U_{\Delta_1\cap\Gamma}(\beta,n)) \subset w^*\mathrm{cl}(U_\Gamma(\alpha,0))$. Moreover, from the inductive hypothesis applied to $\Delta_1\cap\Gamma$, there is a set $\Delta_2 \in A_\infty$ with $\mu(\Delta_2) \geq 1 - \frac{\varepsilon}{2}$ and $X^\perp \cap w^*\mathrm{cl}(U_{\Delta_2\cap\Delta_1\cap\Gamma}) \subset w^*\mathrm{cl}(U_{\Delta_1\cap\Gamma}(\beta,0))$. Let $\Delta \equiv \Delta_1 \cap \Delta_2$. Then $\mu(\Delta) \geq 1 - \varepsilon$. Suppose that

$\nu \in X^{\perp} \cap w^*cl(U_{\Delta\cap\Gamma})$. It will be demonstrated that $\nu \in w^*cl(U_{\Gamma}(\alpha,0))$ which will evidently complete the proof. By definition of Δ_2 it is possible to choose a sequence $\{f_n: n = 1,2,...\}$ in $U_{\Delta_1\cap\Gamma}(\beta,0)$ with weak* $\lim_n f_n = \nu$. Fix a positive integer p and choose $\delta > 0$ so small that $\|g\| \leqq \frac{1}{p}$ whenever $g \in X$, $0 \leqq g \leqq 1$ and $\|g\|_1 \leqq \delta$. (See (40).) Since $\nu \in X^{\perp}$, it is possible to find a set $\Sigma \in A$ wit $\mu(\Sigma) \leqq \delta$ and $\nu(\Omega\setminus\Sigma) \leqq \delta$. Let ν' be the restriction of ν to the Borel subsets of Σ and for each n, let $f'_n \equiv f_n\chi_{\Sigma}$. Since $\Sigma \in A$ it follows that weak* $\lim_n f$ $= \nu'$. Evidently $\|f'_n \wedge 1\|_1 \leqq \mu(\Sigma) \leqq \delta$ for each n. Recall that $f_n \in U_{\Delta_1\cap\Gamma}(\beta,0)$, an hereditary set. Thus $f'_n \in U_{\Delta_1\cap\Gamma}(\beta,0)$ as well and since $\|f'_n \wedge 1\|_1 \leqq \delta$ it foll that $\|f'_n \wedge 1\| \leqq \frac{1}{p}$. Consequently, $f'_n \in U_{\Delta_1\cap\Gamma}(\beta,p)$ for each n. Thus

$$\text{weak* } \lim_n f'_n = \nu' \in w^*cl(U_{\Delta_1\cap\Gamma}(\beta,p)).$$

Since $\delta > 0$ was arbitrary, it follows that $\nu \in w^*cl(U_{\Delta_1\cap\Gamma}(\beta,p))$ as well. But Δ was chosen so that

$$\bigcap_p w^*cl(U_{\Delta_1\cap\Gamma}(\beta,p)) \subset w^*cl(U_{\Gamma}(\alpha,0)).$$

Hence $\nu \in w^*cl(U_{\Gamma}(\alpha,0))$ as was to be shown. The proof of Proposition 7.13.14 is complete. Hence Talagrand's Theorem 7.13.2 has been established.

INDEX OF COMMON NOTATION

$\mathbb{N}$ natural numbers

$\mathbb{N}^+$ positive integers

$\mathbb{R}$ real numbers

$\mathbb{R}_+$ positive real numbers

$\mathbb{R}^n$ Euclidean n space

$A \equiv B$ $A = B$. Moreover, either A or B appears for the first time and is defined by this equation

$a \wedge b$ lattice minimum of a and b

$a \vee b$ lattice maximum of a and b

$\oplus$ direct sum

$\mu|_B$ restriction of measure μ to measurable subsets of B

$f|_B$ restriction of function f to subset B of its domain

$L^p(\Omega,F,P)$; $L^p(\mu)$; c_o ; ℓ_p ; $C[0,1]$; $C(Y)$ (Y compact Hausdorff) Standard notation for Banach spaces as defined in Dunford and Schwartz [1958]

BIBLIOGRAPHY

E.M. Alfsen

[1971] Compact convex sets and boundary integrals, Springer-Verlag, New York.

D. Amir and J. Lindenstrauss

[1968] The structure of weakly compact sets in Banach spaces, Ann. of Math. (2) 88, 35-46.

L. Asimow

[1969] Extremal structure of closed convex sets, Notices Amer. Math. Soc. (16), 503-504.

E. Asplund

[1968] Fréchet differentiability of convex functions, Acta Math. 121, 31-47.

[1969] Boundedly Krein-compact Banach spaces, Proc. of Functional Analysis Week (Aarhus), Mat. Inst. Aarhus Univ., 1-4.

E. Asplund and I. Namioka

[1967] A geometric proof of Ryll-Nardzewski's fixed point theorem, Bull. Amer. Math. Soc. 73, 443-445.

E. Asplund and R.T. Rockafellar

[1969] Gradients of convex functions, Trans. Amer. Math. Soc. 139, 443-467.

A. Bellow

[1978] Uniform amarts: a class of asymptotic martingales for which strong almost sure convergence obtains, Z. Wahrsch. Verw. Gebiete (3) 41, 177-191.

C. Bessaga and A. Pełczynski

[1966] On extreme points in separable conjugate spaces, Israel J. Math. 4, 262-264.

E. Bishop

[1966] Unpublished letter to R.R. Phelps.

E. Bishop and K. de Leeuw

[1959] The representation of linear functionals by measures on sets of extreme points, Ann. Inst. Fourier (Grenoble) 9, 305-331.

E. Bishop and R.R. Phelps

[1961] A proof that every Banach space is subreflexive, Bull. Amer. Math. Soc. 67, 97-98.

[1963] The support functionals of a convex set, Proc. Sympos. Pure Math. VII, Convexity, Amer. Math. Soc., Providence, R.I. 27-35.

K. Blizzard

[1980] A Krein-Milman set without the integral representation property, Abstracts of the eighth winter school of the Czechoslovak Acad. of Sci., Spindlerův, Mlyn.

S. Bochner

[1933a] Integration von funktionen deren werte die elemente eines vectorraumes sind, Fund. Math. 20, 262-276.

[1933b] Absolut-additive abstrakte mengenfunktionen, Fund. Math. 21, 211-213.

S. Bochner and A.E. Taylor

[1938] Linear functionals on certain spaces of abstractly-valued functions, Ann. of Math. (2) 39, 913-944.

J. Bourgain

[1976] Strongly exposed points in weakly compact convex sets in Banach spaces, Proc. Amer. Math. Soc. 58, 197-200.

[1977] On dentability and the Bishop-Phelps property, Israel J. Math. 28, 265-271.

[1978a] A geometric characterization of the Radon-Nikodým property in Banach spaces, Compositio Math. (1) 36, 3-6.
[1978b] A note on extreme points in duals, Bull. Soc. Math. Belg. (1) 30, 89-91.
[1979] Un espace non Radon-Nikodým sans arbre diadique, Seminaire d'analyse fonctionnelle 1978-79, Centre de mathématiques (Palaiseau), Juin, exposé no. 29.
[1980] Sets with the Radon-Nikodým property in conjugate space, Studia Math. 66, 291 297.

J. Bourgain and F. Delbaen
[1980] A class of special $\mathcal{L}_\infty$ spaces, Acta Math. 145, 155-176.

J. Bourgain and H.P. Rosenthal
[1980a] Martingales valued in certain subspaces of L^1, Israel J. Math. (1-2) 37, 54-
[1980b] Geometrical implications of certain finite dimensional decompositions, Bull. Soc. Math. Belg., Ser. B, (1) 32, 57-82.
[1983] Applications of the theory of semi-embeddings to Banach space theory, preprin

J. Bourgain and M. Talagrand
[1981] Dans un espace de Banach reticulé solide, la propriété de Radon-Nikodým et ce de Krein-Milman sont équivalentes, Proc. Amer. Math. Soc. 81, 93-96.

R.D. Bourgin
[1969] Thesis, University of Washington.
[1971] Barycenters of measures on certain noncompact convex sets, Trans. Amer. Math. Soc. 154, 323-340.
[1978] Weak* compact convex sets with separable extremal subsets have the Radon-Niko property, Proc. Amer. Math. Soc, (1) 69, 1978, 81-84.
[1979] Partial orderings for integral representations on convex sets with the Radon-Nikodým property, Pacific J. Math. 81, 29-44.

R.D. Bourgin and G.A. Edgar
[1976] Noncompact simplexes in Banach spaces with the Radon-Nikodým property, J. Functional Analysis (2) 23, 162-176.

L. Breiman
[1968] Probability, Addison-Wesley, Reading, Mass.

C. Carathéodory
[1907] Über den variabilitätsbereich der koeffizienten von potenzreihen, die gegeben werte nicht annehmen, Math. Ann. 64, 95-115.

R.V. Chacon and L. Sucheston
[1975] On convergence of vector-valued asymptotic martingales, Z. Wahrsch. Verw. Geb (1) 33, 55-59.

S.D. Chatterji
[1968] Martingale convergence and the Radon-Nikodým theorem in Banach spaces, Math. Scand. 22, 21-41.

G. Choquet
[1956] Existance et unicité des representations intégrales au moyen des points extr aux dans les cones convexes, Séminaire Bourbaki (Dec. 1956) 139, 15 pp.
[1959] Ensembles K-analytiques et K-Sousliniens cas général et cas métrique, Ann. In Fourier (Grenoble) 9, 75-89.
[1962] Remarques à propos de la démonstration de l'unicité de P.A. Meyer, Séminaire Brelot-Choquet-Deny (Theorie de Potentiel) 6, No. 8, 13 pp.
[1969] Lectures on Analysis, Vol II (Representation theory), W.A. Benjamin, Inc., Ne York.

J.A. Clarkson
[1936] Uniformly convex spaces, Trans. Amer. Math. Soc. 40, 396-414.

D.L. Cohn
[1980] Measure theory, Birkhäuser, Boston, Mass.

J.B. Collier
[1976] The dual of a space with the Radon-Nikodým property, Pacific J. Math. (1) 64, 103-106.

W.J. Davis, T. Figiel, W.B. Johnson and A. Pełczynski
[1974] Factoring weakly compact operators, J. Functional Analysis 17, 311-327.

W.J. Davis, N. Ghoussoub and J. Lindenstrauss
[1981] A lattice renorming theorem and applications to vector-valued processes, Trans. Amer. Math. Soc. (2) 263, 531-540.

W.J. Davis and R.R. Phelps
[1974] The Radon-Nikodým property and dentable sets in Banach spaces, Proc. Amer. Math. Soc. 45, 119-122.

F. Delbaen (see also J. Bourgain)
[1983] Semi-embeddings and Radon-Nikodým spaces. To appear.

K. de Leeuw (see E. Bishop)

J. Diestel
[1975] Geometry of Banach spaces - selected topics, Lecture Notes in Math. #485, Springer-Verlag, Berlin.

J. Diestel and B. Faires
[1974] On vector measures, Trans. Amer. Math. Soc. 198, 253-271.

J. Diestel and J.J. Uhl, Jr.
[1977] Vector measures, Math. Surveys 15, Amer. Math. Soc., Providence, R.I.

N. Dinculeanu
[1967] Vector measures, Pergamon Press, Oxford.

J.L. Doob
[1940] Regularity properties of certain families of chance variables, Trans. Amer. Math. Soc. 47, 455-486.

R.M. Dudley
[1966] Convergence of Baire measures, Studia Math. 27, 251-268.

N. Dunford
[1937] Integration of vector-valued functions, Bull. Amer. Math. Soc. 43, 24. (Abstract)

N. Dunford and A.P. Morse
[1936] Remarks on the preceding paper of James A. Clarkson, Trans. Amer. Math. Soc. 40, 415-420.

N. Dunford and B.J. Pettis
[1940] Linear operations on summable functions, Trans. Amer. Math. Soc. 47, 323-392.

N. Dunford and J.T. Schwartz
[1958] Linear operators Part I: General theory, Interscience Publishers, Inc., New York.

G.A. Edgar (see also R.D. Bourgin)
[1975a]A noncompact Choquet theorem, Proc. Amer. Math. Soc. 354-358.

[1975b] Disintegration of measures and the vector-valued Radon-Nikodým theorem, Duke Math. J. (3) 42, 447-450.
[1976] Extremal integral representations, J. Functional Analysis (2) 23, 145-161.
[1977] Measurability in a Banach space, Indiana Univ. Math. J. 26, 663-677.
[1978a] On the Radon-Nikodým property and martingale convergence, Vector space measu and applications (Proc. conf. Univ. Dublin, Dublin 1977) II, Lecture Notes in Math # 645, Springer-Verlag, Berlin, 62-76.
[1978b] Unpublished manuscript.
[1979] Measurability in a Banach space, II, Indiana Univ. Math. J. 28, 559-579.
[1980] A long James space, Measure Theory, Oberwolfach 1979 (Proc. Conf., Oberwolfac 1979), Lecture Notes in Math. # 794, Springer-Verlag, Berlin, 31-37.

G.A. Edgar and L. Sucheston
[1976] The Riesz decomposition for vector-valued amarts, Z. Wahrsch. Verw. Gebiete 3 85-92.

G.A. Edgar and M. Talagrand
[1980] Liftings of functions with values in a completely regular space, Proc. Amer. Math. Soc. 78, 345-349.

L. Egghe
[1977] On the Radon-Nikodým-Property, and related topics in locally convex spaces, Vector Space Measures and Applications II (Proc. Dublin 1977), Lecture Notes in Ma # 645, 77-90.
[1980] The Radon-Nikodým-Property, σ dentability and martingales in locally convex spaces, Pacific J. Math. (2) 87, 313-322.

I. Ekeland
[1979] Nonconvex minimization problems, Bull. Amer. Math. Soc. (New Series)1, 443-47

I. Ekeland and G. Lebourg
[1976] Generic Fréchet differentiability and perturbed optimization problems in Bana spaces, Trans. Amer. Math. Soc. (2) 224, 193-216.

J. Elton
[1982] The approximate Krein-Milman property does not imply the Krein-Milman propert Unpublished manuscript.

P. Enflo
[1972] Banach spaces which can be given an equivalent uniformly convex norm, Israel Math. 13, 281-288.

B. Faires (see J. Diestel)

T. Figiel (see W.J. Davis)

S. Fitzpatrick
[1977] Weak* compact convex sets with the RNP, Rainwater Seminar, Univ. of Washingtc March 1977, mimeographed notes.
[1978] Monotone operators and dentability, Bull. Austral. Math. Soc. 18, 77-82.
[1980] The differentiability of distance functions and the GSP in Banach spaces, The Univ. of Washington. Chapter 2.

S. Fitzpatrick and J.J. Uhl, Jr.
[1981] Personal communication.

D.H. Fremlin
[1975] Pointwise compact sets of measurable functions, Manuscripta Math. 15, 219-242

D.H. Fremlin and M. Talagrand
[1979] A decomposition theorem for additive set-functions, with applications to Pettis integrals and ergodic means, Math. Z. 168, 117-142.

D.J.H. Garling
[1976] Chatterji's martingale convergence theorem, Durham Symposium on the Relations between Infinite-dimensional and Finite-dimensional Convexity, Bull. London Math. Soc. (1) 8, Ed. by D.G. Larman and C.A. Rogers, p. 9.

R.F. Geitz
[1981] Pettis integration, Proc. Amer. Math. Soc. (1) 82, 81-86.
[1982] Geometry and the Pettis integral, Trans, Amer. Math. Soc, (2) 269, 535-548.

R.F. Geitz and J.J. Uhl, Jr.
[1981] Vector-valued functions as families of scalar-valued functions, Pacific J. Math. (1) 95, 75-83.

I.M. Gelfand
[1938] Abstrakte funktionen und lineare operatoren, Mat. Sb. (N.S.) (4) 46, 235-286.

N. Ghoussoub (see also W.J. Davis)
[1982] Renorming dual Banach lattices, Proc. Amer. Math. Soc. (4) 84, 521-524.
[1983] Riesz-space valued measures and processes, J. Multivariate Anal., to appear.

N. Ghoussoub and E. Saab
[1981] On the weak Radon-Nikodým property, Proc. Amer. Math. Soc. (1) 81, 81-84.

N. Ghoussoub and M. Talagrand
[1979] Order dentability and the Radon-Nikodým property in Banach lattices, Math. Ann. 243, 217-225.

D. Gilliam
[1976] Geometry and the Radon-Nikodým theorem in strict Mackay convergence spaces, Pacific J. Math. (2) 65, 353-364.

G. Godefroy
[1979] Étude des projections de norme 1 de E" sur E. Unicité de certains préduax. Applications., Ann. Inst. Fourier (Grenoble) (4) 29, 53-70.
[1981] Points de Namioka espaces normants applications à la théorie isométrique de la dualité, Israel J. Math. (3) 38, 209-220.

A. Goldman
[1977] Mesures cylindriques, mesures vectorielles et questions de concentration cylindrique, Pacific J. Math. (2) 69, 385-413.

A. Grothendieck
[1955] Produits tensoriels et espaces nucleaires, Mem. Amer. Math. Soc. 16.

J. Hagler and E. Odell
[1978] A Banach space not containing ℓ_1 whose dual ball is not weak* sequentially compact, Illinois J. Math. 22, 290-294.

P.R. Halmos
[1950] Measure theory, Van Nostrand, New York.

R.E. Harrell and L.A. Karlovitz
[1970] Girths and flat Banach spaces, Bull. Amer. Math. Soc. 76, 1288-1291.
[1972] Nonreflexivity and the girth of spheres, Inequalities, III (Proc. Third Symposium, Univ. Calif., L.A. 1969), Academic Press, New York, 121-127.

[1974] The geometry of flat Banach spaces, Trans. Amer. Math. Soc. 192, 209-218.
[1975] On tree structures in Banach spaces, Pacific J. Math. (1) 59, 85-91.

R. Haydon
[1976a] An extreme point criterion for separability of a dual Banach space, and a new proof of a theorem of Corson, Quart. J. Math. Oxford Ser. (2), 27 no. 107, 379-385.
[1976b] Some more characterizations of Banach spaces containing ℓ_1, Math. Proc. Cambridge Phil. Soc. 80, 269-276.

E. Hewitt
[1950] Linear functionals on spaces of continuous functions, Fund. Math. 37, 161-189.

E. Hille
[1972] Methods in classical and functional analysis, Addison-Wesley, Reading, Mass.

E. Hille and R.S. Phillips
[1957] Functional analysis and semi-groups, Amer. Math. Soc. Colloq. Pub. XXXI, Providence, R.I.

A. Ho
[1979] The Radon-Nikodým property and weighted trees in Banach spaces, Israel J. Math. (1) 32, 59-66.
[1982] The Krein-Milman property and complemented bushes in Banach spaces, Pacific J Math. 98, 347-363.

J. Hoffmann-Jørgensen
[1970] The theory of analytic spaces, Aarhus, Various Publication Series No. 10.

R.E. Huff
[1974] Dentability and the Radon-Nikodým property, Duke Math. J. 41, 111-114.
[1980] On nondensity of norm-attaining operators, Rev. Roumaine Math. Pures Appl. 25 239-241.

R.E. Huff and P.D. Morris
[1975] Dual spaces with the Krein-Milman property have the Radon-Nikodým property, Proc. Amer. Math. Soc. 49, 104-108.
[1976] Geometric characterizations of the Radon-Nikodým property in Banach spaces, Studia Math. (2) 56, 157-164.

A. Ionescu Tulcea and C. Ionescu Tulcea
[1969] Topics in the theory of lifting, Springer-Verlag, Ergebnisse der Mathematik und ihrer Grenzgebiete 48, Berlin.

C. Ionescu Tulcea (see A. Ionescu Tulcea)

R.C. James
[1951] A non-reflexive Banach space isometric with its second conjugate space, Proc. Nat. Acad. Sci. U.S.A. 37, 174-177.
[1964] Weakly compact sets, Trans, Amer. Math. Soc. 113, 129-140.
[1972a] Some self-dual properties of normed linear spaces, Symposium on Infinite Dimensional Topology, Ed. by R.D. Anderson, Ann. of Math. Studies 69, 159-175.
[1972b] Super-reflexive Banach spaces, Canad. J. Math. (5) XXIV, 896-904.
[1974] A separable somewhat reflexive Banach space with nonseparable dual, Bull. Amer. Math. Soc. 80, 738-743.
[1981] Subbushes and extreme points in Banach spaces, Proceedings of Research Workshop on Banach Space Theory, Univ. of Iowa, Ed. by Bor-Luh Lin, 59-81.

R.C. James and J.J. Schäffer
[1972] Super-reflexivity and the girth of spheres, Israel J. Math. 11, 398-404.

L. Janicka
[1980] Some measure-theoretical characterization of Banach spaces not containing ℓ_1, Bull. Acad. Polon. Sci. Ser. Sci. Math. (7-8) 27, 561-565.

J.E. Jayne and C.A. Rogers
[1977] The extremal structure of convex sets, J. Functional Analysis (3) 26, 251-288.

F. John
[1948] Extremum problems with inequalities as subsidiary conditions, Courant Anniversary Volume, Interscience, New York, 187-204.

W.B. Johnson (see W.J. Davis)

W.B. Johnson and J. Lindenstrauss
[1980] Examples of $\mathcal{L}_1$ spaces, Ark. Mat. 18, 101-106.

W.B. Johnson and E. Odell
[1974] Subspaces of L_p which embed into ℓ_p, Compositio Math. 28, 37-49.

N.J. Kalton
[1979] An analogue of the Radon-Nikodým property for non locally convex quasi-Banach spaces, Proc. Edinburgh Math. Soc. (2) 22 no. 1, 49-60.

L.A. Karlovitz (see also R.E. Harrell)
[1973] On duals of flat Banach spaces, Math. Ann. 202, 245-250.

J.L. Kelley and I. Namioka
[1963] Linear topological spaces, Van Nostrand, Princeton, N.J.

G. Koumoullis
[1981] On perfect measures, Trans. Amer. Math. Soc. (2) 264, 521-537.

K. Kunen and H. Rosenthal
[1982] Margingale proofs of some geometrical results in Banach space theory, Pacific J. Math 100, 153-175.

T. Kuo
[1975] On conjugate Banach spaces with the Radon-Nikodým property, Pacific J. Math. 59, 497-503.

E.B. Leach and J.H. Whitfield
[1972] Differentiable functions and rough norms on Banach spaces, Proc. Amer. Math. Soc. 33, 120-126.

G. Lebourg (see I. Ekeland)

L. LeCam
[1957] Convergence in distribution of stochastic processes, Univ. Calif. Publ. in Stat. 2, 207-236.

P. Lévy
[1937] Théorie de l'addition des variables aléatoires, Gauthier-Villars, Paris. (Second edition, 1954.)

D. Lewis and C. Stegall
[1973] Banach spaces whose duals are isomorphic to $\ell_1(\Gamma)$, J. Functional Analysis 12, 177-187.

J. Lindenstrauss (see also D. Amir, W.J. Davis, W.B. Johnson)
[1963] On operators which attain their norm, Israel J. Math. 3, 139-148.

[1966] On extreme points in ℓ_1, Israel J. Math. 4, 59-61.
[1967] On complemented subspaces of m, Israel J. Math. 5, 153-156.
[1972] Weakly compact sets- their topological properties and the Banach spaces they generate, Symposium on Infinite Dimensional Toplogy, Louisiana State Univ. Baton Rouge, La. 1967, Ann. of Math. Studies 69, Princeton Univ. Press, Princeton, N.J. 1972, 235-273.

J. Lindenstrauss and H.P. Rosenthal
[1969] The $\mathcal{L}_p$ spaces, Israel J. Math. 7, 325-349.

J. Lindenstrauss and C. Stegall
[1975] Examples of separable spaces which do not contain ℓ_1 and whose duals are non-separable, Studia Math. 54, 81-105.

J. Lindenstrauss and L. Tzafriri
[1973] Classical Banach spaces, Lecture Notes in Math. # 338, Springer-Verlag, Berlin
[1979] Classical Banach spaces II, Springer-Verlag, Ergebnisse der Mathematik und ihre Grenzgebiete 97, New York.

H.P. Lotz, N.T. Peck and H. Porta
[1979] Semi-embeddings of Banach spaces, Proc. Edinburgh Math. Soc. (2) 22 no. 3, 233-240.

M.N. Lusin
[1917] Théorie des fonctions. Sur la classification de M. Baire, Comptes Rendus des Séances de l'Académie des Sciences 164, 91-94.

P. Mankiewicz
[1978] A remark on Edgar's extremal integral representation theorem, Studia Math. (3) 63, 259-265.

E. Marczewski and R. Sikorski
[1948] Measures in non-separable metric spaces, Colloq. Math. 1, 133-139.

H.B. Maynard
[1973] A geometrical characterization of Banach spaces having the Radon-Nikodým property, Trans Amer. Math. Soc. 185, 493-500.

P.W. McCartney
[1980] Neighborly bushes and the Radon-Nikodým property for Banach spaces, Pacific J Math. (1) 87, 157-168.

P.W. McCartney and R.C. O'Brien
[1980] A separable Banach space with the Radon-Nikodým property that is not isomorph to a subspace of a separable dual, Proc. Amer. Math. Soc. (1) 78, 40-42.

P.A. Meyer
[1966] Probability and Potentials, Blaisdell, Waltham, Mass.

E. Michael
[1956] Continuous selection I, Ann. Math. (2) 63, 361-382.

E. Michael and I. Namioka
[1976] Barely continuous functions, Bull. Acad. Polon. Sci. Ser. Sci. Math. Astronom Phys. (10) 24, 889-892.

D.P. Mil'man
[1947] Characteristics of extreme points of regularly convex sets, Dokl. Akad. Nauk SSSR (N.S.) 57, 119-122. (Russian.)

G.J. Minty
[1964] On the monotonicity of the gradient of a convex function, Pacific J. Math. 14, 243-247.

P.D. Morris (see also R.E. Huff)
[1981] Banach spaces which always contain supremum-attaining elements, Proc. Amer. Math. Soc. 83, 496-498.

A.P. Morse (see N. Dunford)

K. Musial
[1979] The weak Radon-Nikodým property in Banach spaces, Studia Math. 64, 151-173.
[1980] Martingales of Pettis integrable functions, Measure Theory, Proc. Conf., Oberwolfach, 1979, Lecture Notes in Math # 794, Springer-Verlag, Berlin, 324-339.

I. Namioka (see also E. Asplund, J.L. Kelley, E. Michael)
[1967] Neighborhoods of extreme points, Israel J. Math. 5, 145-152.
[1981] Personal communication.

I. Namioka and R.R. Phelps
[1975] Banach spaces which are Asplund spaces, Duke Math. J. 42, 735-750.

J. von Neumann
[1949] On rings of operators. Reduction theory., Ann. Math. (2) 50, 401-485.

O.M. Nikodým
[1930] Sur une généralisation des intégrales de M.J. Radon, Fund. Math. 15, 131-179.

R.C. O'Brien (see P.W. McCartney)

E. Odell (see J. Hagler, W.B. Johnson)

E. Odell and H.P. Rosenthal
[1975] A double-dual characterization of separable Banach spaces containing ℓ^1, Israel J. Math. 20, 375-384.

J.C. Oxtoby and S. Ulam
[1939] On the existence of a measure invariant under a transformation, Ann. Math. (2) 40, 560-566.

T. Parthasarathy
[1972] Selection theorems and their applications, Lecture Notes in Mathematics # 263, Springer-Verlag, Berlin.

N.T. Peck (see also H.P. Lotz)
[1971] Support points in locally convex spaces, Duke Math. J. 38, 271-278.

A. Pełczynski (see also C. Bessaga, W.J. Davis)
[1968] On Banach spaces containing $L_1(\mu)$, Studia Math. 30, 231-246.

B.J. Pettis (see also N. Dunford)
[1938] On integration in vector spaces, Trans. Amer. Math. Soc. 44, 277-304.

R.R. Phelps (see also E. Bishop, W.J. Davis, I. Namioka)
[1957] Subreflexive normed linear spaces, Ark. Mat. VIII Fasc 6, 444-450.
[1963] Support cones and their generalizations, Proc. Sympos. Pure Math. VII, Convexity, Amer. Math. Soc., Providence, R.I., 393-401.
[1964] Weak* support points of a convex set in E*, Israel J. Math 2, 177-182.
[1966] Lectures on Choquet's theorem, Van Nostrand, Mathematical Studies # 7, Van Nostrand, Princeton, N.J.

[1974a] Dentability and extreme points in Banach spaces, J. Functional Analysis (1) 17, 78-90.
[1974b] Support cones in Banach spaces and their applications, Advances in Math. 13, 1-19.
[1983] Convexity in Banach spaces: some recent results. To appear in Proc. Conf. on Convexity at Siegen, Germany, July, 1982. To be published by Birkhäuser.

R.S. Phillips (see also E. Hille)
[1940a] Integration in a convex linear topological space, Trans. Amer. Math. Soc. 47, 114-145.
[1940b] A decomposition of additive set functions, Bull. Amer. Math. Soc. 46, 274-277

G. Pisier
[1975] Martingales with values in uniformly convex spaces, Israel J. Math. (3-4) 20, 326-350.

H. Porta (see H.P. Lotz)

J. Radon
[1913] Theorie und andwendungen der absolut additiven mengenfunktionen, Akademie der Wissenschaften Wien, Sitzungsberichte 122-2a, 1295-1438.

J. Rainwater
[unpublished notes] A theorem of Ekeland and Lebourg on Fréchet differentiability of convex functions on Banach spaces.

G. Restrepo
[1964] Differentiable norms in Banach spaces, Bull. Amer. Math. Soc. 70, 413-414.

O.I. Reynov
[1975] Operators of type RN in Banach spaces, Dokl. Akad. Nauk SSSR 220, 3, 528-531. (Russian.)
[1977] Some classes of sets in Banach spaces and the topological characterization of operators of type RN, Zap. Nauk Sem. Lomi An SSSR 73, 224-228. (Russian.)
[1978a] On some classes of linear continuous mappings, Mat. Zametki 23, 2, 285-296. (Russian.)
[1978b] RN-sets in Banach spaces, Funkcional Anal. i Prilozen. 12, 1, 80-81. (Russian
[1978c] On purely topological characterization of operators of type RN, Funkcional. Anal. i Prilozen. 12, 4. (Russian.)
[1979] On two questions in the theory of linear operators, Primen. Funkcional. Anal. v Teorii Priblizenii 9, 102-114. (Russian.)
[1981] On a class of Hausdorff compacts and GSG Banach spaces, Studia Math. LXXI, 113-126.

L.H. Riddle, E. Saab and J.J. Uhl, Jr.
[1983] Sets with the weak Radon-Nikodým property in dual Banach spaces. To appear.

L.H. Riddle and J.J. Uhl, Jr.
[1983] Martingales and the fine line between Asplund spaces and spaces not containing a copy of ℓ_1. To appear.

M.A. Rieffel
[1967] Dentable subsets of Banach spaces, with application to a Radon-Nikodým theorem, Proc. Conf. Irvine, Calif., 1966, B.R. Gelbaum, editor, Functional Analysis, Academic Press, London, Thompson, Washington, D.C., 71-77.

J.W. Roberts
[1977] A compact convex set with no extreme points, Studia Math. 60, 255-266.
[1983] Compact convex sets with no extreme points in the spaces $L^p([0,1])$, $0 \leq p < 1$ To appear.

R.T. Rockafellar (see E. Asplund)

C.A. Rogers (see J.E. Jayne)

H.P. Rosenthal (see also J. Bourgain, K. Kunen, J. Lindenstrauss, E. Odell)
[1970] On relatively disjoint families of measures, with some applications to Banach space theory, Studia Math. 37, 13-36.
[1974] A characterization of Banach spaces containing ℓ^1, Proc. Nat. Acad. Sci. U.S. A. 71, 2411-2413.
[1979] Geometric properties related to the Radon-Nikodým property, Seminaire d'Initiation a l'Analyse, Univ. de Paris VI, 1979-80.
[1983] A subsequence splitting result for L^1-bounded sequences of random variables. To appear.

H.L. Royden
[1968] Real analysis, Macmillan, second edition, New York.

E. Saab (see also N. Ghoussoub, L.H. Riddle)
[1975] Dentabilité, points extrémaux et propriété de Radon-Nikodým, Bull. Sci. Math. (2) 99, 129-134.
[1976] Sur la propriété de Radon-Nikodým dans les espaces localement convexe de type (BM), C.R. Acad. Sci. Paris Ser. A, 899-902.
[1978] On the Radon-Nikodým property in a class of locally convex spaces, Pacific J. Math. (1) 75, 281-291.
[1980a] A characterization of w*-compact convex sets having the Radon-Nikodým property, Bull. Sci. Math. (2) 104, 79-88.
[1980b] Universally Lusin-measurable and Baire-1 projections, Proc. Amer. Math. Soc. (4) 78, 514-518.
[1981a] On measurable projections in Banach spaces, Pacific J. Math. 97, 453-459.
[1981b] Extreme points, separability, and weak K-analyticity in dual Banach spaces, J. London Math. Soc. (2) 23, 165-170.

J. Saint-Raymond
[1975] Représentations intégrale dans certains convexes, Séminaire Choquet, 14e année (1974/75), Initiation à l'Analyse, Exp.2, 11pp., Secrétariat Mathématique, Paris.

V.V. Sazanov
[1965] On perfect measures, Amer. Math. Soc. Transl. (2) 48, 229-254.

W. Schackermayer
[1980] Private communication.

H.H. Schaefer
[1974] Banach lattices and positive operators, Springer-Verlag, Die Grundlehren der mathematischen Wissenschaften in Einzeldarstellungen 215, Berlin.

J.J. Schäffer (see also R.C. James)
[1976] Geometry of spheres in normed spaces, Marcel Dekker, Lecture Notes in Pure and Applied Mathematics 20, New York.
[1980] Girth, superreflexivity, and isomorphic classification of normed spaces and subspaces, Bull. Acad. Polon. Sci. Sér. Sci. Math. 28, 573-584.
[1981] Girth and isomorphic classification of normed spaces: remarks and conjectures, Proc. Research Workshop on Banach Space Theory, Ed. by Bor-Luh Lin, Univ. Iowa, June-July 1981, 153-167

J.T. Schwartz (see N. Dunford)

L. Schwartz
[1973] Radon measures on arbitrary topological spaces and cylindrical measures, Oxford Univ. Press, for Tata Inst. of Fund. Research, London.

J. Shirey
[1980] On the theorem of Helley concerning finite dimensional subspaces of a dual space, Pacific J. Math. 86, 571-577.

W. Sierpinski and E. Szpilrajn
[1936] Remarque sur le problème de la mesure, Fund. Math. 26, 256-261.

R. Sikorski (see E. Marczewski)

R.M. Solovay
[1967] Real-valued measurable cardinals, Axionmatic Set Theory (D.S. Scott, ed.), Amer. Math. Soc., Providence, R.I., 397-428.

C. Stegall (see also D. Lewis, J. Lindenstrauss)
[1975] The Radon-Nikodým property in conjugate Banach spaces, Trans. Amer. Math. Soc 206, 213-223.
[1978a] The duality between Asplund spaces and spaces with the Radon-Nikodým propert Israel J. Math. 29, 408-412.
[1978b] Optimization of functions on certain subsets of Banach spaces, Math. Ann. 23 171-176.
[1981] The Radon-Nikodým property in conjugate Banach spaces. II, Trans. Amer. Math. Soc. (2) 264, 507-519.

V. Strassen
[1965] The existence of probability measures with given marginals, Ann. Math. Statis 36, 423-439.

L. Sucheston (see R.V. Chacon, G.A. Edgar)

F. Sullivan
[1981] Ordering and completeness of metric spaces, Nieuw Arch. Wisk. (3) XXIX, 178-1

E. Szpilrajn (see W. Sierpinski)

M Talagrand (see also J. Bourgain, G.A. Edgar, D.H. Fremlin, N. Ghoussoub)
[1975] Sur une conjecture de H.H. Corson, Bull. Sci. Math. (2) 99, 211-212.
[1977a] Sur la structure Borélienne des espaces analytiques, Bull. Sci. Math. (2) 10 415-422.
[1977b] Espaces de Banach faiblement K-analytiques, C.R. Acad. Sci. Paris Ser. A 284 745-748.
[1977c] Sur les espaces de Banach faiblement K-analytiques, C.R. Acad. Sci. Paris Se A, 119-122.
[1978] Comparaison des Boreliens d'un espace de Banach pour les topologies fortes et faibles, Indiana Univ. Math. J. (6) 27, 1001-1004.
[1979] Espaces de Banach faiblement K-analytiques, Ann. Math. 110, 407-438.
[1980] Sur les mesures vectorielles définies par une application Pettis-intégrable, Bull. Soc. Math. France 108, 475-483.
[1983a] Three convex sets. Preprint.
[1983b] La structure des espaces de Banach reticules ayant la propriété de Radon-Nikodým. Preprint.

A.E. Taylor (see S. Bochner)

E.G.F. Thomas
[1977] Integral representations in conuclear spaces, Vector Space Measures and Appli tions II, Proc. Dublin, Lecture Notes in Math. # 645, 172-179.
[1978] Représentations intégrale dans les cones convexes, C.R. Acad. Sci. Paris Ser. A 286, 515-518.
[1979] A converse to Edgar's theorem, Measure Theory, Oberwolfach 1979, Ed. by D. Kölzow, Lecture Notes in Math. # 794, 497-512.

S.L. Troyanski
[1971] On locally uniformly convex and differentiable norms in certain non-separable Banach spaces, Studia Math. XXXVII, 173-180.

L. Tzafriri (see J. Lindenstrauss)

J.J. Uhl, Jr. (see also J. Diestel, S. Fitzpatrick, R.F. Geitz, L.H. Riddle)
[1969] The range of a vector-valued measure, Proc. Amer. Math. Soc. 23, 158-163.
[1972] A note on the Radon-Nikodým property for Banach spaces, Rev. Roumaine Math. Pures Appl. 17, 113-115.
[1978] Vector valued functions equivalent to measurable functions, Proc. Amer. Math. Soc. (1) 68, 32-36.

S. Ulam (see J.C. Oxtoby)

V.S. Varadarajan
[1965] Measures on topological spaces, Amer. Math. Soc. Transl. 48, 161-228.

H. von Weizsäcker
[1978] Strong measurability, liftings and the Choquet-Edgar theorem, Vector Space Measures and Applications II, Proc. Dublin 1977, Lecture Notes in Math. # 645, Springer-Verlag, Berlin.

H. von Weizsäcker and G. Winkler
[1979] Integral representations in the set of solutions of a generalized moment problem, Math. Ann. 246, 23-32.
[1980] Non-compact extremal integral representations: some probabalistic aspects, Functional Analysis: Surveys and Recent Results II, ed. by K.D. Bierstedt and B. Fuchssteiner, North-Holland, New York, 115-48.

J.H. Whitfield (see E.B. Leach)

G. Winkler (see also H. von Weizsäcker)
[1978] On the integral representation in convex noncompact sets of tight measures, Math. Z. 158, 71-77.

CITATION INDEX

SUBJECT INDEX

345: A. Tannenbaum, Invariance and System Theory: Algebraic Geometric Aspects. X, 161 pages. 1981.

846: Ordinary and Partial Differential Equations, Proceedings. d by W. N. Everitt and B. D. Sleeman. XIV, 384 pages. 1981.

847: U. Koschorke, Vector Fields and Other Vector Bundle hisms – A Singularity Approach. IV, 304 pages. 1981.

848: Algebra, Carbondale 1980. Proceedings. Ed. by R. K. o. VI, 298 pages. 1981.

49: P. Major, Multiple Wiener-Itô Integrals. VII, 127 pages. 1981.

50: Séminaire de Probabilités XV. 1979/80. Avec table générale xposés de 1966/67 à 1978/79. Edited by J. Azéma and M. Yor. 04 pages. 1981.

851: Stochastic Integrals. Proceedings, 1980. Edited by D. ams. IX, 540 pages. 1981.

352: L. Schwartz, Geometry and Probability in Banach Spaces. pages. 1981.

353: N. Boboc, G. Bucur, A. Cornea, Order and Convexity in ntial Theory: H-Cones. IV, 286 pages. 1981.

354: Algebraic K-Theory. Evanston 1980. Proceedings. Edited M. Friedlander and M. R. Stein. V, 517 pages. 1981.

355: Semigroups. Proceedings 1978. Edited by H. Jürgensen, trich and H. J. Weinert. V, 221 pages. 1981.

856: R. Lascar, Propagation des Singularités des Solutions uations Pseudo-Différentielles à Caractéristiques de Multipli-Variables. VIII, 237 pages. 1981.

857: M. Miyanishi. Non-complete Algebraic Surfaces. XVIII, pages. 1981.

58: E. A. Coddington, H. S. V. de Snoo: Regular Boundary Value lems Associated with Pairs of Ordinary Differential Expressions. 5 pages. 1981.

359: Logic Year 1979–80. Proceedings. Edited by M. Lerman, hmerl and R. Soare. VIII, 326 pages. 1981.

60: Probability in Banach Spaces III. Proceedings, 1980. Edited Beck. VI, 329 pages. 1981.

61: Analytical Methods in Probability Theory. Proceedings 1980. d by D. Dugué, E. Lukacs, V. K. Rohatgi. X, 183 pages. 1981.

362: Algebraic Geometry. Proceedings 1980. Edited by A. Lib- and P. Wagreich. V, 281 pages. 1981.

363: Processus Aléatoires à Deux Indices. Proceedings, 1980. d by H. Korezlioglu, G. Mazziotto and J. Szpirglas. V, 274 pages.

364: Complex Analysis and Spectral Theory. Proceedings, '80. Edited by V. P. Havin and N. K. Nikol'skii, VI, 480 pages.

365: R. W. Bruggeman, Fourier Coefficients of Automorphic s. III, 201 pages. 1981.

66: J.-M. Bismut, Mécanique Aléatoire. XVI, 563 pages. 1981.

67: Séminaire d'Algèbre Paul Dubreil et Marie-Paule Malliavin. edings, 1980. Edited by M.-P. Malliavin. V, 476 pages. 1981.

368: Surfaces Algébriques. Proceedings 1976–78. Edited by aud, L. Illusie et M. Raynaud. V, 314 pages. 1981.

69: A. V. Zelevinsky, Representations of Finite Classical Groups. 4 pages. 1981.

70: Shape Theory and Geometric Topology. Proceedings, 1981. d by S. Mardešić and J. Segal. V, 265 pages. 1981.

71: Continuous Lattices. Proceedings, 1979. Edited by B. Bana- wski and R.-E. Hoffmann. X, 413 pages. 1981.

72: Set Theory and Model Theory. Proceedings, 1979. Edited B. Jensen and A. Prestel. V, 174 pages. 1981.

Vol. 873: Constructive Mathematics, Proceedings, 1980. Edited by F. Richman. VII, 347 pages. 1981.

Vol. 874: Abelian Group Theory. Proceedings, 1981. Edited by R. Göbel and E. Walker. XXI, 447 pages. 1981.

Vol. 875: H. Zieschang, Finite Groups of Mapping Classes of Surfaces. VIII, 340 pages. 1981.

Vol. 876: J. P. Bickel, N. El Karoui and M. Yor. Ecole d'Eté de Probabilités de Saint-Flour IX – 1979. Edited by P. L. Hennequin. XI, 280 pages. 1981.

Vol. 877: J. Erven, B.-J. Falkowski, Low Order Cohomology and Applications. VI, 126 pages. 1981.

Vol. 878: Numerical Solution of Nonlinear Equations. Proceedings, 1980. Edited by E. L. Allgower, K. Glashoff, and H.-O. Peitgen. XIV, 440 pages. 1981.

Vol. 879: V. V. Sazonov, Normal Approximation – Some Recent Advances. VII, 105 pages. 1981.

Vol. 880: Non Commutative Harmonic Analysis and Lie Groups. Proceedings, 1980. Edited by J. Carmona and M. Vergne. IV, 553 pages. 1981.

Vol. 881: R. Lutz, M. Goze, Nonstandard Analysis. XIV, 261 pages. 1981.

Vol. 882: Integral Representations and Applications. Proceedings, 1980. Edited by K. Roggenkamp. XII, 479 pages. 1981.

Vol. 883: Cylindric Set Algebras. By L. Henkin, J. D. Monk, A. Tarski, H. Andréka, and I. Németi. VII, 323 pages. 1981.

Vol. 884: Combinatorial Mathematics VIII. Proceedings, 1980. Edited by K. L. McAvaney. XIII, 359 pages. 1981.

Vol. 885: Combinatorics and Graph Theory. Edited by S. B. Rao. Proceedings, 1980. VII, 500 pages. 1981.

Vol. 886: Fixed Point Theory. Proceedings, 1980. Edited by E. Fadell and G. Fournier. XII, 511 pages. 1981.

Vol. 887: F. van Oystaeyen, A. Verschoren, Non-commutative Algebraic Geometry, VI, 404 pages. 1981.

Vol. 888: Padé Approximation and its Applications. Proceedings, 1980. Edited by M. G. de Bruin and H. van Rossum. VI, 383 pages. 1981.

Vol. 889: J. Bourgain, New Classes of $\mathcal{L}^p$-Spaces. V, 143 pages. 1981.

Vol. 890: Model Theory and Arithmetic. Proceedings, 1979/80. Edited by C. Berline, K. McAloon, and J.-P. Ressayre. VI, 306 pages. 1981.

Vol. 891: Logic Symposia, Hakone, 1979, 1980. Proceedings, 1979, 1980. Edited by G. H. Müller, G. Takeuti, and T. Tugué. XI, 394 pages. 1981.

Vol. 892: H. Cajar, Billingsley Dimension in Probability Spaces. III, 106 pages. 1981.

Vol. 893: Geometries and Groups. Proceedings. Edited by M. Aigner and D. Jungnickel. X, 250 pages. 1981.

Vol. 894: Geometry Symposium. Utrecht 1980, Proceedings. Edited by E. Looijenga, D. Siersma, and F. Takens. V, 153 pages. 1981.

Vol. 895: J.A. Hillman, Alexander Ideals of Links. V, 178 pages. 1981.

Vol. 896: B. Angéniol, Familles de Cycles Algébriques – Schéma de Chow. VI, 140 pages. 1981.

Vol. 897: W. Buchholz, S. Feferman, W. Pohlers, W. Sieg, Iterated Inductive Definitions and Subsystems of Analysis: Recent Proof-Theoretical Studies. V, 383 pages. 1981.

Vol. 898: Dynamical Systems and Turbulence, Warwick, 1980. Proceedings. Edited by D. Rand and L.-S. Young. VI, 390 pages. 1981.

Vol. 899: Analytic Number Theory. Proceedings, 1980. Edited by M.I. Knopp. X, 478 pages. 1981.

Vol. 900: P. Deligne, J. S. Milne, A. Ogus, and K.-Y. Shih, Hodge Cycles, Motives, and Shimura Varieties. V, 414 pages. 1982.

Vol. 901: Séminaire Bourbaki vol. 1980/81 Exposés 561–578. III, 299 pages. 1981.

Vol. 902: F. Dumortier, P.R. Rodrigues, and R. Roussarie, Germs of Diffeomorphisms in the Plane. IV, 197 pages. 1981.

Vol. 903: Representations of Algebras. Proceedings, 1980. Edited by M. Auslander and E. Lluis. XV, 371 pages. 1981.

Vol. 904: K. Donner, Extension of Positive Operators and Korovkin Theorems. XII, 182 pages. 1982.

Vol. 905: Differential Geometric Methods in Mathematical Physics. Proceedings, 1980. Edited by H.-D. Doebner, S.J. Andersson, and H.R. Petry. VI, 309 pages. 1982.

Vol. 906: Séminaire de Théorie du Potentiel, Paris, No. 6. Proceedings. Edité par F. Hirsch et G. Mokobodzki. IV, 328 pages. 1982.

Vol. 907: P. Schenzel, Dualisierende Komplexe in der lokalen Algebra und Buchsbaum-Ringe. VII, 161 Seiten. 1982.

Vol. 908: Harmonic Analysis. Proceedings, 1981. Edited by F. Ricci and G. Weiss. V, 325 pages. 1982.

Vol. 909: Numerical Analysis. Proceedings, 1981. Edited by J.P. Hennart. VII, 247 pages. 1982.

Vol. 910: S.S. Abhyankar, Weighted Expansions for Canonical Desingularization. VII, 236 pages. 1982.

Vol. 911: O.G. Jørsboe, L. Mejlbro, The Carleson-Hunt Theorem on Fourier Series. IV, 123 pages. 1982.

Vol. 912: Numerical Analysis. Proceedings, 1981. Edited by G. A. Watson. XIII, 245 pages. 1982.

Vol. 913: O. Tammi, Extremum Problems for Bounded Univalent Functions II. VI, 168 pages. 1982.

Vol. 914: M. L. Warshauer, The Witt Group of Degree k Maps and Asymmetric Inner Product Spaces. IV, 269 pages. 1982.

Vol. 915: Categorical Aspects of Topology and Analysis. Proceedings, 1981. Edited by B. Banaschewski. XI, 385 pages. 1982.

Vol. 916: K.-U. Grusa, Zweidimensionale, interpolierende Lg-Splines und ihre Anwendungen. VIII, 238 Seiten. 1982.

Vol. 917: Brauer Groups in Ring Theory and Algebraic Geometry. Proceedings, 1981. Edited by F. van Oystaeyen and A. Verschoren. VIII, 300 pages. 1982.

Vol. 918: Z. Semadeni, Schauder Bases in Banach Spaces of Continuous Functions. V, 136 pages. 1982.

Vol. 919: Séminaire Pierre Lelong – Henri Skoda (Analyse) Années 1980/81 et Colloque de Wimereux, Mai 1981. Proceedings. Edité par P. Lelong et H. Skoda. VII, 383 pages. 1982.

Vol. 920: Séminaire de Probabilités XVI, 1980/81. Proceedings. Edité par J. Azéma et M. Yor. V, 622 pages. 1982.

Vol. 921: Séminaire de Probabilités XVI, 1980/81. Supplément: Géométrie Différentielle Stochastique. Proceedings. Edité par J. Azéma et M. Yor. III, 285 pages. 1982.

Vol. 922: B. Dacorogna, Weak Continuity and Weak Lower Semicontinuity of Non-Linear Functionals. V, 120 pages. 1982.

Vol. 923: Functional Analysis in Markov Processes. Proceedings, 1981. Edited by M. Fukushima. V, 307 pages. 1982.

Vol. 924: Séminaire d'Algèbre Paul Dubreil et Marie-Paule Malliavin. Proceedings, 1981. Edité par M.-P. Malliavin. V, 461 pages. 1982.

Vol. 925: The Riemann Problem, Complete Integrability and Arithmetic Applications. Proceedings, 1979-1980. Edited by D. Chudnovsky and G. Chudnovsky. VI, 373 pages. 1982.

Vol. 926: Geometric Techniques in Gauge Theories. Proceedings, 1981. Edited by R. Martini and E.M. de Jager. IX, 219 pages. 1982.

Vol. 927: Y. Z. Flicker, The Trace Formula and Base Change GL (3). XII, 204 pages. 1982.

Vol. 928: Probability Measures on Groups. Proceedings 1981. Ed by H. Heyer. X, 477 pages. 1982.

Vol. 929: Ecole d'Eté de Probabilités de Saint-Flour X – 19 Proceedings, 1980. Edited by P.L. Hennequin. X, 313 pages. 19

Vol. 930: P. Berthelot, L. Breen, et W. Messing, Théorie de Dieudo Cristalline II. XI, 261 pages. 1982.

Vol. 931: D.M. Arnold, Finite Rank Torsion Free Abelian Gro and Rings. VII, 191 pages. 1982.

Vol. 932: Analytic Theory of Continued Fractions. Proceedings, 1 Edited by W.B. Jones, W.J. Thron, and H. Waadeland. VI, 240 pa 1982.

Vol. 933: Lie Algebras and Related Topics. Proceedings, 1 Edited by D. Winter. VI, 236 pages. 1982.

Vol. 934: M. Sakai, Quadrature Domains. IV, 133 pages. 1982.

Vol. 935: R. Sot, Simple Morphisms in Algebraic Geometry 146 pages. 1982.

Vol. 936: S.M. Khaleelulla, Counterexamples in Topological Ve Spaces. XXI, 179 pages. 1982.

Vol. 937: E. Combet, Intégrales Exponentielles. VIII, 114 pa 1982.

Vol. 938: Number Theory. Proceedings, 1981. Edited by K. A IX, 177 pages. 1982.

Vol. 939: Martingale Theory in Harmonic Analysis and Ba Spaces. Proceedings, 1981. Edited by J.-A. Chao and W.A. W czyński. VIII, 225 pages. 1982.

Vol. 940: S. Shelah, Proper Forcing. XXIX, 496 pages. 1982.

Vol. 941: A. Legrand, Homotopie des Espaces de Sections. 132 pages. 1982.

Vol. 942: Theory and Applications of Singular Perturbations. ceedings, 1981. Edited by W. Eckhaus and E.M. de Jager. V, pages. 1982.

Vol. 943: V. Ancona, G. Tomassini, Modifications Analytique 120 pages. 1982.

Vol. 944: Representations of Algebras. Workshop Proceed 1980. Edited by M. Auslander and E. Lluis. V, 258 pages. 1982

Vol. 945: Measure Theory. Oberwolfach 1981, Proceedings. Ec by D. Kölzow and D. Maharam-Stone. XV, 431 pages. 1982.

Vol. 946: N. Spaltenstein, Classes Unipotentes et Sous-groupe Borel. IX, 259 pages. 1982.

Vol. 947: Algebraic Threefolds. Proceedings, 1981. Edite A. Conte. VII, 315 pages. 1982.

Vol. 948: Functional Analysis. Proceedings, 1981. Edited by D. ković, H. Kraljević, and S. Kurepa. X, 239 pages. 1982.

Vol. 949: Harmonic Maps. Proceedings, 1980. Edited by R.J. M. Kalka and H.C.J. Sealey. V, 158 pages. 1982.

Vol. 950: Complex Analysis. Proceedings, 1980. Edited by J. E IV, 428 pages. 1982.

Vol. 951: Advances in Non-Commutative Ring Theory. Proceed 1981. Edited by P.J. Fleury. V, 142 pages. 1982.

Vol. 952: Combinatorial Mathematics IX. Proceedings, 1981. E by E. Billington, S. Oates-Williams, and A.P. Street. XI, 443 p 1982.

Vol. 953: Iterative Solution of Nonlinear Systems of Equations. ceedings, 1982. Edited by R. Ansorge, Th. Meis, and W. Tö VII, 202 pages. 1982.